ANNUAL REVIEW OF ENTOMOLOGY

ANNUAL REVIEW OF ENTOMOLOGY

VOLUME 41, 1996

THOMAS E. MITTLER, *Editor*
University of California, Berkeley

FRANK J. RADOVSKY, *Editor*
Oregon State University, Corvallis

VINCENT H. RESH, *Editor*
University of California, Berkeley

ANNUAL REVIEWS INC.   4139 EL CAMINO WAY   P.O. 10139   PALO ALTO, CALIFORNIA 94303-0139

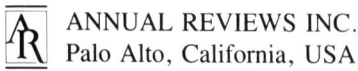 ANNUAL REVIEWS INC.
Palo Alto, California, USA

International Standard Serial Number: 0066-4170
International Standard Book Number: 0-8243-0141-2
Library of Congress Catalog Card Number: A56-5750

Annual Review and publication titles are registered trademarks of Annual Reviews Inc.

♾ The paper used in this publication meets the minimum requirements of
American National Standard for Information Sciences—Permanence of Paper
for Printed Library Materials, ANSI Z39.48-1984.

Annual Reviews Inc. and the Editors of its publications assume no responsibility for the
statements expressed by the contributors to this *Review*.

Typesetting by Kachina Typesetting Inc., Tempe, Arizona; John Olson, President;
Marty Mullins, Typesetting Coordinator; and by the Annual Reviews Inc. Editorial Staff

PRINTED AND BOUND IN THE UNITED STATES OF AMERICA

PREFACE

Perhaps the most enjoyable part of the process that leads to a volume of the *Annual Review of Entomology* is sitting around a table as an editorial committee member discussing potential topics for chapters. At the meeting, we review booklets containing updates of past volumes, correspondence on the progress made by authors, and production and sales information. Over coffee and muffins in the morning and sandwiches at lunch, ideas come and go, votes are taken, and topics are accepted or rejected, tabled, and revisited. Since 1989, Amanda Suver, the production editor, has kept us all organized as we choose a balanced assortment of chapter topics and has kept the authors on track for meeting the deadlines for the annual volume. Amanda is leaving Annual Reviews and moving East. On behalf of the editorial committee, I thank Amanda for her work and patience and acknowledge her excellent editorial skills, as evidenced by the quality of the six volumes she has handled.

Women have always played an important role as production editors of the *Annual Review of Entomology*, but only since 1985 have they served on the editorial board. In the 1970s and 1980s, women were authors of only 5–10% of the chapters of the *Review* (a conservative estimate based on authors for whom first names are given). However, more recently the contributions from women have almost doubled; women are authors of approximately 19% of chapters published during the 1990s. I leave the board with the knowledge that women entomologists will continue to make valuable contributions to their field and that this will be reflected in the *Annual Review of Entomology*. I appreciate having had the opportunity to be involved with this important publication series.

JUDY MYERS
FOR THE EDITORIAL COMMITTEE

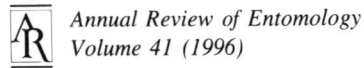
Annual Review of Entomology
Volume 41 (1996)

CONTENTS

(*continued*)   vii

viii   Contents   *(continued)*

For the convenience of readers, a detachable order form/envelope is bound into the back of this volume.

Annu. Rev. Entomol. 1996. 41:1–22

HOST IMMUNITY TO TICKS

Stephen K. Wikel

Department of Entomology, 127 Noble Research Center, Oklahoma State
University, Stillwater, Oklahoma 74078

KEY WORDS: immunomodulation, resistance, vaccination

ABSTRACT

The tick-host-pathogen interface is characterized by complex immunological
interactions. Tick feeding induces host immune regulatory and effector pathways
involving antibodies, complement, antigen-presenting cells, T lymphocytes, and
other bioactive molecules. Acquired resistance impairs tick engorgement, ova
production, and viability. Tick countermeasures to host defenses reduce T-lymphocyte proliferation, elaboration of the T_H1 cytokines interleukin-2 and inter-
feron-γ, production of macrophage cytokines interleukin-1 and tumor necrosis
factor, and antibody responses. The dynamic balance between acquired resistance
and tick modulation of host immunity affects engorgement and pathogen transmission. A thorough understanding of acquired immunity to ticks is essential for
rational development of antitick vaccines.

Perspectives and Overview

A complex array of immunological responses characterizes the tick-host-
pathogen interface. Studies of the interactions of tick-introduced immunogens
with host immune regulatory and effector pathways are vital to an understanding of blood feeding; salivary-gland function; digestion; pathogen acquisition by the tick, transmission to the host, and establishment of tick-borne
pathogens in the host; and vaccine-based tick control. Advances during the
past 25 years in fundamental knowledge, experimental approaches, and instrumentation have allowed us to increase our understanding of the cellular and
molecular bases of host immunity to ticks and tick-induced immunomodulation
of the host.

A vast amount of work remains to be done in the ongoing battle against
ticks and tick-borne diseases. The vectors and pathogens can be most effectively studied by interactive, multidisciplinary research teams. The development of acaricide resistance (66) calls for alternative control strategies for ticks
and tick-borne diseases. The most promising of these alternatives, antitick
vaccines, have advanced from a possibility to a reality in the past ten years.

1

The complexity and variety of tick-borne diseases continues to increase, highlighting the validity of the observations made over a decade ago by Hoogstraal (42). Lyme borreliosis is the most frequently reported vector-borne disease in the United States (94). Ehrlichial diseases are of increasing importance to physicians and veterinarians (84). Human ehrlichiosis was first reported in the United States in 1986, and clinical illness was linked to infection with *Ehrlichia chaffeensis* (8, 39). The recent cases of human granulocytic ehrlichiosis in Minnesota and Wisconsin indicate that this disease is more complicated than previously thought (9). Isolation of an intraerythrocytic protozoan (WA1) from a patient in the State of Washington added another possible member to the list of tick-borne infectious agents of humans (98). How host-derived inflammatory or immunological signals affect activation of the pathogen within the vector and subsequent transmission of tick-borne pathogens merits investigation.

The influence of the tick on host responses to infection with a tick-borne pathogen exemplifies the immunological complexities of the host-tick-pathogen interface. The immune response in hamsters to tick-transmitted *Borrelia burgdorferi* differed from responses to needle inoculation of the same organisms (85). Hamster antibody reactivity to tick-transmitted infection was similar to that of humans; i.e. minimal or absent responses to the spirochete outer surface proteins A and B (OspA and OspB). Immune responses to *B. burgdorferi* were most accurately analyzed after tick transmission of spirochetes (38). These findings indicated the importance of studying immune responses to a tick-borne pathogen after vector transmission of the disease-causing agent. Tick modulation of host immunocompetence must be considered in any study of immune reactivity to infestation and tick-transmitted microorganisms. The tick cannot be simplistically viewed as a crawling hypodermic needle and syringe.

The following topics regarding the tick-host interface have been reviewed: host acquired immunity to ticks (3, 18, 19, 114, 125), tick-induced host immunomodulation (123), salivary-gland physiology and feeding (49, 88), pharmacological properties of tick salivary glands (24, 75, 76, 99), and antitick vaccines (50, 97, 100, 119, 122).

Acquired Resistance

Tick feeding induces host immune regulatory and effector pathways involving antibodies, complement, cytokines, antigen-presenting cells, and T lymphocytes (18, 114, 123, 125). Immunologically acquired host resistance to tick feeding can result in reduced blood-meal volume, decreased engorgement weight, prolonged duration of feeding, diminished production of ova, reduced viability of ova, inhibition of molting, and death of engorging ticks.

Acquired resistance has been most often observed after infestation by female ticks. However, feeding male ixodids also induce acquired resistance, but to a lesser degree than that caused by females alone or together with males (74).

Host grooming is an important factor in reducing tick burden (11). Immunological mediators induced by tick antigens introduced into host skin contribute to the itch sensation, which stimulates grooming (1).

Immunology of acquired resistance to tick feeding has been most extensively studied in associations between ticks and bovines or laboratory animals (18, 26, 61, 114, 125). Cattle and laboratory animals exhibit similar immune responses to tick infestation (35, 72, 114, 121). However, tick–laboratory animal associations not occurring in nature resulted in a greater degree of expression of acquired resistance (76). The availability of reagents for dissection of laboratory-animal immunological pathways has permitted analyses not possible for other host species. On the other hand, probes developed for bovine cytokines and lymphocyte subpopulations have proved valuable tools for the characterization of acquired resistance by cattle (72).

The genetic background of an experimental host must be considered in any study of acquired immunological resistance. For example, bovine breeds differ in their resistance to ticks, a factor recognized and extensively used by cattle breeders (27, 47).

The characterization of acquired resistance to ticks began with Trager's finding (100) that guinea pigs developed resistance after a single infestation with *Dermacentor variabilis* larvae. He observed minimal skin reactivity at tick-attachment sites during primary infestation. Cutaneous reactions at tick-feeding sites on resistant guinea pigs were characterized by epidermal hyperplasia, edema, and granulocytic inflammatory infiltrates.

Allen (2) established that, at the site of feeding, tick-resistant guinea pigs developed cutaneous reactions consisting primarily of basophils and eosinophils. The findings of such histological studies of cutaneous responses of tick-resistant cattle and laboratory animals have been reviewed (3, 18, 124). Basophils appear to be common elements in cutaneous responses of cattle (4) and laboratory animals expressing acquired resistance to ticks (15). Numbers of infiltrating basophils vary with the tick-host association studied. Basophil-rich cutaneous infiltrations [referred to as cutaneous basophil hypersensitivity (CBH)] are T-lymphocyte responses (32), and these delayed hypersensitivity reactions are mediated by the T_H1 helper lymphocyte subpopulation (60).

Basophils apparently play a role in limiting tick feeding. One study revealed that expression of acquired resistance is reduced by administration of antibasophil serum and that eosinophils are also important in establishing resistance (22). Basophils degranulate more readily during repeated infestations in the presence of tick salivary antigens (17), and mast cell degranulation occurs at tick-attachment sites during initial and subsequent exposures (15). Infestation induces synthesis of tick-reactive homocytotropic antibodies that bind Fc receptors on basophils and mast cells (16, 111). Continued introduction of salivary-gland immunogens could result in the formation of antigen-homocytotropic antibody complexes, causing the release of basophil and mast cell granule contents. Tick salivary enzymes could hydrolyze cell membranes, causing the release of vasoactive moieties (88).

BALB/c mice developed resistance to tick feeding, expressing it during the third and fourth infestations (28). Male mice permitted more larvae to engorge during primary infestations than did females. A possible explanation for this difference was the observation that the dorsal and ventral skin surfaces of female mice had higher histamine levels than those of male mice (53). Cutaneous reactions at tick-attachment sites on repeatedly infested animals contained mast cells and eosinophils (28). Mast cell–normal (WB × C57BL/6) F_1-+/+ mice developed resistance to *Haemaphysalis longicornis* larvae, whereas congenic mast cell–deficient W/W^v mice did not acquire infestation resistance (56). Bone marrow transplanted from normal mice rescued congenic mast cell–deficient mice, which then developed resistance to *H. longicornis* larvae (56). Cultured mast cells injected into the skin of W/W^v mice appeared to be essential for expression of resistance (57). Immunoglobulin E levels increased after tick feeding, correlating with an immediate hypersensitivity component to resistance (104).

Ultrastructural analysis of tick-attachment sites on resistant mice revealed the accumulation of large numbers of basophils (95). These mast cell–deficient mice acquired resistance to *D. variabilis* after two infestations (95). Basophils, eosinophils, and neutrophils were observed at tick-attachment sites. The basophil response may compensate for mast cell deficiency. A basophil-rich cutaneous hypersensitivity appears to be expressed by tick-resistant mice, guinea pigs, rabbits, and cattle.

The alternative pathway of complement activation contributes to expression of acquired resistance (112). Salivary-gland antigens, IgG, and complement were localized at the dermal-epidermal junction of resistant guinea pigs (6, 7). Levels of complement component C3 increase during tick infestations (70). In addition, development of tick-specific IgG has been reported for many ixodid-host associations (18, 19, 114). Continuous introduction of saliva during feeding would allow complexes of circulating antibody and antigen to form and be deposited at feeding sites. Subsequent fixation of complement could enhance lesion formation and potentially have an impact on the engorging tick.

Complement activated by either the alternative or classical pathways could contribute to feeding-lesion formation through anaphylatoxin generation and chemotactic activities. C3a and C5a cause degranulation of mast cells and basophils and a concomitant release of eosinophil chemotactic factors (histamine, eosinophil chemotactic factor of anaphylaxis) and vasoactive substances

(34). C5a is chemotactic for neutrophils and monocytes (34). Moreover, C5a- and lymphocyte-derived mediators are chemotactic for guinea pig basophils (106). Digestive tracts of ticks that obtained blood meals from resistant hosts contained intact basophils, intact eosinophils, and granules of both cell types (105). Basophil and eosinophil granules were observed within tick gut cells that displayed membrane damage and other signs of injury. Bioreactive molecules released by basophils, eosinophils, and mast cells may influence tick physiological responses, thus causing cellular injury and behavioral changes. Histamine, leukotrienes, prostaglandins, eosinophil major basic protein, enzymes, and other mediators of inflammatory and immune responses likely contribute to formation of the lesion at the bite site and affect the engorging tick.

In one study, histamine content at tick-attachment sites on animals displaying acquired resistance was significantly elevated, and in vivo administration of both H_1 and H_2 histamine receptor antagonists reduced expression of acquired resistance (115). Administration of H_1 receptor antagonist inhibits expression of rabbit resistance to *Ixodes ricinus* (14). In vitro quantities of histamine similar to those found at bite sites of resistant animals inhibited the ability of *Dermacentor andersoni* to salivate and engorge (5, 69). Elevated histamine levels should accompany the intense accumulations of basophils that occur at tick-attachment sites on resistant animals. In addition to a direct effect upon the feeding tick, vasoactive substances would facilitate the accumulation of immunoglobulins, complement, and cells at the bite site by increasing vascular permeability and other changes in the endothelium.

Cutaneous immune responses at tick-bite sites during acquisition and expression of acquired resistance require the participation of antigen-presenting cells, antigen-specific T and B lymphocytes, and cytokines. Administration of cyclosporin A reduced antibody- (type I) and cell-mediated (type IV) hypersensitivity, further confirming the involvement of T lymphocyte–dependent responses in acquiring resistance (37).

In addition, Langerhans cells are important in acquired resistance (7, 62–65). The association of tick salivary-gland antigens with dendritic cells in the skin of resistant guinea pigs supports the hypothesis that Langerhans cells trap immunogens (6, 7). Langerhans cells decrease in the epidermis during primary infestation and increase at tick-attachment sites during the early phase of secondary infestation, when resistance is expressed (62).

Short-wavelength ultraviolet radiation C depleted the guinea pig epidermal supply of Langerhans cells for 6 days after treatment, with little residual inflammation (63). This treatment prior to an initial tick infestation of guinea pig ears significantly reduced the acquisition of tick resistance, and in a resistant animal, treatment before a challenge infestation reduced the expression of antitick immunity (64). In vitro studies confirmed that Langerhans cells function as antigen-presenting accessory cells in the expression of acquired resistance to ticks (65).

The skin and draining lymph nodes of *I. ricinus*–infested BALB/c mice were examined using in situ hybridization for interferon-γ (IFN-γ), interleukin-2 (IL-2), and interleukin-4 (IL-4) mRNA expression (58). Infestation with *I. ricinus* nymphs resulted in strong expression of IFN-γ and IL-2 mRNA and lower levels of expression of IL-4 mRNA. Repeated infestations resulted in higher levels of expression of IFN-γ and IL-2 in the skin. These observations suggested the presence of a cutaneous T_H1 response elicited by tick immunogens (58).

Localized cutaneous immunity to tick feeding should be the subject of continued study using both in vivo and in vitro approaches. Particular attention needs to be focused upon local cytokine and T-lymphocyte responses. The initiation of tick feeding in a region drained by an accessible lymph node would allow the study of cytokine elaboration, T-lymphocyte responses, and local antibody production during the course of primary and repeated infestations. An in vitro culture system of syngeneic Langerhans cells and lymphocytes would prove valuable in assessing local immune interactions of tick and host.

Bovine resistance to tick infestation consists of innate and acquired components (27, 83). *Bos indicus* cattle exhibit the strongest innate resistance (27, 96). Natural and acquired resistance of Hereford calves to *B. microplus* seem unrelated, and individual animals differ in natural susceptibility to infestation (10). Innate resistance, which partly reflects the ability to mount a more intense immune response to infestation, appears to be linked to breed differences in immune response capabilities (73). As for acquired resistance, upon exposure to tick salivary-gland immunogens, *B. indicus* had heightened immune responsiveness compared with *Bos taurus,* as measured by macrophage elaboration of interleukin-1 (IL-1) and in vitro proliferation of B and T lymphocytes to mitogens (72). The tropics and subtropics have placed strong selection pressures on *B. indicus* cattle to enable them to withstand harsh environmental factors in addition to parasites and other disease-causing agents (27). Selection of animals with the ability to limit tick infestation by heightened immune responsiveness results in a survival advantage.

Studies to date provide an increasingly clear picture of the immunology of acquired resistance. A gap in achieving a more complete understanding of these interactions was the limited amount of information available regarding immunogens responsible for specific responses. Antibody responses to tick salivary-gland immunogens have been reviewed (18, 19, 114, 124, 125). However, little information was obtained about the role of these molecules in acquisition and expression of resistance. Tick salivary gland–derived molecules responsible for stimulating in vitro proliferation of lymphocytes from infested laboratory animals (120) and cattle (35, 121) have not been identified.

Changes in tick salivary-gland proteins during the course of feeding (59) point to an array of tick immunogens and responses to host factors. In addition to other components in saliva, immunogens unique to attachment cement could contribute to host reactivity. Attachment cement likely serves to expose saliva immunogens to host skin and allow them to be adsorbed during feeding and, possibly, be a site of continued exposure after detachment. Similar antigenic determinants were detected in attachment cement and salivary glands of several ixodid species (46).

During infestations of laboratory animals (110) and cattle (73), tick-reactive antibody titers increased. A negative correlation was reported between mean engorgement weight of *Boophilus decoloratus* and serum γ globulin levels of infested bovines (74). Sera derived from resistant animals were used to screen for reactive molecules by means of immunoprecipitation (23) and immunoblotting (20, 86, 92, 109, 119).

In guinea pigs, resistance to *Rhipicephalus appendiculatus* appears to be mediated by antibodies (92). The presence and concentration of reactive immunogens varied with the tick feeding cycle. A 90-kDa molecule derived from *R. appendiculatus* salivary glands was implicated in induction of acquired resistance by rabbits (91). A 20-kDa salivary gland polypeptide reportedly induces resistance to *Amblyomma americanum* (21). The roles of these molecules in acquisition and expression of resistance should be further evaluated.

Antibodies stimulated by tick feeding reacted with a variety of proteins and polypeptides in salivary-gland extracts (92, 119). Sera collected from rabbits infested with *Hyalomma anatolicum anatolicum* were used to probe electrophoretograms of saliva and salivary-gland extracts (36). All saliva antigens, and 12 of 17 reactive molecules in the salivary-gland extract, were glycoconjugates. Although similar immunogens were present throughout different stages of feeding, the concentrations of salivary-gland immunogens changed during feeding (40). In addition, antigenic composition of *Amblyomma hebraeum* saliva changed with the size of the tick during engorgement (30). *A. americanum* and *D. andersoni* share immunogenic determinants (124). *A. americanum, D. variabilis,* and *Ixodes dammini* all express cross-reactive 45.0- and 91.0-kDa proteins and polypeptides (45). Antibodies obtained from rabbits and rats infested with *I. dammini* reacted with glycoproteins obtained from *I. dammini, Ixodes scapularis,* and *D. variabilis* (108).

The salivary-gland extract of *D. andersoni* stimulated in vitro proliferation of lymph-node cells collected from guinea pigs during primary and secondary infestations (120). Antigen-specific responses occurred from 2 to 4 days after termination of an initial infestation through the end of a second infestation. The most intense responsiveness occurred with lymphocytes collected 24 h after initiation of a second exposure. Peripheral blood lymphocytes collected from rabbits infested with *I. ricinus* proliferated in vitro upon culture with salivary-gland extract and integumental antigens of adult female *I. ricinus* (90). The most intense responses came at the end of a second infestation of animals given the greatest number of ticks.

By the third or fourth infestation, *B. taurus* cows and calves repeatedly infested with *D. andersoni* adults developed peripheral blood lymphocytes reactive in vitro with salivary-gland extracts of the same ixodid species (121). Purebred *B. indicus* calves infested with *A. americanum* developed in vitro lymphocyte responsiveness to this tick's salivary-gland extract after the first, second, and third infestations, whereas the reactivity of cells from crossbred animals was considerably less (35).

The use of antibodies alone to detect immunogens responsible for induction and expression of acquired resistance could be potentially misleading. Antibodies contribute to expression of resistance. However, many antibodies are likely not reactive with molecules involved in processes that influence tick biology. Attention should also be focused upon immunogens of salivary-gland origin that stimulate T lymphocytes during infestation. The relative importance of T_H1 and T_H2 lymphocyte responses during acquisition and expression of resistance needs to be determined. Attention should be focused upon host defenses modulated by infestation and identification of tick-derived molecules responsible for induction of immunosuppression. Tick-mediated immunosuppression is likely directed toward those elements of the immune system that limit successful acquisition of the blood meal (123).

Pharmacological Properties of Tick Saliva

The characterization of pharmacological and immunological properties of tick saliva provides a better understanding of the interrelated elements involved in acquisition and expression of host immunity to ticks. Advances in studies of tick salivary-gland physiology allow new insights into the biological properties of tick salivas (49, 88, 89). Salivary gland–derived molecules have antihemostatic, vasodilatory, antiinflammatory, and immunosuppressive properties (24, 75–77, 99, 123). Several tick salivary-gland factors appear to have more than one biological activity. For example, molecules inhibiting coagulation and enhancing vasodilation contribute to formation of the feeding site. Antiinflammatory and immunosuppressive molecules reduce host defenses that impair tick engorgement. Tick-mediated host immunosuppression appears to achieve a balance between reduction of immune defenses that limit engorgement and maintenance of sufficient immunocompetence for host survival. In addition, tick-induced host immunosuppression likely contributes to successful establishment of infection by tick-borne pathogens.

Tick salivary glands contain apyrase, which inhibits platelet aggregation by hydrolyzing adenosine triphosphate (ATP) and adenosine diphosphate (ADP) to adenosine monophosphate (AMP) and orthophosphate (75, 99). Prostaglandin E_2 (PGE_2), which is produced by tick salivary glands, inhibits platelet aggregation and causes vasodilation (24, 79). Salivary apyrase may prevent aggregation of neutrophils and mast cell degranulation (79). *A. americanum* saliva contains very high concentrations of PGE_2 (78), and *I. dammini* saliva contains prostacyclin, which blocks platelet aggregation, inhibits mast cell degranulation, and induces vasodilation (80). Intrinsic and extrinsic pathways of coagulation are inhibited by elements of *D. andersoni* saliva that act upon factors V and VII (41). Salivary-gland extracts of *R. appendiculatus* contain a 65-kDa anticoagulant that inhibits the activity of factor Xa or other components of the prothrombinase complex (54).

PGE_2 potentiates pain produced by bradykinin; however, *I. dammini* saliva contains a kininase capable of destroying bradykinin (79). Kininase activity could reduce cutaneous irritation caused by bradykinin, thereby decreasing the likelihood that a host will remove feeding ticks by grooming.

Tick Modulation of Host Immune Function

A common theme throughout host-parasite relationships appears to be the ability of the infectious agent to alter host immune defenses by modulating various regulatory and/or effector pathways (29). The discovery of host immunosuppressive factors associated with ticks was not unexpected (71, 72, 79, 116, 123). Over evolutionary time, blood-feeding arthropods likely developed strategies for suppression of host defenses that directly affect the arthropods' survival. Salivary-gland immunosuppressive molecules are not limited to long-term blood feeders. For example, factors capable of modulating host immune responses occur in salivary glands of *Simulium vittatum* (25) and *Aedes aegypti* (13). Vector-mediated immunosuppression of the host helps in both blood-meal acquisition as well as in effective transmission of pathogens. In addition, the responses to both arthropods and pathogens involve common elements of the immune system.

I. dammini saliva contains an inactivator of complement anaphylatoxins (81) and inhibits activation of the alternative complement pathway, thus blocking deposition of C3b and release of C3a (75).

Tick feeding also suppressed the generation of a primary IgM response to a thymic-dependent immunogen (117). In this study, reduced antibody responsiveness to sheep red blood cells remained evident for several days after termination of infestation. Likewise, *R. appendiculatus* infestation reduced rabbit antibody responses to bovine serum albumin (33). Immunosuppression was attributed to lymphocytotoxic factors in tick salivary glands. Reduced lymphocyte proliferative responses and cytokine elaboration resulting from salivary-gland extracts of other tick species were not caused by cytotoxicity (71, 116, 123).

In vitro, lymphocytes obtained from guinea pigs infested with adult *D. andersoni* exhibited significantly reduced proliferative responses to the T-lymphocyte mitogens concanavalin A (Con A) and phytohemagglutinin (PHA), whereas reactivity of B lymphocytes to *Escherichia coli* lipopolysaccharide (LPS) remained normal (116). The in vitro responsiveness to PHA of peripheral blood lymphocytes from *B. taurus* cows and calves subjected to one to four experimental infestations with ten female and five male *D. andersoni* was significantly reduced during third and fourth exposures (121). Rabbits infested with *I. ricinus* had reduced in vitro responsiveness to Con A (90).

Boophilus microplus infestation of purebred *B. taurus* caused a marginal decrease in numbers of T lymphocytes, beginning with the second infestation and lasting until the end of a fourth exposure (44). Peripheral blood lymphocytes collected from *B. microplus*–infested hosts were less responsive to in vitro stimulation with PHA than were similar cells of tick-free animals. Differences were attributed to a direct effect on T lymphocytes rather than the slight decrease in T-cell numbers. *B. microplus* saliva and salivary glands reduced the in vitro responsiveness to PHA of T lymphocytes collected from uninfested cattle (44). In this report (44), immunosuppression induced by *B. microplus* salivary-gland factors was linked to factors other than PGE_2.

These authors subsequently concluded that PGE_2 in *B. microplus* saliva suppressed the in vitro responses of bovine peripheral blood mononuclear cells to PHA (43). Indeed, the importance of tick salivary PGE_2 as a potential immunosuppressant had already been proposed (75, 79). The *B. microplus* saliva studied was collected in capillary tubes containing heparinized blood, plasma, or washed red blood cells in plasma, or in feeding chambers containing plasma (43). Capillary-tube collections were performed at 35°C for 16 h, and feeding-chamber collections of saliva at 35°C for 24 h, before PGE_2 levels were determined via radioimmunoassay. The saliva used to suppress PHA responses contained 33 ng PGE_2 ml^{-1} and 1.30 mg protein ml^{-1}. The primary half-life of PGE_2 in blood is less than one minute (101). In this study (43), the immunosuppressive effect of standard PGE_2 differed from that of *B. microplus* saliva. The role of PGE_2 in tick-mediated host immunosuppression remains to be established.

In the future, the immunosuppressive properties of tick saliva need to be determined for samples of whole saliva and saliva depleted of PGE_2 by immunoprecipitation. The relative importance of PGE_2 and salivary proteins in modulation of host immunity must be determined. Both prostaglandins and proteins introduced into the host during feeding might contribute to host immunosuppression.

Tick salivary-gland extracts prepared daily from engorging female *D. andersoni* were cultured in vitro with lymphocytes derived from normal (uninfested) mice to determine how the extracts affected host responsiveness to Con

A and LPS (71). Day 0 salivary-gland extracts (from unfed ticks) suppressed Con A responses by 6.8%; whereas suppression was 44.1% for the day 1 extracts and 68.4% for the day 9 extract. Exposure to salivary-gland extracts, prepared on days 0–9 of engorgement, enhanced B-lymphocyte responsiveness to LPS. In a similar manner, peripheral blood lymphocytes collected from purebred *B. indicus* and purebred *B. taurus* were suppressed in response to Con A and enhanced in reactivity to LPS upon in vitro cultivation with *D. andersoni* salivary-gland extracts prepared on days 0–9 of feeding (72). The overall in vitro responsiveness of *B. indicus* cells was greater than that of similar *B. taurus* cells, although the percent suppression of Con A responses was similar in both groups.

T lymphocytes of infested laboratory animals and cattle had reduced in vitro responses to T-cell polyclonal activators. Salivary-gland extracts and saliva suppressed normal T-cell reactivity upon coculture in vitro. Reduced Con A responses were not attributed to salivary gland–extract cytotoxicity for T cells or to shifts in dose-response curves for polyclonal stimulators. Mitogen-induced proliferation substituted for activation by immunogen-major histocompatibility molecule complex (MHC) (93), and Con A acted in part by binding the T-lymphocyte receptor for antigen (48). Tick salivary gland–derived molecules reduce host T-lymphocyte function, which could suppress regulatory and effector pathways involved in acquired resistance.

Salivary-gland extracts of *Dermacentor reticulatus* reduced the activity of natural killer cells collected from healthy human blood donors (52). Natural killer cells participate in the innate immune defenses that lyse tumor and virus-infected target cells in a MHC-unrestricted manner (102). Natural killer cells do not develop secondary responses, and IFN-γ enhances their activity (102). These cells have yet to be implicated in acquired resistance. However, reduced natural killer cell function could impair defenses against tick-borne pathogens, particularly viruses.

Cytokines orchestrated all aspects of immunoregulation (51), and pathogens have developed strategies for suppressing the complex web of cytokine interactions (55). Salivary-gland extracts prepared daily from engorging female *D. andersoni* were assessed for their ability to influence cytokine elaboration by normal murine macrophages and T lymphocytes (71). The macrophage cytokines evaluated were IL-1 and tumor necrosis factor–α (TNF-α), while the T-lymphocyte cytokines studied were IL-2, IL-4, and IFN-γ. Cytokine levels were monitored using bioassays, because transcription of cytokine mRNA increased in the absence of protein release (127).

Suppression of IL-1 elaboration by salivary-gland extracts prepared on days 0–5 ranged from 89.9 to 61.6%, whereas extracts from days 0–9 significantly reduced TNF-α release by 62.5–94.6% (71). T-lymphocyte production of IL-2 was suppressed by 14.1–31.9%, whereas IFN-γ levels decreased by 8.7–57.0%

(71). Subsequent studies in this laboratory revealed that IL-4 levels were not altered by tick salivary-gland extracts. Trypsin sensitivity and lack of suppressive activity in a total lipid extract indicated that one or more proteins caused immunosuppression.

Murine spleen cells suffered a 77.4% reduction in their ability to elaborate IL-2 in the presence of *I. dammini* saliva (103). These authors attributed the observed suppression of spleen cell in vitro responsiveness to Con A and PHA to inhibition of IL-2 production. In addition, *I. dammini* saliva dramatically reduced nitric oxide production by LPS-stimulated macrophages (103). Finally, the suppressive effects observed with saliva dilutions of 1:100 to 1:1000 appeared to result from a protein of 5.0 kDa or greater rather than from PGE_2 (103).

Other investigators found that in *B. indicus* and *B. taurus,* macrophage elaboration of IL-1 was suppressed to a similar degree by salivary-gland extracts prepared from female *D. andersoni* on days 0–5 of engorgement (72). However, *B. indicus* macrophages produced more IL-1 than *B. taurus* macrophages. Salivary-gland extracts prepared on days 0–9 suppressed TNF production by cells from either type of cattle (72). A factor contributing to innate tick resistance of cattle of *B. indicus* genetic composition appeared to be cytokine-response capabilities more vigorous than those of *B. taurus* in the presence of tick salivary-gland molecules.

Tick modulation of host immunity inhibits regulatory and effector pathways involved in acquisition and expression of acquired resistance. Inhibition of the alternative pathway of complement activation, antianaphylatoxin activity, and reduction of natural killer cell function suppresses the innate response pathways of the host immune system. Reduced antibody responses facilitate feeding by reducing host immunoglobulins reactive with tick salivary-gland molecules. Analysis of host cytokine networks reveals that macrophage and T_H1-lymphocyte function are suppressed by tick salivary-gland extracts (71). Macrophages participate in the critical process of antigen presentation and cytokine regulation of B and T lymphocytes in the immune response. Tick-mediated suppression of T_H1-lymphocyte reactivity may act to inhibit expansion of antigen specific T-lymphocyte clones (IL-2), differentiation of B lymphocytes (IL-2), activation of macrophages (IFN-γ), enhancement of natural killer cell activity (IFN-γ), and delayed hypersensitivity responses. Tick countermeasures appear to target the major pathways involved in acquired host immunuity.

In experiments using *Francisella tularenisis* (113) and *B. burgdorferi* (31), host acquired resistance induced through feeding of pathogen-free ticks impaired subsequent pathogen transmission by these species. Host resistance to tick feeding may result in an unfavorable environment for pathogen development at the tick-attachment site, and/or host immune effector elements such as antibodies may neutralize tick immunosuppressive factors that facilitate pathogen transmission. The basis for blocking pathogen transmission needs to be characterized.

Tick modulation of host immunity appears to be mediated by salivary gland–derived proteins and possibly by PGE_2. PGE_2 inhibits T_H1, but not T_H2, production of cytokines (12). Efforts to identify, isolate, and characterize tick salivary gland–derived molecules that suppress host immune responses are under way. An intriguing approach, first proposed by Titus & Ribeiro (99), to the control of tick-borne pathogens and tick infestation is the development of a vaccine that neutralizes host immunosuppressive factors introduced by the tick. A vaccine would allow more complete development and expression of host immunity to the vector and any introduced pathogens.

Acquired Resistance: a Synthesis of Current Knowledge

The findings of studies described above can be used to develop a general model for events occurring during acquisition and expression of immunological resistance to ticks. The following model accounts for recognized host defenses and tick countermeasures to these responses. Obviously, the responses of different tick and host species will vary.

Initiation of tick feeding on a previously uninfested animal presents salivary immunogens to cells in the epidermis and dermis at the bite site. The types of immunogens introduced into the host change during the course of engorgement. Proteins and other immunogenic molecules in tick saliva can be processed and presented to antigen-specific T lymphocytes in the epidermis (Langerhans cells), dermis (macrophages), and draining lymph nodes (macrophages, dendritic cells). Thus, helper T lymphocytes are presented tick immunogens in the context of class II MHC molecules on the antigen-presenting cell surface. T_H1 and T_H2 lymphocytes provide immunoregulatory signals for generation of cell-mediated and antibody responses. In addition, T_H1 lymphocytes are effector cells of delayed hypersensitivity reactions, including CBH responses elicited by feeding ticks. Immunogens, antigen-presenting cells, T lymphocytes, and cytokines contribute to the activation and differentiation of B lymphocytes, which produce tick-reactive circulating and homocytotropic antibodies. The tick-specific primary delayed hypersensitivity and antibody responses require days to weeks to develop. Additional information is needed regarding the nature of specific immunogens that induce and subsequently interact with host antitick defenses.

The introduction of tick saliva into the skin of an unsensitized host causes degranulation of mast cells, possibly via enzymatic breakdown of plasma membranes. Chemoattractants and vasoactive factors released in this manner could contribute to formation of the modest leukocytic influx observed at tick-attachment sites during primary exposure. Generation of C5a by activation of the alternative complement pathway would contribute to cellular influx at the bite site.

Ticks can modulate both innate and primary immune responses of hosts. Tick saliva contains inhibitors of the alternative pathway of complement activation, anaphylatoxins, and natural killer cells, which are all innate defenses. Tick saliva also reduces macrophage cytokine elaboration, which impairs the earliest steps in development of antitick immunity by altering signals to T and B lymphocytes. T-lymphocyte proliferative capacity is impaired, as shown by Con A responses. Furthermore, T_H1-lymphocyte production of IL-2 and IFN-γ is reduced, while T_H2-lymphocyte cytokine elaboration of IL-4 is normal. Reduced T_H1 lymphocyte function would diminish delayed type hypersensitivity to tick immunogens, which contributes to the cellular influx at attachment sites on resistant hosts. Reduced IFN-γ would impair macrophage activation.

Tick feeding induces many antibodies that vary in specificity. The roles of these antibodies in immunity to ticks remain to be defined. Reduced antibody responses to thymic-dependent antigens could result from altered function of macrophages, T lymphocytes, or other yet-to-be-described interactions. Normal IL-4 levels support the observed generation of homocytotropic antibodies. Circulating antibodies are capable of activating the classical pathway of complement, thus generating a diverse array of biological activities.

Repeated, or continuous, exposure brings feeding ticks into contact with the immune effector elements induced by primary infestation. Primary introduction of tick saliva stimulates generation of memory T and B lymphocytes, which assure a more vigorous immune response upon reinfestation. Langerhans cells, macrophages, and lymph-node dendritic cells remain important in antigen processing and presentation. Mast cells residing in the dermis and basophils become armed with surface homocytotropic antibodies capable of reacting with introduced saliva immunogens.

In resistant animals, basophils appear to be attracted to attachment sites by soluble mediators and T lymphocytes. C5a, generated by activation of complement through either the alternative or classical pathways, is chemotactic. Lymphocytes are present at the bite site within hours of attachment. The development of tick-reactive circulating antibodies capable of fixing complement via the classical pathway results in higher levels of C5a, thus contributing to a greater accumulation of basophils. At tick attachment sites, basophils and mast cells likely degranulate when tick salivary antigens complex with homocytotropic antibodies occupying cellular Fc receptors. Tick salivary antigens introduced over the course of engorgement could result in an equally long-term release of bioactive molecules from mast cells already present as well as from infiltrating mast cells.

Eosinophils act as feedback regulators of basophil and mast cell–derived bioactive molecules. Basophil and mast cell–derived histamine, leukotriene B_4, and the eosinophil chemotactic factor of anaphylaxis attract eosinophils to the bite site. Moreover, C5a is chemotactic for eosinophils. Neutrophils are attracted to the attachment site by a high-molecular-weight chemotactic factor derived from basophils and mast cells and activated by soluble factors. Macrophages observed in feeding lesions are probably involved in the elimination of antigen, as well as in antigen processing and presentation to any influx of immunocompetent cells.

Very little is known about the specific mechanisms that disrupt tick feeding, impair egg production, and reduce viability. Histamine, eosinophil basic protein, prostaglandins, leukotrienes, enzymes, and other biologically active molecules released by basophils, eosinophils, and mast cells might all be contributing factors. Histamine, basophil granules, and eosinophil granules are implicated in expression of acquired resistance. Although antibodies, lymphocytes, complement, and other elements of the immune and inflammatory responses participate in acquired resistance, their specific roles must still be established. For example, how do tick-reactive immunoglobulins affect the feeding ixodid? What is the importance, if any, of prostaglandins, leukotrienes, eosinophil basic protein, and other constituents of cells attracted to bite sites on resistant hosts? Do cytokines and acute-phase proteins have a direct role in expression of acquired resistance?

Cytotoxic T lymphocytes and natural killer cells have not been implicated in antitick resistance. However, this topic remains to be fully investigated. The immunogens involved in acquisition and expression of acquired resistance must be identified and characterized. Certainly, a considerable body of interesting work awaits us.

Antitick Vaccines

Acaricide resistance is a significant threat to effective control of ticks and tick-borne diseases (66, 107). The development of antitick vaccines represents one of the most promising alternatives to chemical control and has the advantages of target-species specificity, environmental safety, lack of human health risk, ease of administration, and cost. Dramatic advances in immunology, molecular biology, and biotechnology have made the development of antitick vaccines a reality. However, the effectiveness of such vaccines will depend on a thorough understanding of acquired host immunity to ticks. This section cites selected examples of antitick vaccine research. For more information, the reader is referred to other reviews (50, 97, 114, 119, 122, 126).

Numerous investigators have used whole-tick extracts and salivary-gland homogenates as vaccine immunogens, which induced variable levels of protection (122). These investigators selected logical immunogens based on knowledge of host immunity to ticks; however, the most promising vaccines contain defined immunogens (50, 122). Particular attention has focused on concealed or novel immunogens not normally introduced into the host during feeding (122, 126), and tick gut immunogens have received the most attention.

An elegant series of studies resulted in development of a recombinant vaccine to limit *B. microplus* infestation (97). A recombinant protein, derived from a well-characterized membrane glycoprotein of digest cells, Bm86, was expressed in various systems, with a eukaryotic expression system thought to be optimal (97). Antibodies to Bm86 inhibit digest-cell endocytosis. An additional *B. microplus* immunogen, Bm91, was identified and purified for use as an antitick vaccine immunogen (82). Multiple immunogens may produce an even more effective vaccine.

Opdebeeck and coworkers (68) induced immunity by vaccination with *B. microplus* midgut membranes. Subsequent studies revealed that carbohydrate-dependent determinants were essential for maintaining integrity of the immunogen (67).

Sera derived from mice infested with *I. ricinus* nymphs and rabbits infested with adults of the same species reacted on immunoblots with a 25-kDa soluble integumental protein of adult females, which was also present in extracts of engorged nymphs (86). This protein, which was detected in fed nymphs, appeared to be associated with integument formation (18). The 25-kDa antigen was detected in the salivary glands of *I. ricinus* (18). Immunoglobulins reactive with the 25-kDa salivary gland integument protein reacted with a 20-kDa molecule in the integument of *R. appendiculatus* (86). In turn, cattle vaccinated with the *R. appendiculatus* 20-kDa molecule were resistant to challenge with adults of the same species (87).

Primary tissue-culture cells from developing larvae of *A. americanum* induced a significant level of immunity to infestations with adult *A. americanum* and *D. andersoni* (118). Preparations of *A. americanum* digestive tract cell brush borders were effective vaccine immunogens (119). In vitro cultures of immortalized tick gut cells derived from genetically engineered digestive tract cells of several ixodid species (SK Wikel, RN Ramachandra & DK Bergman, unpublished observation), have been successfully used to induce immunity to tick infestation.

Careful attention must be given to the characterization of cell-mediated and humoral responses induced by immunization. Vaccines will be used in genetically diverse populations of livestock. Consequently, the immunogens selected must contain epitopes capable of being processed and presented to immunocompetent cells in the context of varied MHC antigens. Research should not be limited to the development of recombinant vaccines, which have the potential for containing epitopes recognized by a limited repertoire of MHC molecules.

ACKNOWLEDGMENTS

Research of the author is supported by USDA Grants 88-34116-3759 and 88-34116-4206, the Oklahoma Center for Advancement of Science and Technology, Pfizer Animal Health, and the Oklahoma Agricultural Experiment Station (project number OKL02174). Appreciation is expressed to Douglas K Bergman, Rangappa N Ramachandra, John R Sauer, and Nancy E Wikel for their review of the manuscript and countless helpful discussions.

Literature Cited

1. Alexander JO'D. 1986. The physiology of itch. *Parasitol. Today* 2:345–51
2. Allen JR. 1973. Tick resistance: basophils in skin reactions of resistant guinea pigs. *Int. J. Parasitol.* 3:195–200
3. Allen JR. 1989. Immunology of interactions between ticks and laboratory animals. *Exp. Appl. Acarol.* 7:5–13
4. Allen JR, Doube BM, Kemp DH. 1977. Histology of bovine skin reactions to *Ixodes holocyclus*, Neuman. *Can. J. Comp. Med.* 41:26–35
5. Allen JR, Kemp DH. 1982. Observations on the behavior of *Dermacentor andersoni* larvae infesting normal and resistant guinea pigs. *Parasitology* 84:195–204
6. Allen JR, Khalil HM, Graham JE. 1979. The location of tick salivary gland antigens, complement and immunoglobulin in the skin of guinea pigs infested with *Dermacentor andersoni* larvae. *Immunology* 38:467–72
7. Allen JR, Khalil HM, Wikel SK. 1979. Langerhans cells trap tick salivary gland antigens in tick-resistant guinea pigs. *J. Immunol.* 122:563–65
8. Anderson BE, Dawson JE, Jones, DC, Wilson KH. 1991. *Ehrlichia chaffeensis,* a new species associated with human ehrlichiosis. *J. Clin. Microbiol.* 29:2838–42
9. Bakken JS, Dumler JS, Chen SM, Eckeman MR, Van Etta LC, Walker DH. 1994. Human granulocytic ehrlichiosis in the upper midwest United States. *J. Am. Med. Assoc.* 272:212–18
10. Barriga OO, da Silva SS, Azevedo JSC. 1993. Inhibition and recovery of tick function in cattle repeatedly infested with *Boophilus microplus. J. Parasitol.* 79:710–15

11. Bennett GF. 1969. *Boophilus microplus* (Acari: Ixodidae): experimental infestations on cattle restrained from grooming. *Exp. Parasitol.* 26:323–28
12. Betz M, Fox BS. 1991. Prostaglandin E$_2$ inhibits production of Th1 lymphokines but not of Th2 lymphokines. *J. Immunol.* 146:108–13
13. Bissonnette EY, Rossignol PA, Befus AD. 1993. Extracts of mosquito salivary gland inhibit tumor necrosis factor α release from mast cells. *Parasite Immunol.* 15:27–33
14. Brossard M. 1982. Rabbits infested with adult *Ixodes ricinus* L.: effects of mepyramine on acquired resistance. *Experientia* 38:702–4
15. Brossard M, Fivaz V. 1982. *Ixodes ricinus* L.: mast cells, basophils and eosinophils in the sequence of cellular events in the skin of infested or reinfested rabbits. *Parasitology* 85:583–92
16. Brossard M, Girardin P. 1979. Passive transfer of resistance in rabbits infested with adult *Ixodes ricinus* L.: humoral factors influence feeding and egg laying. *Experientia* 35:1395–96
17. Brossard M, Monneron JP, Papatheodorou V. 1982. Progressive sensitization of circulating basophils against *Ixodes ricinus* L. antigens during repeated infestations of rabbits. *Parasite Immunol.* 4:355–61
18. Brossard M, Rutti B, Haug T. 1991. Immunological relationships between host and ixodid ticks. In *Parasite-Host Associations: Coexistence or Conflict,* ed. CA Toft, A Aeschliman, L Bolic, pp. 177–200. Oxford: Oxford Univ. Press
19. Brown SJ. 1985. Immunology of ac-

quired resistance to ticks. *Parasitol. Today* 1:165–71

20. Brown SJ. 1988. Western blot analysis of *Amblyomma americanum*–derived stage-specific and shared antigens using serum from guinea pigs expressing resistance. *Vet. Parasitol.* 28:163–71

21. Brown SJ, Askenase PW. 1986. *Amblyomma americanum:* physiochemical isolation of a protein derived from the tick salivary gland that is capable of inducing immune resistance in guinea pigs. *Exp. Parasitol.* 62:40–50

22. Brown SJ, Galli SJ, Gleich GJ, Askenase PW. 1982. Ablation of immunity to *Amblyomma americanum* by anti-basophil serum: cooperation between basophils and eosinophils in expression of immunity to ectoparasites (ticks) in guinea pigs. *J. Immunol.* 129: 790–96

23. Brown SJ, Shapiro SZ, Askenase PW. 1984. Characterization of tick antigens inducing host immune resistance. I. Immunization of guinea pigs with *Amblyomma americanum* derived salivary gland extracts and identification of an important salivary gland protein antigen with guinea pig anti-tick antibodies. *J. Immunol.* 133:3319–25

24. Champagne DE. 1994. The role of salivary vasodilators in bloodfeeding and parasite transmission. *Parasitol. Today* 10:430–33

25. Cross MC, Cupp MS, Galloway AL, Enriquez FJ. 1993. Modulation of murine immunological responses by salivary gland extract of *Simulium vittatum* (Diptera: Simuliidae). *J. Med. Entomol.* 30: 928–35

26. de Castro JJ, Cunningham MP, Dolan TT, Dransfield RD, Newson RM, Young AS. 1985. Effects on cattle of artificial infestations with the tick *Rhipicephalus apppendiculatus. Parasitology* 90:21–33

27. de Castro JJ, Newson RM. 1993. Host resistance in cattle tick control. *Parasitol. Today* 9:13–17

28. den Hollander N, Allen JR. 1985. *Dermacentor variabilis:* acquired resistance to ticks in BALB/c mice. *Exp. Parasitol.* 59:118–29

29. Dessaint J-P, Capron AR. 1993. Survival strategies of parasites in their immunocompetent hosts. In *Immunology and Molecular Biology of Parasitic Infections,* ed. KS Warren, pp. 87–99. Oxford: Blackwell

30. Dharampaul S, Kaufman WR, Belosevic M. 1993. Differential recognition of saliva antigens from the ixodid tick *Amblyomma hebraeum* (Acari: Ixodidae)

by sera from infested and immunized rabbits. *J. Med. Entomol.* 30:262–66

31. Dizij A, Arndt S, Seitz HM, Kurtenbach K. 1994. *Clethrionomys glareolus* acquired resistance to *Ixodes ricinus:* a mechanism to prevent spirochete inoculation? In *Advances in Lyme Borreliosis Research,* ed. R Cevenini, V Sambri, M LaPlaca, pp. 228–31. Bologna

32. Dvorak HF, Dvorak AM, Simpson BA, Richardson HB, Leskowitz S, Karnovsky MJ. 1970. Cutaneous basophil hypersensitivity. II. A light and electron microscopic description. *J. Exp. Med.* 132:558–82

33. Fivaz BH. 1989. Immune suppression induced by the brown ear tick *Rhipicephalus appendiculatus* Neuman, 1901. *J. Parasitol.* 75:946–52

34. Frank MM, Fries LF. 1989. Complement. In *Fundamental Immunology,* ed. WE Paul, pp. 679–701. New York: Raven

35. George JE, Osburn RL, Wikel SK. 1985. Acquisition and expression of resistance by *Bos indicus* and *Bos indicus* × *Bos taurus* calves to *Amblyomma americanum* infestation. *J. Parasitol.* 71:174–82

36. Gill HS, Boid R, Ross CA. 1986. Isolation and characterization of salivary gland antigens from *Hyalomma anatolicum anatolicum. Parasite Immunol.* 8: 11–25

37. Girardin P, Brossard M. 1989. Effects of cyclosporin A on humoral immunity to ticks and on cutaneous immediate and delayed hypersensitivity reactions to *Ixodes ricinus* L. salivary-gland antigens in re-infested rabbits. *Parasitol. Res.* 75:657–62

38. Golde WT, Burkot TR, Sviat S, Keen MG, Mayer LW, et al. 1993. The major histocompatibility complex–restricted response of recombinant inbred strains of mice to natural tick transmission of *Borrelia burgdorferi. J. Exp. Med.* 177: 9–17

39. Goldman DP, Artenstein AW, Bolan CD. 1992. Human ehrlichiosis: a newly recognized tick-borne disease. *Am. Fam. Phys.* 46:199–208

40. Gordon JR, Allen JR. 1987. Isolation and characterization of salivary gland antigens from the female tick, *Dermacentor andersoni. Parasite Immunol.* 9: 337–52

41. Gordon JR, Allen JR. 1991. Factor V and VII anticoagulant activities in the salivary glands of feeding *Dermacentor andersoni* ticks. *J. Parasitol.* 77:167–70

42. Hoogstraal H. 1981. Changing patterns

of tick-borne diseases in modern society. *Annu. Rev. Entomol.* 26:75–99

43. Inokuma H, Kemp DH, Willadsen P. 1994. Prostaglandin E_2 production by the cattle tick (*Boophilus microplus*) into feeding sites and its effect on the response of bovine mononuclear cells to mitogen. *Vet. Parasitol.* 53:293–99

44. Inokuma H, Kerlin RL, Kemp DH, Willadsen P. 1993. Effects of cattle tick (*Boophilus microplus*) infestation on the bovine immune system. *Vet. Parasitol.* 47:107–18

45. Jaworski DC, Muller MT, Simmen FA, Needham GR. 1990. *Amblyomma americanum:* identification of tick salivary gland antigens from unfed and early feeding females with comparisons to *Ixodes dammini* and *Dermacentor variabilis. Exp. Parasitol.* 70:217–26

46. Jaworski DC, Rossell R, Coons LB, Needham GR. 1992. Tick (Acari: Ixodidae) attachment cement and salivary gland cells contain similar immunoreactive polypeptides. *J. Med. Entomol.* 29:305–9

47. Johnston TH, Bancroft MJ. 1918. A tick-resistant condition in cattle. *R. Soc. Queensland Proc.* 30:219–317

48. Kanellopoulos JM, DePetris S, Leca G, Crumpton MJ. 1985. The mitogenic lectin from *Phaseolus vulgaris* does not recognize the T3 antigen of human T-lymphocytes. *Eur. J. Immunol.* 15:478–86

49. Kaufman WR. 1989. Tick-host interactions: a synthesis of current concepts. *Parasitol. Today* 5:47–56

50. Kay BH, Kemp DH. 1994. Vaccination against arthropods. *Am. J. Trop. Med. Hyg.* 50:87–96 (Suppl.)

51. Kroemer G, de Alboran IM, Gonzalo JA, Martinez AC. 1993. Immunoregulation by cytokines. *Crit. Rev. Immunol.* 13:163–91

52. Kubes M, Fuchsberger N, Labuda M, Zuffova E, Nuttall P. 1994. Salivary gland extracts of partially fed *Dermacentor reticulatus* ticks decrease natural killer cell activity *in vitro. Immunology.* 82:113–16

53. Lebel B, Scheinmann P, Canu P, Burtin C. 1980. Histamine levels in mouse tissues of different strains: influence of sex. *Agents Act.* 10:149–50

54. Limo MK, Voigt WP, Tumbo-Oeri AG, ole-Moi Yoi OK. 1991. Purification and characterization of an anticoagulant from the salivary glands of the ixodid tick *Rhipicephalus appendiculatus. Exp. Parasitol.* 72:418–29

55. Marrack P, Kappler J. 1994. Subversion of the immune system by pathogens. *Cell* 76:323–32

56. Matsuda H, Fukui K, Kiso Y, Kitamura Y. 1985. Inability of genetically mast cell-deficient w/w^v mice to acquire resistance against larval *Haemaphysalis longicornis* ticks. *J. Parasitol.* 71:443–48

57. Matsuda H, Nakano T, Kiso Y, Kitamura Y. 1987. Normalization of antitick response of mast cell-deficient w/w^v mice by intracutaneous injection of cultured mast cells. *J. Parasitol.* 73:155–60

58. Mbow ML, Rutti B, Brossard M. 1994. IFN-G, IL-2 and IL-4 mRNA expression in the skin and draining lymph nodes of BALB/c mice repeatedly infested with nymphal *Ixodes ricinus* ticks. *Cell. Immunol.* 156:254–61

59. McSwain JL, Essenberg RC, Sauer JR. 1982. Protein changes in the salivary glands of the female lone star tick, *Amblyomma americanum*, during feeding. *J. Parasitol.* 68:100–6

60. Mossman TR, Coffman RL. 1989. TH1 and TH2 cells: differential patterns of lymphokine secretion lead to different functional properties. *Annu. Rev. Immunol.* 7:145–73

61. Newson RM, Chiera JW. 1989. Development of resistance in calves to nymphs of *Rhipicephalus appendiculatus* (Acarina: Ixodidae) during test feeds. *Exp. Appl. Acarol.* 6:19–27

62. Nithiuthai S, Allen JR. 1984. Significant changes in epidermal Langerhans cells of guinea-pigs infested with ticks (*Dermacentor andersoni*). *Immunology* 51:133–41

63. Nithiuthai S, Allen JR. 1984. Effects of ultraviolet irradiation on epidermal Langerhans cells in guinea-pigs. *Immunology* 51:143–51

64. Nithiuthai S, Allen JR. 1984. Effects of ultraviolet irradiation on the acquisition and expression of tick resistance in guinea-pigs. *Immunology* 51:153–59

65. Nithiuthai S, Allen JR. 1985. Langerhans cells present tick antigens to lymph node cells from tick sensitized guinea-pigs. *Immunology* 55:157–63

66. Nolan J. 1990. Acaricide resistance in single and multi-host ticks and strategies for control. *Parassitologia* 32:145–53

67. Opdebeeck JP, Lee RP, Wong JYM, Jackson LA. 1992. Vaccination of cattle against *Boophilus microplus*. In *Proc. First Int. Conf. Tick-borne Pathogens at the Host Vector Interface: An Agenda for Research*, ed. UG Munderloh, TJ Kurtti. pp. 233–39. St. Paul: Univ. Minn. Coll. Agric.

68. Opdebeeck JP, Wong JYM, Jackson,

LA, Dobson, C. 1988. Vaccines to protect Hereford cattle against the cattle tick, *Boophilus microplus. Immunology* 63:363–67

69. Paine SH, Kemp DH, Allen JR. 1983. In vitro feeding of *Dermacentor andersoni* (Stiles): effects of histamine and other mediators. *Parasitology* 86:419–28

70. Papatheodorou V, Brossard M. 1987. C3 levels in the sera of rabbits infested and reinfested with *Ixodes ricinus* L. and in midguts of fed ticks. *Exp. Appl. Acarol.* 3:53–59

71. Ramachandra RN, Wikel SK. 1992. Modulation of host-immune responses by ticks (Acari: Ixodidae): effect of salivary gland extracts on host macrophages and lymphocyte cytokine production. *J. Med. Entomol.* 29:818–26

72. Ramachandra RN, Wikel SK. 1995. Effects of *Dermacentor andersoni* (Acari: Ixodidae) salivary gland extracts on *Bos indicus* and *Bos taurus* lymphocytes and macrophages: in vitro cytokine elaboration and lymphocyte blastogenesis. *J. Med. Entomol.* 32:338–45

73. Rechav Y. 1987. Resistance of Brahman and Hereford cattle to African ticks with reference to serum gamma globulin levels and blood composition. *Exp. Appl. Acarol.* 3:219–32

74. Rechav Y, Clarke FC, Dauth J. 1991. Acquisition of immunity in cattle against the blue tick, *Boophilus decoloratus. Exp. Appl. Acarol.* 11:51–56

75. Ribeiro JMC. 1987. Role of saliva in blood-feeding by arthropods. *Annu. Rev. Entomol.* 463–78

76. Ribeiro JMC. 1989. Role of saliva in tick/host interactions. *Exp. Appl. Acarol.* 7:15–20

77. Ribeiro JMC. 1989. Vector saliva and its role in parasite transmission. *Exp. Parasitol.* 69:104–6

78. Ribeiro JMC, Evans PM, McSwain, JL, Sauer J. 1992. *Amblyomma americanum:* characterization of salivary prostaglandins E_2 and F_{2a} by RP-HPLC/bioassay and gas chromatography-mass spectrometry. *Exp. Parasitol.* 74:112–16

79. Ribeiro JMC, Makoul GT, Levine J, Robinson DR, Spielman A. 1985. Antihemostatic, antiinflammatory and immunosuppressive properties of the saliva of a tick, *Ixodes dammini. J. Exp. Med.* 161:332–44

80. Ribeiro JMC, Makoul GT, Robinson DR. 1988. *Ixodes dammini:* evidence for salivary prostacyclin secretion. *J. Parasitol.* 74:1068–69

81. Ribeiro JMC, Spielman A. 1986. *Ixodes dammini:* salivary anaphylatoxin inactivating activity. *Exp. Parasitol.* 62:292–97

82. Riding GA, Jarmey J, McKenna RV, Pearson R, Cobon GS, Willadsen P. 1994. A protective "concealed" antigen from *Boophilus microplus.* Purification, localization and possible function. *J. Immunol.* 153:5158–66

83. Riek RF. 1962. Studies on the reactions of animals to infestation with ticks. VI. Resistance of cattle to infestation with the tick *Boophilus microplus. Aust. J. Agric. Res.* 13:532–50

84. Rikihisa Y. 1991. The tribe *Ehrlichieae* and ehrlichial diseases. *Clin. Microbiol. Rev.* 4:286–308

85. Roehrig JT, Piesman J, Hunt AR, Keen MG, Happ CM, Johnson BJB. 1992. The hamster immune response to tick-transmitted *Borrelia burgdorferi* differs from the response to needle inoculated, cultured organisms. *J. Immunol.* 149:3648–53

86. Rutti B, Brossard M. 1989. Repetitive detection by immunoblotting of an integumental 25-kDa antigen in *Ixodes ricinus* and a corresponding 20-kDa antigen in *Rhipicephalus appendiculatus* with sera from pluriinfested mice and rabbits. *Parasitol. Res.* 75:325–29

87. Rutti B, Brossard M. 1992. Vaccination of cattle against *Rhipicephalus appendiculatus* with detergent solubilized tick tissue proteins and purified 20 kDa protein. *Ann. Parasitol. Hum. Comp.* 67:50–54

88. Sauer JR, McSwain JL, Bowman AS, Essenberg RC. 1995. Tick salivary gland physiology. *Annu. Rev. Entomol.* 40:245–67

89. Sauer JR, McSwain JL, Essenberg RC. 1994. Cell membrane receptors and regulation of cell function in ticks and blood-sucking insects. *Int. J. Parasitol.* 24:33–52

90. Schorderet S, Brossard M. 1993. Changes in immunity to *Ixodes ricinus* by rabbits infested at different levels. *Med. Vet. Entomol.* 7:186–92

91. Shapiro SZ, Buscher G, Dobbelaere DAE. 1987. Acquired resistance to *Rhipicephalus appendiculatus* (Acari: Ixodidae): identification of an antigen eliciting resistance in rabbits. *J. Med. Entomol.* 24:147–54

92. Shapiro SZ, Voigt WP, Fujisaki K. 1986. Tick antigens recognized by serum from a guinea pig resistant to infestation with the tick *Rhipicephalus appendiculatus. J. Parasitol.* 72:454–63

93. Sharon N. 1983. Lectin receptors as lymphocyte surface markers. *Adv. Immunol.* 34:213–98

94. Spach DH, Liles WC, Campbell GL, Quick RE, Anderson DE Jr, Fritsche TR. 1993. Tick-borne diseases in the United States. *New Engl. J. Med.* 329: 936–47
95. Steeves EBT, Allen JR. 1990. Basophils in skin reactions of mast cell–deficient mice infested with *Dermacentor variabilis. Int. J. Parasitol.* 20:655–67
96. Strother GR, Burns EC, Smart LI. 1974. Resistance of purebred Brahman, Hereford, and Brahman × Hereford crossbred cattle to the lone star tick, *Amblyomma americanum* (Acarina: Ixodidae). *J. Med. Entomol.* 11:559–63
97. Tellam RL, Smith D, Kemp DH, Willadsen P. 1992. Vaccination against ticks. In *Animal Parasite Control Utilizing Biotechnology,* ed. WK Yong. pp. 303–31. Boca Raton, FL: CRC
98. Thomford JW, Conrad PA, Telford SR III, Mathiesen D, Bowman BH, et al. 1994. Cultivation and phylogenetic characterization of a newly recognized pathogenic protozoan. *J. Inf. Dis.* 169: 1050–56
99. Titus RG, Ribeiro JMC. 1990. The role of vector saliva in transmission of arthropod-borne disease. *Parasitol. Today* 6:157–60
100. Trager W. 1939. Acquired immunity to ticks. *J. Parasitol.* 25:57–81
101. Trang LE. 1980. Prostaglandins and inflammation. *Sem. Arthritis Rheum.* 9: 153–90
102. Trichieri G. 1989. Biology of natural killer cells. *Adv. Immunol.* 47:187–376
103. Urioste S, Hall LR, Telford SR III, Titus RG. 1994. Saliva of the Lyme disease vector, *Ixodes dammini,* blocks cell activation by a nonprostaglandin E_2–dependent mechanism. *J. Exp. Med.* 180: 1077–85
104. Ushio H, Watanabe N, Kiso Y, Higuchi, S, Matsuda H. 1993. Protective immunity and mast cell and eosinophil responses in mice infested with larval *Haemaphysalis longicornis* ticks. *Parasitol. Immunol.* 15:209–14
105. Voss-McCowan M. 1991. *Changes in the midgut of female* Amblyomma americanum *as a result of feeding upon hosts expressing different levels of acquired anti-tick resistance.* PhD thesis. Univ. N. Dak. 310 pp.
106. Ward PA, Dvorak HF, Cohen S, Yoshida T, Data R, Selvaggio SS. 1975. Chemotaxis of basophils by lymphocyte-dependent and lymphocyte independent mechanisms. *J. Immunol.* 114: 1523–27
107. Wharton RH, Roulston WJ. 1970. Resistance of ticks to chemicals. *Annu. Rev. Entomol.* 15:381–404
108. Wheeler CM, Coleman JL, Benach JL. 1991. Salivary gland antigens of *Ixodes dammini* are glycoproteins that have interspecies cross-reactivity. *J. Parasitol.* 77:965–73
109. Whelen AC, Richardson LK, Wikel SK. 1984. Ixodid tick antigen recognized by the infested host: immunoblotting studies. *IRCS Med. Sci.* 12:910–1
110. Whelen AC, Richardson LK, Wikel SK. 1986. Dot-ELISA assessment of guinea pig antibody responses to repeated *Dermacentor andersoni* infestations. *J. Parasitol.* 72:155–62
111. Whelen AC, Wikel SK. 1993. Acquired resistance of guinea pigs to *Dermacentor andersoni* mediated by humoral factors. *J. Parasitol.* 79:908–12
112. Wikel SK. 1979. Acquired resistance to ticks. Expression of resistance by C4 deficient guinea pigs. *Am. J. Trop. Med. Hyg.* 28:586–90
113. Wikel SK. 1980. Host resistance to tick-borne pathogen by virtue of resistance to tick infestation. *Ann. Trop. Med. Parasitol.* 74:103–4
114. Wikel SK. 1982. Immune responses to arthropods and their products. *Annu. Rev. Entomol.* 27:21–48
115. Wikel SK. 1982. Histamine content of tick attachment sites and the effects of H_1 and H_2 histamine antagonists on the expression of resistance. *Ann. Trop. Med. Parasitol.* 76:179–85
116. Wikel SK. 1982. Influence of *Dermacentor andersoni* infestation on lymphocyte responsiveness to mitogens. *Ann. Trop. Med. Parasitol.* 76:627–32
117. Wikel SK. 1985. Effects of tick infestation on the plaque-forming cell response to a thymic dependent antigen. *Ann. Trop. Med. Parasitol.* 79: 195–98
118. Wikel SK. 1985. Resistance to ixodid tick infestation induced by administration of tick tissue culture cells. *Ann. Trop. Med. Parasitol.* 79:513–18
119. Wikel SK. 1988. Immunological control of hematophagous arthropod vectors: utilization of novel antigens. *Vet. Parasitol.* 29:235–64
120. Wikel SK, Graham JE, Allen JR. 1978. Acquired resistance to ticks. IV. Skin reactivity and *in vitro* lymphocyte responsiveness to salivary gland antigen. *Immunology* 32:257–63
121. Wikel SK, Osburn RL. 1982. Immune responsiveness of the bovine to repeated low-level infestations with *Dermacentor andersoni. Ann. Trop. Med. Parasitol.* 76:405–14

122. Wikel SK, Ramachandra RN, Bergman DK. 1992. Immunological strategies for suppression of vector arthropods: novel approaches in vector control. *Bull. Soc. Vector Ecol.* 17:10–19

123. Wikel SK, Ramachandra RN, Bergman DK. 1994. Tick-induced modulation of the host immune response. *Int. J. Parasitol.* 24:59–66

124. Wikel SK, Whelen AC. 1986. Ixodid-host immune interaction. Identification and characterization of relevant antigens and tick-induced host immunosuppression. *Vet. Parasitol.* 20: 149–74

125. Willadsen P. 1980. Immunity to ticks. *Adv. Parasitol.* 18:293–313

126. Willadsen P, McKenna RV. 1991. Vaccination with 'concealed' antigens: myth or reality? *Parasite Immunol.* 13: 605–16

127. Yao Q, Cua D, Sensintaffer J, Dimacalli M, Stohlman S. 1994. Increased TNF transcription in the absence of protein release by mouse hepatitis virus infected macrophages. *FASEB J.* 8:A252

Annu. Rev. Entomol. 1996. 41:23–43

CULICOIDES VARIIPENNIS AND BLUETONGUE-VIRUS EPIDEMIOLOGY IN THE UNITED STATES[1]

Walter J. Tabachnick

Arthropod-Borne Animal Diseases Research Laboratory, USDA, ARS, University Station, Laramie, Wyoming 82071

KEY WORDS: arbovirus, livestock, vector capacity, vector competence, population genetics

ABSTRACT

The bluetongue viruses are transmitted to ruminants in North America by *Culicoides variipennis*. US annual losses of approximately $125 million are due to restrictions on the movement of livestock and germplasm to bluetongue-free countries. Bluetongue is the most economically important arthropod-borne animal disease in the United States. Bluetongue is absent in the northeastern United States because of the inefficient vector ability there of *C. variipennis* for bluetongue. The vector of bluetongue virus elsewhere in the United States is *C. variipennis sonorensis*. The three *C. variipennis* subspecies differ in vector competence for bluetongue virus in the laboratory. Understanding *C. variipennis* genetic variation controlling bluetongue transmission will help identify geographic regions at risk for bluetongue and provide opportunities to prevent virus transmission. Information on *C. variipennis* and bluetongue epidemiology will improve trade and provide information to protect US livestock from domestic and foreign arthropod-borne pathogens.

INTRODUCTION

Arthropod-borne pathogens cause significant mortality and morbidity to humans and animals. The bluetongue viruses, which cause bluetongue disease in ruminants, are among the most important arthropod-borne animal pathogens in the United States. The primary North American arthropod vector of the bluetongue viruses is a biting midge, *Culicoides variipennis* (Diptera: Cera-

---

[1]The US Government has the right to retain a nonexclusive royalty-free license in and to any copyright covering this paper.

topogonidae). International regulations prohibit the movement of livestock and their germplasm from countries harboring animals with bluetongue viruses to countries with livestock considered virus free. Many US livestock populations are infected with bluetongue viruses or are located in areas endemic for the disease. Consequently, the US livestock industry has suffered estimated annual losses of $125 million because of lost trade in cattle, sheep, or their germplasm to bluetongue-free countries, such as those in the European Union (24, 88).

This paper focuses on the role of *C. variipennis* in bluetongue disease epidemiology. We must understand the mechanisms controlling *C. variipennis'* ability to vector the bluetongue viruses if we are to reduce the effects of bluetongue disease in North America. Investigations of other arthropod-borne, disease-causing pathogens involve the same issues discussed here: (*a*) the importance of the vector in disease epidemiology, (*b*) the importance of the vector in pathogen biology, and (*c*) features of the vector that provide opportunities for controlling the disease.

BLUETONGUE DISEASE

Bluetongue disease was first described in 1902 as malarial catarrhal fever in South African sheep (39). Soon afterwards, the disease agent was recognized as filterable (109). In 1944, South African species in the genus *Culicoides* were recognized as vectors of bluetongue virus (15). From their origins in Africa, bluetongue viruses have spread to the Middle East, Asia, the Americas, and Australia (25). Bluetongue disease in the United States was first described as "soremuzzle" in Texas (33), and bluetongue virus was isolated from sheep with soremuzzle in California in 1952 (64). *C. variipennis* was subsequently identified as a vector in the United States (94).

Bluetongue Pathogenesis

Several reviews discuss bluetongue-virus pathogenesis in ruminants (62, 89, 90). Bluetongue viruses can infect several domestic and wild ruminant species, but the most significant diseases occur in sheep. Clinical signs include a rise in temperature lasting 5–7 days; hyperemia and swelling of the buccal and nasal mucosa; profuse salivation; swollen tongue; hemorrhage in the mucosal membranes of the mouth; oral erosions; and hemorrhage in the coronary bands of the hoof, which produces lameness. Sheep may vomit because of lesions in the esophagus and pharynx, which can lead to aspiration of the ruminal contents, pneumonia, and frequently death. The severity of the disease varies according to virus serotype and is less drastic in indigenous than in introduced sheep (18, 40, 89, 90, 128). Although sheep mortality may range from 5 to

50%, bluetongue-virus infections in many regions of the world produce no overt disease (90).

Clinical bluetongue disease in cattle is rare. Cattle develop a prolonged viremia lasting several weeks, which may allow the virus to survive during winter or other times when vector populations are absent or small. Controversy has surrounded the extent of clinical signs. Currently, cattle are thought to develop signs of bluetongue disease rarely («5% of infected animals), and the virus is considered to have little if any effect on reproduction. However, early prenatal infection may lead to embryonic death. Fetuses infected at later stages of gestation survive without persistent infection, and infected animals develop specific antibodies (62, 89–92).

Bluetongue Viruses

Bluetongue viruses are double-stranded RNA viruses in the genus *Orbivirus,* family Reoviridae. They have 24 serotypes worldwide. The bluetongue genome (molecular weight 12×10^6) can be resolved via polyacrylamide electrophoresis into 10 gene segments, which encode ten mRNAs for seven structural and three nonstructural proteins. The viruses are icosahedral particles with the RNA genome encapsidated in a double-layered protein coat (38, 59, 101, 102). The outer coat contains two major proteins, VP2 and VP5, while the inner coat consists of two major proteins, VP3 and VP7. Serotype specificity resides in VP2. VP7 contains epitopes that react with group-specific bluetongue antibodies. The roles of the minor core proteins (VP1, VP4, and VP6), as well as of the nonstructural proteins (NS1, NS2, NS3), have been described (17).

The bluetongue-virus proteins function differently in mammalian and insect cells. Treatment of viruses with trypsin or chymotrypsin results in cleavage of VP2 from the outer capsid, producing an infectious subparticle. Further treatment uncoats the inner core (71). Although inner-core particles are not as infectious as intact virus to mammalian cells, all three particles are equally infective to susceptible *C. variipennis.* (68).

The genetic diversity among bluetongue serotypes and related orbiviruses, e.g. African horse sickness viruses and epizootic hemorrhagic disease viruses, is known (28–30, 100–102). Nucleotide sequences for VP3, VP5, VP7, NS1, and NS3 reveal close genetic relationships between orbiviruses from the same geographic region (29, 30). Bluetongue viruses from Australia form a distinct topotype. Within each region, topotypes contain similar serotypes. For instance, VP3 sequences show that BLU-1 in Australia is related to the Australian topotype consisting of serotypes 3, 9, 16, 20, 21, and 23, whereas BLU-1 in South Africa is closer to South African serotypes 3 and 9 (29). A close phylogenetic relationship, based on VP3 gene-sequence data, between a US

BLU-2 isolate and serotypes 1, 6, and 12 from Jamaica and Honduras supports proposals that BLU-2 was introduced into the United States from the Caribbean (95, 106).

The relationships between viral diversity and the different *Culicoides* vectors present in different regions are unknown. Studies of an RNA arbovirus, vesicular stomatitis virus, suggested that arboviruses evolved in a punctuated fashion as they entered new environments and were transmitted by new arthropod vectors (84). New variants of bluetongue virus could also result from gene segment reassortment between serotypes. Reassortment frequencies are higher in *Culicoides* vectors than in sheep (103). The influence of reassortment on the population biology of the bluetongue viruses is unknown (27). However, *Culicoides* vectors clearly influence bluetongue-virus variation and biology (123).

Bluetongue Epidemiology

WORLDWIDE   Bluetongue viruses are distributed worldwide between latitudes 40°N and 35°S. They infect ruminants wherever suitable *Culicoides* vectors are present. There are regional differences in the viruses, in species of *Culicoides,* and in clinical signs in infected animals. Clinical bluetongue disease is not generally seen in the Central American–Caribbean Basin, where BLU-1, -3, -4, -6, -8, -12, -14, and -17 have been observed, presumably vectored by *C. insignis* (124, 127). Similarly clinical disease is not generally observed in Australia, where BLU-1, -3, -9, -15, -16, -20, -21, and -23 are transmitted by *C. wadai, C. brevitarsis, C. fulvus,* and *C. actoni* (110, 111). In Asia, various *Culicoides* spp. are the suspected vectors for BLU-1, -2, -3, -9, -12, -14–19, -20, -21, and -23 (34, 112, 126). Clinical disease does appear in Africa and the Middle East, where serotypes 1–19, 22, and 24 are found. There, the primary vector is *C. imicola,* although *C. tororoensis, C. milnei, C. obsoletus,* and *C. schultzei* may play minor roles (68). In many regions of the world, the vectors of the bluetongue viruses are unknown (123).

Bluetongue viruses caused disease outbreaks between 1956 and 1960 in Spain and Portugal (97). The principal vector, *C. imicola,* was also the vector of the related African horse sickness viruses in Spain (70). The potential for bluetongue virus in Europe has resulted in animal health restrictions to ensure bluetongue-free animal imports. The range of *C. imicola* in Europe does not extend beyond the Iberian Peninsula because of inhospitable climate (97). However, *C. obsoletus* and *C. pulicaris,* capable vectors of bluetongue in the laboratory (43), are common in Northern Europe (69). Without an understanding of the vector ability of European *Culicoides* spp. in the field, bluetongue incursions into Europe remain a concern.

NORTH AMERICA   US bluetongue serotypes are 2, 10, 11, 13, and 17 (3, 26). In a serologic survey for bluetongue-virus antibody in US cattle, Metcalf et al (72) examined ~20,000 sera and found that 17.8% were positive for bluetongue antibody. Bluetongue antibody prevalence ranged from 0 to 79% in different states. It was highest in southern and western states and lowest in northern states, where prevalence was ≤2%. These values were confirmed in several serologic surveys conducted during the ensuing two decades involving thousands of additional samples of cattle sera (93). Seropositive animals have been traced to origin to determine the contribution of animal movement to the presence of bluetongue-positive animals in northern states. Of more than 32,000 cattle tested in New York State, only 14 were seropositive, and all of these were originally from high seroprevalence regions of the United States (58).

Bluetongue viruses were found in the Okanagan Valley of British Columbia in 1976 and 1987 (113). The dire implications for the Canadian livestock trade prompted a serosurvey of more than 175,000 cattle between 1976 and 1992 (14, 113). BLU-11 was identified (13), but virus was not observed in collections of *C. variipennis* in the Okanagan Valley (65), and evidence of bluetongue-virus infection outside the valley has not been observed (14). Bluetongue viruses may have been transmitted to animals in this region after being introduced from the United States. However, despite this example of sporadic transmission, these viruses are apparently not endemic to Canada (14).

CULICOIDES VARIIPENNIS

Much evidence suggests that *C. variipennis* is the primary North American vector of the bluetongue viruses: (*a*) The species is widespread. (*b*) Many studies show it feeds on wild and domestic ruminants. (*c*) In the laboratory, feeding on susceptible ruminant hosts has resulted in infection, and under laboratory conditions, it transmits virus to susceptible hosts. (*d*) Bluetongue viruses have been isolated on numerous occasions from field-collected *C. variipennis* (2, 8, 10, 20–23, 41, 46, 48, 50, 51, 60, 61, 63, 66, 67, 82, 85, 86, 103, 108, 114, 116, 122, 125, 131, 134). *C. insignis* vectors the bluetongue viruses in South and Central America and through the northern extension of its range in southern Florida (31, 55, 124). Little evidence points to a major role for other *Culicoides* species in North America. *C. venustus* (53), *C. debilipalpis,* and *C. stellifer* (74) support little or no infection in the laboratory, and other species have not been incriminated in bluetongue epizootics. *C. brookmani* or *C. boydi* might serve as bluetongue vectors for desert bighorn sheep in areas of California, as indicated by their abundance and the near absence of *C. variipennis* in these habitats (78). Various models attempting to predict bluetongue-virus transmission are based on climactic variables relating to insect activity (44, 135).

Systematics

C. variipennis is in the subgenus *Monoculicoides*. Based on morphologic variations in collections throughout the United States, *C. variipennis* was divided into five subspecies: *C. variipennis variipennis, C. variipennis sonorensis, C. variipennis occidentalis, C. variipennis australis,* and *C. variipennis albertensis* (132). Other workers believed these forms were species because of the absence of morphologic hybrids in regions where they were sympatric (12, 54). However, studies with laboratory-reared *C. variipennis* suggested that some of the morphologic characters were modified by the environment and thus invalid for use in classification (35). The difficulty in defining subspecies relationships resulted in a single grouping known as the *C. variipennis* complex (133).

Isozyme electrophoretic analyses of ~200 *C. variipennis* populations from the United States have helped define population and subspecies genetic relationships (36, 37, 115, 118). Populations, analyzed for genetic variation using 11–21 different isozyme-encoding loci, confirmed only three members in the *C. variipennis* complex (36, 37, 118): *C. variipennis variipennis* (northern regions of the United States), *C. variipennis sonorensis* (from Florida to California; north to Virginia and Ohio; and in the west as far north as Washington and British Columbia), and *C. variipennis occidentalis* (Arizona to California, north to Washington and British Columbia).

Limited gene flow was found between California *C. variipennis sonorensis* and *C. variipennis occidentalis* populations (36, 118). *C. variipennis sonorensis* larvae resided in highly polluted organic habitats, while *C. variipennis occidentalis* larvae inhabited highly saline habitats, e.g. Borax Lake, California (36). Collections from approximately 100 New England sites yielded only *C. variipennis variipennis* populations (37). No isozyme genes diagnostic for subspecies have been identified, although other molecular markers can be used (96). However, gene frequencies, and genetic similarities based on gene-frequency differences, showed that populations within a subspecies are more closely related to each other than to populations from other subspecies, regardless of geographic proximity. Populations classified morphologically as *C. variipennis australis* are genetically *C. variipennis sonorensis* (FR Holbrook & WJ Tabachnick, unpublished observations). Populations from the Gulf Coast of the United States in which *C. variipennis variipennis* and *C. variipennis sonorensis* occurred in the same larval habitat lacked any genetic hybrids (FR Holbrook & WJ Tabachnick, unpublished observations). This observation suggests that the three subspecies are indeed separate species. However, pending formal descriptions, they should continue to be referred to using the subspecies designation.

Population Genetics

Active *C. variipennis sonorensis* adults are not present during the winter in Colorado. Populations overwinter predominantly as larvae in permanent aquatic habitats (5, 6). In one study in this region (117), the gene frequencies of permanent populations remained stable through two seasons at all but two loci. *C. variipennis sonorensis* populations collected from temporary larval sites, which did not persist through the winter, showed genetic changes each summer that resulted from chance effects when these habitats were colonized each spring (117). Migration, at a rate of ~2.15 *C. variipennis sonorensis* per generation (regardless of population size), allowed temporary populations to differentiate from permanent populations through chance and prevented permanent populations from differentiating from one another. This study defined the major features of Colorado *C. variipennis sonorensis* population genetics (117): (*a*) Permanent larval populations maintain genetic stability; (b) no migration occurs between permanent populations during the winter; and (*c*) temporary populations are founded each spring and differ from permanent populations owing to chance (117). Although *C. variipennis* may disperse several kilometers (56; FR Holbrook, personal communication), as well as longer distances via wind (105–107), such dispersal did not affect population differentiation. Temporary populations, separated by only a few hundred meters, were not panmictic and were genetically differentiated (117). Weather is the major factor shaping the genetic structure of Colorado *C. variipennis sonorensis* populations.

Differentiation among other US populations provides additional evidence for the effect of weather on *C. variipennis* population dynamics and genetic structure. One measure of genetic variation between two populations is the average (av) genetic distance (*D*) based on allele-frequency differences. The av *D* among all populations in a region compared with the av *D* in another region quantifies the differences in regional genetic diversity. The av $D \pm$ SE (n = number of pairwise population comparisons) showed significant differences (36, 37): among New England *C. variipennis variipennis* populations, av $D = 0.046 \pm 0.002$ ($n = 276$); among Colorado *C. variipennis sonorensis,* av $D = 0.040 \pm 0.010$ ($n = 21$); among California *C. variipennis sonorensis,* av $D = 0.010 \pm 0.007$ ($n = 171$); and between California *C. variipennis occidentalis,* av $D = 0.132 \pm 0.017$ ($n = 10$). *C. variipennis sonorensis* populations within a single Colorado county exhibited levels of genetic diversity similar to those of New England *C. variipennis variipennis.* Both of these populations have significantly higher genetic diversity than California *C. vari-*

ipennis sonorensis collected throughout the state. In Colorado and New England, temporary populations arising each spring probably generate temporally differentiated populations and thus greater genetic diversity. Populations of *C. variipennis sonorensis* in California, which enjoy longer seasons and have active adult migration, experience greater gene flow and less genetic differentiation (36). In contrast, California *C. variipennis occidentalis* populations showed the highest genetic differentiation, as a result of their wide geographic separation from one another and a lack of gene flow with nearby *C. variipennis sonorensis* (36).

In summary, the three major groups in the *C. variipennis* complex share only limited gene flow; their population genetics are influenced primarily by weather; and their distributions are associated with the North American distribution of the bluetongue viruses. The distribution of bluetongue in the United States has been stable for more than 20 years, despite potential change resulting from animal movement within the United States and into Canada, and from migration of infected *C. variipennis* between regions. However, infected exotic *Culicoides* spp. could still enter the United States (105), as could foreign livestock carrying exotic forms of the bluetongue viruses, particularly those from the Central American–Caribbean Basin. *Culicoides* spp. do not respect national, regional, or political boundaries (69).

Vector Capacity

Traits associated with arthropod ability to transmit pathogens, such as host preference, biting or feeding rates, gonotrophic cycle, population densities, and vector longevity, determine vector capacity. Vector capacity also depends on vector competence, which involves the ability of the vector to be infected with the pathogen, the ability of the vector to infect progeny by transovarial transmission, and the ability of the vector to transmit the pathogen to a suitable host (see 9, 11, 32, 73, 98, 120).

Information on *C. variipennis* vector capacity is limited. Population densities vary throughout geographic regions. Although *C. variipennis variipennis* larvae and adults are often found on dairies throughout New York State (80, 104), *C. variipennis sonorensis* larvae and adults on southern California dairies are even more abundant—by one to two orders of magnitude (75, 76). If biting rates are related to adult density, this relationship would explain why bluetongue viruses are not endemic in the northeastern United States, where biting rates are probably substantially reduced (77). Flight activity may also influence biting rates. *C. variipennis* flight activity depends on light intensity and temperature, and most flight occurs at dusk and dawn (6, 57). Population variation regarding flight has not been studied.

Another component of vector capacity is the daily survivorship of adult females. Infected females must survive the incubation period to allow the pathogen to replicate so that transmission to an animal may follow. The extrinsic incubation period is 10–14 days at 23°C in *C. variipennis* (21) but varies substantially with temperature (82). Survivorship estimates of *C. variipennis* in the field are based on parity rates, determined by observing pigment granules deposited in the abdominal cuticle after blood feeding (1, 16), and on estimates of the gonotrophic cycle determined by examining degenerative relics in the ovariole pedicel (5, 81). The daily survivor rate in New York State was estimated at 0.62–0.88 (80), and a similar survivor rate was estimated in the western United States (83, 134). Since these data are based on estimates of gonotrophic cycles that are governed by temperature (79), more field studies are needed to assess population variation.

The limited information on vector capacity indicates population and regional variations in *C. variipennis* biting rates, extrinsic incubation time, and perhaps—although more data are needed—daily survivorship. Clearly more results must be gathered about variation in vector capacity in field populations and about the effects of this variation on bluetongue transmission (77).

Vector Competence

C. variipennis transmits bluetongue viruses, African horse sickness virus, akabane virus, and epizootic hemorrhagic disease viruses (7, 19, 22, 42, 52, 60, 61). Laboratory studies have not provided evidence for transovarial transmission of bluetongue viruses from infected *C. variipennis* females to their progeny (47, 86). Therefore it is unlikely that transovarial transmission is a major overwintering mechanism for the virus when adult vectors are not active. Information concerning variation in the ability of infected *C. variipennis* to transmit bluetongue virus is limited. *C. variipennis sonorensis,* containing 2.7–5.1 $\log_{10}$ $TCID_{50}$ (tissue culture infectious doses) per fly regularly transmitted bluetongue virus, while flies with ≤ 2.5 $\log_{10}$ $TCID_{50}$ did not (41). However, the difficulties in determining transmission rates in the laboratory have prevented evaluations of transmission variation among field populations, using different serotypes and viruses.

SUSCEPTIBILITY TO INFECTION   More information is available on *C. variipennis sonorensis* susceptibility to infection with the bluetongue viruses than for other vector-competence traits. However, most of this information is based on studies of a single laboratory colony, known as the 000, sonora strain, or AA colony (49). The transmission studies cited above used this strain. Studies of AA colony flies showed that bluetongue viruses adsorb to host red blood cells and can be observed inside red blood cells up to two days after a *C. variipennis sonorensis* blood meal (108). The *C. variipennis sonorensis* peritrophic membrane did not prevent virus infection of the midgut epithelium, which may occur in the first few hours after ingestion. Bluetongue-virus replication occurs in midgut cells, and the viral particles exit these cells through the basolateral extracellular membrane into extracellular spaces. Virus infection did not result in *C. variipennis sonorensis* cytopathology (108). In tests of three different bluetongue serotypes infecting another colony of *C. variipennis sonorensis* (2), virus first appeared in midgut cells and then in secondary target tissues, e.g. hindgut, fat body, salivary gland, thoracic muscle, and ovarian tissues, excluding follicles and eggs. Salivary gland involvement is particularly important, since this organ delivers the pathogen to a susceptible host during subsequent blood feedings. Within four days after blood feeding, bluetongue virus can be detected in *C. variipennis sonorensis* salivary glands, in salivary gland cytoplasm, in plasma membranes of acinar cells, extracellularly, and within cisternae of vacuoles and endoplasmic reticulum (87). Similar information on bluetongue-virus replication in other subspecies, populations, and other *Culicoides* spp. is lacking.

A key feature of *C. variipennis* infection with bluetongue virus is the likely interference via the mesenteron or gut barrier. *C. variipennis* that were intrathoracically inoculated with bluetongue virus showed higher infection rates than those fed virus via a blood meal (85). Infection rates for intrathoracically inoculated *C. variipennis* have approached 100%, even in colonies that exhibited only 30% infection following ingestion of an infected blood meal (45, 116). In all likelihood, a gut barrier prevents some *C. variipennis* from becoming infected through blood meals. In contrast, inoculation bypasses the midgut, so the infection rates of inoculated insects are higher. The nature of this gut barrier or its role in determining infection of *C. variipennis* field populations is unknown.

In addition to environmental circumstances, *C. variipennis* susceptibility to infection with bluetongue viruses depends on several factors: the subspecies, the population, and the strain of *C. variipennis;* the strain of the virus; any circumstances that may alter the physiologic condition of the insect; the temperature of extrinsic incubation; and the numbers of infectious virions in the blood meal. Virogenesis proceeds much faster, and individual bluetongue-infected *C. variipennis sonorensis* tended to have more virus antigen when incubated at higher temperatures (82). More *C. variipennis sonorensis* females fed with blood meals containing $\geq 10^6$ pfu/ml were infected than those fed on lower concentrations, and no flies were infected from blood meals containing $\leq 10^{-4}$ pfu/ml (46). Susceptibility to infection varies with nutritional status of the larvae: Poor larval nutrition and crowding resulted in small *C. variipennis sonorensis* females that were more susceptible than larger females (WJ Tabachnick, unpublished observations). Different groups from the same generation of the AA colony showed significant variation in BLU-4 infection rates, which casts doubt on the accuracy of laboratory assessments of vector competence (41). AA colony showed stable infection rates of ~30% with BLU-11 and -17 for several years (48, 49). These observations are consistent with findings that *C. variipennis* infection rates depend on the insect strain, virus isolate, and serotype (50). The infection rates of two *C. variipennis* colonies differed with each of the US serotypes. However, because differences depended on the serotype, the response of either colony to one serotype could not be predicted based on the response to another (67).

Different factors influencing *C. variipennis*–bluetongue virus interactions probably cause infection rates to vary among different insect strains and different viruses. The absence of exotic bluetongue viruses, e.g. Central American–Caribbean bluetongue serotypes, from North America may result from different vector competence and capacity levels of *C. variipennis* for these serotypes— these characteristics need to be evaluated in North American vectors. An understanding of the variety of factors influencing vector competence will require information on the underlying mechanisms, e.g. functional interactions between insect and virus proteins and the effects of environmental factors.

POPULATION HETEROGENEITY   Although the specific factors that influence susceptibility to bluetongue infection remain unknown, susceptibility certainly

Table 1 Susceptibility to infection by bluetongue viruses in *Culicoides variipennis* populations from different US states, 1978–1990[a]

State	Number of populations	Av. no. insects tested/ population ± SE	Av. % infected insects/population ± SE	Subspecies
New York	5	313 ± 2.2	2.7 ± 0.4	*variipennis*
New Jersey	2	154 ± 95.0	0.6 ± 0.6	*variipennis*
Maryland	3	175 ± 75.0	1.0 ± 1.0	*variipennis*
Virginia	1	617	0.8	??
Montana	1	123	4.1	??
Missouri	2	530 ± 189.0	3.2 ± 0.8	??
Nebraska	4	33 ± 9.8	24.2 ± 6.9	*sonorensis*
Colorado	12	192 ± 55.9	8.9 ± 1.6	*sonorensis*
Oregon	1	29 ±	27.6 ±	*sonorensis*
California	14	157 ± 21.2	22.5 ± 2.9	*sonorensis*
California	2	308 ± 143.0	1.2 ± 1.1	*occidentalis*
Nevada	2	24 ± 13.0	11.3 ± 2.2	*sonorensis*
New Mexico	1	44	54.6	*sonorensis*
Utah	1	74	14.8	*sonorensis*
Texas	2	70 ± 58.0	30.1 ± 11.6	*sonorensis*
Florida	3	34 ± 26.7	26.1 ± 13.5	*sonorensis*

[a] RH Jones & WJ Tabachnick, unpublished observations.

varies among different subspecies and populations of *C. variipennis* (50). Table 1 shows infection rates with US bluetongue serotypes 2, 10, 11, 13, and 17 for field populations sampled throughout the United States (RH Jones & WJ Tabachnick, unpublished observations). *C. variipennis variipennis* and *C. variipennis occidentalis* are less susceptible than *C. variipennis sonorensis*. Infection of *C. variipennis sonorensis* populations varied from 1.6% in Weld County, Colorado, to 54.6% in Eddy County, New Mexico (RH Jones, unpublished observations). These rates differ for different viruses, and infection with any given serotype does not generally correlate to the rate for other serotypes.

The average infection rate of *C. variipennis* populations from a given state and the seroprevalence of bluetongue antibody in cattle from that state appear strongly correlated (121). This observation is consistent with the hypothesis that the presence of competent *C. variipennis sonorensis* determines bluetongue distribution in the United States. Although vector competence varies greatly within *C. variipennis sonorensis*, *C. variipennis variipennis* exhibits consistently low susceptibility to infection with US bluetongue-virus serotypes. Of *C. variipennis occidentalis*, only populations from Borax Lake and the Salton Sea in California have been tested for bluetongue-virus susceptibility to infection, and these groups were generally refractory. Other *C. variipennis occidentalis* populations must be tested to determine whether any show higher infection rates. *C. insignis* in southern Florida has infection rates of 20.0–60.5% and can transmit bluetongue virus in the laboratory. This species is likely the predominant bluetongue vector in south Florida (124).

Culicoides variipennis and Bluetongue Epidemiology

Several factors relating bluetongue epidemiology in North America to *C. variipennis* distributions are apparent: (*a*) Bluetongue-virus transmission is virtually absent in the northern United States despite the presence of *C. variipennis variipennis*. (*b*) *C. variipennis sonorensis* is the subspecies in endemic regions of the United States. (*c*) Only BLU-2, -10, -11, -13, and -17 have been observed in the United States, and BLU-2 occurs in isolated instances in the south. (*d*) Bluetongue-virus transmission has been virtually absent from Canada, except in the Okanagan Valley, despite the presence of *C. variipennis* sensu lato (*e*) The epidemiology has been stable despite animal movement and the potential for migration of infected *Culicoides* spp.

C. variipennis sonorensis is the primary North American vector of the bluetongue viruses. *C. variipennis variipennis* should not be considered a vector of bluetongue viruses because (*a*) it has a low susceptibility to infection in the laboratory; (*b*) no viruses have been isolated from field-collected insects; (*c*) in regions where it is the only *C. variipennis* subspecies, bluetongue transmission to ruminants has not occurred; and (*d*) environmental conditions in regions where it occurs reduce vector capacity for long periods—for example, low temperatures increase extrinsic incubation period and prolong the gonotrophic cycle, and lower densities reduce biting rates. The stable 20-year absence of bluetongue in the northeastern United States can only be explained by the nonvector status of *C. variipennis variipennis*. The distributions of *C. variipennis sonorensis* are critical for determining North American regions at risk for bluetongue. In addition, *C. variipennis sonorensis* populations probably sporadically reside in dynamic transition regions, where the fly may extend its range owing to temporary environmental conditions. These populations may cause the low levels and irregular instances of bluetongue transmission seen in such states as Indiana, Ohio, and Virginia (FR Holbrook, personal communication).

The vector status of *C. variipennis occidentalis* is less certain. The Borax Lake population has low susceptibility in the laboratory (Table 1), and bluetongue viruses have not been isolated from this population. However, this subspecies exists sympatrically with *C. variipennis sonorensis* in areas of the western United States where the bluetongue viruses are endemic. Until molecular genetic markers became available, identifying sources of viral isolates from members of the *C. variipennis* complex in western field collections was difficult. *C. variipennis occidentalis* is probably not a major North American vector of the bluetongue viruses, but this supposition must by confirmed by studies using genetic markers to identify field populations and vectors during epizootics.

Canadian bluetongue epizootics likely resulted from incursions of infected *C. variipennis sonorensis* into the Okanagan Valley from the United States or from the importation of viremic ruminants, from which resident *C. variipennis sonorensis* obtained viruses they transmitted to indigenous cattle. Little information is available regarding *C. variipennis* distributions in Canada. *C. variipennis variipennis* occurs in Ontario (118), and probably in the eastern provinces, where *C. variipennis sonorensis* is unlikely. Although *C. variipennis sonorensis* has been collected in the Okanagan Valley of British Columbia and in southern Alberta (FR Holbrook, personal communication), its distribution to the east and north is unknown. The distribution of *C. variipennis sonorensis* in Canada is important because, although some *C. variipennis sonorensis* populations may not be efficient vectors, the current limited information indicates that any region with *C. variipennis sonorensis* is at risk for bluetongue-virus transmission during the insect season.

Our ability to evaluate, predict, and perhaps interrupt the vector potential of *C. variipennis,* as well as to determine regions at risk for bluetongue-virus transmission, depends on the following: (*a*) valid distributions of the subspecies, (*b*) knowledge of genetic control mechanisms responsible for vector capacity and competence, (*c*) ability to analyze populations for genes controlling vector capacity and vector competence, and (*d*) information on environmental factors that contribute to variation in vector capacity (119–121).

Genetics of Culicoides variipennis Vector Competence

Investigations on the genetic control of *C. variipennis* susceptibility to infection with bluetongue virus demonstrated the presence of a single controlling gene in two laboratory colonies. Strains of highly susceptible and resistant *C. variipennis sonorensis* were selected from the AA and AK colonies (48, 116). Crosses between susceptible and resistant lines provided evidence for a major locus and a modifier controlling susceptibility (48). Similar studies of colony lines showed that the major controlling locus acted via a maternal effect and paternal imprinting. That is, the mother's genotype determined the progeny phenotype, and the paternal gene was always dominant in offspring (116). This inheritance pattern allowed construction of isogenic pools of flies and identification of a candidate controlling protein that was used to isolate a cDNA clone for sequencing to determine function (KE Murphy & WJ Tabachnick, unpublished observations). Once the candidate gene is identified, further studies will be necessary to determine its role in controlling vector competence variation in field populations. Vector competence is a complex trait, and consequently, it is likely that several genes and various interactions with environmental factors control variation within the species (120).

Genetic mapping studies using DNA molecular markers may in future help us identify other *C. variipennis* vector-competence and vector-capacity genes (99, 120). The long-term goals of these studies are the identification and analysis not only of these genes, but also of the environmental factors that influence them in different *C. variipennis* populations. This information will allow us to (*a*) identify the conditions enabling bluetongue-virus transmission; (*b*) interrupt transmission using releases of genetically altered, resistant *C. variipennis;* and (*c*) reduce vector capacity by changing environmental conditions that affect vector phenotypes.

CONCLUSIONS

The absence of bluetongue virus from the north and northeastern regions of the United States, temporary incursions into Canada, and the presence of only five serotypes in the United States are consistent with the predominant role of the members of the *C. variipennis* complex in transmission. We are only beginning to appreciate the complexities of arthropod-pathogen interactions. Indeed, the study of *C. variipennis* and bluetongue epidemiology in North America involves many issues common to studies of human and animal arthropod-borne pathogens. Investigations of *C. variipennis* and bluetongue virus show the critical nature of vector-virus interaction that must be understood to predict vector populations and geographic regions at risk for disease. The results may lead to novel biological control strategies, as opposed to chemical measures, to interrupt pathogen transmission and reduce the effects of disease on animal populations.

Bluetongue in North America largely depends on the distribution of *C. variipennis sonorensis*. Based on bluetongue epidemiology and current vector distributions, it is unlikely that the northern United States and large portions of eastern Canada are at risk for bluetongue-virus transmission.

Regulators establishing policies for animal movement should consider the current situation but must also consider the potential for changes in epidemiology. Caution is warranted. For instance, *C. variipennis variipennis* is probably not a bluetongue-virus vector. However, we do not understand genetic and environmental control mechanisms and the effects of new serotypes or viral variants on vector competence and capacity. Thus, vector status could change. The United States must continue to monitor animal populations in its bluetongue-free areas (130). Once we know the factors responsible for transmission, we can assess the competence of vectors in other regions. For example, we could determine the actual risk for bluetongue-virus transmission by European *Culicoides* spp.

The future for using vector-pathogen information for more effective control of arthropod-borne pathogens is promising. The information reviewed in this chapter can serve as the foundation for efforts to reduce the effect of bluetongue disease on national economies. Regionalization within countries according to the presence of bluetongue virus vectors can reduce unnecessary animal-health regulations and increase opportunities for international trade.

Molecular-biology experiments are under way to determine the mechanisms of action of arthropod traits influencing pathogen transmission. The arthropod-pathogen interactions between *C. variipennis* and bluetongue viruses, *Aedes aegypti* and dengue or yellow fever viruses, or *Anopheles* spp. and the malaria pathogen may even share similarities that will afford opportunities for general control strategies. The first step is to identify the underlying vector-pathogen interactions for different systems, the controlling genes, and the effects of the environment.

ACKNOWLEDGMENTS

I appreciate the constructive comments by S Brodie, F Holbrook, G Hunt, B Mullens, S Narang, R Nunamaker, E Schmidtmann, and W Wilson on earlier drafts of the manuscript.

Literature Cited

1. Akey DH, Potter HW. 1979. Pigmentation associated with oogenesis in the biting fly *Culicoides variipennis* (Diptera: Ceratopogonidae): determination of parity. *J. Med. Entomol.* 16:67–70
2. Ballinger ME, Jones RH, Beaty BJ. 1987. The comparative virogenesis of three serotypes of bluetongue virus in *Culicoides variipennis* (Diptera: Ceratopogonidae). *J. Med. Entomol.* 24:61–65
3. Barber TL. 1979. Temporal appearance, geographic distribution and species of origin of bluetongue virus serotypes in the United States. *Am. J. Vet. Res.* 40:1654–56
4. Barber TL, Jochim MM, eds. 1985. *Bluetongue and Related Orbiviruses.* New York: Liss
5. Barnard DR, Jones RH. 1980. *Culicoides variipennis:* seasonal abundance, overwintering, and voltinism in northeastern Colorado. *Environ. Entomol.* 9:709–12
6. Barnard DR, Jones RH. 1980. Diel and seasonal patterns of flight activity of Ceratopogonidae in northeastern Colorado. *Environ. Entomol.* 9:446–51
7. Boorman J, Mellor PS, Penn M, Jennings M. 1975. The growth of African Horse Sickness in embryonated hen eggs and the transmission by *Culicoides variipennis* Coquillet (Diptera: Ceratopogonidae). *Arch. Virol.* 47:343–49
8. Bowne JG, Jones RH. 1966. Observations on bluetongue virus in the salivary glands of an insect vector, *Culicoides variipennis*. *Virology* 30:127–33
9. Chamberlain RW, Sudia WD. 1961. Mechanism of transmission of viruses by mosquitoes. *Annu. Rev. Entomol.* 6:371–90
10. Chandler LJ, Ballinger ME, Jones RH, Beaty BJ. 1985. The virogenesis of bluetongue virus in *Culicoides variipennis.* See Ref. 4, pp. 245–53
11. DeFoliart GR, Grimstad PR, Watts DM. 1987. Advances in mosquito-borne arbovirus/vector research. *Annu. Rev. Entomol.* 32:479–505
12. Downes JA. 1978. *Culicoides variipennis* complex: a necessary re-alignment of nomenclature (Diptera: Ceratopogonidae). *J. Med. Entomol.* 12:379:63–69
13. Dulac GC, Dubuc C, Myers DJ, Taylor EA, Ward D, Sterritt WG. 1989. Incursion of bluetongue virus type 11 and epizootic hemorrhagic disease of deer type 2 for two consecutive years in the Okanagan Valley. *Can. Vet. J.* 30:351–54
14. Dulac GC, Sterritt WG, Dubuc C, Afshar A, Myers EA, et al. 1992. Incursions of orbiviruses in Canada and their serologic monitoring in the native animal population between 1962 and 1991. See Ref. 129, pp. 120–27
15. Dutoit R. 1944. The transmission of bluetongue and horse-sickness by *Culicoides.* Onderstepoort. *J. Vet. Sci. Anim. Ind.* 19:7–16
16. Dyce AL. 1969. The recognition of nulliparous and parous *Culicoides* (Diptera: Ceratopogonidae) without dissection. *J. Aust. Entomol. Soc.* 8:11–15
17. Eaton BT, Hyatt AD, Brookes SM. 1990. The replication of bluetongue virus. *Curr. Top. Microbiol. Immunol.* 162:89–118
18. Erasmus BJ. 1975. Bluetongue in sheep and goats. *Aust. Vet. J.* 51:165–70
19. Foster NM, Beckon RD, Luedke AJ, Jones RH, Metcalf HE. 1977. Transmission of two strains of epizootic hemorrhagic disease virus in deer by *Culicoides variipennis. J. Wildl. Dis.* 13:9–16
20. Foster NM, Jones RH. 1973. Bluetongue virus transmission with *Culicoides variipennis* via embryonated chicken eggs. *J. Med. Entomol.* 10:529–32
21. Foster NM, Jones RH. 1979. Multiplication rate of bluetongue virus in the vector *Culicoides variipennis* (Diptera: Ceratopogonidae) infected orally. *J. Med. Entomol.* 15:302–3
22. Foster NM, Jones RH, Luedke AJ. 1968. Transmission of attenuated and virulent bluetongue virus with *Culicoides variipennis* infected orally via sheep. *Am. J. Vet. Res.* 29:275–79
23. Foster NM, Jones RH, McCrory BR. 1963. Preliminary investigations on insect transmission of bluetongue virus in sheep. *Am. J. Vet. Res.* 24:1195–200
24. Gibbs EPJ. 1983. Bluetongue—an analysis of current problems, with particular reference to importation of ruminants to the United States. *J. Am. Vet. Med. Assoc.* 182:1190–94
25. Gibbs EPJ, Greiner EC. 1988. Bluetongue and epizootic hemorrhagic disease. In *The Arboviruses: Epidemiology and Ecology,* ed. TP Monath, pp. 39–70. Boca Raton: CRC
26. Gibbs EPJ, Greiner EC, Taylor WP, Barber TL, House JA, Pearson JE. 1983. Isolation of bluetongue serotype 2 from cattle in Florida: a serotype of blue-

tongue virus hitherto unrecognized in the western hemisphere. *Am. J. Vet. Res.* 44:2226–28

27. Gorman BM. 1990. The bluetongue viruses. *Curr. Top. Microbiol. Immunol.* 162:1–19
28. Gorman BM. 1992. An overview of the orbiviruses. See Ref. 129, pp. 335–47
29. Gould AR, McColl KA, Pritchard LI. 1992. Phylogenetic relationships between bluetongue viruses and other orbiviruses. See Ref. 129, pp. 452–60
30. Gould AR, Pritchard LI. 1990. Relationships amongst bluetongue viruses revealed by comparisons of capsid and outer protein coat nucleotide sequences. *Virus Res.* 17:31–39
31. Greiner EC, Barber TL, Pearson JE, Kramer WL, Gibbs EPJ. 1985. Orbiviruses from *Culicoides* in Florida. See Ref. 4, pp. 195–200
32. Hardy JL, Houk EJ, Kramer LD, Reeves WC. 1983. Intrinsic factors affecting vector competence of mosquitoes for arboviruses. *Annu. Rev. Entomol.* 28:229–62
33. Hardy WT, Price DA. 1952. Soremuzzle of sheep. *J. Am. Vet. Med. Assoc.* 120:23–25
34. Hassan H. 1992. Status of bluetongue in the Middle East and Asia. See 129, pp. 38–41
35. Hensleigh DA, Atchley WR. 1977. Morphometric variability in natural and laboratory populations of *Culicoides variipennis* (Diptera: Ceratopogonidae). *J. Med. Entomol.* 14:379–86
36. Holbrook FR, Tabachnick WJ. 1995. The *Culicoides variipennis* (Diptera: Ceratopogonidae) in California. *J. Med. Entomol.* In press
37. Holbrook FR, Tabachnick WJ, Brady RC. 1995. The *Culicoides variipennis* complex (Diptera: Ceratopogonidae) in the six New England States. *Vet. Med. Entomol.* In press
38. Huismans H, van Dijk AA. 1990. Bluetongue virus structural components. *Curr. Top. Microbiol. Immunol.* 162:21–42
39. Hutcheon D. 1902. Malarial catarrhal fever of sheep. *Vet. Rec.* 14:629–33
40. Jeggo MH, Gumm ID, Taylor WP. 1985. Clinical and serological response of sheep to serial challenge with different bluetongue serotypes. *Res. Vet. Sci.* 34:205–11
41. Jennings DM, Mellor PS. 1987. Variation in the responses of *Culicoides variipennis* (Diptera: Ceratopogonidae) to oral infection with bluetongue virus. *Arch. Virol.* 95:107–82
42. Jennings DM, Mellor PS. 1989. *Culicoides:* biological vectors of Akabane virus. *Vet. Microbiol.* 21:125–31
43. Jennings DM, Mellor PS. 1988. The vector potential of British *Culicoides* species for bluetongue virus. *Vet. Microbiol.* 17:1–10
44. Johnson BG. 1992. An overview and perspective on orbivirus disease prevalence and occurrence of vectors in North America. See Ref. 129, pp. 58–64
45. Jones RH, Foster NM. 1966. The transmission of bluetongue virus to embryonated chicken eggs by *Culicoides variipennis* (Diptera: Ceratopogonidae) infected by intrathoracic inoculation. *Mosq. News* 26:185–89
46. Jones RH, Foster NM. 1971. The effect of repeated blood meals infective for bluetongue on the infection rate of *Culicoides variipennis. J. Med. Entomol.* 8:499–501
47. Jones RH, Foster NM. 1971. Transovarial transmission of bluetongue virus unlikely for *Culicoides variipennis. Mosq. News* 31:434–37
48. Jones RH, Foster NM. 1974. Oral infection of *Culicoides variipennis* with bluetongue virus: development of susceptible and resistant lines from a colony population. *J. Med. Entomol.* 11:316–23
49. Jones RH, Foster NM. 1978. Relevance of laboratory colonies of the vector in arbovirus research—*Culicoides variipennis* and bluetongue. *Am. J. Trop. Med. Hyg.* 27:168–77
50. Jones RH, Foster NM. 1978. Heterogeneity of *Culicoides variipennis* field populations to oral infection with bluetongue virus. *Am. J. Trop. Med. Hyg.* 27:178–83
51. Jones RH, Foster NM. 1979. *Culicoides variipennis:* threshold to infection for bluetongue virus. *Ann. Parasitol. Hum. Comp.* 54:250
52. Jones RH, Roughton RD, Foster NM, Bando BM. 1977. *Culicoides,* the vector of epizootic hemorrhagic disease in white-tailed deer in Kentucky in 1971. *J. Wildl. Dis.* 13:2–8
53. Jones RH, Schmidtmann ET, Foster NM. 1983. Vector competence studies for bluetongue and epizootic hemorrhagic disease viruses with *Culicoides venustus* (Ceratopogonidae). *Mosq. News* 26:185–89
54. Jorgensen NM. 1979. The systematics, occurrence and host preference of *Culicoides* (Diptera: Ceratopogonidae) in southeastern Washington. *Melandria* 3:1–47
55. Kline DL, Greiner EC. 1992. Field observations on the ecology of adult and immature stages of *Culicoides* spp. as-

sociated with livestock in Florida USA. See Ref. 129, pp. 297–305

56. Lillie TH, Marquardt WC, Jones RH. 1981. The flight range of *Culicoides variipennis* (Diptera: Ceratopogonidae). *Can. Entomol.* 113:419–26

57. Linhares AX, Anderson JR. 1990. The influence of temperature and moonlight on flight activity of *Culicoides variipennis* (Coquillet) (Diptera: Ceratopogonidae)) in northern California. *Pan-Pac. Entomol.* 66:199–207

58. Lopez JW, Dubovi EJ, Cupp EW, Lein DH. 1992. An examination of the bluetongue virus status of New York State. See Ref. 129, pp. 140–46

59. Loudon PT, Liu HM, Roy P. 1992. Genes to complex structures of bluetongue viruses: structure-function relationships of bluetongue virus proteins. See Ref. 129, pp. 383–89

60. Luedke AJ, Jones RH, Jochim MM. 1967. Transmission of bluetongue between sheep and cattle by *Culicoides variipennis*. *Am. J. Vet. Res.* 28:457–60

61. Luedke AJ, Jones RH, Jochim MM. 1976. Serial cyclic transmission of bluetongue virus in sheep and *Culicoides variipennis*. *Cornell Vet.* 66:536–50

62. MacLachlan NJ, Barratt-Boyes SM, Brewer AW, Stott JL. 1992. Bluetongue virus infection of cattle. See Ref. 129, pp. 723–36

63. MacLachlan NJ, Nunamaker RA, Katz JB, Sawyer MM, Akita GY, et al. 1994. Detection of bluetongue virus in the blood of inoculated calves: comparison of virus isolation, PCR assay, and in vitro feeding of *Culicoides*. *Arch. Virol.* 136:1–8

64. McKercher DG, McGowan B, Howarth JA, Saito JK. 1953. A preliminary report on the isolation and identification of the bluetongue virus from sheep in California. *J. Am. Vet. Med. Assoc.* 122:300–11

65. McMullen BA. 1978. *Culicoides* (Diptera Ceratopogonidae) of the south Okanagan area of British Columbia. *Can. Entomol.* 110:1053–57

66. Mecham JO, Dean VC, Wigington JG, Nunamaker RA. 1990. Detection of bluetongue virus in *Culicoides variipennis* by an antigen capture enzyme-linked immunosorbant assay. *J. Med. Entomol.* 27:602–6

67. Mecham JO, Nunamaker RA. 1994. Complex interactions between vectors and pathogens: *Culicoides variipennis* (Diptera: Ceratopogonidae) with bluetongue virus. *J. Med. Entomol.* 31:903–7

68. Mellor PS. 1990. The replication of bluetongue virus in *Culicoides* vectors.

Curr. Top. Microbiol. Immunol. 162:143–61

69. Mellor PS. 1993. *Culicoides:* Do vectors respect national boundaries? *Br. Vet. J.* 149:5–8

70. Mellor PS, Boned J, Hamblin C, Graham S. 1990. Isolations of African horse sickness virus from insects made during the 1988 epizootic in Spain. *Epidemiol. Infect.* 105:447–54

71. Mertens PPC, Burroughs JN, Anderson J. 1987. Purification and properties of virus particles, infectious sub-viral particles, and cores of bluetongue virus serotypes 1 and 4. *Virology* 157:375–86

72. Metcalf HE, Pearson JE, Klingsporn AL. 1981. Bluetongue in cattle: a serologic survey of slaughter cattle in the United States. *Am. J. Vet. Res.* 44:1057–61

73. Mitchell CJ. 1983. Mosquito vector competence and arboviruses. *Curr. Top. Vector Res.* 1:63–92

74. Mullen GR, Jones RH, Bravernan Y, Nusbaum KE. 1985. Laboratory infections of *Culicoides debilipalpis* and *C. stellifer* (Diptera: Ceratopogonidae) with bluetongue virus. See Ref. 4, pp. 239–43

75. Mullens BA. 1985. Age-related activity and sugar feeding by *Culicoides variipennis* (Diptera: Ceratopogonidae) in southern California. *J. Med. Entomol.* 22:32–37

76. Mullens BA. 1989. A quantitative survey of *Culicoides variipennis* (Diptera: Ceratopogonidae) in dairy wastewater ponds in southern California. *J. Med. Entomol.* 26:559–65

77. Mullens BA. 1992. Integrated management of *Culicoides variipennis:* a problem of applied ecology. See Ref. 129, pp. 896–905

78. Mullens BA, Dada CE. 1992. Spatial and seasonal distribution of potential vectors of hemorrhagic disease viruses to peninsular bighorn sheep in the Santa Rosa mountains of southern California. *J. Wildl. Dis.* 28:192–205

79. Mullens BA, Holbrook FR. 1991. Temperature effects on the gonotrophic cycle of *Culicoides variipennis* (Diptera: Ceratopogonidae). *J. Am. Mosq. Control Assoc.* 7:588–91

80. Mullens BA, Rutz DA. 1984. Age structure and survivorship of *Culicoides variipennis* (Diptera: Ceratopogonidae) in central New York State, USA. *J. Med. Entomol.* 21:194–203

81. Mullens BA, Schmidtmann ET. 1982. The gonotrophic cycle of *Culicoides variipennis* (Diptera: Ceratopogonidae) and its implications in age-grading field

populations in New York State. *J. Med. Entomol.* 19:340–49

82. Mullens BA, Tabachnick WJ, Holbrook FR, Thompson LH. 1995. Effects of temperature on virogenesis of bluetongue virus serotype 11 in *Culicoides variipennis sonorensis. Vet. Med. Entomol.* 9:71–76

83. Nelson RL, Scrivani RP. 1972. Isolations of arboviruses from parous midges of the *Culicoides variipennis* complex, and parous rates in biting populations. *J. Med. Entomol.* 9:277–81

84. Nichol ST, Rowe JE, Fitch WM. 1993. Punctuated equilibrium and positive Darwinian evolution in vesicular stomatitis virus. *Proc. Natl. Acad. Sci. USA* 90:10424–28

85. Nunamaker RA, Mecham JO. 1989. Immunodetection of bluetongue virus and epizootic hemorrhagic disease virus in *Culicoides variipennis* (Diptera: Ceratopogonidae). *J. Med. Entomol.* 26:256–59

86. Nunamaker RA, Sieburth PJ, Dean VC, Wigington JG, Nunamaker CE, Mecham JO. 1990. Absence of transovarial transmission of bluetongue virus in *Culicoides variipennis:* immunogold labelling of bluetongue virus antigen in developing oocytes from *Culicoides variipennis* (Coquillett). *Comp. Biochem. Physiol. Ser. A* 96:19–31

87. Nunamaker RA, Wick BC, Nunamaker CE. 1988. Immunogold labelling of bluetongue virus in cryosections from *Culicoides variipennis* (Coquillett) salivary gland. *Proc. Annu. Meet. Electr. Microsc. Soc. Am., 46th,* pp. 372–73

88. Osburn BI. 1985. Economics of bluetongue in the United States. *Proc. Symp. Arbovirus Res. Australia, 4th,* p. 245. Brisbane: CSIRO

89. Parsonson IM. 1990. Pathology and pathogenesis of bluetongue infections. *Curr. Top. Microbiol. Immunol.* 162: 119–42

90. Parsonson IM. 1992. Overview of bluetongue virus infection in sheep. See Ref. 129, pp. 13–24

91. Parsonson IM. 1992. Vertical transmission of the orbiviruses. See Ref. 129, pp. 996–1000

92. Parsonson IM. 1993. Bluetongue virus infection of cattle. *Proc. US Anim. Health Assoc.* 97:120–25

93. Pearson JE, Gustafson GA, Shafer AL, Alstad AD. 1992. Distribution of bluetongue in the United States. See Ref. 129, pp. 128–39

94. Price DA, Hardy WT. 1954. Isolation of the bluetongue virus from Texas sheep: *Culicoides* shown to be a vector. *J. Am. Vet. Med. Assoc.* 121:255

95. Pritchard LI, Gould AR, Mertens PPC, Wilson WC, Thompson LH. 1994. Complete nucleotide sequence of bluetongue virus serotype 2 RNA segment 3 (Ona-A). Phylogenetic analyses reveal the probable origins of this virus. *Virus Res.* 35:247–61

96. Raich T, Archer JL, Robertson MA, Tabachnick WJ, Beaty BJ. 1993. Polymerase chain reaction approaches to *Culicoides* (Diptera: Ceratopogonidae) identification. *J. Med. Entomol.* 30:228–32

97. Rawlings P. 1993. Bluetongue virus vectors: a European perspective. *Proc. Soc. Vet. Epidemiol. Prev. Med.* pp. 134–44

98. Reeves WC. 1990. *Epidemiology and Control of Mosquito-Borne Arboviruses in California, 1943–1987.* Sacramento, CA: Calif. Mosq. Vector Control Assoc.

99. Robertson MA, Tabachnick WJ. 1992. Molecular genetic approaches *Culicoides variipennis* vector competence for bluetongue virus. See Ref. 129, pp. 271–77

100. Roy P. 1989. Bluetongue virus genetics and genome structure. *Virus Res.* 13: 179–206

101. Roy P. 1992. Bluetongue virus proteins. *J. Gen. Virol.* 73:3051–64

102. Roy P, Marshall JA, French TJ. 1990. Structure of the bluetongue virus genome and its encoded proteins. *Curr. Top. Microbiol. Immunol.* 162:43–88

103. Samal SK, El Hussein A, Holbrook FR, Beaty BJ, Ramig RF. 1987. Mixed infection of *Culicoides variipennis* with bluetongue virus serotypes 10 and 17: evidence for high frequency of reassortment in the vector. *J. Gen. Virol.* 68: 2319–29

104. Schmidtmann ET, Mullens BA, Schwager SJ, Spear S. 1983. Distribution, abundance, and a probability model for *Culicoides variipennis* on dairy farms in New York State. *Environ. Entomol.* 12:768–73

105. Sellers RF. 1992. Weather, *Culicoides,* and the distribution and spread of bluetongue and African horse sickness viruses. See Ref. 129, pp. 284–90

106. Sellers RF, Maaroof AR. 1989. Trajectory analysis and bluetongue virus serotype 2 in Florida 1982. *Can. J. Vet. Res.* 53:100–202

107. Sellers RF, Maaroof AR. 1991. Possible introduction of epizootic hemorrhagic disease of deer virus (serotype 2) and bluetongue virus (serotype 11) into British Columbia in 1987 and 1988 by

infected *Culicoides* carried on the wind. *Can. J. Vet. Res.* 55:367–70

108. Sieburth PJ, Nunamaker CE, Ellis J, Nunamaker RA. 1991. Infection of the midgut of *Culicoides variipennis* (Diptera: Ceratopogonidae) with bluetongue virus. *J. Med. Entomol.* 2:74–85

109. Spreull J. 1905. Malarial catarrhal fever (bluetongue) of sheep in South Africa. *J. Comp. Pathol. Ther.* 18:321–37

110. St George TD. 1992. Occupation of an ecological niche in the Pacific Ocean countries by bluetongue and related viruses. See Ref. 129, pp. 76–84

111. Standfast HA, Muller MJ. 1989. Bluetongue in Australia—an entomologist's view. *Aust. Vet. J.* 66:396–97

112. Standfast HA, Muller MJ, Dyce AL. 1992. An overview of bluetongue virus vector biology and ecology in the Oriental and Australasian regions of the western Pacific. See Ref. 129, pp. 253–61

113. Sterritt WG, Dulac GC. 1992. Evolving perceptions of bluetongue: a challenge for government and industry. *Can. Vet. J.* 33:109–11

114. Stott JL, Osburn BI, Bushnell R, Loomis KRE Squire. 1985. Epizootiological study of bluetongue virus infection in California livestock: an overview. See Ref. 4, pp. 571–82

115. Tabachnick WJ. 1990. Genetic variation in laboratory and field populations of the vector of bluetongue disease, *Culicoides variipennis* (Diptera: Ceratopogonidae). *J. Med. Entomol.* 27:24–30

116. Tabachnick WJ. 1991. Genetic control of oral susceptibility to infection of *Culicoides variipennis* for bluetongue virus. *Am. J. Trop. Med. Hyg.* 45:666–71

117. Tabachnick WJ. 1992. Microgeographic and temporal patterns of genetic variation in natural populations of *Culicoides variipennis*. *J. Med. Entomol.* 29:384–95

118. Tabachnick WJ. 1992. Genetic differentiation among North American populations of *Culicoides variipennis*. *Ann. Entomol. Soc. Am.* 85:140–47

119. Tabachnick WJ. 1992. Genetics, population genetics, and evolution of *Culicoides variipennis:* implications for bluetongue virus transmission in the USA and its international impact. See Ref. 129, pp. 262–70

120. Tabachnick WJ. 1994. The role of genetics in understanding insect vector competence for arboviruses. *Adv. Dis. Vector Res.* 10:93–108

121. Tabachnick WJ, Holbrook FR. 1992. The *Culicoides variipennis* complex and the distribution of the bluetongue viruses in the United States. *Proc. US Anim. Health Assoc.* 96:207–12

122. Tabachnick WJ, MacLachlan NJ, Thompson LH, Hunt GJ, Patton JF. 1995. Infection of *Culicoides variipennis* using PCR detectable bluetongue virus in cattle blood. *Am. J. Trop. Med. Hyg.* Submitted

123. Tabachnick WJ, Mellor PS, Standfast HA. 1992. Recommendations for research on *Culicoides* vector biology. See Ref. 129, pp. 977–81

124. Tanya VN, Greiner EC, Gibbs EPJ. 1992. Evaluation of *Culicoides insignis* (Diptera: Ceratopogonidae) as a vector of bluetongue virus. *Vet. Microbiol.* 32:1–14

125. Tanya VN, Greiner EC, Shroyer DA, Gibbs EP. 1993. Vector competence parameters of *Culicoides variipennis* (Diptera: Ceratopogonidae) for bluetongue virus serotype 2. *J. Med. Entomol.* 30:204–8

126. Taylor WP. 1986. The epidemiology of bluetongue. *Rev. Sci. Technol. Off. Int. Epizoot.* 5:351–56

127. Thompson LH, Mo CL, Oviedo MT, Homan EJ, Interamerican Bluetongue Team. 1992. Prevalence and incidence of bluetongue viruses in the Caribbean Basin: serologic and virologic findings. See Ref. 129, pp. 106–13

128. Uren MF, St George TD. 1985. The clinical susceptibility of sheep to four Australian serotypes of bluetongue virus. *Aust. Vet. J.* 62:175–76

129. Walton TE, Osburn BI, eds. 1992. *Bluetongue, African Horsesickness and Related Orbiviruses.* Boca Raton, FL: CRC

130. Walton TE, Tabachnick WJ, Thompson LH, Holbrook FR. 1992. An entomologic and epidemiologic perspective for bluetongue regulatory changes for livestock movement from the United States and observations on bluetongue in the Caribbean Basin. See Ref. 129, pp. 952–60

131. Wieser-Schimpf L, Wilson WC, French DD, Baham A, Foil LD. 1993. Bluetongue virus in sheep and cattle and *Culicoides variipennis* and *C. stellifer* (Diptera: Ceratopogonidae) in Louisiana. *J. Med. Entomol.* 30:719–24

132. Wirth WW, Jones RH. 1957. The North American subspecies of *Culicoides variipennis* Diptera (Heleidae). *US Dep. Agric. Tech. Bull.* 1170:1–35

133. Wirth WW, Morris C. 1985. The taxonomic complex, *Culicoides variipennis.* See Ref. 4, pp. 165–75

134. Work TM, Sawyer MM, Jessup DA, Washino RK, Osburn BI. 1990. Effects

of anesthetization and storage temperature on bluetongue virus recovery from *Culicoides variipennis* (Diptera: Ceratopogonidae) and sheep blood. *J. Med. Entomol.* 27:331–33

135. Wright JC, Getz RR, Powe TA, Nusbaum KE, Stringfellow DA, et al. 1993. Model based on weather variables to predict seroconversion to bluetongue virus in Alabama cattle. *Prev. Vet. Med.* 16:217–78

Annu. Rev. Entomol. 1996. 41:45–73

INSECT PESTS OF BEANS IN AFRICA: Their Ecology and Management

T. Abate

Institute of Agricultural Research, Nazareth Research Center, PO Box 436, Nazareth, Ethiopia

J. K. O. Ampofo

SADC/CIAT Regional Bean Programme, PO Box 2704, Arusha, Tanzania

KEY WORDS: bean entomology, *Phaseolus vulgaris,* subsistence farming, IPM

ABSTRACT

Damage by insect pests, inter alia, is considered the liming factor of bean production in Africa. This paper reviews the current status of insect pests of beans, focusing on their ecology and management, as well as the potential for integrated pest management (IPM) approaches in subsistence farming conditions, under which most beans are grown in Africa. Although numerous insect pests attack all parts of beans, bean stem maggots and bruchids are the most important field and storage pests, respectively. Foliage beetles, flower thrips, pollen beetles, pod borers, pod bugs, and sap suckers such as aphids also inflict significant damage. Control of bean pests in Africa is achieved through the use of a traditional IPM approach that consists of appropriate sowing dates, optimum plant density, varietal mixtures, intercropping, good crop husbandry, and locally available materials. Research should focus on low-input IPM approaches that encompass farmers' current practices, host-plant resistance, and natural biological control.

INTRODUCTION

Beans (*Phaseolus vulgaris*) form an important food and cash crop in Africa, particularly in the eastern, southern, and Great Lakes regions of the continent. Of all agricultural commodities produced in these parts of Africa, beans are considered the second most important source of human dietary protein (after maize) and the third most important source of calories (after maize and cassava)

0066-4170/96/0101-0045$08.00

(137). Beans are grown in the majority of African countries, but most production is in Burundi, Rwanda, Kenya, Uganda, Tanzania, Zaire, Malawi, and Mozambique (195), with an annual estimated total area of approximately 3.9 million hectares (52). The major producers (metric tons) are Kenya (414,000), Uganda (395,000), Tanzania (283,000), Zaire (269,000), Rwanda (220,000), Burundi (169,000), Ethiopia (120,000), and South Africa (104,000). Together, African nations produce over 2.5 million metric tons (~25% of the global production) of dry beans annually (137) with an average per capita consumption of 14.4 kg per year. Although most beans are produced by small-scale farmers for home consumption (193), high-quality dry and snap beans exported to markets in Europe and elsewhere constitute a significant proportion of export crops in many countries, notably Kenya, Ethiopia, Zimbabwe, and Tanzania (75, 84). The value of these crops in terms of foreign exchange is considerable.

Beans originated in the highlands of Central and South America (68, 74) and were introduced into Africa some 400 years ago (111, 145). Because the crop arrived without many of its field pests, the pest spectrum for beans in Africa differs significantly from that attacking the crop in its ancestral region. The major exceptions are storage pests, which tend to be seed borne and therefore are cosmopolitan in their distribution. However, many indigenous pests of other legumes, notably the cowpea, *Vigna* spp., and its close relatives (Leguminosae) have adapted to the crop, and every part of the plant—from roots to the mature pods—is attacked. These pests are widely distributed and cause damage to the crop wherever it is grown, even though their individual status varies with location.

Bean production systems and farmer practices vary widely across geographic regions ranging from low-latitude, subhumid eastern African highlands with high potential productivity to mid-latitude, semiarid areas at mid-altitudes with acidic soils and low potential for productivity. Depending on the environment, beans can be grown either as a sole crop or in association with other crops (32, 40, 51, 91, 110, 132, 133, 192, 196). In addition to sole crops, the major bean cropping systems consist of beans intercropped with cereals (such as maize and sorghum), root and tuber crops (e.g. potato, yam, cassava, taro, enset), field pea, and banana (195). For instance, in Malawi nearly 94% of beans are grown in association with other foods, notably maize (61); in eastern Kenya, only 40% of the crop is grown in monoculture; and in Uganda beans are rarely planted in pure stands (111, 159). In Rwanda, Burundi, Malawi, Zaire, and parts of Zambia, crops of mixed bean varieties are also common. Farmers in parts of Uganda plant beans during seven months of the year, although they know that few of their plantings are agronomically suitable, in order to ensure against crop failure (104). Such practices have implications for the incidence of pest populations and their natural enemies.

Yield potentials in the range of 3–4 tons/ha are frequently reported (8, 10,

125), but bean yields in traditional cropping systems remain low (0.6 tons/ha) because of a lack of improved varieties, poor management practices, and the ravages of pests and diseases. Farmers generally do not apply fertilizer to beans (61) nor actively control pests or diseases in the bean crop. Research on improved management strategies for increased bean productivity is under way in many national agricultural research systems (NARS) in the bean-growing regions and through regional collaborative research networks between the NARS and with CIAT (the International Center for Tropical Agriculture) with support from CIDA (the Canadian International Development Agency), SDC (the Swiss Development Cooperation), and USAID (the United States Agency for International Development), as well as many other international donor agencies.

Here, we review the current status of insect pests of beans, focusing on their importance, distribution, biology, ecology, and current management strategies, as well as the potential for integrated pest management (IPM) of bean pests in Africa.

MAGNITUDE OF THE PEST PROBLEM

Table 1 lists estimates of yield losses in beans attributed to insect pests. Despite the usefulness of such estimates as indicators of the potential ability of a particular pest to inflict damage under the worst-case scenario, these figures have to be interpreted with great caution for several reasons. First, most yield estimates are obtained from on-station experiments and therefore do not represent the usual practices and cropping systems of the African farmer. Second, such experiments are conducted in artificial environments with increased pest pressure. For example, planting is synchronized with periods of peak pest population; susceptible cultivars are most often used as the test crop; and only monocropping systems are studied. None of these experiment designs reflect the ingenuity and accumulated experience of the farmer: The farmer knows when to plant; bean cultivars used cannot be classified as susceptible; and most cropping systems in small-scale bean production in Africa are not monocrops either in time or in space. Third, most, if not all, data are obtained from limited areas, and information gathered at a specific location cannot be accurately extrapolated to the national or regional situation. Finally, the current yield-loss estimates do not take into account contributions of other factors such as vegetational diversity, diseases, and edaphic and climatic conditions. Beans in Africa are grown under diverse vegetational conditions that may profoundly affect arthropod pest and natural-enemy populations (8, 9, 12, 16, 34, 36, 142, 143). Wortmann (194) discusses in more detail the problems of estimating yield loss from pests and diseases on beans in Africa.

Table 1 Major field pests of beans in Africa, losses attributed to them, and the nature of damage

Pest	Nature of damage	Yield loss (%)	Country	Reference
Bean stem maggot *Ophiomyia* spp. (Diptera: Agromyzidae)	Stem mining and girdling	60	Mozambique	58
		50	Central Africa	40
		30–100	Kenya	59
		33–100	Tanzania	95, 189
		8–37	Ethiopia	14
		100	Uganda	72
		50–100	Zimbabwe	180
Foliage beetle *Ootheca* sp. (Coleoptera: Chrysomelidae)	Defoliation	18–31	Tanzania	98
Pod borers *Helicoverpa armigera* (Lepidoptera: Noctuidae) *Maruca testulalis* (Lepidoptera: Pyralidae)	Removal of flowers and tunneling of pods	15–25	Kenya	59
		33–53	Tanzania	92
		12–16	Ethiopia	21
Aphids *Aphis fabae* (Homoptera: Aphididae)	Sap sucking and transmission of bean common mosaic virus	90	Uganda	130
		50	Burundi	42
		37	Tanzania	178
General	—	27	Tanzania	96
Tobacco whitefly *Bemisia tabaci*	Sap sucking	14–86	Sudan	161

THE PESTS

In African fields, numerous insect pests attack all parts of common bean during all stages of growth, from seedling to stored product (3, 5, 17, 20, 22, 94, 174). Only a few of these are recognized as major pests. Their economic importance varies from one environment to another, but bean researchers generally agree that bean stem maggots (*Ophiomyia* spp.) and bruchids are the most important field and storage pests, respectively. Other important pests include aphids, flower thrips, pod borers, and pod bugs. Locally significant pests include the tobacco whitefly in Sudan, pollen beetles in southern Africa, and the so-called le cigar [*Apoderus humeralis* (Coleoptera: Curculionidae)] as well as caterpillars of the painted lady butterfly in Madagascar.

Bean pests can be classified into five broad categories according to the plant growth stage or plant part they primarily attack: seedlings, foliage, flowers, pods, and stored seeds.

SEEDLING PESTS

Bean Stem Maggots

DISTRIBUTION AND IMPORTANCE   Bean stem maggots (BSM, Diptera: Agro-myzidae), also known as bean flies (*Ophiomyia* spp.), are the most important field pests of beans in Africa. *Ophiomyia phaseoli* had been thought to be the only BSM species that attacked beans in Africa until studies by Greathead (72) revealed the presence of two other species, i.e. *Ophiomyia spencerella* and *Ophiomyia centrosematis*, in East Africa. Until recently, *O. phaseoli* was regarded as the principal and most widespread BSM species in Africa (8–10, 14, 15, 56, 74, 94, 102, 105). Recent reports (18, 19, 41, 42, 58, 69, 76, 85, 91, 113, 124, 136, 144, 175–177, 180) indicate that at least one of these three BSM species is common in each of the major bean-growing countries in Africa. *O. phaseoli* and *O. spencerella* are the most important of the three species; *O. centrosematis* usually occurs rarely and in small numbers (18, 19, 42, 72). The distribution of both *O. phaseoli* and *O. centrosematis* ranges throughout tropi-cal and subtropical Africa, Asia, and Australia, but *O. spencerella* has not been recorded outside of Africa.

BIOLOGY   Adults of the three BSM species have similar external morphologi-cal features, and male aedeagal characters are used to distinguish them. Puparia are distinguished by their color and by the characteristics of posterior spiracles (33, 72, 85, 179). Puparia of *O. phaseoli* and *O. centrosematis* are brown, whereas those of *O. spencerella* are shiny black, with a gray ventral surface. The posterior spiracles in *O. phaseoli* are somewhat bifurcated and have eight to nine lobes; in *O. centrosematis* they are blunt with three lobes (see also 94).

The eggs of BSM are white, slender, and about 1 mm long. They are laid singly on different parts of the seedling. *O. phaseoli* oviposits mostly on the newly emerging leaf; some eggs may also be laid on the stem and petioles in older plants. *O. spencerella* deposits its eggs mostly in the hypocotyl, although sometimes it may lay them on the stem as well. *O. centrosematis* prefers the stem for egg laying. Each female can lay approximately 100 eggs in her lifetime. Incubation takes 2–4 days, depending on the temperature. On hatch-ing, the small white maggots mine in the leaf and then bore into the stem, which produces cracks near soil level. The larvae pass through three instars over a period of 6–14 days. The mature larvae pupate mostly in the stem and sometimes in petioles or in the soil.

The puparia are barrel-shaped and about 3 mm long. The pupal period lasts about 7 days, after which the shiny black adult flies, measuring approximately 2 mm, emerge. Mating takes place 2–6 days after emergence, and egg laying begins 2–4 days after mating. Thus, depending on the temperature, the life cycle is completed in 20–30 days. Further details on BSM biology can be found elsewhere (26, 27, 29, 39, 67, 72, 86, 116, 178).

Larvae damage seedlings by feeding on the stem, which results in characteristic swelling and cracking just above ground level. Damaged plants become yellowish, wilted (signs often confused with symptoms of moisture stress or root diseases), and stunted and may eventually die unless they form adventitious (secondary) roots. The success of adventitious roots depends on the amount of moisture available to the plant (17). More than 30 species of cultivated and wild plants in the family Leguminosae reportedly serve as hosts of BSM. Sunn hemp (*Crotalaria juncea*) and several species of wild *Crotalaria* are important sources of infestation (8, 15, 17, 72).

ECOLOGY   The prevalence of BSM species is heavily influenced by several climatic and edaphic factors and agronomic practices. Reports from many countries indicate that *O. phaseoli* and *O. centrosematis* are more prevalent at lower altitudes and warmer climatic conditions, whereas *O. spencerella* is more abundant at higher altitudes and cooler, wetter environments (18, 42, 175, 176). In most instances, *O. phaseoli* is more common in early-sown beans, whereas *O. spencerella* is the dominant species in late-sown beans (18, 58, 91, 136, 176). In studies by Letourneau (113) in Malawi, the responses of the above two species to the effects of varietal mixtures, soil fertility, and cropping patterns varied; BSM numbers increased with increases in phosphorus levels in the soil, whereas intercropping with maize did not affect pest numbers. By contrast, several reports (8, 10, 91, 122, 192, 196) suggest that pest damage is reduced in mixed-cropped beans. Studies by Abate (18) suggest that BSM infestation levels and species composition are influenced by one or more combinations of several environmental factors and cultural practices, including temperature, crop variety, growth stage of the host plant, soil fertility, and cropping patterns.

MANAGEMENT   BSM management by farmers begins with cultural practices. Management decisions include sowing time, optimum plant population, varietal mixtures, and intercropping. Good crop husbandry is also a crucial success factor. Although researchers have identified host plant–resistance factors (1, 2, 8, 13, 22, 35), this information has not yet reached the farmer.

Numerous pupal parasitoids attack BSM (8, 11, 15, 42, 72) and cause significant mortality, thus providing good natural biological control. Pupal predation also occurs. *Opius phaseoli* (Hymenoptera: Braconidae) and *Eucoilidea* sp. (Hymenoptera: Cynipidae) are the major parasitoids. The former mainly attacks *Ophiomyia phaseoli,* whereas the latter is the dominant parasitoid of *O. spencerella* and *O. centrosematis.*

Effective seed-dressing insecticides have been identified (8, 14, 33, 186), but their use in subsistence farming is limited. Integrated pest management of BSM might best be achieved through the use of resistant cultivars, coupled with appropriate cultural practices (10, 35).

OTHER SEEDLING PESTS   Several other insect species attack common bean seedlings (17, 45, 46, 162, 181). Notable among these are cutworms, mainly *Agrotis* spp. (Lepidoptera: Noctuidae); termites, *Microtermes* spp. (Isoptera: Termitidae); and the two-spotted cricket, *Gryllus bimaculatus* (Orthoptera: Gryllidae). These insects may have economic significance in some countries, but in most instances, they are considered minor pests.

FOLIAGE PESTS

Foliage Beetles

DISTRIBUTION AND IMPORTANCE   Many members of the family Chrysomelidae attack and defoliate bean plants during the seedling stages. The more important foliage beetles (Coleoptera: Chrysomelidae) include *Ootheca bennigseni, Ootheca mutabilis,* and *Medythia quaterna* (syn. *Luperodes lineata*). *Ootheca* spp. are widely distributed in eastern and southern Africa. Although specific yield loss estimates are not available, these beetles are reportedly major pests in Zambia (175) and important pests in Kenya, Burundi, and Rwanda (94). They are considered minor pests in Zimbabwe (69). In Tanzania, yield losses in the range of 18–31% are attributed to *O. bennigseni* (98). The striped foliage beetle, *M. quaterna,* is found in many bean-growing regions in Africa (166).

BIOLOGY   The biology of *O. mutabilis* in Nigeria has been studied (131), and the biology of *O. bennigseni* is believed to be similar. The adults of bean foliage beetles are oval, 6 mm long, and shiny. Their color most often ranges from orange to light brown, although darker brown or blackish forms also occur in the same population. Eggs are yellow, elliptical, and translucent and are laid in the soil in batches of approximately 60 eggs glued together. Larvae emerge after approximately 2 weeks and feed on roots of beans and other plant species.

Attack by the early instars may go unnoticed, but the older larvae remove lateral roots and cause wilting and premature senescence in bean plants. The larvae go through three instars: The first two last about 6 days each and the third lasts nearly 10 days, followed by a prepupal stage, which lasts approximately 15 days. The pupal stage, which occurs in an earthen cell within the soil, lasts approximately 16 days. The duration of the development stages depends on soil temperature, among other factors, and varies between 60 and

250 days. The teneral adults of *Ootheca* spp. may undergo an obligatory diapause until the onset of the rainy season, when they emerge. This pattern of emergence coincides with the presence of young beans in the field.

Bean foliage beetles chew holes between the veins of the leaf and may cause extensive defoliation of young bean seedlings. Damage to the growing point of a young seedling kills the plant, and heavy infestation may completely destroy a crop. Infestation and damage is usually most severe on young seedlings, but sometimes the infestation may persist through the postflowering stages. In addition, the adults occasionally feed on floral parts. The adults also serve as effective vectors of various viruses of cowpea in Africa (31), but their ability to transmit viruses in beans has not been established.

The biology of *M. quaterna* is not fully understood, but the egg, larval, and pupal stages are found in the soil. The adult is about 4 mm long with white and light brown longitudinal stripes on the elytra. Bean damage is caused by the adult, which feeds at the margins of newly opened leaves and the walls of developing pods. *M. quaterna* is also reported to be a vector of cowpea mottle virus (31, 191), but its status as a vector of bean viruses is uncertain. Other chrysomelids of minor importance on beans belong to *Monolepta* spp.

ECOLOGY   The two *Ootheca* species are distinguishable by their aedeagal characteristics. Records kept at the Kenya Museum of Natural History, Entomology Section, suggest that *O. mutabilis* may be more predominant in the lower-altitude environments, while *O. bennigseni* dominates in the medium-altitude zones (1000–1500 m). *O. bennigseni* is not reported from West Africa.

Various weevils also attack and cause defoliation in bean plants. Their damage is usually minor and rarely warrants control. However, "le cigar," *Apoderus humeralis,* is the principal defoliator of bean plants (148); it is not reported from mainland Africa. *A. humeralis* adults chew circular holes between the leaf veins. In addition, the gravid female lays single eggs at the tip of trifoliate leaves and rolls them in the shape of a cigar, wherein the emerging larva feeds and grows through pupation. Adults of *Alcidodes leucogrammus, Systates* spp., and *Nematocerus* spp. feed on leaf margins of bean plants, where they chew out circular holes also. Eggs of these species are laid in the soil near the base of the plant. Emerging larvae bore into the stem, which causes a cankerous swelling and renders the plants prone to lodging. Pupation takes place within the stem.

Several thrips species (e.g. *Frankliniella* spp. and *Sericothrips* sp.) also feed on leaves. On the island of Mauritius, *Thrips palmi* is the principal thrips species attacking beans (147). This species is also not reported on beans in mainland Africa and appears to be a recent arrival to the island. Another locally important foliage pest in Madagascar is the so-called painted lady, *Pyrameis* (syn. *Vanessa* or *Cynthia*) *cardui* (Lepidoptera: Nymphalidae). This species commonly occurs on the mainland but does not damage beans.

MANAGEMENT   Foliage-pest management generally relies on chemical appli-cations. However, Karel & Rweyemamu (99) found moderate levels of resis-tance to *O. bennigseni* in several CIAT breeding lines and local materials in Tanzania. Karel (93), in greenhouse studies, found that application of a 1% aqueous extract of powder from seeds of the neem tree (*Azadirachta indica*) inhibited feeding of *O. bennigseni*. Chemical control of species such as *Liriomyza trifolii* can induce large population buildups and result in greater severity of pest damage (167). Fagoonee & Toory (63) reported on work to develop resistance to *L. trifolii*. Several other foliage-pest infestations do not warrant control because the plants can grow out of the damage.

FLOWER PESTS

Bean Flower Thrips

DISTRIBUTION AND IMPORTANCE   Various species of thrips (Thysanoptera: Thripidae) attack bean plants in Africa, but *Megalurothrips sjostedti* (syn. *Taeniothrips sjostedti*), the cowpea flower thrips, is perhaps the most impor-tant, occurring widely throughout the continent (83, 129, 183). Other thrips reported from beans include *Taeniothrips nigricornis* and *Frankliniella dampfi* (83, 97, 166), as well as *Sericothrips occipitalis*.

BIOLOGY   Observations on the biology of *M. sjostedti* on beans were made in Uganda (83, 107). The adult *M. sjostedti* is a tiny black insect, about 2 mm long. Males have not been observed, so parthenogenesis is assumed to be the reproductive mechanism. Eggs are laid in leaves, stems, and calyces of flower buds. The pale orange nymphs reside in flower buds or open flowers. The first two instars inhabit the plant, but the third instar enters the soil where it turns into a prepupa. The pupal stage lasts 8–10 days.

Flower malformation, distortion, and discoloration characterize thrips in-jury. In severe infestations, flower buds do not open and may abort prema-turely. Thrips infestation and damage is usually considered minor in dry bean production. However, in snap beans, infestation and damage by *M. sjostedti* and *Caliothrips* sp. can significantly reduce pod quality (37, 126).

ECOLOGY   Many members of the thrips species complex that attack beans also attack several other crops and weed species, and the severity of their infestation on beans relates to the proximity of these other crops to the bean field. For instance, *M. sjostedti* survives on *Cajanus cajan* in the absence of beans, and *T. palmi* and *Frankliniella* spp. also have various alternate hosts (173) from which they can invade a bean crop. Both nymphs and adults attack bean plants. Feeding occurs at the base of petals and stigma, where feeding punctures can be observed easily with the aid of a hand lens. Damage is more severe in drier areas.

MANAGEMENT   Using chemical control, Ingram (83) achieved neither pest reduction nor a yield increase, and Karel et al (97) observed that thrips infestations in general rarely warrant chemical control. Current information on natural enemies is scarce, although Salifu (161) and Letourneau & Altieri (114) did observe *Orius* spp. (Hemiptera: Anthocoridae) preying on *M. sjostedti* and *Frankliniella* sp. Kyamanywa and coworkers (109, 110) observed that bean-maize intercropping reduced *M. sjostedti* infestations significantly, and Kyamanywa & Ampofo (108) demonstrated that this species preferred open areas to shaded areas, such as those prevailing in bean-maize intercrops. Virtually no research results regarding the development of thrips resistance in beans have been reported.

Pollen Beetles

DISTRIBUTION AND IMPORTANCE   The major pollen and flower feeders (Coleoptera: Meloidae) belong to two genera, *Coryna* and *Mylabris,* which are widely distributed throughout sub-Saharan Africa. The more common species are *Coryna apicicornis, Coryna kersteni, Mylabris amplectens, Mylabris aperta, Mylabris dicincta, Mylabris farquharsoni, Mylabris temporalis,* and *Mylabris tristigma* (47, 66, 78, 85, 112, 166). These beetles are generally strikingly colored, with red or yellow bands or spots on black elytra and orange to yellow antennae. Body length ranges between 15 and 30 mm. When alarmed, the adult may exude a fluid that can irritate human skin. Therefore, these beetles are also called blister beetles. Adults feed on floral parts of bean plants, and large populations may reduce fertilization and pod set drastically (173).

BIOLOGY   The complex biology of pollen beetles is similar among species. Oviposition takes place within the soil, and the eggs hatch into active triungulin larvae that feed on egg pods of the Orthoptera (71). The larvae undergo hypermetamorphosis, during which each instar looks different. Older larvae are sluggish and euriciform in shape with a cylindrical body, a well-developed head, and the presence of both thoracic and some abdominal prolegs. Pupation occurs in the soil.

ECOLOGY   Flower and pollen beetles feed on various other crops, notably maize, sorghum, and flowers of weed species. Intercropping beans with these cereals often leads to high levels of infestation. The larvae are also predators of grasshopper egg pods, and pollen beetle populations tend to surge after grasshopper outbreaks (71). The adults are strong fliers and migrate to bean fields from their breeding sites.

MANAGEMENT   The meloids are generally difficult to control. Severe outbreaks may warrant repeated application of commercial insecticides. Otherwise hand-picking may suffice to control low infestations.

Legume Pod Borer (Lepidoptera: Pyralidae)

DISTRIBUTION AND IMPORTANCE   The legume pod borer, *Maruca testulalis* (Lepidoptera: Pyralidae), attacks beans and other grain legumes throughout the tropical and subtropical world (173). It occurs in nearly all countries in sub-Saharan Africa. On beans, it attacks both flowers and green pods. In Tanzania, yield loss in the range of 30% is attributed to this pest (92).

BIOLOGY   The biology of *M. testulalis* has been studied extensively on cowpea (28, 87, 134, 182, 184). This moth has light brown forewings with white markings. The hind wings are pearly white with brown markings at the lateral edge. It rests with wings outspread; the wingspan is 15–28 mm. Although the moth is chiefly nocturnal, it may also be seen during daytime. The translucent eggs are laid singly on terminal shoots, flower buds, and pods and hatch in 2–3 days. Larvae go through five instars in 8–14 days. Early instars are dull white, but later instars are black-headed, with irregularly shaped brown or black spots on the dorsal, lateral, and ventral surfaces of each body segment. Pupation, lasting 5–10 days, takes place within a cocoon in the soil. The early larvae, in the absence of flower buds and flowers, feed on young tender shoots, peduncles, and stems (134). Otherwise, the preferred feeding sites are floral parts and pods. Because attack by the early instars on flower buds and flowers is internal, there is usually very little sign of damage until the flower wilts and drops. The larvae prefer concealment when feeding and characteristically attack pods at the junction between two pods or between a pod and a leaf or stem. Also, to stay hidden they frequently web together flowers, pods, and leaves. They hide during the day in these feeding nests and wander during the night in order to invade fresh flowers and pods.

ECOLOGY AND MANAGEMENT   *M. testulalis* is generally managed through the application of chemical pesticides (92, 97). However, Okeyo-Owuor (135) studied the ecology of *M. testulalis* on cowpea in Kenya and observed only one generation of the pest on a crop. He also found that pathogens (e.g. *Nosema* sp. and *Bacillus* spp.) significantly reduced *M. testulalis* populations. Host-plant resistance to *M. testulalis* occurs in several cowpea varieties (88, 134), but little research has focused on bean resistance to this pest.

POD AND SEED FEEDERS

African Bollworm (Lepidoptera: Noctuidae)

IMPORTANCE AND DISTRIBUTION   The African bollworm (ABW), *Helicoverpa* (syn. *Heliothis*) *armigera,* is widely distributed in tropical and subtropical Africa and as well as in Europe, Asia, and Australia. In many countries, it is a major pest of beans and other important crops. Johnson et al (90) provide a comprehensive review of *H. armigera* and related species, and Abate (17) reviewed the economic importance of this pest in Africa. Although much has been said about the importance of ABW, few estimates of crop losses attributable to this pest are based on sound research findings. Abate & Negasi (21) reported a 12–16% (reaching up to 46% under severe conditions) yield reduction of common bean caused by ABW in Ethiopia. *H. armigera* is extremely polyphagous, occurring on approximately 60 species of cultivated plants and about 67 wild hosts in 39 families across its geographical distribution range (152). Cotton, sorghum, maize, beans, chickpea, groundnut, sunflower, and tomato are among the most frequently attacked crops; wild hosts include *Abutilon* spp. (Malvaceae), *Datura stramonium* and *Solanum nigrum* (Solanaceae), *Cleome gynandra* (Caparidaceae), and *Guizotia scabra* (Compositae).

BIOLOGY   Several investigators (55, 77, 140, 150, 156, 178) have studied the biology of *H. armigera*. Fecundity is heavily influenced by environmental factors. For example, the average number of eggs per female during the rainy season was 1226, while that in the dry season was 198 (see 150). Eggs are nearly spherical and are whitish yellow in color when first laid (turning brownish later). Measuring about 0.5 mm in diameter, the eggs are characterized by their sculptured, longitudinal ribs; they are flattened at the base and look rather dome shaped. Incubation takes 1–5 days, with peak hatching at 2–3 days. On emergence, the larva consumes all of the egg shell except for the base, which adheres to the leaf surface. The first instar moves over the plant by crawling and hanging from a thread. The first two instars feed on the tender leaves or young, developing parts of the plant. Although the color of the *H. armigera* larvae varies among mixtures of green, straw yellow, pink, reddish brown, and black, they are easily distinguished from similar caterpillars by the long, alternating dark and pale longitudinal bands that run along the back. The larvae normally undergo six instars, but seven have been recorded under cold weather conditions. Each instar can be distinguished by head-capsule size.

The larval period lasts 18–25 days. A fully grown larva measures about 40 mm in length and undergoes a prepupal period of about 3 days, burying itself in the ground prior to pupation. Pupation takes place in the soil and lasts an average of 12–23 days. Thus, two to eight generations per year are possible, depending on temperature, food supply, and availability of host plants. The shiny, mahogany brown pupa measures 14–18 mm in length and has two tapering parallel spines at the posterior tip. A facultative diapause may occur during the pupal stage before the moth emerges. Under warm tropical conditions in the field, the number of individuals undergoing diapause is usually less than 1% of the population (77, 157).

The adult is a stout-bodied moth measuring 18–19 mm in length, with a wingspan of about 40 mm. The body is light brown, and the hindwings have dark marginal spots. Upon eclosion, the adult female feeds for 1–3 days, during which time she also mates. Oviposition takes place between 20:00 and 23:00 hours and usually coincides with flowering of the host plant. Some egg-laying may also occur during the vegetative stage on such host plants as chickpea and lablab bean, but even in these crops, peak egg laying coincides with peak flowering. Each female lays an average of 750–1000 eggs over a period of 10–11 days. Eggs are laid singly and are usually randomly distributed within the field. Higher numbers of eggs are laid on the top two-thirds of the plant. On leaves, eggs are deposited on the upper surface.

ECOLOGY   Several authors have given detailed accounts of the ecology and behavior of *H. armigera* and related species (64, 140, 141, 149, 151, 156, 157, 185). *H. armigera* is adapted to numerous environmental conditions, and adult females have a strong aptitude for migration (141, 149). Population levels are influenced by several factors, including cropping patterns and seasonal changes in climatic conditions and locations (6, 12, 16, 55). For example, Coaker (55) found that *H. armigera* does not achieve pest status in areas where the climatic conditions allowed both the pest and its natural enemies to survive all year round. Observations in Ethiopia indicate that damage by *H. armigera* on beans is more serious in warmer areas of the rift valley, where beans are commonly grown. Small plots under subsistence farming suffer much less damage. Like the damage caused by many other insects (34, 36, 142, 143, 192), *H. armigera* damage to intercropped bean is less than damage to sole crops of beans.

MANAGEMENT   Farmers' methods of managing *H. armigera* rely on cultural practices (including intercropping, use of varietal mixtures, timely sowing, optimum plant density, etc). Small-scale growers do not apply insecticides to control this insect. Although host-plant resistance may have greater potential against *H. armigera* under low-input production systems (115), research in this area has been weak. A few experiments in Africa (1, 2) and elsewhere

(154) have revealed significant differences in the response of bean varieties to *H. armigera* damage, but the use of host-plant resistance in the management of *H. armigera* has advanced little. The high migration ability of the larvae from neighboring crops necessitates a high degree of host-plant resistance.

Numerous natural enemies attack various stages of *H. armigera* in Africa (11, 12, 73, 128, 139, 151, 186a), resulting in varying levels of natural biological control. Furthermore, at least four groups of indigenous pathogens have been identified or tested in parts of southern Africa (17, 54, 57, 73, 121, 155, 158, 190), and promising results have been reported with the nuclear polyhedrosis virus. However, use of such pathogens has not yet been implemented. Although vacant niches are available for introduced natural enemies (73, 103), there is little prospect for classical biological control of *H. armigera* in Africa, especially on a subsistence crop such as beans.

Pod-Sucking Bugs

DISTRIBUTION AND IMPORTANCE   The major pod-sucking bugs (Hemiptera: Coreidae) are the spiny brown bugs *Clavigralla tomentosicollis* (syn. *Acanthomia tomentosicollis*), *Clavigralla schadabi* (syn. *Acanthomia horrida*), *Clavigralla hystricodes* (syn. *Acanthomia hystricodes*), and *Anoplocnemis curvipes*. These pests frequently severely damage developing bean pods and seeds, especially in the hotter, drier lowlands of Africa. They have a wide range of other hosts such as cowpea, soybean, green and black grams, sesame, cotton, and sorghum, and their occurrence on beans may be related to the abundance of these hosts in the vicinity of the bean field.

Members of the Pentatomidae are often recorded as minor bean pests of low economic importance (97, 178). The cosmopolitan green stink bug, *Nezara viridula*, and *Piezodorus guildini* are widespread examples. Mating and oviposition occurs within the bean field, and both adults and the often multicolored nymphs suck sap from young pods, causing necrosis and deformation, yellowing, and premature drying of pods with poor seed formation.

BIOLOGY   Materu (117–119) studied the biology and population dynamics of *C. schadabi* and *C. tomentosicollis* in Tanzania, and Egwuatu & Taylor (62) and Ochieng (131) also studied the biology of *A. curvipes* in Nigeria. Adults of *C. tomentosicollis* are about 10 cm long, hairy, and brown, whereas *C. schadabi* is smaller and has a pair of elongated spines on the anterior end of the thorax. *C. tomentosicollis* lays eggs in batches of 10–70; on average, a single female may lay 200 eggs. The eggs hatch in approximately 6 days. Each instar lasts about 2 days, except for the fifth and final instar, which lasts approximately 6 days. The total nymphal period is about 14 days. Nymphs are

very sluggish and form colonies on pods and peduncles. The adults feed singly
or in mating pairs and drop off the plant when disturbed.

Eggs of *C. schadabi* are laid singly, and about 250 eggs may be laid by a
single female. In other respects, the biology and nature of damage are similar
to those of *C. tomentosicollis.* Incubation lasts approximately 6 days, and the
nymph goes through five instars. The nymphal stage lasts about 25–35 days.

The giant coreid bug (*A. curvipes*) is a large black bug approximately 3 cm
long. The adults are strong fliers and escape to nearby trees when disturbed.
The males have a single large spine on each hind leg and can be distinguished
from the female by this character. Eggs are dark gray and are laid in batches
(chains) of about 10–40 on leguminous trees and weeds but seldom on beans.
However, a related species, *Anoplocnemis madagascariensis,* which is restricted to Madagascar, completes its entire life cycle on beans. The newly
hatched nymphs are bright red and later turn black. Nymphs go through five
instars and during the first two resemble ants. Adults and nymphs of pod bugs
suck sap from green pods and cause dimpling and wrinkling of the seed coat
as well as browning and shriveling of the seeds. Jones (89) observed that the
browning and shriveling were caused by the transmission of the fungus *Ne-
matospora coryli,* but Materu (117) could not establish the link between the
fungus and these specific symptoms. Although he did confirm an association
of *N. coryli* with damage by *Clavigralla* spp., he concluded that the wrinkling
probably resulted from a toxic effect of the bug's saliva on the cotyledons.
Wallace (189) also reported that *N. viridula* is a vector of *N. coryli.*

The viability of coreid bug–damaged seeds was reduced by 65–85% compared with that of healthy seeds (117). *A. curvipes* damage is similar to that
by *Clavigralla* spp. In addition, *A. curvipes* sometimes sucks the sap out of
young shoots and causes them to wilt.

Several species of coreid bugs (e.g. *Riptortus dentipes, Riptortus tenui-
cornis,* and *Riptortus longipes*) reportedly also damage bean plants (66, 112).
The adults of these bugs are slender, about 17 mm long, and light brown with
white or yellow lateral lines. Like the adults of *A. curvipes, Riptortus* spp.
adults are strong fliers and prefer to oviposit on weeds and other legumes
outside bean fields. They are therefore difficult to control with insecticides.

MANAGEMENT   Several natural enemies attack the coreid bugs, including
Clara magnifica (Diptera: Tachinidae), which parasitizes the adults; *Pheidole*
spp. (Hymenoptera: Formicidae), which are egg predators; *Gryon charon* and
Protelenomus anoplocnemis (Hymenoptera: Scelionidae); and *Ooencyrtus* spp.
(Hymenoptera: Encyrtidae) (120, 131). Various insecticides such as endosulfan
and dimethoate also effectively control pod-sucking bugs (130, 178), but
Matteson (120) observed that spiny brown bug populations could rise as a
result of continued insecticide use.

SAP SUCKING PESTS

Aphids

DISTRIBUTION AND IMPORTANCE   Of the aphids (Homoptera: Aphididae), the black bean aphid, *Aphis fabae,* is the principal aphid pest directly damaging beans in Africa (153). In lowland areas, the cowpea aphid, *Aphis craccivora,* may also colonize and damage bean plants. They form colonies around the stem, leaves, and growing points; suck sap from plants; and cause seedlings to wilt and die. All parts of the plant may be damaged, and older plants may be stunted as a result of aphid attack. Moreover, the direct damage caused by aphids may not be as economically important as their ability to transmit bean common mosaic virus (BCMV) and bean aphids. In fact, several aphid species do not colonize bean plants but are frequently found in the bean crop environment and may transmit such viruses as BCMV. Suspected species include *Aphis gossypii, Aphis spiraecola, Myzus persicae,* and *Rhopalosiphum maidis.*

BIOLOGY   *A. fabae* is dull black with black cornicles and cauda. The third antennal segment bears 9–20 irregularly arranged sensoria. *A. craccivora* has 3–8 sensoria arranged in a more regular manner. The femur bears many fine hairs on all surfaces, and the cauda has 10–19 hairs. Adults of *A. fabae* are approximately 2 mm long and very often carry a powdery white secretion on the abdominal segments. In the warm tropics of Africa, reproduction is only by parthenogenesis. Hosts in this region include multiple leguminous species. Wingless forms are produced when food is abundant and climatic conditions are optimal. However, when food is in short supply or the colony becomes overcrowded, winged forms develop. The winged aphids disperse to colonize new fields or plants and may invade bean fields soon after seedlings have emerged; however, damage to the preflowering stage is more critical (101). The aphid life cycle takes 11–13 days, and adult longevity varies from 6 to 15 days.

ECOLOGY AND MANAGEMENT   Aphid populations usually build up during dry weather, and thus infestations are more severe during the dry season. In humid weather, large colonies are often eliminated by members of the Entomophtoraceae (e.g. *Erynia neoaphidis* and *Neozygites fresenii*) (43). *A. fabae* is also parasitized by various Hymenoptera from the family Aphidiidae (e.g. *Aphidius colemani*). Other Aphidiidae—*Lysiphlebus fabarum, Lysiphlebus confusus, Lysiphlebus cardui, Lysiphlebus testaceipes* (introduced from Cuba), *Ephedrus plaiator,* and *Trioxys angelicae* (from Europe)—introduced into Burundi appear to have become established and promise effective control (43).

   With regard to cultural practices, A'Brook (24, 25) observed that increased space around groundnut plants increased the incidence of *A. craccivora* and the spread of groundnut rosette virus. Ogenga-Latigo et al (132, 133) reported reduced aphid infestation and damage when beans were intercropped with densely populated or older maize.

Tobacco Whitefly

DISTRIBUTION AND IMPORTANCE   The tobacco whitefly (TWF), *Bemisia tabaci* (Homoptera: Aleyrodidae), is the only whitefly species recorded on beans in Africa. It is widely distributed in bean-producing regions. Although it is considered a minor pest in many countries (7, 69), it has recently achieved major pest status in some countries. For example, it is the most important bean pest in Sudan, where its damage causes yield losses of 14–86%, depending on the season (160). In Kenya, it is an important pest of snap bean (126). The excessive use of pesticides and consequent loss of predators and parasites may have caused the recent upsurge in the importance of this insect. The direct damage caused by whiteflies on beans is usually not as important as the transmission of viruses such as cowpea mild mottle virus (30).

BIOLOGY   The pear-shaped eggs, measuring about 0.2 mm in diameter, are laid on the leaf; they are white when first laid but turn brownish later. Incubation lasts approximately 7 days. Upon hatching, the young move a small distance and attach themselves to the leaf, where they feed until adult emergence. The adult is a minute insect measuring approximately 2 mm in length. The life cycle is completed in 15–30 days, depending on the temperature.

ECOLOGY   TWF is a cosmopolitan pest widely distributed in the tropics and subtropics. It is an extremely polyphagous insect, recorded in 63 families of plants (123). Its attack is more severe during the dry season, but it prefers humid conditions for reproduction and development.

MANAGEMENT   Under small-scale production, no control measures are required, but severe infestations in commercial fields necessitate the use of insecticides. Hymenopterous parasitoids of the genus *Encarsia* are believed to give good natural biological control when no insecticides are used.

Leafhoppers

Two species of leafhoppers (Homoptera: Cicadellidae), *Empoasca lybica* and *Empoasca dolichi,* are widely distributed in the major African bean-growing areas. They are generally considered minor pests compared with *Empoasca kraemeri,* the principal pest of beans in subtropical and tropical America.

Red Spider Mite

DISTRIBUTION AND IMPORTANCE   The red spider mite (RSM) (Acari: Tetrany-chidae), *Tetranychus cinnabarinus,* is widely distributed in Africa and else-where. Although yield-loss data are not available, this insect is considered a major pest of common bean in Zimbabwe (69). In some areas of Ethiopia, it is a serious pest of beans grown under irrigation (7), and in Kenya, it is a serious pest of snap bean (126). Its infestation of beans in Sudan is reportedly slight and localized (166).

BIOLOGY   The pale pinkish eggs are laid singly on the lower surface of the leaf, and 15–20 are laid by each female. Incubation takes 1–3 days. The emerging larvae feed and produce excessive webs on infested plants, which may be weakened or wilt and dry up under severe conditions. Depending on temperature, completion of the life cycle takes about 7–10 days. RSM is a very polyphagous pest, infesting many plants from different families. RSM feeding on bean leaves causes a characteristic brownish spotting on the under-side of the leaves.

MANAGEMENT   No control is needed in small-scale bean production. In snap-bean production, however, high populations tend to curtail the harvest period, so control is often necessary.

STORAGE PESTS

Bean Bruchids

DISTRIBUTION AND IMPORTANCE   Stored beans suffer heavy losses in terms of both quality and quantity, and these losses are caused mostly by bean bruchids (Coleoptera: Bruchidae), although other pests such as the flour mite, *Acarus siro* (Acari: Acaridae), may be important in some African countries. Beans in Africa fall under attack by half a dozen species of bruchids: the bean bruchid or common bean weevil, *Acanthoscelides obtectus;* the cowpea bruchids *Callosobruchus chinensis* and *Callosobruchus maculatus;* the Rho-desian bean weevil, *Callosobruchus rhodesianus;* the Mexican bean weevil or spotted bean weevil, *Zabrotes subfasciatus;* and unidentified species of *Bruchidius.* Of these, *A. obtectus* and *Z. subfasciatus* are the most important (70, 127) and occur together in many instances.

A. *obtectus* and the two *Callosobruchus* spp. occur throughout Africa (79), whereas *Z. subfasciatus* has a limited distribution, perhaps because it was only recently introduced to the continent; *C. rhodesianus* occurs in southern Africa. Although bean bruchids cause substantial crop losses, little quantitative data on actual losses under field conditions are available. Negasi (127) reported that in Ethiopia stored bean damage by *A. obtectus* and *Z. subfasciatus* reached up to 38%, with a corresponding bean weight loss of about 3.2%.

BIOLOGY   The two major species, *A. obtectus* and *Z. subfasciatus,* have similar biologies (60, 81, 188). After hatching from the eggs, the first-instar larvae penetrate the seed coat, form a cell, and proceed to develop inside the seed. They pass through four instars before pupation. Feeding by the last instar produces the characteristic circular "windows" that become visible externally as insect development progresses. The newly formed adults may remain in the cell for several days before pushing out the window and exiting the seed. Mating occurs immediately, and egg laying follows. The life cycles of *A. obtectus* and *Z. subfasciatus* are completed in about 28 and 24 days, respectively. Respective adult longevity for the two species is approximately 12 and 11 days. The adult female of *A. obtectus* lays an average of 60 eggs, and a *Z. subfasciatus* female will lay approximately 35 eggs in her lifetime. Adult bruchids do not feed; only the larvae cause damage.

The behavior and characteristics of the two species differ in the following ways (169, 188):

1. The eggs of *Z. subfasciatus* are glued to the seed coat, whereas those of *A. obtectus* are scattered between the seeds.
2. *Z. subfasciatus* infests only harvested seeds, whereas *A. obtectus* infestation may start in the field on growing pods.
3. Upon hatching, *Z. subfasciatus* larvae bore through the seed directly. Those of *A. obtectus* wander about before they penetrate the seed.
4. *Z. subfasciatus* adults are smaller than those of *A. obtectus.*
5. The larvae of *Z. subfasciatus* are white, whereas those of *A. obtectu*s are dirty white or pale yellow.
6. *Z. subfasciatus* males and females are easily distinguishable, but those of *A. obtectu*s are not.

ECOLOGY   Studies in Latin America showed that the prevalence of the two species is influenced by the ambient temperature regimes (188). For example, *A. obtectus* prefers cooler climates at higher altitudes, where it is the dominant species, whereas *Z. subfasciatus* prefers warmer climates in the lower altitudes and therefore is more important in the tropical and subtropical regions (79). Recent studies in southern and eastern Africa (70, 127) show that these preferences do not necessarily hold in this continent. The African strain of *Z. subfasciatus* may be different from the Latin American strain, and in Africa, factors other than altitude and temperature differences, such as time of year, may influence abundance of both African species.

MANAGEMENT   Recent studies in eastern and southern Africa of farmers' bruchid-management methods (70, 127, 168) revealed a dozen means of keeping the pest population below damaging levels. These include solar heating, varietal selection, removal of infested grain at harvest, admixing grain with ash and botanicals, enrobing the seed with mud, use of different storage methods, and granary hygiene. A few farmers store unthreshed bean to protect against *Z. subfasciatus,* which, as mentioned above, infests only harvested beans. Research on tumbling containers (146) and repeated sieving (170) as control methods has produced encouraging results, but the applicability of these techniques to large-scale production remains to be proven. Information on the use of vegetable oils (80, 187, 187a) and insecticidal control (4) is available but is of little use under subsistence farming conditions. Little or no information is available on biological control of bruchids, natural or otherwise, in Africa.

Research on host-plant resistance has identified promising cultivars and their resistance mechanisms (38, 48–50, 53, 106, 121a, 127, 138, 171, 172, 187b, 188a). The resistance factor arcelin (an antimetabolite) was recently incorporated into several germplasm lines at CIAT and distributed to NARS, but the benefits of these experiments have yet to reach the farmer. Tests with botanicals such as seeds of the neem tree (*A. indica*) and the pepper tree (*Schinus molle*) in Ethiopia have produced promising results (23, 127), and the possibility of using other materials available in Zimbabwe (53) should be vigorously pursued.

FUTURE FOCUS

Beans are typically produced on a small scale, usually in association with other crops. In such production systems, farmers rarely apply purchased inputs such as fertilizers and pesticides to the beans; as a result only some 20% of the potential yield is realized in small-scale fields. Current recommendations for bean pest management emphasize chemical pesticides, but these are often not available to the farmers because of high cost and infrequent supply. However, research and technology offer various alternatives for the development of more sustainable bean-pest management in Africa. These studies include exploration of indigenous farmers' practices and their pest-management knowledge. Efforts focused on improving the efficiency of collaborations between researchers and farmers will lead to strategies that are easier for African growers to adopt.

Research emphasis should be on low-input, integrated pest management (IPM) approaches that encompass farmers' current practices (intercropping, varietal mixtures, optimum sowing date and plant population), biological control, and use of locally available materials as the major control components.

Advances have already been made in the identification of host-plant resistance, the implementation of cultural practices, and the use of botanical pesticides and other locally available material, as well as natural enemy complexes of the major pests. Further research is needed to establish the role of natural enemies in bean-pest management.

Our experience disproves the argument that IPM may not be applicable to small-scale farming in Africa. The African farmer has a wealth of indigenous techniques. In fact, African farmers were practicing IPM long before the term was coined. IPM applied in the African smallholder concept should not be interpreted in "simple financial terms" (65) such as savings on pesticides: Instead, these cropping systems traditionally use a combination of several practices such that pest outbreaks of the magnitude experienced in the temperate regions are uncommon. Indeed, recent high incidences of whiteflies in farmers' fields in various regions may have resulted from the increased use of pesticides in these areas. Thus, IPM in Africa should emphasize an understanding of cropping systems in order to increase their efficiency. This aim will involve a multidisciplinary approach requiring the combined efforts of entomologists, agronomists, soil experts, plant pathologists, and breeders.

The fact that estimates of actual crop losses caused by bean pests in Africa are inadequate suggests a need to generate meaningful crop-loss data that relate to the existing farming practices. This goal can be achieved by setting up a network of trials in selected environments across the regions. The existing bean research networks could facilitate this approach, which is the only one that will establish the status of particular pest species and allow priorities to be set for efficient use of the meager resources available for research in Africa.

Plant quarantine services in Africa must be strengthened to guard against the dangers of inadvertently introduced new pests from outside or within the continent. The vegetable leafminer, *Liriomyza trifolii* (5, 100), was accidentally introduced to East Africa from North America in the 1970s. It has become widespread in this region and is already a major pest of beans in Mauritius. A related species, *Liriomyza huidobrensis,* from South America has also become an important pest of beans in this country (147). Furthermore, the major bruchid bean pests were also introduced from South America.

Many pest control technologies have been developed over the years, but few have reached the end-users. In collaboration with farmers, these techniques should be refined. Collaboration with social scientists should also be sought in this effort.

Literature Cited

1. Abate T. 1983. *Screening of haricot bean varieties against bean fly (BNF) and African bollworm (ABW).* Nursery I, Nazret, 1982/1983. 7 pp. (Typescript)
2. Abate T. 1983. *Screening of haricot bean varieties against bean fly and African bollworm.* Nursery II, Nazret, Jima, Kobo, Mekele. 3 pp. (Typescript)
3. Abate T. 1984. Further notes on insect pests of grain legumes in Ethiopia. *Comm. Ethiop. Entomol. Newsl.* 3(2):4–5
4. Abate T. 1985. Evaluation of some insecticides for the control of the bean bruchid, *Callosobruchus chinensis* (L.), on stored haricot bean. *Ethiop. J. Agric. Sci.* 7:109–18
5. Abate T. 1987. New records of arthropod pests of grain legumes in Ethiopia. *Bean Improv. Coop. Annu. Rep.,* pp. 62–63. NY State Agric. Exp. Stn., Geneva, NY
6. Abate T. 1988. Experiments with trap crops against African bollworm, *Heliothis armigera,* in Ethiopia. *Entomol. Exp. Appl.* 48:135–40
7. Abate T. 1988. *Insect and Mite Pests of Horticultural and Miscellaneous Plants in Ethiopia. IAR Handbook,* No. 1. Addis Ababa, Ethiopia: Inst. Agric. Res. (Ethiopia). 115 pp.
8. Abate T. 1990. *Studies on genetic, cultural and insecticidal controls against the bean fly,* Ophiomyia phaseoli *(Tryon) (Diptera: Agromyzidae), in Ethiopia.* PhD thesis. Simon Fraser Univ., Burnaby, BC, Canada. 177 pp.
9. Abate T. 1990. Bean entomology. See Ref. 84, pp. 54–57
10. Abate T. 1990. Prospects for integrated management of bean fly (*Ophiomyia phaseoli*). See Ref. 177a, pp. 190–97
11. Abate T. 1991. *Entomophagous Arthropods of Ethiopia: a Catalog.* Tech. Man. No. 4. Addis Ababa: IAR. 50 pp.
12. Abate T. 1991. Intercropping and weeding: effects on some natural enemies of African bollworm, *Heliothis armigera* (Hbn.) (Lep., Noctuidae), in bean fields. *J. Appl. Entomol.* 112:38–42
13. Abate T. 1991. Research methods in host plant resistance against bean stem maggots. In *Bean Stem Maggot Research Methods. Occas. Publ. Ser.* No. 7, ed. JKO Ampofo, pp. 3–13. Cali, Colombia: CIAT
14. Abate T. 1991. Seed dressing insecticides for the control of bean fly [*Ophiomyia phaseoli* (Tryon) (Diptera: Agromyzidae)] in Ethiopia. *Trop. Pest Manage.* 37:334–38
15. Abate T. 1991. The bean fly *Ophiomyia phaseoli* (Tryon) (Dipt., Agromyzidae) and its parasitoids in Ethiopia. *J. Appl. Entomol.* 111:278–85
16. Abate T. 1992. Sowing date and plant density effects on pest and predator numbers in haricot bean fields. *Ethiop. J. Agric. Sci.* 13:30–36
17. Abate T. 1993. Control of other pests and diseases. In *Dryland Farming in Africa,* ed. JRJ Rowland, 9:188–204. London: Macmillan
18. Abate T, Ayalew G, Girma A. 1995. *Ecology of bean stem maggots in Ethiopia.* Presented at 2nd Int. Crop Sci. Conf. East. and South. Afr., Blantyre, Malawi, Feb. 19–24
19. Abate T, Ayalew G, Wale M. 1990. New species of bean flies in Ethiopia. *Comm. Ethiop. Entomol. Newsl.* 11(2): 6–9
20. Abate T, Gebremedhin T, Ali K. 1982. *Arthropod Pests of Grain Legumes in Ethiopia: Their Importance and Distribution.* Addis Ababa: IAR. 56 pp.
21. Abate T, Negasi A. 1981. Chemical control of African bollworm (*Heliothis armigera*) (Hubner) with ultra-low-volume sprays. *Ethiop. J. Agric. Sci.* 3:49–55
22. Abate T, Negasi F, Ali K. 1986. A review of grain legume pest management research in Ethiopia. In *A Review of Crop Protection Research in Ethiopia,* ed. T Abate, pp. 327–44. Addis Ababa: IAR. 685 pp.
23. Abate T, Negasi F, Ayalew G. 1992. *Botanicals in pest management.* Presented at CEE/EPC Joint Conf., Addis Ababa
24. A'Brook J. 1964. The effect of planting date and spacing on the incidence of groundnut rosette disease and of the vector, *Aphis craccivora* Koch, at Mokwa, northern Nigeria. *Ann. Appl. Biol.* 54:199–208
25. A'Brook J. 1968. The effect of plant spacing on the numbers of aphids trapped over the groundnut crop. *Ann. Appl. Biol.* 61:289–94
26. Abul-Nasr S, Assem MAH. 1968. Studies on the biological processses of the bean fly, *Melanagromya phaseoli* (Tryon). *Bull. Soc. Entomol. Egypte* 52: 283–95
27. Agarwal NS, Pandey ND. 1962. Bionomics of *Melanagromyza phaseoli*

Coq. (Diptera: Agromyzidae). *Indian J. Entomol.* 23(4):293–98

28. Akinfewa S. 1975. *Bioecological studies of* Maruca testulalis Geyer *(Lepidoptera: Pyralidae) in the Zaria area of northern Nigeria.* MS thesis. Ahmadu Bello Univ., Zaria, Nigeria

29. Ali AM. 1957. On the bionomics and control of the bean fly, *Agromyza phaseoli* Coq. *Bull. Soc. Entomol. Egypte* 41:551–54

30. Allen DJ. 1987. *Principal Diseases of Beans in Africa: Study Guide.* Cali, Colombia: CIAT. 32 pp.

31. Allen DJ, Anno-Nyako FO, Ochieng RS, Ratinam M. 1981. Beetle transmission of cowpea mottle and southern bean mosaic viruses in West Africa. *Trop. Agric. Trinidad* 58:267–74

32. Allen DJ, Edje OT. 1990. Common bean in African farming systems. See Ref. 177a, pp. 20–31

33. Allen DJ, Smithson JB, eds. 1986. *Proc. Bean Fly Workshop: Afr. Workshop Ser.* No. 1. Cali, Colombia: CIAT. 29 pp.

34. Altieri MA, Francis CA, van Schoonhoven A, Doll JD. 1978. A review of insect prevalence in maize (*Zea mays* L.) and bean (*Phaseolus vulgaris* L.) polycultural systems. *Field Crops Res.* 1(1):33–49

35. Ampofo JKO. 1993. Host plant resistance and cultural strategies for bean stem maggot management. In *Proc. 2nd Meet. Pan-Afr. Work. Group Bean Entomol. CIAT Afr. Workshop Ser.* No. 25, pp. 4–73. Cali, Colombia: CIAT

36. Andow DA. 1991. Vegetational diversity and arthropod population response. *Annu. Rev. Entomol.* 36:561–86

37. Anyango JJ, Ochiel GRS. 1990. Survey and identification of insect pests of French bean with emphasis on control of bean thrips (*Taeniothrips* sp. and *Caliothrips* sp.) in Kenya. *Natl. Agric. Res. Lab. Annu. Rep.*, pp. 86–87. Nairobi: Kenya Agric. Res. Inst.

38. Applebaum SW, Guez M. 1972. Comparative resistance of *Phaseolus vulgaris* beans to *Callosbruchus chinensis* and *Acanthoscelides obtectus* (Coleoptera: Bruchidae): the differential digestions of soluble heteropolysaccharide. *Entomol. Exp. Appl.* 15:203–7

39. Asian Vegetable Research and Development Center. 1984. *Agromyzid flies of some native legume crops in Java.* AVRDC Publ. 84–126. Shanhua, Taiwan. AVRDC. 98 pp.

40. Assefa H. 1994. *Epidemiology of bean rust in Ethiopia.* PhD thesis. Wageningen Agric. Univ., Wageningen, The Netherlands. 172 pp.

41. Autrique A. 1985. *Compte-rendu de seminaire CIAT-ISABU-IRAZ sur la production et l' amelioration du haricot dans les pays des Grand Lacs. 20–25 Mai,* pp. 44–53. Bujumbura, Burundi: ISABU

42. Autrique A. 1989. Bean pests in Burundi: their status and prospects for control. In *Proc. Meet. Pan-Afr. Work. Group Bean Entomol., 1st, Aug. 6–9. CIAT Afr. Workshop Ser.* No. 11, pp. 1–9. Cali, Colombia: CIAT

43. Autrique A, Stary P, Ntahimpera L. 1989. Biological control of pest aphids by hymenopterous parasitoids in Burundi. *FAO Plant Prot. Bull.* 37(2):71–76

44. Deleted in proof

45. Bigger M. 1965. The biology and control of termites damaging field crops in Tanganyika. *Bull. Entomol. Res.* 56:417–44

46. Buckmire KU. 1978. Pests of grain legumes in the Commonwealth Caribbean. See Ref. 174a, 179–84

47. Buyckx EJE. 1962. *Precis des Maladies et des Insectes Nuisibles Rencontres sur les Plantes Cultivees au Congo, au Rwanda et au Burundi. Publ. Hors Ser.* pp. 592–610. Brussels: INEAC

48. Cardona C, Dick K, Posso CE, Ampofo K, Nadhy SM. 1992. Resistance of a common bean (*Phaseolus vulgaris* L.) cultivar to postharvest infestations by *Zabrotes subfasciatus* (Boheman) (Coleoptera: Bruchidae). II. Storage tests. *Trop. Pest Manage.* 38:173–75

49. Cardona C, Kornegay J, Posso CE, Morales F, Ramires H. 1990. Comparative value of four arcelin variants in the development of dry bean lines resistant to the Mexican bean weevil. *Entomol. Exp. Appl.* 56:197–206

50. Cardona C, Posso CE, Kornegay J, Valor J, Serano M. 1989. Antibiosis effects of wild dry bean accessions on the Mexican bean weevil and the bean weevil (Coleoptera: Bruchidae). *J. Econ. Entomol.* 82:310–15

51. Centro Internacional de Agricultura Tropical. 1987. Improving bean yields in Africa's Great Lakes region. *CIAT Int.* 6(2):3–6

52. Centro Internacional de Agricultura Tropical. 1993. *Trends in CIAT Commodities.* Work. Doc. No. 8. Cali, Colombia: CIAT

53. Chinwada P. 1994. *The bean bruchids,* Acanathoscelides obtectus *(Say) and* Zabrotes subfasciatus *(Boheman): distribution patterns in Zimbabwe, resistance levels in* Phaseolus vulgaris L. *(common bean) lines and control.* Mas-

ters thesis. Univ. Zimbabwe, Harare. 159 pp.

54. Coaker TH. 1958. Experiments with a virus disease of the cotton bollworm *Heliothis armigera* (Hbn). *Ann. Appl. Biol.* 46:536–41

55. Coaker TH. 1959. Investigations on *Heliothis armigera* (Hb.) in Uganda. *Bull. Entomol. Res.* 50:487–506

56. Commonwealth Institute of Biological Control. 1978. *Biological Control of Major Pests of Legumes, Pod Borers and Bean Flies.* Status Pap. No. 12. London: CAB. 8 pp.

57. Daoust RA, Roome RE. 1974. Bioassay of a nuclear-polyhedrosis virus and *Bacillus thuringiensis* against the American bollworm, *Heliothis armigera,* in Botswana. *J. Invertebr. Pathol.* 23:318–21

58. Davies G. 1990. Progress in research on bean stem maggots (*Ophiomyia* spp.) in Mozambique. See Ref. 177a, pp. 208–19

59. De Lima CPF. 1983. Management of pests of subsistence crops: legumes and pulses. See Ref. 197, pp. 246–48

60. Deny JE, Credland PE. 1991. Development, fecundity and egg dispersion of *Zabrotes subfasciatus. Entomol. Exp. Appl.* 59:9–17

61. Edje OT, Mughogho LK, Rao YP, Msuku WAB. 1981. Bean production in Malawi. In *Potential for Field Beans in Eastern Africa: Proc. Reg. Workshop, Lilongwe, Malawi, March 9–14, 1980,* pp. 55–97. CIAT Ser. No. 03EB. Cali, Colombia: CIAT

62. Egwuatu RI, Taylor TA. 1977. Studies on the biology of *Acanthomia tomentosicollis* (Stal) (Hemiptera: Coreidae) in the field and insectary. *Bull. Entomol. Res.* 67:249–57

63. Fagoonee I, Toory V. 1984. Preliminary investigations of host selection mechanisms by the leaf miner *Liriomyza trifolii. Insect Sci. Appl.* 4:337–41

64. Fitt GP. 1989. The ecology of *Heliothis* species in relation to agroecosystems. *Annu. Rev. Entomol.* 34:17–52

65. Food and Agriculture Organization of the United Nations. 1993. A global strategy for integrated pest management: results of an international meeting. *FAO Crop Prot. Bull.* 41(3–4):151–54

66. Forsyth J. 1966. *Agricultural Insects of Ghana.* Accra: Ghana Univ. Press. 163 pp.

67. Gangrade GA, Kogan M. 1980. Sampling stem flies in soybean. In *Sampling Methods in Soybean Entomology,* ed. M Kogan, DC Herzog, pp. 394–403. New York: Springer-Verlag

68. Gepts P, Debouck D. 1991. Origin, domestication, and evolution of the common bean (*Phaseolus vulgaris* L.). See Ref. 188b, pp. 7–53

69. Giga DP. 1989. Constraints to bean production in Zimbabwe. In *Proc. Meet. Pan-Afr. Work. Group Bean Entomol., 1st, Aug. 6–9. CIAT Afr. Workshop Ser.* No. 11, pp. 21–25. Cali, Colombia: CIAT

70. Giga DP, Ampofo JKO, Silim MN, Negasi F, Nahimana M, Nchimbi Msolla S. 1992. *On-Farm Storage Losses Due to Bean Bruchids and Farmers' Control Strategies: a Report on a Travelling Workshop in Eastern and Southern Africa. Occas. Publ. Ser.* No. 8. Cali, Colombia: CIAT

71. Greathead DJ. 1963. A review of the insect enemies of Acridoidea (Othoptera). *Trans. R. Entomol. Soc. London* 114:437–514

72. Greathead DJ. 1969. A study in East Africa of the bean flies (Dipt., Agromyzidae) affecting *Phaseolus vulgaris* and of their natural enemies, with the description of a new species of *Melanagromyza* Hend. *Bull. Entomol. Res.* 59:541–56

73. Greathead DJ, Girling DJ. 1982. Possibilities for natural enemies in *Heliothis* management and the contribution of the Commonwealth Institute of Biological Control. See Ref. 151a, pp. 147–58

74. Greenway P. 1945. The origin of some East African food plants. III. *E. Afr. Agric. J.* 10:177–80

74a. Grisley W, ed. 1991. *Proc. Workshop on National Research Planning for Bean Production Uganda. CIAT Afr. Workshop Ser.* No. 10. Cali, Colombia: CIAT

75. Grisley W, Mwesigwa D. 1991. A report on the socio-economics of bean production and marketing in Uganda: information for research planning. See Ref. 74a, pp. 75–81

76. Guled AA. 1989. Current status of grain legume pests in Somalia. In *Proc. Meet. Pan-Afr. Work. Group Bean Entomol., 1st, Aug. 6–9. CIAT Afr. Workshop Ser.* No. 11, p. 12. Cali, Colombia: CIAT

77. Hackett DS, Gatehouse AG. 1982. Studies on the biology of *Heliothis* spp. in Sudan. See Ref. 151a, pp. 29–38

78. Hall MJR. 1985. The blister beetle: a pest of man, his animals and crops. *Zimbabwe Sci. Newsl.* 19:11–15

79. Hill DS. 1983. In *Agricultural Insect Pests of the Tropics and Their Control,* pp. 454–55. London: Cambridge Univ. Press. 746 pp. 2nd ed.

80. Hill J, van Schoonhoven A. 1981. Effectiveness of vegetable oil fractions in

controlling the Mexican bean weevil on stored beans. *J. Econ. Entomol.* 74:478–79

81. Howe RW, Curie JE. 1964. Some laboratory observations on the rates of development, mortality and oviposition of several species of Bruchidae breeding in stored pulses. *Bull. Entomol. Res.* 55:437–77

82. Deleted in proof

83. Ingram WR. 1969. Observations on the pest status of bean flower thrips in Uganda. *E. Afr. Agric. For. J.* 34:482–84

84. Institute of Agricultural Research (Ethiopia), spons. org. 1990. *Research on Haricot Bean in Ethiopia: an Assessment of Status, Progress, Priorities and Strategies.* Addis Ababa: IAR. 114 pp.

85. Ismay J. 1989. Bean stem maggot identification demonstration. In *Proc. Meet. Pan-Afr. Work. Group Bean Entomol., 1st, Aug. 6–9. CIAT Afr. Workshop Ser.* No. 11, pp. 35–40. Cali, Colombia: CIAT

86. Jack RW. 1913. The bean stem maggot. *Rhod. Agric. J.* 10:545–53

87. Jackai LEN. 1981. Relationship between cowpea crop phenology and field infestation by the legume pod borer, *Maruca testulalis. Ann. Entomol. Soc. Am.* 74:402–8

88. Jackai LEN. 1982. A field screening technique for resitance of cowpea (*Vigna unguiculata*) to the pod borer *Maruca testulalis* (Geyer) (Lepidoptera: Pyralidae). *Bull. Entomol. Res.* 72:145–56

89. Jones AJ. 1953. *Annual Report of the Entomologist, Arusha, for the Season 1952–53.* Part 2. Tanganyika Dep. Agric.

90. Johnson SJ, King EG, Bradley JR Jr, eds. 1986. Theory and tactics of *Heliothis* population management. I. Cultural and biological control. *South. Coop. Ser. Bull.* No. 316. Okla. State Univ., Stillwater. 161 pp.

91. Karamura EB. 1991. Survey of bean fly (*Ophiomyia* spp.) in Kabale district. See Ref. 74a, pp. 42–48

92. Karel AK. 1985. Yield losses from and control of bean pod borers *Maruca testulalis* (Lepidoptera: Pyralidae) and *Heliothis armigera* (Lepidoptera: Noctuidae). *J. Econ. Entomol.* 77:761–65

93. Karel AK. 1989. Response of *Ootheca bennigseni* (Coleoptera: Chrysomelidae) to extracts from neem. *J. Econ. Entomol.* 82:1799–803

94. Karel AK, Autrique A. 1989. Insects and other pests in Africa. In *Bean Production Problems in the Tropics*, HF Schwartz, MA Pastor-Corrales, pp. 455–504. Cali, Colombia: CIAT. 2nd ed.

95. Karel AK, Matee JJ. 1986. Yield losses in common beans following damage by beanfly, *Ophiomyia phaseoli* Tryon (Diptera: Agromyzidae). *Bean Improv. Coop. Annu. Rep.*, pp. 115–16. NY State Agric. Exp. Stn., Geneva, NY

96. Karel AK, Mghogho RMK. 1985. Effects of insecticide and plant populations on the insect pests and yield of common bean (*Phaseolus vulgaris*). *J. Econ. Entomol.* 78:917–21

97. Karel AK, Ndunguru BJ, Price M, Semuguruka SH, Singh BB. 1981. Bean production in Tanzania. In *Potential for Field Beans in Eastern Africa: Proc. Reg. Workshop, Lilongwe, Malawi, March 9–14, 1980*, pp. 122–54. CIAT Ser. No. 03EB. Cali, Colombia: CIAT

98. Karel AK, Rweyemamu CL. 1984. Yield losses in field beans following foliar damage by *Ootheca bennigseni* (Coleoptera: Chrysomelidae). *J. Econ. Entomol.* 77:761–65

99. Karel AK, Rweyemamu CL. 1985. Resistance to foliar beetle, *Ootheca bennigseni* (Coleoptera: Chrysomelidae), in common beans. *Environ. Entomol.* 14:662–64

100. Katundu JM. 1980. Agromyzid leaf-miner: a new pest to Tanzania. *Trop. Grain Legume Bull.* 20:8–10

101. Khaemba BM, Ogenga-Latigo MW. 1985. Effects of the interaction of two levels of the black bean aphid, *Aphis fabae* Scopoli (Homoptera: Aphididae), and four stages of plant growth and development performance of the common bean, *Phaseolus vulgaris*, under greenhouse conditions in Kenya. *Insect Sci. Appl.* 6:645–48

102. Khamala CPM. 1978. Pests of grain legumes in Kenya. See Ref. 174a, pp. 127–34

103. King EG, Powell JE, Smith JW. 1982. Prospects for utilization of parasites and predators for management of *Heliothis* spp. See Ref. 151a, pp. 103–22

104. Kisakye J, Nabasrye M, Tushemereirwe W, Bakamwangiraki C, Kavuma JB. 1987. *A diagnostic survey of Kabale district.* Presented at the Reg. Bean Workshop, Kampala, Uganda, June. 16 pp.

105. Kornegay J, Cardona C. 1991. Breeding for insect resistance in beans. See Ref. 188b, pp. 619–48

106. Kornegay JL, Cardona C. 1991. Inheritance of resistance to *Acanthoscelides obtectus* in a wild common bean accession crossed to commercial bean cultivars. *Euphytica* 52(2):103–11

107. Kyamanywa S. 1988. *Ecological factors governing* Megalurothrips sjostedti *(Trybom) populations on cowpea/maize intercropped systems.* PhD thesis. Makerere Univ. Kampala, Uganda. 253 pp.
108. Kyamanywa S, Ampofo JKO. 1988. Effect of cowpea/maize mixed cropping on the incident light at the cowpea canopy and flower thrips (Thysanoptera: Thripidae) population density. *Crop Prot.* 7(3):186–89
109. Kyamanywa S, Baliddawa CW, Ampofo JKO. 1993. Effect of maize plants on colonization of cowpea plants by bean flower thrips *Megalurothrips sjostedti. Entomol. Exp. Appl.* 69:61–68
110. Kyamanywa S, Tukahirwa EM. 1988. The effect of mixed cropping beans, cowpeas and maize on population densities of bean flower thrips, *Megalurothrips sjostedti* (Trybom) (Thripidae). *Insect Sci. Appl.* 9:255–59
111. Leakey CLA. 1970. The improvement of beans (*Phaseolus vulgaris* L.) in East Africa. In *Crop Improvement in East Africa*, ed. CLA Leakey, pp. 99–28. Farnham Royal: CABI
112. Le Pelley RH. 1959. *Agricultural Insects of East Africa.* Nairobi: E. Afr. High Comm. 307 pp.
113. Letourneau DK. nd. Low-input pest control in Malawian subsistence agriculture: bean flies, cropping patterns, fertilizers and mulch. Board Environ. Studies. Univ. Calif. Santa Cruz. 24 pp.
114. Letourneau DK, Altieri MA. 1983. Abundance patterns of a predator, *Orius tristicolor* (Hemiptera Anthocoridae) and its prey *Frankliniella occidentalis* (Thysanoptera: Thripidae): habitat attraction in polyculture versus monoculture. *Environ. Entomol.* 12:1464–69
115. Lukefahr MJ. 1982. A review of the problems, progress, and prospects for host-plant resistance to *Heliothis* species. See Ref. 151a, pp. 223–32
116. Manohar S, Balasubramanian M. 1980. Note on the ovipostion behaviour of agromyzid stem fly *Ophiomia phaseoli* Tryon (Diptera Agromyzidae) in blackgram. *Madras Agric. J.* 67:470–71
117. Materu MEA. 1970. Damage caused by *Acanthomia tomentosicollis* Stål and *A. horrida* Germ. (Hemiptera: Coreidae). *E. Afr. Agric. For. J.* 35:429–35
118. Materu MEA. 1971. Population dynamics of *Acanthomia* spp. (Hemiptera; Coreidae) on beans and pigeon peas in the Arusha area of Tanzania. *E. Afr. Agric. For. J.* 36:361–83
119. Materu MEA. 1972. Morphology of adults and description of the young stages of *Acanthomia tomentosicollies* Stål and *A. horrida* Germ. (Hemiptera, Coreidae). *J. Nat. Hist.* 6:427–50
120. Matteson PC. 1981. Egg parasitoids of hemipteran pests of cowpea in Nigeria and Tanzania, with special reference to *Ooencyrtus particiae* Subba Rao (Hymenoptera: Encyrtidae) attacking *Clavigralla tomentosicollis* Stål. (Hemiptera: Coreidae). *Bull. Entomol. Res.* 71:547–54
121. McKinley DJ. 1982. The prospects for the use of nuclear polyhedrosis virus in *Heliothis* management. See Ref. 151a, pp. 123–34
121a. Minney BHP, Gatehouse AMR, Dobie P, Deny J, Cardona C, Gatehouse JA. 1990. Biochemical bases of seed resistance to *Zaborotes subfasciatus* (bean weevil) in *Phaseolus vulgaris* (common bean): a mechanism for arcelin toxicity. *J. Insect Physiol.* 36:757–67
122. Mohamed RA, Teri JM. 1989. Farmers' strategies of insect pest and disease management in small-scale bean production systems in Mgeta, Tanzania. *Insect Sci. Appl.* 10:821–25
123. Mound LA, Halsey SH. 1978. In *Whitefly of the World*, pp. 118–24. Chichester, UK: Br. Mus. Nat. Hist./Wiley. 340 pp.
124. Mushi CS, Youngquist WC, eds. 1992. *Proc. Workshop Bean Res. Planning in Tanzania. CIAT Afr. Workshop Ser.* No. 24. Cali, Colombia: CIAT
125. Mwalyego F. 1991. Effect of some cultivar mixtures on disease management and yield of beans. In *Bean Research*, ed. RB Mabagala, N Mollel, 6:100–4. Cali, Colombia: CIAT
126. Nderitu JH, Anyango JJ. nd. *Survey on the pests and their current control measures on French beans in Mwea division, Kirinyaga district, Kenya.* 23 pp. (Typescript)
127. Negasi F. 1994. *Studies on the economic importance and control of bean bruchids in haricot bean.* MS thesis. Alemaya Univ. Agric., Alemaya, Ethiopia. 103 pp.
128. Nyambo BT. 1990. Efect of natural enemies on the cotton bollworm, *Heliothis armigera* (Hubner) (Lepidoptera: Noctuidae) in western Tanzania. *Trop. Pest Manage.* 36:50–58
129. Nyiira ZM. 1973. Pest status of thrips and lepidopterous species on vegetables in Uganda. *E. Afr. Agric. For. J.* 39:131–35
130. Nyiira ZM. 1978. Pests of grain legumes in Uganda. See Ref. 174a, pp. 117–21
131. Ochieng RS. 1977. *Studies on the bionomics of two major pests of cowpea* Vigna unguiculata *L. (Walp.): Ootheca mutabilis Sahlb. (Coleoptera: Chrysomelidae) and* Anoplocnemis curvipes *F.*

(Hemiptera: Coreidae). PhD thesis. Univ. Ibadan, Nigeria. 267 pp.

132. Ogenga-Latigo MW, Ampofo JKO, Baliddawa CW. 1992. Influence of maize row spacing on infestation and damage of inter-cropped beans by the bean aphid (*Aphis fabae* Scop.). I. Incidence of aphids. *Field Crops Res.* 30:111–21

133. Ogenga-Latigo MW, Baliddawa CW, Ampofo JKO. 1992. Influence of maize row spacing on infestation and damage to intercropped beans by the bean aphid *Aphis fabae* (Scop.). II. Reduction of bean yields. *Field Crops Res.* 30:123–30

134. Okech SHO. 1991. Colonizing responses of *Maruca testulalis* (Geyer) (Lepidoptera: Pyralidae) to different cowpea cultivars in relation to their resistance/susceptibility. In *Abstracts of PhD and MPhil Theses Submitted by ARPPIS Scholars 1983–1987 Classes*, p. 34. Nairobi: ICIPE Sci.

135. Okeyo-Owuor JB. 1991. In *Abstracts of PhD and MPhil Theses Submitted by ARPPIS Scholars 1983–1987 Classes*, p. 35. Nairobi: ICIPE Sci.

136. Oree A, Slumpa S, Ampofo JKO. 1990. *Effect of environment and location on the species composition and populations of beanfly* (Ophiomyia *spp.: Diptera, Agronomyzidae) in Tanzania*. Presented at 2nd Reg. Bean Res. Workshop in East. Afr., Nairobi, Kenya, March 4–10

137. Pachico D. 1993. The demand for bean technology. In *Trends in CIAT Commodities 1993*, pp. 60–73. Cali: CIAT

138. Padgham J, Pike V, Dick K, Cardona C. 1992. Resistance of a common bean (*Phaseolus vulgaris* L.) cultivar to postharvest infestation by *Zabrotes subfaciates* (Boheman) (Coleoptera: Bruchidae). 2. Storage tests. *Trop. Pest Manage.* 38:167–72

139. Parsons FS. 1940. Investigations on the cotton bollworm, *Heliothis armigera*, Hubn. (*obsoleta* Fabr.). Part II. The incidence of parasites in quantitative relation to bollworm populations in South Africa. *Bull. Entomol. Res.* 31: 89–109

140. Parsons FS. 1940. Investigations on the cotton bollworm, *Heliothis armigera*, Hubn. Part III. Relationship between oviposition and the flowering curves of food plants. *Bull. Entomol. Res.* 31:147–77

141. Pedgley DE. 1986. Windborne migration in the Middle East by the moth *Heliothis armigera* (Lepidoptera: Noctuidae). *Ecol. Entomol.* 11: 467–70

142. Perrin RM. 1977. Pest management in multiple cropping systems. *Agro-Ecosystems* 3:93–118

143. Perrin RM, Phillips ML. 1978. Some effects of mixed cropping on the population dynamics of insect pests. *Entomol. Exp. Appl.* 24:385–93

144. Pomela ML. 1989. Bean insect pests of Lesotho. In *Proc. 1st Meet. Pan-Afr. Work. Group Bean Entomol. CIAT Afr. Workshop Ser. No. 11*, pp. 10–11. Cali, Colombia: CIAT

145. Purseglove JW. 1976. The origins and migrations of crops in tropical Africa. In *Origins of African Plant Domestication*, ed. JR Harlan, JM de Wet, ABL Stemler, pp. 293–309. The Hague: Monton

146. Quentin ME, Spencer JL, Milles JR. 1991. Bean tumbling as a control measure for the common bean weevil *Acanthoscelides obtectus*. *Entomol. Exp. Appl.* 60:105–9

147. Rajabalee A. 1993. Integrated pest management (IPM) development and practice in Mauritius. In *Proc. 2nd Meet. Pan-Afr. Work. Group on Bean Entomol. CIAT Afr. Workshop Ser. No. 25*, pp. 78–82. Cali, Colombia: CIAT

148. Randianandrianina L. 1988. Mise en point d'un milieu artificiel pour l'elevage de *Cynthia cardui* L. (= *Vanessa* = *Pyrameis*) Lepidoptera Nymphalidae, ravageur du soja (*Gyhycine max* Meril). In *Les Legumineuses a Graines*, ed. Y Demarly, pp. 357–63. Stockholm: FIS (Fond. Int. Sci.)

149. Raulston JR, Wolf WW, Lingren PD, Sparks AN. 1982. Migration as a factor in *Heliothis* management. See Ref. 197, pp. 61–64

150. Reed W. 1965. *Heliothis armigera* (Hb.) (Noctuidae) in western Tanganyika. I. Biology, with special reference to the pupal stage. *Bull. Entomol. Res.* 56:117–25

151. Reed W. 1965. *Heliothis armigera* (Hb.) (Noctuidae) in western Tanganyika. II. Ecology and natural and chemical control. *Bull. Entomol. Res.* 56:127–40

151a. Reed W, Kumble V, eds. 1982. *International Workshop on* Heliothis *Management*. Patancheru, India: ICRISAT

152. Reed W, Pawar CS. 1982. *Heliothis:* a global problem. See Ref. 151a, pp. 9–14

153. Remaudiere G, Amonyin G, Autrique A. 1985. Les plantes hotes des pucerons africains. In *Contribution a l'Ecologie des Aphides Africains. Plant Production and Protection Papers*, ed. G Remaudiere, A Autrique, 64:103–9. Rome: FAO

154. Rogers DJ. 1982. Screening legumes for

resistance to *Heliothis*. See Ref. 151a, pp. 277–87

155. Roome RE. 1975. Field trials with a nuclear polyhedrosis virus and *Bacillus thuringiensis* against larvae of *Heliothis armigera* (Hb.) (Lepidoptera: Noctuidae), in Botswana. *Bull. Entomol. Res.* 65:507–14

156. Roome RE. 1975. Activity of adult *Heliothis armigera* (Hb.) (Lepidoptera, Noctuidae) with reference to the flowering of sorghum and maize in Botswana. *Bull. Entomol. Res.* 65:523–30

157. Roome RE. 1979. Pupal diapause in *Heliothis armigera* Hubner (Lepidoptera: Noctuidae) in Botswana: its regulation by environmental factors. *Bull. Entomol. Res.* 69:149–60

158. Roome RE, Daoust RA. 1976. Survival of the nuclear polyhedrosis virus of *Heliothis armigera* on crops and in soil in Botswana. *J. Invertebr. Pathol.* 27:7–12

159. Rubaihayo PR, Mulindwa D, Sengooba T, Kamugira F. 1981. Bean production in Uganda. In *Potential for Field Bean Production in Eastern Africa,* pp. 155–86. Ser. No. 03EB–1. Cali, Colombia: CIAT

160. Salifu AB. 1986. *Studies on apsects of the biology of flower thrips,* Megalurothrips sjostedti *(Trybon), with particular reference to resistance in its host cowpea,* Vigna unguiculata *(L.) Walp.* PhD thesis. Wye College, Univ. London, London, UK. 269 pp.

161. Salih SH, Bushara AG, Ali MEK. 1990. Common bean (*Phaseolus vulgaris*) production and research in the Sudan. *Proc. 2nd Workshop on Bean Res. East. Afr. CIAT Afr. Workshop Ser.* No. 7, ed. JB Smithson, pp. 130–36. Cali, Colombia: CIAT

162. Sands WA. 1973. Termites as pests of tropical food crops. *Pest Art. News Summ.* 19:167–77

163. Deleted in proof

164. Deleted in proof

165. Deleted in proof

166. Schmutterer H. 1969. *Pests of Crops in North-East and Central Africa.* Stuttgart: Gustav Fischer. 296 pp.

167. Schreiner I, Nafus D, Bjork C. 1986. Control of *Liriomyza trifolii* (Burgess) (Dipt.: Agromyzidae) on yardlong (*Vigna unguiculata*) and pole beans (*Phaseolus vulgaris*) on Guam: effect on yield loss and parasite numbers. *Trop. Pest Manage.* 32:333–37

168. Silim MN. 1991. Current research on bruchids in Uganda. In *Potential for Field Beans in Eastern Africa: Proc. Reg. Workshop, Lilongwe, Malawi,*

March 9–14, 1980, pp. 49–55. CIAT Ser. No. 03EB. Cali, Colombia: CIAT

169. Silim MN. 1994. An additional character for sexing the adults of the dried bean beetle *Acanthoscelides obtectus* (Say) (Coleoptera: Bruchidae). *J. Stored Prod. Res.* 30(1):61–63

170. Silim MN. 1994. Bean sieving, a possible control measure for the dried bean beetles, *Acanthoscelides obtectus* (Say) (Coleoptera: Bruchidae). *J. Stored Prod. Res.* 30(1):65–69

171. Silim MN, Agona A. 1993. Studies on the control of the bean bruchids *Acanthoscelides obtectus* (Say) and *Zabrotes subfasciatus* (Boheman) (Coleoptera: Bruchidae) in the East African region. In *Proc. 2nd Meet. Pan-Afr. Work. Group on Bean Entomol. CIAT Afr. Workshop Ser.* No. 25, pp. 50–59. Cali, Colombia: CIAT

172. Sindibona JMV, Kayitare J. 1987. Contribution a l'etude de la resistance varietale du haricot (*Phaseolus vulgaris* L.) aux bruches (*Acanthoscelides obtectus* Say et *Zabrotes subfasciatus* Boh). *Bull. Agric. Rwanda* 20(2):120–28

173. Singh SR, Taylor TA. 1978. Pests of grain legumes in Nigeria and their control. See Ref. 174a, pp. 99–101

174. Singh SR, van Emden HF. 1979. Insect pests of grain legumes. *Annu. Rev. Entomol.* 24:255–78

174a. Singh SR, van Emden HF, Taylor TA, eds. 1978. *Pests of Grain Legumes: Ecology and Control.* London: Academic. 454 pp.

175. Sithanantham S. 1989. Status of bean entomology in Zambia. In *Proc. 1st Meet. Pan-Afr. Work. Group Bean Entomol. CIAT Afr. Workshop Ser.* No. 11, pp. 17–20. Cali, Colombia: CIAT

176. Slumpa S, Ampofo JKO. 1990. Recent advances in bean stem maggot research in northern Tanzania. See Ref. 177a, pp. 220–27

177. Slumpa S, Kabungo D. 1989. Status of bean entomology research in Tanzania. In *Proc. 1st Meet. Pan-Afr. Work. Group Bean Entomol. CIAT Afr. Workshop Ser.* No. 11, pp. 13–16. Cali, Colombia: CIAT

177a. Smithson JB, ed. 1990. *Bean Research,* Vol. 5, *Progress in Improvement of Common Bean in Eastern and Southern Africa, CIAT Afr. Workshop Ser.* No. 12. Cali, Colombia: CIAT

178. Swaine G. 1969. Studies on the biology and control of pests of seed beans (*Phaseolus vulgaris*) in northern Tanzania. *Bull. Entomol. Res.* 59:323–38

179. Talekar NS, Chen BS. 1986. The beanfly pest complex of tropical soybean. In

Soybean in Tropical and Subtropical Cropping Systems, ed. S Shanmugasundaram, EW Sulzberg, BT McLeod, pp. 257–71. Shanhua, Taiwan: AVRDC

180. Taylor CE. 1958. The bean stem maggot. *Rhod. Agric. J.* 55:634–36

181. Taylor DE. 1982. Cutworms. *Zimbabwe Agric. J.* 79:19–20

182. Taylor TA. 1967. The bionomics of *Maruca testulalis* Geyer (Lepidoptera: Pyralidae), a major pest of cowpeas in Nigeria. *J. West Afr. Sci. Assoc.* 12:111–29

183. Taylor TA. 1969. On the population dynamics and flight activity of *Taeniothrips sjostedti* (Trybom) (Thysanoptera: Thripidae) on cowpea. *Bull. Entomol. Soc. Nigeria* 2:60–71

184. Taylor TA. 1978. *Maruca testulalis:* an important pest of tropical legumes. See Ref. 174a, pp. 193–200

185. Topper CP. 1987. Nocturnal behaviour of adults of *Heliothis armigera* (Hubner) in the Sudan Gezira and pest control implications. *Bull. Entomol. Res.* 77:541–44

186. Trutmann P, Paul KB, Chisahayo D. 1992. Seed treatments increase yield of farmer varietal field bean mixtures in the central African highlands through multiple disease and bean fly control. *Crop Prot.* 11:458–64

186a. van den Berg H. 1993. *Natural control of* Helicoverpa armigera *in smallholder crops in East Africa.* PhD thesis. Wageningen Agric. Univ. 233 pp.

187. Van Rheenen HA, Pere WM, Magoya JK. 1983. Protection of stored seed bean seeds against the bean bruchid. *FAO Plant Prot. Bull.* 31:121–25

187a. van Schoonhoven A. 1978. Use of vegetable oils to protect stored beans from bruchid attack. *J. Econ. Entomol.* 78:254–56

187b. van Schoonhoven A, Cardona C. 1982. Low levels of resistance to the Mexican bean weevil in dry beans. *J. Econ. Entomol.* 75:567–69

188. van Schoonhoven A, Cardona C, Valencia CA. 1986. *Main Insect Pests of Stored Beans and Their Control: Study Guide.* Ser. 04EB–05.03. Cali, Colombia: CIAT. 40 pp.

188a. van Schoonhoven A, Cardona C, Valor J. 1983. Resistance to the bean weevil and the Mexican bean weevil (Coleoptera: Bruchidae) in non-cultivated common bean accessions. *J. Econ. Entomol.* 76:1255–59

188b. van Schoonhoven A, Voysest O, eds. 1991. *Common Beans: Research for Improvement.* Wallingford: CABI

189. Wallace GB. 1939. French bean diseases and beanfly in East Africa. *E. Afr. Agric. J.* 7:170–75

190. Whitlock VH. 1977. Simultaneous treatments of *Heliothis armigera* with a nuclear polyhedrosis and a granulosis virus. *J. Invertebr. Pathol.* 29:297–303

191. Whitney WK, Gilmer RM. 1974. Insect vectors of cowpea viruses in Nigeria. *Ann. Appl. Biol.* 77:17–21

192. Woolley J, Davis JHC. 1991. The agronomy of intercropping with beans. See Ref. 188b, pp. 707–35

193. Woolley J, Lepiz RI, de Aquino TPC. 1991. See Ref. 188b, pp. 679–706

194. Wortmann CS. 1992. *Assessment of Yield Loss Caused by Biotic Stress on Beans in Africa.* Occas. Publ. Ser. No. 4. Cali, Colombia: CIAT. 17 pp.

195. Wortmann CS, Allen DJ. 1994. *African bean production environments: their characteristics, constraints and opportunities.* Occas. Publ. Ser. No. 11. Cali, Colombia: CIAT. 47 pp.

196. Wortmann CS, Bosch C, Mukandala L. 1993. *The Banana-Bean Intercropping System in Kagera Region of Tanzania: Results of a Diagnostic Survey.* Occas. Publ. Ser. No. 6. Cali, Colombia: CIAT

197. Youdowei A, Service MW, eds. 1983. *Pest and Vector Management in the Tropics.* London: Longman

Annu. Rev. Entomol. 1996. 41:75–100
Copyright © 1996 by Annual Reviews Inc. All rights reserved

ECOLOGY OF INSECT COMMUNITIES IN NONTIDAL WETLANDS

Darold P. Batzer

Department of Biology, Canisius College, Buffalo, New York 14208

Scott A. Wissinger

Departments of Biology and Environmental Science, Allegheny College, Meadville, Pennsylvania 16335 and Rocky Mountain Biological Laboratory, Crested Butte, Colorado 81224

KEY WORDS: aquatic ecology, colonization, plant-insect interactions, predation, wetland management

ABSTRACT

Published research about wetland insects has proliferated, and a conceptual foundation about how wetland insect populations and communities are regulated is being built. Here we review and synthesize this new body of work. Our review begins with a summary of insect communities found in diverse wetland types, including temporary pools, seasonally flooded marshes, perennially flooded marshes, forested floodplains, and peatlands. Next, we critically discuss research on the population and community ecology of wetland insects, including the importance of colonization strategies and insect interactions with the physical environment, plants, predators, and competitors. Results from many of the experimental studies that we review indicate that some commonly held beliefs about wetland insect ecology require significant reevaluation. We then discuss the importance of wetland insect ecology for some applied concerns such as efforts to manage wetland insect resources as waterfowl food and development of ecologically sound strategies to control pest mosquitoes. We conclude with a discussion of wetland conservation, emphasizing insect aspects.

PERSPECTIVES AND OVERVIEW

The amount of published research on wetland insect communities has grown steadily since the mid 1980s. One reason for this increase is that insect communities in small wetlands have proven to be useful systems for testing

75

ecological paradigms. Both community structure of insects and the environment of small wetlands can be readily manipulated or mimicked for experimental investigation (e.g. 11, 81, 115, 180, 184, 195). The newfound interest in the community ecology of wetland insects also is evident in the applied sector. Endangered wetland ecosystems are receiving a great deal of public attention, and the importance of insects in these systems is being recognized. In addition, waterfowl and pest management concerns about wetlands now place a high priority on information about insect ecology. In the 1960s and 1970s, wildlife workers discovered that insects were important waterfowl foods (120), and much of the recent research on wetland insects has been sponsored by wildlife interests. In addition, although mosquitoes have always been a research focus in wetlands, investigators currently take an increasingly holistic approach and study wetlands in a community context. Much of our understanding about how wetland insect communities are regulated comes from this applied work.

Although research on wetland insects has proliferated worldwide, recent reviews have focused on particular regions and specific wetland types (55, 90, 119, 154). Reviews of Canadian wetlands have been some of the most thorough (55, 154). Reviews of individual insect groups provide additional sources of information about wetland insects (33, 72, 99, 141, 171). Much of the early work on wetland insects is covered by Wiggins et al (189).

The theoretical foundations of insect population and community regulation in wetlands remain in the initial stages of development. One finds little mention of the important insect fauna in general reviews of wetland ecology (e.g. 22, 113) or in standardized field procedures to evaluate the biotic integrity of wetland habitats (3, 174). Researchers of wetland insects have typically used ecological models developed in other habitats as the basis for their work (e.g. 103, 119), and only recently have they begun to experimentally test and develop paradigms that specifically apply to wetlands. Those efforts are the major focus of our review, and we contend that many commonly held beliefs about wetland insect ecology require significant reevaluation.

We begin by reviewing the characteristics of insect communities found in diverse wetland types. We restrict our coverage to temporary pools, nontidal marshes, forested floodplains, and peatlands. Large wetlands, such as the Florida Everglades, include several of these wetland categories (143). We do not address tidal marshes because of space limitations. We then discuss the population and community ecology of wetland insects, including patterns of colonization and insect interactions with the physical environment, plants, predators, and competitors. We develop these themes using work conducted specifically in wetlands and refrain from forcing ecological paradigms developed in other habitats on wetland insect communities. We conclude with a discussion of applied aspects of wetland insect ecology such as insects and waterfowl management, pest-insect control, and insect aspects of wetland conservation.

WETLAND INSECT COMMUNITIES

Temporary Pools

The smallest and often the most ephemeral of the wetlands are temporary pools that fill from seasonal rains or snow melt (131, 189, 190). Some flood and dry repeatedly (131); others fill for a few months (92, 131, 189); and some remain flooded for much of the year (131, 189). Most of the research on temporary-pool insects focuses on their adaptations for and interactions with the short-term flooding regimes. Another major area of research addresses biotic interactions.

Insects use various life-history strategies to survive in temporary pools. Many have methods for resisting drought (83, 189, 190), whereas others combine rapid aerial colonization and larval development to effectively avoid drought (189, 190, 194). However, drought is often a major mortality factor for insects in temporary pools (81, 150). Many species are habitat generalists that also flourish in more permanent wetlands (70, 124).

Temporary pools have especially diverse beetle and midge communities (4, 70, 92, 129, 160, 189). Among the different wetland types, temporary pools are unique in that mosquito larvae are often the numerically dominant residents (117, 150). Specific community structures are strongly influenced by the flooding regimes of individual pools (44, 131, 189) (see section on hydrology, below). In addition to wet-dry cycles, other abiotic factors that commonly affect insects include pool size (8, 130, 156) and water temperature (8, 130). Insect productivity can be very high in temporary pools (124, 189), possibly because decomposition during waterless stages enhances detrital food quality (189).

Biotic interactions often regulate wetland insect populations in temporary pools. This observation is somewhat surprising given the relatively short periods available for insects to interact with each other. High densities can lead to competitive interactions (82, 115, 150, 180). In these fishless habitats, predaceous insects can become very abundant (92), and they are often the top aquatic predators (see section on predation).

Seasonally Flooded Marshes

Seasonally flooded meadows or marshes are also highly ephemeral wetlands. Extensive stands of moist-soil annual plants and hydrophilic perennials are a unique feature of these habitats (147). Seasonally flooded marshes were historically common on river floodplains, but most have disappeared because of river regulation (1, 147). This type of natural habitat still occurs over large portions of the Florida Everglades (143). Because seasonally flooded marshes are productive habitats for ducks, similar types of habitats are often constructed (e.g. moist-soil management units) and maintained artificially on waterfowl management areas (147). These managed habitats are where most insect research in seasonally flooded marshes has been done, and much of this research focuses on the impact of human manipulations on insect communities (12, 49, 52, 101, 147).

In managed seasonally flooded marshes, insects typically dominate the macroinvertebrate communities (12, 49, 52, 101, 125). Perhaps because of the intensive management, opportunists flourish. For example, rat-tailed maggots were very abundant in a managed California habitat (12), but reference to their occurrence elsewhere is scant. Typically, a few insect families dominate the communities of managed seasonal marshes, with the midges being particularly prominent (12, 49, 52, 125, 152). A beneficial aspect of such dominance in terms of waterfowl management is that midges in the genus *Chironomus* thrive in managed habitats (10, 12, 49, 52, 102), and they are often preferred duck food (10, 52, 102). Predatory insects are often very abundant (10, 12, 13, 49, 101, 125, 152), and like in temporary pools, they can become important community regulators (e.g. 11).

Perennially Flooded Marshes

Extensive stands of emergent aquatic plants are distinctive features of perennial-water marshes, and submersed macrophytes and algae are also conspicuous. As a result, much of the research on marsh insects involves their interactions with plants. Emergents are particularly important resources for marsh insects. Terrestrial or semiaquatic forms feed on above-water portions of emergent plants (27, 56) or use that area as habitat (51). Decaying plant material provides food for aquatic detritivores (28, 126). Submersed portions of living and dead emergent plants serve as habitat for numerous aquatic insects (9, 28, 51, 121). The habitat templates created by interspersions of emergent plants and open water are particularly productive areas for insects (107, 127).

Submersed or floating weed beds also support many insects (15, 39, 106, 159, 198). Numerous collembolans, aphids, and water striders can be found at the air-water interface of weed beds (15). Many aquatic insects use submersed plant surfaces as habitat (15, 39, 136, 137, 159). Midge larvae can be particularly numerous on these surfaces, even more so than on emergent plants (15, 198). An important role of macrophytes to marsh insects may be in providing substrates for growth of periphytic algal foods (28, 68). Algae is another important food for many marsh insects (28, 59, 68, 126), and it is also used by midge larvae to construct tube cases (11).

Not all marsh insects intimately associate with plants. Mud substrates are productive areas of many marshes, particularly for benthic dipteran larvae (9, 101, 122). Benthos and certain nektonic insects can be more abundant in open water than emergent vegetation zones (9, 14, 90, 159). Deep, artificially created channels can support numerous marsh insects (23, 143). Some species migrate between open and emergent areas (35, 118), although the transitional zone itself is seldom a preferred habitat (118, 159).

Forested Floodplains

Forested wetlands or swamps typically occur on river or stream floodplains. In addition to the flooded woodlands, these habitats contain temporary pools, semipermanent backwater ponds, and shallow lakes (1, 62, 165). Depending on rainfall and elevation, floodplain habitats can be flooded year-round or for only short durations (38, 61, 165), and specific patterns of inundation influence invertebrate communities (165). Invertebrate productivities can be very high in such habitats (165), perhaps because aqueous nutrients from flowing channels (62) and leaf fall from trees (38, 105) enrich forested floodplains. The secondary productivity from adjacent wetlands probably has important implications for riverine habitat ecologies, given that the floodplains cover a larger surface area than the associated channels (61).

In terms of faunal diversity, insects, especially midges, can dominate the invertebrate communities of forested floodplains (61, 62, 102). Midges are the only insects that consistently reach high densities in such habitats (78, 100, 101, 105, 165). A few predatory insects can be important in terms of biomass and productivity (38, 61, 105, 165), but noninsects such as aquatic worms, mollusks, and crustaceans tend to dominate (38, 49, 61, 78, 100, 165). Unexpectedly, insect shredders adapted to consume tree leaves are often rare (38, 105).

Floodplains often share insect taxa with their associated channels (38, 61). Those taxa tolerant of low levels of dissolved oxygen prosper most in the wetland portion of these systems (38, 165). Lotic insects use floodplain wetlands to exploit the rich food resources and warmer water (1), to reproduce (1, 38), or to find a haven from predation (168). Predatory fish themselves move into floodplains to feed on aquatic invertebrates and trapped terrestrials (148). Many different terrestrial insects live on dry floodplains (65) or in tree foliage (64).

Forested floodplains are a prevalent wetland type in the southern latitudes, yet their invertebrate communities are understudied relative to other wetlands (49, 78, 114, 158, 165). Most research focuses on insects closely associated with channels, so even less is known about the fauna of difficult-to-access backswamps (165). Learning more about the ecology of invertebrate communities in these wetlands should become a research priority.

Peatlands

Peat develops in wetlands where decomposition does not keep pace with plant production (181). Peatlands are categorized as fens or bogs based on hydrology, with fens usually being dominated by sedges or other monocots and bogs by *Sphagnum* mosses (181). Peat accumulation causes peatland surfaces to gradually rise above the water table (181), and thus many of the insects found in these habitats are of terrestrial origin. Aquatic species can occur in the water around peatland vegetation itself (e.g. 155), although most of the research addressing the aquatic fauna has been conducted in peatland pools.

Finnamore & Marshall (55) recently edited an extensive review of the terrestrial arthropod fauna of Canadian peatlands; therefore, we emphasize research conducted elsewhere. Some of the terrestrial insects of peatlands are invaders from surrounding upland habitats (86). However, others are unique to peatlands (36, 98), and particular assemblages of terrestrial species can be useful in characterizing peatland habitats (36, 74). Ants are especially important ecologically because mound-building and hummock-dwelling species can influence vegetation patterns and habitat microtopography (30, 98).

A 1987 review by Rosenberg & Danks covers the truly aquatic insects of Canadian peatlands (154); this treatment also addressed research outside of Canada. The more recently published work indicates that, similar to the terrestrial fauna, the aquatic insect communities of peatlands consist of a mixture of invaders from other habitats and species unique to peatlands (155). Insects occur in diverse aquatic microhabitats, including saturated peat itself (138, 155), shallow pools (79, 138), and deeper ponds (25, 94). A unique habitat for some aquatic insects is the water that accumulates inside pitcher plants (54 and references therein). The deeper ponds of peatlands can be particularly productive habitats for insects (138). The relative size of pools (94), water temperature (25, 173), dissolved oxygen level (25), and water stability (79, 94) have all been identified as important abiotic regulators of aquatic peatland insects. Although many peatland pools are acidic, the primary influence of acid on insects is probably indirect. The lack of fish in acidic habitats may be a more important influence on insects than the water chemistry itself (73).

POPULATION AND COMMUNITY ECOLOGY

Colonization and Life-History Strategies

Many wetlands dry, either seasonally or unpredictably, during droughts, and invertebrates that thrive in these habitats exhibit two general strategies for rapid colonization during and after inundation. The first is desiccation resistance, which is best documented in flightless invertebrates (protists, rotifers, crustaceans, annelids, mollusks) that deposit resistant eggs or spores in the substrate (189, 190). Many wetland insects, including dragonflies, caddisflies, beetles, and various dipterans, also have desiccation-resistant eggs (189). Many dipteran eggs are similar to the resistant stages of flightless invertebrates in that they are often deposited just before basins dry and, in their diapausal state, can resist harsh conditions (189, 190). In some mosquitoes and a few midges, egg maturation within clutches is staggered, thus providing an egg bank that remains viable over several dry-wet cycles (2, 190). Several wetland-inhabiting damselflies in wetlands deposit eggs endophytically in the stems of emergent plants, where the eggs undergo diapause until habitats reflood (48 and references therein). Insects that oviposit in dried wetlands typically enclose eggs in a protective gelatinous mass. Dry-basin oviposition has been especially well studied for several species of *Sympetrum* dragonflies from northern wetlands (177) and for limnephilid caddisflies that oviposit under rocks or wood in or near dry basins (16; SA Wissinger & WW Brown, unpublished data).

Desiccation resistance in adult aquatic insects is best documented for beetles that survive drying by burying themselves in the mud or crawling under rocks or logs in or near wetland basins (81, 92, 101, 189). Some odonate nymphs also resist drying and/or freezing by burying themselves in the substrate of dry habitats (33, 83). Desiccation resistance during the larval stage has been documented for most families of aquatic Diptera (101, 141, 189). Of these, the midges are often most abundant and widespread in moist soils under dried wetland basins (79, 189). Given that both the egg and larvae of midges can be desiccation resistant and adults are such good dispersers (see below), it is perhaps not surprising that they are often the most diverse and abundant insects in ephemeral habitats (26, 44, 61, 189, 190).

A second strategy that aquatic insects use to colonize wetlands involves adult immigration and oviposition. The most rapid aerial colonizers are opportunists such as mosquitoes and midges, which often deposit eggs within a few days after inundation (43, 89, 101, 150, 153, 183, 191). In several wetland insects, adult longevity facilitates colonization. For example, in limnephilid caddisflies, adults emerge just before ponds dry and diapause for a month or more in terrestrial habitats before returning to lay eggs (16, 99; SA Wissinger & WW Brown, unpublished data). Similarly, long-lived dragonfly adults survive the dry phase of wetland hydroperiods in the terrestrial environment and deposit eggs when basins refill (32, 94). Dragonfly colonization of newly inundated habitats is also aided by the general tendency to disperse beyond the limits of their natal habitats (111).

The most common strategy for the aerial colonization of ephemeral wetlands appears to involve cycles of migration between permanent and temporary aquatic habitats (189, 194). For example, in a seasonally flooded marsh that was managed as waterfowl habitat, all of the numerically dominant insects (various dipterans, a beetle, and a water boatman) colonized by immigrating from nearby aquatic habitats rather than by means of desiccation-resistant stages (12). Our review of the literature suggests that this general pattern of "cyclic colonization" (sensu 194) is important for the maintenance of insect populations in many natural and managed wetlands, including seasonally and perennially flooded marshes (9, 12, 29, 70, 76, 122, 125, 128, 143, 144), temporary pools in larger wetland complexes (81, 94, 129), and tropical and temperate swamps (61, 101, 114, 158).

Cyclic colonization is best documented for adult hemipterans and beetles, which exhibit a characteristic life-history syndrome that often involves flight polymorphism (cf 72, 93, 171, 176). For beetles and hemipterans that inhabit temporary waters, long-lived adults typically overwinter in permanent habitats and then fly to newly inundated habitats in spring. Upon arrival, wing musculature histolyzes in the females, and the developing ovaries expand into the vacated space. Depending on the duration of the wet phase, one or more short-winged or wingless generations are completed, and then, as habitats dry, long-winged individuals with immature gonads emigrate and return to permanent refugia. Flight polymorphisms facilitate dispersal to and from permanent refugia and allows for the rapid establishment of large populations in the ephemeral habitat (72, 171). The importance of cyclic colonization for the maintenance of beetle populations in temporary habitats is emphasized by Svensson (176), who suggests that their absence in those habitats near the edges of geographical ranges is related to the relatively low density of source populations.

Although most cyclic colonization occurs between local habitats, several large dragonflies colonize ephemeral wetlands by migrating over long distances (34). In the tropics, these strong flyers complete several generations per year by tracking shifting monsoonal rains and exploiting temporary habitats across a latitudinal gradient (32). In North America, these dragonflies migrate to temperate wetlands and other fishless habitats where they complete one summer generation before emigrating south in autumn (193).

Influence of Physicochemical Variables

Wetland insects can be affected by several abiotic factors (e.g. 66), but only the influences of hydrology, acidity, or nutrient levels have been studied in detail. The periodic flooding and drying of wetland habitats is probably the most important single influence on insect communities, and the way wetland insects deal with drought is a key to their success (189). The research interest in acidity stems from concerns about acid pollution and curiosity about what impact low natural pH has on insects. Phosphorous and nitrogen levels are also a concern in terms of pollutants leading to eutrophication; conversely, research has examined how the natural nutrient-poor status of some wetlands influences insects (28, 59, 123). Other physicochemical variables that can influence wetland insect communities are habitat sizes (8, 94, 156), high or low water temperatures (8, 25, 143, 173), and dissolved oxygen levels (25, 143, 144).

HYDROLOGY   Water-level stability of wetlands is a major influence on insect community structure and productivity (9, 61, 81, 94, 131, 189). Highly ephemeral habitats tend to be dominated by beetles and mosquitoes, while midges and odonates predominate in habitats that are flooded longer (12, 70, 81, 94, 131, 159). Insects dependent on a drought-resistant stage are virtually eliminated when ephemeral habitats remain flooded (125). If wetlands are flooded infrequently or for short durations, their invertebrate productivities may be low (61). A long and intense drought phase can also have a negative impact on wetland invertebrates (44, 81).

Field observations indicate that invertebrates in temporary pools grow more rapidly than those in perennial water wetlands (142, 189). However, in the few reports that directly compare insects from each of these habitats, the efficiency of growth and feeding is actually higher for insects from perennial habitats. For example, Pritchard & Berté (142) found that a caddisfly adapted for perennially flooded habitats grew more efficiently than a species adapted for life in ephemeral pools, and they believed the more rapid growth of the latter species in the field was simply related to differences in habitat temperature. In addition, when populations of the same midge species living in temporary and perennial pools were compared, their feeding rates, feeding efficiencies, and productivities were all higher in the perennial habitat (124). The general belief that temporary pools are intrinsically more productive habitats for insects than perennially flooded habitats needs further evaluation.

ACIDITY   Although some insects select wetland habitats with specific acidities (4, 66), experimental acidification of wetland habitats often has shown little impact on insects. For example, damselfly nymphs were not affected by pH regulation unless levels became very low (63). Experimental reduction of the pH of a fen peatland had no detectable influence on midge emergence (155). In contrast, low pH may actually benefit insects in some cases because insectivorous fish cannot survive in many acidic habitats (17, 73, 77, 139). However, low pH may harm certain insects. In studies contrasting acid and circumneutral wetlands, the acidic habitats supported fewer mayfly nymphs (17) or had detrital food resources of lower quality (89).

NUTRIENTS   Descriptive studies indicate that nutrient levels may influence wetland insect communities (62, 66, 74, 107). However, excessive nutrient loading in the Florida Everglades increased invertebrate diversity (144), contrary to expectation (62, 66). A series of experiments in a nutrient-poor Canadian marsh found that naturally low phosphorous and nitrogen levels can limit midge productivity (28, 59, 123).

Primary Consumers

Many terrestrial insects in wetlands consume the foliage of living macrophytes (27, 56, 64, 103, 162). However, aquatic forms rarely eat living vascular plants (119, 167); instead they consume detritus from dead plants. Decomposing leaves and stems from wetland macrophytes (28, 119) and trees (38, 105) are important sources of detritus, as are inputs from riparian habitats and upland run-off (124, 133, 160). Algae are another food important to some wetland insects, particularly midge larvae (11, 28, 59, 68, 119).

Which of the above sources of plant carbon are most important to the aquatic food webs in wetlands is under debate (119). Recent studies that use molecular techniques (stable isotope analysis) indicate that aquatic insect and other invertebrate consumers commonly eat algal periphyton or emergent macrophytes (in the form of detritus) but rarely eat filamentous algae or detritus from submersed macrophytes (24, 112, 126). A surprising result of these studies was that unknown sources of carbon, rather than the conspicuous forms tested above, were important to some consumers (24, 126).

MACROPHYTE CONSUMERS   Most wetland insects that consume living macrophytes occur on the above-water portions of the plants or bore inside them, and they tend to be members of primarily terrestrial groups (aphids, leaf beetles, weevils, moth caterpillars) (15, 56, 103, 162). Some of these herbivores can have a great impact on plant growth and survival (64, 103). Foote et al (56) list several published accounts of insect herbivores removing from 25 to 56% of the standing crop of wetland macrophytes. Certain moth larvae can completely defoliate swamp trees (64). A beetle that consumed only 6% of the available biomass of water lily leaves still had a marked impact on detrital inputs into a wetland (162). Thus, although the subject is largely unstudied, the impact of terrestrial insect herbivores on overall wetland ecosystem function is probably very significant.

DETRITUS CONSUMERS   As aquatic macrophytes or tree leaves decompose, aquatic invertebrate numbers on this material tend to increase (28, 41, 89, 101, 128, 153, 167). This pattern may be related to the increasing palatability and protein content of the decaying plant material (166, 167), as indicated by experimental evidence from some wetlands. A comparison of natural and artificial litter showed that both types initially supported similar numbers of midges, but as the natural litter decayed, it became the preferred substrate (28).

However, field experimentation often fails to link wetland invertebrate abundance with detritus quality. Invertebrate colonization rates can be similar on plant litters of differing nutritional qualities (89, 128), and fewer invertebrates occurred on litter from fast-decaying *Equisetum fluviatile* than slower decaying *Carex rostrata* (41). In some cases, temporal changes in invertebrate numbers on decaying litter may simply reflect their life-history patterns (128) or natural colonization cycles (153).

An issue separate from detritus quality is its quantity. Our survey of the literature uncovered little experimental evidence that the available supply of detrital food in wetlands can be limiting to detritivores. Where wetland macrophytes were cut and the detritus removed, detritivore populations either were unaffected or showed unexpected increases (122, 125). In another experiment, albeit nonreplicated, detritus was added to a marsh enclosure, but the expected increase in detritivore numbers failed to materialize (5). In a detailed descriptive study, a midge that dominated the benthos of two different wetland pools consumed only 4.6–13.0% of the available detritus over the course of a year (124). In certain cases, the low oxygen conditions created by decomposing plant matter may reduce rather than increase invertebrate numbers, at least temporarily (59). However, some detritivorous mosquitoes can reach high enough densities in temporary pools that they compete for space, which leads to food shortages despite the presence of abundant supplies of detritus (150).

In some cases, however, the availability of plant litter and detritus resources may limit insect populations. As mentioned previously, the substitution of a nonnutritive substrate for plant litter can reduce midge numbers (28). Increases in plant litter biomass caused by above-normal flooding of a marsh resulted in more midges and other consumers, although whether the detritus served as food or habitat was unclear (121). Finally, in a small pond study, midge growth rates, but not densities, increased as a result of enriched microdetrital food supplies (145).

In some wetland habitats, invertebrate detritivores play an important role in litter decomposition (89). However, in many wetlands, shredders adapted to process plant litter are rare (38, 105). If invertebrate consumers are excluded from decaying plant litter, decomposition rates can remain unchanged (153). In wetlands, most plant decomposition is probably accomplished by leaching and through microbial pathways rather than via invertebrate processing (41, 89, 105, 153, 189).

ALGAL CONSUMERS   In contrast to detritus, a relatively strong case can be made that algal food supplies may become limiting in wetland food webs. Experiments in a seasonally flooded marsh (11), a perennially flooded marsh (68), and simulated temporary pools (115) indicate that midge larvae can overgraze their algal food supplies, which may lead to intra- or interspecific competition for food. Enrichment of marshes with nitrogen and phosphorous, either from experimental introductions or pollutants (28, 59, 144), can increase the algal food supply and with it numbers of invertebrate consumers (except see 123). Herbicide-induced reductions in algal biomass can limit numbers of insect consumers (45). The importance of algae to wetland food webs has probably been underestimated (119).

Wetland Plants as Habitat

Apart from providing food for insects, wetland plants also serve as structural habitat. Distributions of some nonherbivorous species have been predicted accurately from information about which plants dominate wetland habitats (e.g. 149). In marshes, some of the most productive areas for insects are where open water is interspersed with emergent and submersed plants (107, 127, 143). Experimentally creating such interspersions can increase numbers of many insect taxa (12, 87, 122). However, some taxa such as epiphytic fly larvae and free-swimming mosquito larvae prefer denser plant stands (12, 137). Although living emergent plants rarely support high densities of insects directly on their surfaces (50, 198), the litter that accumulates from them is valuable habitat for epiphytic and mining midges that live on or in this material (121). The substrates lying beneath emergent plant litter can also support high densities of benthic insects (50).

In contrast to emergent plants, submersed plants support large numbers of epiphytic insects on their surfaces (50, 198). However, aquatic insect or other invertebrate use of submersed plant habitats varies among plant species, and it has generally been assumed that differences in plant morphology and architecture are responsible (i.e. plants with more colonizable substrate would support the most invertebrates) (39, 50, 71, 80, 198). If all factors other than plant morphology are held constant, this hypothesis appears to be valid (80). However, in natural plant beds, the mechanisms controlling wetland invertebrate numbers are considerably more complex. Plants with the most finely dissected leaves do not necessarily support the most invertebrates (39). Submersed plants with similar morphologies but different growth cycles harbor different invertebrate communities (71), and communities also change as individual plant beds grow and senesce (15). In addition to the insects that occur directly on submersed plant surfaces, free-swimming insects find these plant beds to be valuable refuges from predators (14, 137).

Predator-Prey Interactions and Food Webs

Fish are typically abundant only in those wetlands that remain flooded over long periods of time and are deep enough to prevent winterkill. They also occur as transients that move in and out of lacustrine marshes and floodplains with flood waters (61, 148). Numerous studies suggest that wetlands with fish will have lower insect diversity and lower biomass than those that are fishless (46, 76, 77, 104). Many of the taxa that fish prey upon most heavily are also important in the diets of migratory waterfowl (10). Thus, fish are usually considered unwanted predators by managers of waterfowl habitat (12, 52, 77, 104).

The insect communities in fish and fishless wetlands also differ because of the size-selective nature of fish predation. In wetlands without fish, large-bodied active predators such as odonate nymphs, beetle larvae, and hemipterans replace fish as the top aquatic predators in the food web (14, 20, 101, 180). The few species of large predatory insects that can coexist with fish exhibit various antipredator morphologies and behaviors that are not observed in congeneric species that thrive in fishless habitats (14, 20, 73, 84, 109). Fishless habitats that are perennially flooded tend to support the highest diversity and overall abundance of these large predators (49, 61, 94, 159, 171, 176, 192). The lack of an increase in detritivore numbers following enhanced detritus levels in fishless wetlands suggests that large invertebrate predators can control prey abundance (5, 145, 146).

All of the top insect predators in fishless habitats are generalists that will prey on nearly all other invertebrates, including each other (11, 20, 84, 129, 171, 195). When species share common prey and also eat each other [i.e. intraguild predation (see 140)], food web connections will be complex. For example, nymphs of the large dragonfly *Tramea lacerata* prey on smaller dragonflies and on various other taxa such as damselfly nymphs that are also eaten by the smaller dragonflies. The addition of *Tramea* to a community will have a direct negative impact on damselfly nymphs but also an indirect positive effect through reduced numbers of other dragonfly predators. The net impact of *Tramea* predation in a community context is much lower than would be predicted from experiments involving only *Tramea* and damselflies (195). In wetlands with many predatory insects, the community-level impacts of removing or adding a predator will be difficult to understand in the absence of manipulative experimentation.

Most large predatory insects in fishless habitats are also notoriously cannibalistic (6, 84, 130, 140, 171, 175, 180), which may play an important role in regulating their populations (180). A generalized diet, and specifically cannibalism, may be important for colonizing predators that arrive in temporary habitats before preferred prey are abundant (194). Cannibalism in ephemeral wetlands may ensure that at least some individuals complete development before habitats dry (196).

Many of the predatory insects consume zooplankton (84, 197). Because of the high surface area-to-volume ratio, linkages between the benthic and zooplankton communities in shallow lentic habitats like wetlands presumably should be stronger than those in lakes. Although Richardson et al (151) found little evidence for this hypothesis, more data from different wetland types are needed to determine if benthic and planktonic communities are tightly coupled.

Various adult and larval amphibians can be seasonally abundant in fishless marshes and forested wetlands. The eggs and/or larvae of these amphibians are eaten by predatory insects that thrive in ephemeral wetlands (19, 60, 88, 110, 157, 178, 179). Conversely, the tadpoles of many salamanders and a few frogs and toads are predatory and, in some habitats, can regulate the abundance of insects and influence the size distribution of their communities (18, 75, 188). Interactions among fish, amphibians, and predatory insects can be complex. For example, Werner & McPeek (187) found that bullfrog tadpoles, which coexist with fish in permanent habitats, are quite vulnerable to predation by dragonfly nymphs and salamander larvae, both of which are excluded by fish from permanent habitats (187). In contrast, green frog tadpoles, which are most abundant in ephemeral habitats, are more vulnerable to fish than to dragonfly or salamander predation (187). Clearly, hydroperiod will have an important impact on the assemblage of top aquatic predators in wetland communities.

Although wetlands are often described as being situated at the ecotone between aquatic and terrestrial ecosystems, surprisingly little information is available about the movement of energy across that ecotone. Much of the evidence about this phenomenon addresses bird consumption of wetland insects. Many dabbling ducks readily prey on wetland insects (120) (see section on insects and waterfowl management, below). Several other groups of birds, such as swallows, martins, and blackbirds, also rely heavily on wetland insects as prey, especially during the breeding season (17, 120, 135). Orians (134) concluded that the emergence of particularly important prey items such as damselflies can play an important role in the timing of reproduction in marsh-nesting blackbirds. During the breeding season, the impact of these birds on insect mortality can be extremely high. For example, while feeding nestlings, red-wing blackbirds captured 50% of all emerging dragonflies as they transformed on stems of emergent plants (192).

The degree to which aquatic insect communities are regulated by top-down (by predatory insects, fish, amphibians, or birds) or bottom-up processes (abundance of detritus and/or algae) has not been investigated for most wetlands. Several recent experimental studies, however, suggest that the relative importance of top-down and bottom-up control varies between and even within wetlands (11, 69, 125). In a seasonally flooded marsh, Batzer & Resh (11) manipulated vegetation density and found that the densities of beetle larvae (the top predator) were considerably lower in 100% than in

50% vegetation plots. Beetle predation in the 100% vegetation plots was not sufficient to regulate midge populations, which were eventually limited by periphyton abundance (bottom-up control). In adjacent plots with 50% plant cover, the large numbers of beetle larvae were sufficient to control midge densities (top-down control), and in the absence of midges, periphyton abundance increased. Later when beetles emerged, midge populations in 50% plots rebounded in the presence of abundant periphyton. These results emphasize that regulation in wetland food webs can vary temporally depending on insect life-cycle dynamics and spatially depending on vegetation. Dense stands of emergent or submersed vegetation, by acting as a refuge and/or by reducing predator-foraging rates, mediate predator-prey interactions among aquatic insects (11, 37, 199), between fish and aquatic insects (46, 137), and between waterfowl and aquatic insects (122). In summary, the relative importance of top-down and bottom-up mechanisms in regulating wetland insect communities should vary both spatially and temporally depending on vegetation characteristics, hydroperiod, and the seasonal dynamics of predators and prey.

Succession

Above we discussed how life-history strategies affect the initial colonization of newly inundated wetlands. Here we discuss how communities subsequently change following that colonization phase. After inundation, species richness and overall density of the invertebrate community steadily increase for one to six months and then level off for that year (92, 95, 183). Densities can continue to increase during a second year of continuous flooding (101, 102). Multivoltine species that produce population peaks followed by mass emergence will cause temporary deviations in this pattern (95).

In addition to density changes, species compositions can also change (12, 42, 49, 81, 92, 95, 102). In terms of numbers, various crustacean zooplankton and dipteran larvae tend to dominate soon after inundation and are then replaced by beetles, hemipterans, mayflies, odonates, and finally macrocrustaceans and other wingless macrofauna (12, 42, 49, 81, 92, 95, 183). The initially abundant taxa are desiccation-resistant species and rapid cyclic colonizers; they are followed by permanent water species that either disperse slowly or enter the habitat through flood waters (see colonization section).

Some changes in communities reflect trophic status. Among the midges, early colonizers are often detrital feeders, followed by algal feeders, and finally predators (102); *Chironomus* midges are noted early colonizers (12, 49, 95, 102). Densities of many insect predators, as well as other predaceous wetland taxa, tend to increase the longer the wetland remains flooded (12, 49, 92, 95, 183).

APPLIED ECOLOGY

Insects and Waterfowl Management

INSECTS CONSUMED BY WATERFOWL   Invertebrates are readily consumed by
waterfowl on breeding grounds and in wintering areas (47, 120). Waterfowl
often congregate in habitats rich in invertebrates (96, 120). Of the insects,
midge larvae and emerging adults are of most importance to ducks (10, 47,
52, 102, 120, 164), with caddisfly larvae, odonate nymphs, water beetle larvae
and adults, and water boatmen comprising other important foods (10, 21, 120,
132). However, the swimming agility of some of these latter groups makes
them difficult for waterfowl to catch unless densities are high (10, 87).

Large duck species typically consume the larger insect species while smaller
ducks consume both large and small insect taxa (10, 120, 132). Even within
an individual midge species, some ducks may consume only the larger, more
mature larvae (10). This pattern is related in part to the relative coarseness of
the bill lamellae among different kinds of ducks (120, 132). Thus, abundance
of the larger invertebrates, rather than simply overall abundance, may be the
better predictor of waterfowl foraging behavior (87), and factors that influence
invertebrate size may have important management implications (7). Because
waterfowl consume only certain invertebrate taxa, the common practice of
classifying insects only to order or family rather than to species may be
inadequate for an accurate description of the interactions between waterfowl
and insects (10, 52).

MANAGEMENT OF INSECT RESOURCES   Techniques to manage insect populations in waterfowl habitats are being developed. Of the habitats managed for
waterfowl, insects are particularly abundant in seasonally flooded and semipermanent marshes. Midges, beetles, water boatmen, and mosquitoes can dominate the macroinvertebrate fauna of these habitats (12, 23, 49, 87, 101, 106,
122, 125, 153). In managed swamps, many insect groups are less abundant
than in marshes (with the exception of certain midges), but molluskan and
macrocrustacean foods can be more abundant in swamps (49, 100, 101).

Several studies have contrasted relative insect abundances among different
water depths. Some of the insect taxa known to be waterfowl food items are
most abundant in shallow water (95, 100, 101, 152); others flourish in deep
water (13, 23, 95, 143); and still others show no significant depth preference
(12, 101, 106). Even the same benthic midge species can be concentrated in
deep water zones of some marshes and shallow zones of others (10). Given
the mixed responses of insects to water depth, management strategies should
provide a diversity of depths. Although benthic insect larvae in deep water
zones may be inaccessible to dabbling ducks, they become particularly vul-

nerable to them during emergence (164). Seasonal drawdowns also make insects in deep water more available to dabbling ducks (100, 152).

Several studies found that numbers of water boatmen, beetles, and benthic midges can be increased by mowing dense marsh vegetation (12, 13, 87, 122, 125). However, the enhanced numbers of predatory insects in mown habitats can reduce numbers of some prey (11). Rototilling (87) or herbicide treatments (169) are less desirable alternatives to mowing for reducing vegetative cover because they may harm insects.

Adding detrital food to wetlands does not seem to be an effective way to increase numbers of invertebrates for ducks (5, 125). Using fertilizer to increase algal food supplies can increase invertebrate numbers in some cases (59) but not others (123). As discussed previously, the impact of predators may reduce any response of consumers to food additions (5). Several descriptive studies suggest that reducing predation pressure on wetland insects may increase their numbers and thus benefit waterfowl (12, 69, 76, 77, 104).

Pest Management

Because of the environmentally sensitive nature of wetlands, the development of mosquito-control strategies that do not harm nontarget species has become a priority. In tidal marshes, several innovative and safe techniques to control mosquitoes are already being used (40). Unfortunately, in nontidal wetlands, mosquito control has not yet reached this level of sophistication, and pesticide applications are still the primary means of control. However, several studies indicate that natural and ecologically safe strategies could be implemented to control mosquitoes in nontidal wetlands.

In many cases, native predators effectively control mosquito populations. In temporary pools, phantom midge larvae (117), tadpole shrimp (58), or beetles (183) can control mosquitoes. In more permanent habitats, native fishes may be important mosquito predators (97). The presence of some predators may control mosquitoes indirectly by deterring mosquito oviposition (18, 58). Numbers of some predatory beetles in wetlands can be managed by manipulating vegetative cover (12, 13).

Mosquito larvae are usually associated with particular plant growth forms (149), and the cover provided by macrophytes can reduce the susceptibility of mosquito larvae to predation (136). Experimental reductions of plant cover have reduced mosquito densities (12, 13, 136, 137), and some of these manipulations also improve wetland habitats for waterfowl (12, 13). Other techniques to reduce mosquito problems while maintaining or improving habitats for wildlife are listed or reviewed by Batzer & Resh (13).

Herbivorous wetland insects may be useful for controlling noxious plants. For example, in North America, native wetland plants are being displaced by purple loosestrife (*Lythrum salicaria*), an exotic invader from Europe. Malecki et al (103) review an on-going coordinated effort in the United States to implement classical biological-control methodology to manage purple loosestrife using insect herbivores introduced from European wetlands.

Wetland Conservation

Drainage and water regulation can eliminate or degrade wetlands and their associated insect communities (e.g. 1, 36, 65, 91). Grazing of livestock is an additional stress on wetlands that has negative impacts on insects (74, 185). Increased sedimentation rates into wetlands from numerous anthropogenic activities may affect wetland insects indirectly by damaging wetland vegetation (85). Although some wetland insects are threatened with extinction (57, 185), insects rapidly recolonize restored wetlands (91), and the recovery of an endangered damselfly was attributed to creation of new artificial wetlands (57). In efforts to restore or preserve wetlands, insect communities are now being used by scientists as indicators of biotic integrity, particularly in Europe (1, 36, 65, 74, 91).

Pesticides in wetlands affect insects (182), and the potential loss of insect foods for wildlife is of special concern (31, 67). Some insecticides or herbicides are intentionally applied to wetlands for mosquito or pest-plant control (31, 163, 169, 172). However, many pesticides enter wetlands unintentionally via drift or run-off (45, 53, 67, 116, 186). Insecticides pose the gravest threat to wetland insects; for example, organophosphate mosquito larvicides can kill a broad spectrum of wetland insects (163). Alternatively, the microbial mosquito larvicide *Bacillus thuringiensis israelensis* has minimal impact on nontarget wetland insects (31). Although laboratory studies indicate that this insecticide can kill nontarget midge larvae, the water temperatures, water depths, and plant cover typically present in the field help protect naturally occurring midges (31). Insecticides used to control crop or forest pests can drift into wetlands and kill resident insects, particularly if they are applied aerially (53, 116, 186). Unfortunately, some of these insecticides may be unusually persistent in wetland environments (53).

Herbicides also pose a threat to wetland insects. Some herbicides are intentionally applied to wetlands to regulate plant cover or kill noxious weeds (103, 169, 172), but run-off from agriculture is a particularly serious problem (45). The primary influence of herbicides on wetland insects is indirect in that they kill the macrophytes and algae that wetland insects require (45, 172).

Neither eutrophication nor acidification of wetlands appears to strongly influence resident insects, at least directly. Introductions of massive amounts of agricultural wastewater laden with phosphorous and nitrogen into the Florida Everglades did not harm insects (144). However, those insects may eventually be affected indirectly as overgrowth by cattail chokes productive open-water habitats (144). Acid or acidified wetlands typically support healthy insect communities (17, 63, 77, 108, 139, 155). Apparently, many wetland insects can tolerate broad ranges of environmental pH (63, 155, 161), so they may be poor indicators of environmental stress from acid pollutants.

ACKNOWLEDGMENTS

We thank L Harklau and F de Szalay for comments and V Resh and B Peckarsky for encouragement. The National Science Foundation provided support for both DB (DEB-9303248) and SW (BSR-8958253 and DEB-9407856).

Any *Annual Review* chapter, as well as any article cited in an *Annual Review* chapter, may be purchased from the Annual Reviews Preprints and Reprints service. 1-800-347-8007; 415-259-5017; email: arpr@class.org

---

Literature Cited

1. Adamek Z, Sukop I. 1992. Invertebrate communities of former southern Moravian floodplains (Czechoslovakia) and impacts of regulation. *Regul. Rivers Res. Manage.* 7:181–92
2. Adams A. 1984. Crytobiosis in Chironomidae (Diptera)—two decades on. *Antenna* 8:58–61
3. Adamus PR, Clairain EJ, Smith RD, Young RE. 1991. *Wetland Evaluation Techniques (WET)*, Vol. 2, *Methodology. Wetland Res. Prog. Tech. Rep. WRP-DE-3.* US Army Corps Engineers, Waterways Exp. Stn., Vicksburg, Miss.
4. Alarie Y, Leclair R. 1988. Water beetle records from shallow pools in southern Québec (Coleoptera: Dytiscidae). *Coleopt. Bull.* 42:353–58
5. Andersson A, Danell K. 1982. Response of freshwater macroinvertebrates to addition of terrestrial plant litter. *J. Appl. Ecol.* 19:319–25
6. Anholt BR. 1994. Cannibalism and early instar survival in a larval damselfly. *Oecologia* 99:60–65
7. Armstrong DP, Nudds TD. 1985. Factors influencing invertebrate size distributions in prairie potholes and implications for coexisting duck species. *Freshw. Invert. Biol.* 4:41–47
8. Aubin A, Leblanc A. 1986. Effet des variables sur la biomasse animale de sept mares temporaires d'eau douce du Québec méridional. *Hydrobiologia* 139: 143–52
9. Bataille KJ, Baldassarre GA. 1993. Dis-

tribution and abundance of aquatic macroinvertebrates following drought in three prairie pothole wetlands. *Wetlands* 13:260–69
10. Batzer DP, McGee M, Resh VH, Smith RR. 1993. Characteristics of invertebrates consumed by mallards and prey response to wetland flooding schedules. *Wetlands* 13:41–49
11. Batzer DP, Resh VH. 1991. Trophic interactions among a beetle predator, a chironomid grazer, and periphyton in a seasonal wetland. *Oikos* 60:251–57
12. Batzer DP, Resh VH. 1992. Macroinvertebrates of a California seasonal wetland and responses to experimental habitat manipulation. *Wetlands* 12:1–7
13. Batzer DP, Resh VH. 1992. Wetland management strategies that enhance waterfowl habitats can also control mosquitoes. *J. Am. Mosq. Control Assoc.* 8:117–25
14. Bennett DV, Streams FA. 1986. Effects of vegetation on *Notonecta* (Hemiptera) distribution in ponds with and without fish. *Oikos* 46:62–69
15. Bergey EA, Balling SF, Collins JN, Lamberti GA, Resh VH. 1993. Bionomics of invertebrates within an extensive *Potamogeton pectinatus* bed of a California marsh. *Hydrobiologia* 234: 15–24
16. Berté SB, Pritchard G. 1986. The life histories of *Limnephilus externus* Hagen, *Anabolia bimaculata* (Walker) and *Nemotaulius hostilis* (Hagen)

(Trichoptera, Limnephilidae) in a pond in southern Alberta, Canada. *Can. J. Zool.* 64:2348–56

17. Blancher PJ, McNicol DK. 1991. Tree swallow diets in relation to wetland acidity. *Can. J. Zool.* 69:2629–37
18. Blaustein L, Kotler BP. 1993. Oviposition habitat selection by the mosquito, *Culiseta longiareolata:* effects of conspecifics, food and green toad tadpoles. *Ecol. Entomol.* 18:104–8
19. Blaustein L, Margalit J. 1994. Mosquito larvae (*Culiseta longiareolata*) prey upon and compete with tadpoles of *Bufo viridis*). *J. Anim. Ecol.* 63:841–50
20. Blois-Heulin C, Crowley PH, Arrington M, Johnson DM. 1990. Direct and indirect effects of predators on the dominant invertebrates of two freshwater littoral communities. *Oecologia* 84:295–306
21. Botero JE, Rusch DH. 1994. Foods of blue-winged teal in two neotropical wetlands. *J. Wildl. Manage.* 58:561–65
22. Brinson MM, Lugo AE, Brown S. 1981. Primary productivity, decomposition and consumer activity in freshwater wetlands. *Annu. Rev. Ecol. Syst.* 12:123–61
23. Broschart MR, Linder RL. 1986. Aquatic invertebrates in level ditches and adjacent emergent marsh in a South Dakota wetland. *Prairie Nat.* 18:167–78
24. Bunn SE, Boon PI. 1993. What sources of organic carbon drive food webs in billabongs? A study based on stable isotope analysis. *Oecologia* 96:85–94
25. Butler MG, Anderson DH. 1990. Cohort structure, biomass, and production of a merovoltine *Chironomus* population in a Wisconsin bog lake. *J. North Am. Benthol. Soc.* 9:180–92
26. Butler MG, Miller MC, Mozley S. 1980. Macrobenthos. In *Limnology of Tundra Ponds,* ed. JE Hobbie, pp. 297–339. Stroudsburg, PA: Dowden, Hutchinson & Ross
27. Cahoon DR, Stevenson JC. 1986. Production, predation, and decomposition in a low-salinity *Hibiscus* marsh. *Ecology* 67:1341–50
28. Campeau S, Murkin HR, Titman RD. 1994. Relative importance of algae and emergent plant litter to freshwater marsh invertebrates. *Can. J. Fish. Aquat. Sci.* 51:681–92
29. Cantrell MA. 1979. Invertebrate communities in the Lake Chilwa swamp in years of high level. In *Lake Chilwa: Studies of Change in a Tropical Ecosystem,* ed. M Kalk, AJ Mclachlan, C Howard-Williams, pp. 161–73. The Hague: Junk
30. Carpenter QJ, DeWitt CB. 1993. The

31. Charbonneau CS, Drobney RD, Rabeni CF. 1994. Effects of *Bacillus thuringiensis* var. *israelensis* on nontarget benthic organisms in a lentic habitat and factors affecting the efficacy of the larvicide. *Environ. Toxicol. Chem.* 13:267–79
32. Corbet PS. 1962. *A Biology of Dragonflies.* Chicago: Quadrangle Books
33. Corbet PS. 1980. Biology of Odonata. *Annu. Rev. Entomol.* 25:189–217
34. Corbet PS. 1984. Orientation and reproductive condition of migrating dragonflies. *Odonatologica* 13:81–88
35. Corkum LD. 1984. Movements of marsh-dwelling invertebrates. *Freshw. Biol.* 14:89–94
36. Coulson JC, Butterfield JEL, Henderson E. 1990. The effect of open drainage ditches on the plant and invertebrate communities of moorland and on the decomposition of peat. *J. Appl. Ecol.* 27:549–61
37. Crowder LB, Cooper WE. 1982. Habitat structural complexity and the interaction between bluegill and their prey. *Ecology* 1802–13
38. Cuffney TF, Wallace JB. 1987. Leaf litter processing in coastal plain streams and floodplains of southeastern Georgia, U.S.A. *Arch. Hydrobiol. Suppl.* 76:1–24
39. Cyr H, Downing JA. 1988. The abundance of phytophilous invertebrates on different species of submerged macrophytes. *Freshw. Biol.* 20:365–74
40. Dale PER, Hulsman K. 1990. A critical review of salt marsh management methods for mosquito control. *Crit. Rev. Aquat. Sci.* 3:281–311
41. Danell K, Sjöberg K. 1979. Decomposition of *Carex* and *Equisetum* in a northern Swedish lake: dry weight loss and colonization by macro-invertebrates. *J. Ecol.* 67:191–200
42. Danell K, Sjöberg K. 1982. Successional patterns of plants, invertebrates and ducks in a man-made lake. *J. Appl. Ecol.* 19:395–409
43. Davies BR. 1974. The dispersal of Chironomidae: a review. *J. Entomol. Soc. South Afr.* 39:39–62
44. Delettre YR. 1989. Effect of the duration and intensity of drought on abundance and phenology of adult Chironomidae (Diptera) in a shallow temporary pool. *Arch. Hydrobiol.* 114:383–99
45. DeNoyelles F, Kettle WD, Fromm CH, Moffett MF, Dewey SL. 1989. Use of experimental ponds to assess the effect

of a pesticide on the aquatic environment. See Ref. 182, pp. 41–56

46. Diehl S. 1992. Fish predation and benthic community structure: the role of omnivory and habitat complexity. *Ecology* 73:1646–61

47. DuBowy PJ. 1988. Waterfowl communities and seasonal environments: temporal variability in interspecific competition. *Ecology* 69:1439–53

48. Duffy WG. 1994. Demographics of *Lestes disjunctus disjunctus* (Odonata: Zygoptera) in a riverine wetland. *Can. J. Zool.* 72:910–17

49. Duffy WG, LaBar DJ. 1994. Aquatic invertebrate production in Southeastern USA wetlands during winter and spring. *Wetlands* 14:88–97

50. Dvorák J. 1987. Production-ecological relationships between aquatic vascular plants and invertebrates in shallow waters and wetlands—a review. *Arch. Hydrobiol. Beih. Ergeb. Limnol.* 27:181–84

51. Escher RL, Lounibos LP. 1993. Insect associates of *Pistia stratiotes* (Arales: Araceae) in southeastern Florida. *Fla. Entomol.* 76:473–500

52. Euliss NH, Grodhaus G. 1987. Management of midges and other invertebrates for waterfowl wintering in California. *Calif. Fish Game* 73:238–43

53. Fairchild WL, Eidt DC. 1993. Perturbation of the aquatic invertebrate community of acidic bog ponds by the insecticide fenitrothion. *Arch. Environ. Contam. Toxicol.* 25:170–83

54. Farkas MJ, Brust RA. 1986. Phenology of the mosquito *Wyeomyia smithii* (Coq.) in Manitoba and Ontario. *Can. J. Zool.* 64:285–90

55. Finnamore AT, Marshall SA, eds. 1994. *Terrestrial Arthropods of Peatlands, With Particular Reference to Canada.* *Mem. Entomol. Soc. Can.* 169:1–289

56. Foote AL, Kadlec JA, Campbell BK. 1988. Insect herbivory on an inland brackish wetland. *Wetlands* 8:67–74

57. Fox AD, Cham SA. 1994. Status, habitat use and conservation of the scarce blue-tailed damselfly *Ishnura pumilio* (Charpentier) (Odonata: Coenagrionidae) in Britain and Ireland. *Biol. Conserv.* 68: 115–22

58. Fry LL, Mulla MS, Adams CW. 1994. Field introductions and establishment of the tadpole shrimp, *Triops longicaudatus* (Notostraca: Triopsidae), a biological control agent of mosquitoes. *Biol. Control* 4:113–24

59. Gabor TS, Murkin HR, Stainton MP, Boughen JA, Titman RD. 1994. Nutrient additions to wetlands in the Interlake region of Manitoba, Canada: effects of a single pulse addition in spring. *Hydrobiologia* 279/280:497–510

60. Gascon C. 1992. Aquatic predators and tadpole prey in central Amazonia: field data and experimental manipulations. *Ecology* 13:971–80

61. Gladden JE, Smock LA. 1990. Macroinvertebrate distribution and production on the floodplains of two lowland headwater streams. *Freshw. Biol.* 24:533–45

62. Goonan PM, Beer JA, Thompson TB, Suter PJ. 1992. Wetlands of the River Murray flood plain, South Australia. 1. Preliminary survey of the biota and physico-chemistry of ten wetlands from Chowilla to Mannum. *Trans. R. Soc. South Aust.* 116:81–94

63. Gorham CT, Vodopich DS. 1992. Effects of acidic pH on predation rates and survivorship of damselfly nymphs. *Hydrobiologia* 242:51–62

64. Goyer RA, Lenhard GJ, Smith JD. 1990. Insect herbivores of a bald-cypress/tupelo ecosystem. *For. Ecol. Manage.* 33/34:517–21

65. Greenwood MT, Bickerton MA, Castella E, Large ARG, Petts GE. 1991. The use of Coleoptera (Arthropoda: Insecta) for floodplain characterization on the River Trent, U.K. *Regul. Rivers Res. Manage.* 6:321–32

66. Growns JE, Davis JA, Cheal F, Schmidt LG, Rosich RS, Bradley SJ. 1992. Multivariate pattern analysis of wetland invertebrate communities and environmental variables in Western Australia. *Aust. J. Ecol.* 17:275–88

67. Grue CE, DeWeese LR, Mineau P, Swanson GA, Foster JR, et al. 1986. Potential impacts of agricultural chemicals on waterfowl and other wildlife inhabiting prairie wetlands: an evaluation of research needs and approaches. *Trans. North Am. Wildl. Nat. Res. Conf.* 51:357–83

68. Hann BJ. 1991. Invertebrate grazer-periphyton interactions in a eutrophic marsh pond. *Freshw. Biol.* 26:87–96

69. Hanson MA, Butler MG. 1994. Responses to food web manipulation in a shallow waterfowl lake. *Hydrobiologia* 279/280:457–66

70. Hanson BA, Swanson GA. 1989. Coleoptera species inhabiting prairie wetlands of the Cottonwood Lake Area, Stutsman County, North Dakota. *Prairie Nat.* 21:49–57

71. Hargeby A. 1990. Macrophyte associated invertebrates and the effect of habitat permanence. *Oikos* 57:338–46

72. Harrison RG. 1980. Dispersal polymorphisms in insects. *Annu. Rev. Ecol. Syst.* 11:95–118

73. Henrikson BI. 1988. The absence of antipredator behaviour in the larvae of *Leucorrhinia dubia* (Odonata) and the consequences for their distribution. *Oikos* 51:179–83

74. Holmes PR, Boyce DC, Reed DK. 1993. The ground beetle (Coleoptera: Carabidae) fauna of Welsh peatland biotopes: factors influencing the distribution of ground beetles and conservation implications. *Biol. Conserv.* 63:153–61

75. Holomuzki JR, Collins JP. 1987. Trophic dynamics of a top predator, *Ambystoma tigrinum nebulosum* (Caudata: Ambystomatidae) in a lentic community. *Copeia* 1987:949–57

76. Huener JD, Kadlec JA. 1992. Macroinvertebrate response to marsh management strategies in Utah. *Wetlands* 12:72–78

77. Hunter ML, Jones JJ, Gibbs KE, Moring JR. 1986. Duckling responses to lake acidification: do black ducks and fish compete? *Oikos* 47:26–32

78. Irmler U, Junk WJ. 1982. The inhabitation of artificially exposed leaf samples by aquatic macro-invertebrates at the margin of Amazonian inundation forests. *Trop. Ecol.* 23:64–75

79. Jackson JM, Mclachlan AJ. 1991. Rainpools on peat moorland as island habitats for midge larvae. *Hydrobiologia* 209:59–65

80. Jeffries M. 1993. Invertebrate colonization of artificial pondweeds of differing fractal dimension. *Oikos* 67:142–48

81. Jeffries MJ. 1994. Invertebrate communities and turnover in wetland ponds affected by drought. *Freshw. Biol.* 32:603–12

82. Johansson F. 1993. Intraguild predation and cannibalism in odonate larvae: effects of foraging behaviour and zooplankton availability. *Oikos* 66:80–87

83. Johansson F, Nilsson AN. 1991. Freezing tolerance and drought resistance of *Somatochlora alpestris* (Selys) larvae in boreal temporary pools (Anisoptera: Corduliidae). *Odonatologica* 20:245–52

84. Johnson DM. 1991. Behavioral ecology of larval dragonflies and damselflies. *Trends Ecol. Evol.* 6:8–13

85. Jurik TW, Wang SC, van der Valk AG. 1994. Effects of sediment load on seedling emergence from wetland seed banks. *Wetlands* 14:159–65

86. Kaminski M, Krzysztofiak L. 1992. Intensity of invertebrate penetration of the near-shore zone of a polyhumic lake in summer. *Ekol. Polska* 40:127–44

87. Kaminski RM, Prince HH. 1981. Dabbling duck activity and foraging responses to aquatic macroinvertebrates. *Auk* 98:115–26

88. Kehr AI, Schnack JA. 1991. Predator-prey relationship between giant water bugs (*Belostoma oxyurum*) and larval anurans (*Bufo arenarum*). *Alytes* 9:61–69

89. Kok CJ, van der Velde G. 1994. Decomposition and macroinvertebrate colonization of aquatic and terrestrial leaf material in alkaline and acid still water. *Freshw. Biol.* 31:65–75

90. Krieger KA. 1992. The ecology of invertebrates in Great Lakes coastal wetlands: current knowledge and research needs. *J. Great Lakes Res.* 18:634–50

91. LaGrange TG, Dinsmore JJ. 1989. Plant and animal community responses to restored Iowa wetlands. *Prairie Nat.* 21:39–48

92. Lake PS, Bayly IAE, Morton DW. 1989. The phenology of a temporary pond in western Victoria, Australia, with special reference to invertebrate succession. *Arch. Hydrobiol.* 115:171–202

93. Landin J. 1980. Habitats, life histories, migration, and dispersal by flight of two water beetles, *Helophorus brevipalpis* and *H. strigifrons* (Hydrophilidae). *Holarct. Ecol.* 3:190–201

94. Larson DJ, House NL. 1990. Insect communities of Newfoundland bog pools with emphasis on the Odonata. *Can. Entomol.* 122:469–501

95. Layton RJ, Voshell JR. 1991. Colonization of new experimental ponds by benthic macroinvertebrates. *Environ. Entomol.* 20:110–17

96. Lillie RA, Evrard JO. 1994. Influence of macroinvertebrates and macrophytes on waterfowl utilization of wetlands in the Prairie Pothole region of northwestern Wisconsin. *Hydrobiologia* 279/280:235–46

97. Lounibos L, Nishimura N, Dewald LB. 1992. Predation of *Mansonia* (Diptera: Culicidae) by native mosquitofish in southern Florida. *J. Med. Entomol.* 29:236–41

98. Luken JO, Billings WD. 1986. Hummock-dwelling ants and the cycling of microtopography in an Alaskan peatland. *Can. Field-Nat.* 100:69–73

99. Mackay RJ, Wiggins GB. 1979. Ecological diversity in Trichoptera. *Annu. Rev. Entomol.* 24:185–208

100. Magee PA, Fredrickson LH, Humburg DD. 1993. Aquatic macroinvertebrate association with willow wetlands in northeastern Missouri. *Wetlands* 13:304–10

101. Maher M. 1984. Benthic studies of waterfowl breeding habitat in south-west-

ern New South Wales. I. The fauna. *Aust. J. Mar. Freshw. Res.* 35:85–96

102. Maher M, Carpenter SM. 1984. Benthic studies of waterfowl breeding habitat in Southwestern New South Wales. II. Chironomid populations. *Aust. J. Mar. Freshwater Res.* 35:97–110

103. Malecki RA, Blossey B, Hight SD, Schroeder D, Kok LT, Coulson JR. 1993. Biological control of purple loosestrife—a case for using insects as control agents, after rigorous screening, and for integrating release strategies with research. *Bioscience* 43:680–86

104. Mallory ML, Blancher PJ, Weatherhead PJ, McNicol DK. 1994. Presence or absence of fish as a cue to macroinvertebrate abundance in boreal wetlands. *Hydrobiologia* 279/280:345–51

105. McArthur JV, Aho JM, Rader RB, Mills GL. 1994. Interspecific leaf interactions during decomposition in aquatic and floodplain ecosystems. *J. North Am. Benthol. Soc.* 13:57–67

106. McCrady JW, Wentz WA, Linder RL. 1986. Plants and invertebrates in a prairie wetland during duck brood-rearing. *Prairie Nat.* 18:23–32

107. McLaughlin DB, Harris HJ. 1990. Aquatic insect emergence in two Great Lakes marshes. *Wetlands Ecol. Manage.* 1:111–21

108. McNicol DK, Bendell BE, McAuley DG. 1987. Avian trophic relationships and wetland acidity. *Trans. North Am. Wildl. Nat. Res. Conf.* 52:619–27

109. McPeek MA. 1990. Behavioral differences between *Enallagma* species (Odonata) influencing differential vulnerability to predators. *Ecology* 71: 1714–26

110. Miaud C. 1993. Predation on newt eggs (*Triturus alpestris* and *T. helveticus*): identification of predators and protective role of oviposition behaviour. *J. Zool. Lond.* 231:575–82

111. Michiels NK, Dhondt AA. 1991. Characteristics of dispersal in sexually mature dragonflies. *Ecol. Entomol.* 16: 449–59

112. Mihuc T, Toetz D. 1994. Determination of diets of alpine insects using stable isotopes and gut analysis. *Am. Midl. Nat.* 131:146–55

113. Mitsch W, Gosselink JG, 1993. *Wetlands.* New York: Van Norstrand Reinhold. 2nd ed.

114. Mizuno T, Gose K, Lim RP, Furtado JI. 1982. Benthos and attached animals. In *Tasek Bera: The Ecology of a Freshwater Swamp,* ed. JI Furtado, S Mori, pp. 286–306. The Hague: Junk

115. Morin PJ, Lawler SP, Johnson EA.

1988. Competition between aquatic insects and vertebrates: interaction strength and higher order interactions. *Ecology* 69:1401–9

116. Morrill PK, Neal BR. 1990. Impact of deltamethrin insecticide on Chironomidae (Diptera) of prairie ponds. *Can. J. Zool.* 68:289–96

117. Morrison A, Andreadis TG. 1992. Larval population dynamics in a community of nearctic *Aedes* inhabiting a temporary vernal pool. *J. Am. Mosq. Control Assoc.* 8:52–57

118. Murkin EJ, Murkin HR, Titman RD. 1992. Nektonic invertebrate abundance and distribution at the emergent vegetation–open water interface in the Delta Marsh, Manitoba, Canada. *Wetlands* 12: 45–52

119. Murkin HR. 1989. The basis for food chains in prairie wetlands. In *Northern Prairie Wetlands,* ed. AG van der Valk, pp. 316–38. Ames: Iowa State Univ. Press

120. Murkin HR, Batt BDJ. 1987. The interactions of vertebrates and invertebrates in peatlands and marshes. See Ref. 154, pp. 15–30

121. Murkin HR, Kadlec JA. 1986. Responses by benthic macroinvertebrates to prolonged flooding of marsh habitat. *Can. J. Zool.* 64:65–72

122. Murkin HR, Kaminski RM, Titman RD. 1982. Responses by dabbling ducks and aquatic invertebrates to an experimentally manipulated cattail marsh. *Can. J. Zool.* 60:2324–32

123. Murkin HR, Pollard BJ, Stainton MP, Boughen JA, Titman RD. 1994. Nutrient additions to wetlands in the Interlake region of Manitoba, Canada: effects of periodic additions throughout the growing season. *Hydrobiologia* 279/280: 483–95

124. Muthukrishnan J, Palavesam A. 1992. Secondary production and energy flow through *Kiefferulus barbitarsis* (Diptera: Chironomidae) in tropical ponds. *Arch. Hydrobiol.* 125:207–26

125. Neckles HA, Murkin HR, Cooper JA. 1990. Influences of seasonal flooding on macroinvertebrate abundance in wetland habitats. *Freshw. Biol.* 23:311–22

126. Neill C, Cornwell JC. 1992. Stable carbon, nitrogen, and sulfur isotopes in a prairie marsh food web. *Wetlands* 12: 217–24

127. Nelson JW, Kadlec JA. 1984. A conceptual approach to relating habitat structure and macroinvertebrate production in freshwater wetlands. *Trans. North Am. Wildl. Nat. Res. Conf.* 49: 262–70

128. Nelson JW, Kadlec JA, Murkin HR. 1990. Responses by macroinvertebrates to cattail litter quality and timing of litter submergence in a northern prairie marsh. *Wetlands* 10:47–60

129. Nilsson AN. 1986. Community structure in the Dytiscidae (Coleoptera) of a northern Swedish seasonal pond. *Ann. Zool. Fenn.* 23:39–47

130. Nilsson AN, Söderström O. 1988. Larval consumption rates, interspecific predation, and local guild composition of egg-overwintering *Agabus* (Coleoptera, Dytiscidae) species in vernal ponds. *Oecologia* 76:131–37

131. Nolte U. 1989. Observations on neotropical rainpools (Bolivia) with emphasis on Chironomidae (Diptera). *Stud. Neotrop. Fauna Environ.* 24:105–20

132. Nummi P. 1993. Food-niche relationships of sympatric mallards and green-winged teals. *Can. J. Zool.* 71:49–55

133. Oertli B. 1993. Leaf litter processing and energy flow through macroinvertebrates in a woodland pond (Switzerland). *Oecologia* 96:466–77

134. Orians GH. 1980. *Some Adaptations of Marsh-Nesting Blackbirds.* Princeton, NJ: Princeton Univ. Press

135. Orians GH, Wittenberger JF. 1991. Spatial and temporal scales in habitat selection. *Am. Nat.* 137(Suppl.):S29–S49

136. Orr BK, Resh VH. 1989. Experimental test of the influence of aquatic macrophyte cover on the survival of *Anopheles* larvae. *J. Am. Mosq. Control Assoc.* 5:579–85

137. Orr BK, Resh VH. 1992. Influence of *Myriophyllum aquaticum* cover on *Anopheles* mosquito abundance, oviposition, and larval microhabitat. *Oecologia* 90:474–82

138. Paasivirta L, Lahti T, Perätie T. 1988. Emergence phenology and ecology of aquatic and semi-terrestrial insects on a boreal raised bog in central Finland. *Holarct. Ecol.* 11:96–105

139. Parker GR, Petrie MJ, Sears DT. 1992. Waterfowl distribution relative to wetland acidity. *J. Wildl. Manage.* 56:268–74

140. Polis GA, Myers C, Holt RD. 1989. The ecology and evolution of intraguild predation: potential competitors that eat each other. *Annu. Rev. Ecol. Syst.* 20:297–330

141. Pritchard G. 1983. Biology of Tipulidae. *Annu. Rev. Entomol.* 28:1–22

142. Pritchard G, Berté SB. 1987. Growth and food choice by two species of limnephilid caddis larvae given natural and artificial foods. *Freshw. Biol.* 18:529–35

143. Rader RB. 1994. Macroinvertebrates of the northern Everglades: species composition and trophic structure. *Fla. Sci.* 57:22–33

144. Rader RB, Richardson CJ. 1994. Response of macroinvertebrates and small fish to nutrient enrichment in the northern Everglades. *Wetlands* 14:134–46

145. Rasmussen JB. 1985. Effects of density and microdetritus enrichment on the growth of chironomid larvae in a small pond. *Can. J. Fish. Aquat. Sci.* 42:1418–22

146. Rasmussen JB, Downing JA. 1988. The spatial response of chironomid larvae to the predatory leech *Nephelopsis obscura. Am. Nat.* 131:14–21

147. Reid F, Kelley JR, Taylor TS, Fredrickson LH. 1989. Upper Mississippi Valley wetlands—refuges and moist-soil impoundments. In *Habitat Management for Migrating and Wintering Waterfowl in North America,* ed. LM Smith, RL Pederson, RM Kaminski, pp. 181–202. Lubbock: Texas Tech Univ. Press

148. Reimer G. 1991. The ecological importance of floodplains for fish at the river March (Austria). *Arch. Hydrobiol.* 121:355–63

149. Rejmankova E, Savage HM, Rodriguez MH, Roberts DB, Rejmanek M. 1992. Aquatic vegetation as a basis for classification of *Anopheles albimanus* Weideman (Diptera: Culicidae) larval habitats. *Environ. Entomol.* 21:598–603

150. Renshaw M, Service MW, Birley MH. 1993. Density-dependent regulation of *Aedes cantans* (Diptera: Culicidae) in natural and artificial populations. *Ecol. Entomol.* 18:223–33

151. Richardson WB, Wickham SA, Threkleld ST. 1990. Foodweb response to experimental manipulation of a benthivore (*Cyprinus carpio*), zooplanktivore (*Menidia beryllina*) and benthic insects. *Arch. Hydrobiol.* 119:143–65

152. Riley TZ, Bookhout TA. 1990. Response of aquatic macroinvertebrates to early-spring drawdown in nodding smartweed marshes. *Wetlands* 10:173–85

153. Riley TZ, DeRoia DM. 1989. Early decomposition and invertebrate colonization of nodding smartweed leaves. *Wetlands* 9:219–25

154. Rosenberg DM, Danks HV, eds. 1987. *Aquatic Insects of Peatlands and Marshes of Canada. Mem. Entomol. Soc. Can.* 140:1–174

155. Rosenberg DM, Wiens AP, Bilyj B. 1988. Chironomidae (Diptera) of peatlands in northwestern Ontario, Canada. *Holarct. Ecol.* 11:19–31

156. Roth AH, Jackson JF. 1987. The effect

of pool size on recruitment of predatory insects and on mortality in a larval anuran. *Herpetologica* 43:224–32

157. Rowe CL, Sadinski WJ, Dunson WA. 1994. Predation on larval and embryonic amphibians by acid-tolerant caddisfly larvae (*Ptilostomis postica*). *J. Herpetol.* 28:357–64

158. Rzoska J. 1974. The upper Nile swamps, a tropical wetland study. *Freshw. Biol.* 4:1–30

159. Schalles JF, Shure DJ. 1989. Hydrology, community structure, and productivity patterns of a dystrophic Carolina Bay wetland. *Ecol. Monogr.* 59:365–85

160. Schleuter A. 1986. The chironomid communities of permanent and temporary pools depending on water stability and leaf litter input. *Arch. Hydrobiol* 105:471–87

161. Scudder G. 1987. Aquatic and semi-aquatic Hemiptera of peatlands and marshes in Canada. See Ref. 154, pp. 65–98

162. Setällä H, Mäkelä I. 1991. *Galerucella nymphaeae* (Col., Chrysomelidae) grazing increases *Nuphar* leaf production and affects carbon and nitrogen dynamics in ponds. *Oecologia* 86:170–76

163. Siefert RE, Lozano SJ, Brazner JC, Knuth ML. 1989. Littoral enclosures for aquatic field testing of pesticides: effects of chlorpyrifos on a natural system. See Ref. 182, pp. 57–73

164. Sjöberg K, Danell K. 1982. Feeding activity of ducks in relation to diel emergence of chironomids. *Can. J. Zool.* 60:1383–87

165. Sklar FH. 1985. Seasonality and community structure of the backswamp invertebrates in a Louisiana Cypress-Tupelo wetland. *Wetlands* 5:69–86

166. Smock L, Harlowe KL. 1983. Utilization and processing of freshwater wetland macrophytes by the detritivore *Asselus forbesi*. *Ecology* 64:1556–65

167. Smock LA, Stoneburner DL. 1980. The response of macroinvertebrates to aquatic macrophyte decomposition. *Oikos* 35:397–403

168. Söderström O, Nilsson AN. 1987. Do nymphs of *Parameletus chelifer* and *P. minor* (Ephemeroptera) reduce mortality from predation by occupying temporary habitats? *Oecologia* 74:39–46

169. Solberg KL, Higgins KF. 1993. Effects of glyphosate herbicide on cattails, invertebrates, and waterfowl in South Dakota wetlands. *Wildl. Soc. Bull.* 21:299–307

170. Deleted in proof

171. Spence JR, Anderson NM. 1994. Biology of water striders: interactions between systematics and ecology. *Annu. Rev. Entomol.* 39:101–28

172. Stephenson M, Mackie GL. 1986. Effects of 2,4-D treatment on natural benthic macroinvertebrate communities in replicate artificial ponds. *Aquat. Toxicol.* 9:243–51

173. Sternberg K. 1994. Temperature stratification in bog ponds. *Arch. Hydrobiol.* 129:373–82

174. Stiehl RB. 1994. *Habitat Evaluation Procedures Workbook.* Fort Collins, CO: Natl. Biol. Surv., Natl. Ecol. Res. Center

175. Streams FA. 1992, Intrageneric predation by *Notonecta* (Hemiptera: Notonectidae) in the laboratory and in nature. *Ann. Entomol. Soc. Am.* 85:265–73

176. Svensson BW. 1992. Changes in occupancy, niche breadth and abundance of three *Gyrinus* species as their respective range limits are approached. *Oikos* 63:147–56

177. Tai LCC. 1967. *Biosystematic study of Sympetrum (Odonata: Libellulidae).* PhD thesis. Purdue Univ., West Lafayette, IN

178. Tejedo M. 1993. Size-dependent vulnerability and behavioral responses of tadpoles of two anuran species to beetle larvae predators. *Herpetologica* 49:287–94

179. Van Buskirk J. 1988. Interactive effects of dragonfly predation in experimental pond communities. *Ecology* 69:857–67

180. Van Buskirk J. 1989. Density-dependent cannibalism in larval dragonflies. *Ecology* 70:1442–49

181. Vitt D. 1994. An overview of factors that influence the development of Canadian peatlands. See Ref. 55, pp. 7–20

182. Voshell JR, ed. 1989. *Using Mesocosms to Assess the Aquatic Ecological Risks of Pesticides: Theory and Practice. Entomol. Soc. Am. Misc. Publ.* 75:1–88

183. Walton WE, Tietze NS, Mulla MS. 1990. Ecology of *Culex tarsalis* (Diptera: Culicidae): factors influencing larval abundance in mesocosms in Southern California. *J. Med. Entomol.* 27:57–67

184. Walton WE, Tietze NS, Mulla MS. 1991. Consequences of tadpole shrimp predation on mayflies in some Californian ponds. *Freshw. Biol.* 25:143–54

185. Warren MS. 1994. The UK status and suspected metapopulation structure of a threatened European butterfly, the marsh fritillary *Eurodryas aurinia*. *Biol. Conserv.* 67:239–49

186. Wayland M. 1991. Effect of carbofuran on selected macroinvertebrates in a prai-

rie parkland pond: an enclosure approach. *Arch. Environ. Contam. Toxicol.* 21:270–80

187. Werner EE, McPeek MA. 1994. Direct and indirect effects of predators on two anuran species along an environmental gradient. *Ecology* 75:1368–82

188. Whiteman HH, Wissinger SA, Bohonak AJ. 1994. Seasonal movement patterns and diet in a subalpine population of the tiger salamander, *Ambystoma tigrinum nebulosum. Can. J. Zool.* 72:1780–87

189. Wiggins GB, Mackay RJ, Smith IM. 1980. Evolutionary and ecological strategies of animals in annual temporary pools. *Arch. Hydrobiol. Suppl.* 58:97–206

190. Williams DD. 1987. *The Ecology of Temporary Waters.* Portland: Timber

191. Williams DD, Tavares-Cromar A, Kushner DJ, Coleman JR. 1993. Colonization patterns and life-history dynamics of *Culex* mosquitoes in artificial ponds of different character. *Can. J. Zool.* 71:568–78

192. Wissinger SA. 1988. Spatial distribution, life history, and estimates of larval survivorship in a 14-species assemblage of larval dragonflies (Odonata: Anisoptera). *Freshw. Biol.* 20:329–40

193. Wissinger SA. 1989. Life history and size structure of larval dragonfly populations. *J. North Am. Benthol. Soc.* 7:13–28

194. Wissinger SA. 1995. Cyclic colonization and predictable disturbance: a template for biological control in ephemeral crop habitats. *Biol. Control.* 5: In press

195. Wissinger SA, McGrady J. 1993. Intraguild predation and competition between larval dragonflies: direct and indirect effects on shared prey. *Ecology* 74:207–18

196. Wissinger SA, Sparks GB, Brown WW, Rouse GL. Intraguild predation and cannibalism among larvae of detritivorous caddisflies in subalpine ponds. *Ecology.* Submitted

197. Woodward BD, Kiesecker J. 1994. Ecological conditions and the notonectid–fairy shrimp interactions. *Southwest Nat.* 39:160–64

198. Wrubleski DA, Rosenberg DM. 1990. The Chironomidae (Diptera) of Bone Pile Pond, Delta Marsh, Manitoba, Canada. *Wetlands* 10:243–73

199. Zimmerman M, Spence JR. 1989. Prey use of the fishing spider *Dolomedes triton* (Pisauridae, Araneae): an important predator of the neuston community. *Oecologia* 80:187–94

Annu. Rev. Entomol. 1996. 41:101–14

LIVING ON LEAVES: Mites, Tomenta, and Leaf Domatia

David Evans Walter

Department of Entomology, University of Queensland and CRC for Tropical Pest Management, St. Lucia, Queensland 4072 Australia

KEY WORDS: Acari, mutualism, biological control, Phytoseiidae, phylloplane

ABSTRACT

Structures on the surfaces of leaves strongly affect phylloplane mites. Glandular trichomes defend against some plant parasites but can also mire predators. However, leaves with tomenta of nonglandular trichomes are often inhabited by large populations of predatory mites. Tufts of hairs and other minute structures in the vein axils are called leaf domatia. Comparative observations and experimental data demonstrate that leaves with domatia have enhanced levels of predatory mites. By accumulating predatory mites, leaf domatia act as a kind of constitutive defense against herbivores. Mites benefit from leaf domatia by securing a safe place for oviposition and molting. Like several other plant structures, leaf domatia are the manifestation of a long-term and mutually beneficial interaction between plants and arthropods.

PERSPECTIVES AND OVERVIEW

Our human view of the world is biased towards organisms that are easily noticed. Very small or obscure animals fall out of biodiversity inventories, fail to be represented in food web analyses or community studies, and generally escape our notice unless they have an effect on their environment disproportionate to their size. Mites that attack our crops have just such a disproportionate effect. Most mites are a fraction of a millimeter in length, but many of these minute relatives of spiders and scorpions are among our most debilitating agricultural pests (18, 25).

Recent research has revealed a staggering diversity and abundance of mites inhabiting, like arboreal plankton, the canopies of rain forests. Unlike agricultural systems, in which pest species predominate, most rain-forest canopy mites are predators or species that graze on fungi and other microbes that grow on plant surfaces (64). Most of these animals are less than half a millimeter in

101

0066-4170/96/0101-0101$08.00

length, and to them a single leaf must seem a vast area. The smooth, simple surfaces of many leaves offer nowhere for a small arthropod to hide. However, other leaves have numerous hiding places formed by large, raised veins; dense coverings of hair; tunnel-like rolled margins; or small, cave-like structures in the vein axils. Structures restricted to vein axils are called leaf domatia. They have no known function but often house mites. More than a century ago, a Swedish scientist suggested that the association between mites and leaf domatia was analogous to that between ants and formicaria on ant plants (31); however, evidence supporting a mite-plant mutualism has been amassed only in the past few years.

In this review, I try to develop a mite's perspective on the phylloplane (i.e. the leaf surface), and I use this view to answer some basic questions about mite-plant associations: How important are leaf surface structures to interactions among plants and mites? Is the mite-leaf domatia association mutually beneficial? Can a consideration of leaf-surface architecture give us a better understanding of acarine biological-control agents?

MITES ON PLANTS

Mites are the smallest of terrestrial arthropods, so small that even those inhabiting well-studied systems are often overlooked (32, 64). However, in Australian rain forests, from cool-temperate Tasmania to the wet tropics of northern Queensland, much evidence demonstrates that trees, shrubs, and vines are inhabited by an extraordinary abundance and diversity of mites (5, 60, 64, 67). Densities sometimes exceed 10,000 m^{-2} of foliage, and even small trees house hundreds of thousands of mites (67). Although individual mites are small, the large numbers found on forest plants suggest that interactions important both to mites and plants could have developed over evolutionary time.

Age and Extent of the Association

Associated fossil mites and land plants are known from the Devonian (36), and collections from amber (30, 69) and fossil leaf packs (i.e. compressed layers of leaves) (38) indicate that mites have lived on trees at least since the Cretaceous. Most fossil mites, including the earliest plant associates, belong to taxa whose extant descendants feed on fungi, scavenge dead plant matter, and graze on green algae. In humid forests, much of the arboreal plankton comprises these scavenging microbivores (64).

Several lineages of microbivores (and some lineages of predators) have given rise to mites parasitic on higher plants (29). Most of these minute herbivores use stylet-like mouthparts to stab individual plant cells and suck out their contents, providing a ready source of moisture and allowing the mites to avoid many plant chemical defenses. In agricultural systems, plant-parasitic mites are extremely damaging pests with short generation times, high fecundity, and a tendency to overexploit their hosts (6, 18, 25, 29). Most conifers, monocots, and dicots, and many ferns, are attacked by one or more species of plant-parasitic mite (25, 28). Predators of plant-parasitic mites have followed their prey onto plants, and many, especially the Phytoseiidae, are important biological-control agents (19–21). These predators spend their entire lives on plants and tend to remain on some plants even in the absence of prey (2, 6).

Tritrophic-Level Interactions

The realization that three trophic levels are involved in many predator-prey interactions (47) has resulted in the discovery that predatory mites inhabit a complex chemical landscape composed of cues from plants, pests, competitors, and conspecifics (6, 7, 34, 58). This chemical environment has become the focus of much of the research on tritrophic-level interactions. Mites alter their behavior in response to infochemicals, and these changes influence their effectiveness as biological-control agents (58).

Foliar mites also inhabit a landscape of forms—leaf surface structures, epiphylls, and debris (e.g. webbing)—that they must navigate to locate food and mates. These foliar facades influence many aspects of leaf microclimate and chemistry (26) and can also directly affect the behaviors of the small arthropods that hunt on the leaf surface (57). Below, I review the effects of leaf-surface structures on mite abundance, diversity, and trophic interactions.

MITES AND LEAF SURFACE STRUCTURES

From the perspective of an animal as minute as a mite, the leaf surface is highly variable. Mites crossing smooth, rain-shedding leaves or sclerophylls with thick, glossy surfaces face featureless plains with no protection from sun, wind, rain, and larger predators. Other leaves are covered by forests of hairs (tomenta), are dissected by canyons formed by raised veins, or contain cave-like structures offering protection to tiny arthropods. Only recently have the responses of mites to leaf architecture been considered, but topography is central to an understanding of the relationship between phytophagous mites and plants because surface structures clearly alter mite abundance and influence predator-prey interactions.

The tomenta that coat the leaves of many plants come in various forms and densities that produce very different surface habitats (24). Peltate scales, appressed hairs, and dense felts exclude mites from the leaf surface; however, a space of only a few tenths of a millimeter distance will permit the invasion of most foliar mites. For example, wheat curl mite, a minute eriophyid mite that vectors wheat curl virus, disperses on the wind but has much higher landing rates on wheat varieties with pubescent surfaces, presumably because the pubescence induces a boundary layer of lower wind speeds (16, 17).

Foliar mite populations vary widely between adjacent trees that have differences in pubescence. Tropical rain-forest trees with erect hairs on their leaves possessed on average nearly three times as many species and five times as many individual mites as those with glabrous leaves or those coated with appressed scales (64). All three foliar mite trophic guilds—plant-parasites, scavenging microbivores, and predators—were far more abundant on hairy leaves.

Leaf Hairs and the Distribution of Predatory Mites

Spider mites (Tetranychidae) are major pests of grapes in most of the world's vineyards, but outbreaks are often prevented or controlled by phytoseiid mites (18, 19). Duso (9, 10) has demonstrated that two important phytoseiid predators, *Typhlodromus pyri* and *Amblyseius aberrans,* respond to leaf surface structure. A comparison of grape varieties having either glabrous or densely hairy leaf undersurfaces showed that both predators were consistently more abundant on the hairy varieties. This difference was especially evident under conditions of prey scarcity and in hot, dry seasons when grape vines with hairy leaves acted as humid refugia for large populations of predatory mites (9). Similarly, windrows of *Eucalyptus torelliana,* an unusual gum tree with hairy leaves, acted as reservoirs for populations of the phytoseiid *Euseius victoriensis* and allowed this predator to invade adjacent citrus orchards and to control rust mite populations (51).

The dominant predators on Italian grapes with hairy leaves, *T. pyri* and *A. aberrans,* are both relatively small mites; a larger *Amblyseius* species predominated on glabrous grape leaves (9). Most species of *Phytoseius* are found almost exclusively on hairy leaves (59). These mites are also small, and they have relatively narrow bodies, long legs, and enlarged dorsal setae that enclose a space above the dorsum—characteristics that may help them to penetrate tomenta (59). Other Phytoseiidae also prefer hairy leaves (8, 43). These are typically relatively small, narrow mites, e.g. *Metaseiulus occidentalis* and *Neoseiulus caudiglans.*

The penetration of spider-mite webbing and navigation of hairy leaf surfaces may be analogous problems for predatory mites. Like spider-mite silk, dense hairs on leaves can be difficult to move through, even by small phytoseiid mites (43). Similarly, larger predators such as anthocorid bugs cannot penetrate dense tomenta and control spider mites (10). Most phytoseiid predators that effectively control *Tetranychus* species that produce dense webbing have elongate setae, especially along the margin of the dorsal shield (50). These setae presumably help the predator slide through the webbing. Although being small seems to be a prerequisite for success on hairy leaves, effective spider-mite predators vary greatly in size. For example, of the two phytoseiid mites often available commercially for control of the two-spotted mite, *Phytoseiulus persimilis* is more than twice the size of *M. occidentalis.*

Glandular Hairs

In contrast to the dry tomenta on most leaves, assemblages of hairs that produce sticky glandular secretions often negatively affect mite numbers. Ticks that climb certain tropical legumes to quest for hosts may be gummed and killed by glandular hairs (53, 54). Resistance of tomato varieties and other solanaceous plants to attack by spider mites (1, 11, 12, 14, 44, 52) and the broad mite (13) is based on such plants' ability to entangle (and perhaps poison) these small plant-parasites with the exudates of glandular hairs on their leaves and stems. Spider mites produce silk that provides some protection against entanglement, but larger, more active mites such as *P. persimilis* become trapped by glandular hairs on tomato, which considerably reduces their effectiveness as biological-control agents (35, 56).

Extrafloral Nectaries

Glands that produce sugary secretions outside of flowers are referred to as extrafloral nectaries (27). Many plant-inhabiting arthropods (27), including both predatory and fungivorous mites (45), take advantage of the moisture, carbohydrates, and amino acids produced by nectaries. By adding extrafloral nectar (or a honey substitute) to cassava plants with nectaries removed or blocked off, Baker & Klein (2) demonstrated that the phloem exudates at the base of the petioles of young leaves influenced the interaction between cassava green spider mite, *Mononychellus tanajoa,* and the predator mite *Typhlodromalus limonicus.* In the absence of spider mites and the presence of the exudate, the predatory mite disappeared from cassava plants at a slower rate. If spider-mite prey were available, predator populations would be higher and herbivore populations lower when honey was applied to plants with excised nectaries. The carbohydrate source appeared to increase predator reproduction and survival and to reduce emigration. Similarly in a laboratory test, leaves of a garden shrub with nectaries left intact retained seven times as many phytoseiid mites as those with nectaries excised (64).

ACARODOMATIA

In 1887, Lundström (31) suggested that structures found in the vein axils on the underside of the leaves of many trees were produced to shelter mites. He called these tiny structures acarodomatia (mite houses) and compared them to the much larger cavities that housed ants. In contrast to the formicaria (ant domatia) on leaves of many Melastomataceae (23), which are large enough to house colonies of small ants, acarodomatia are typically only a millimeter or two in diameter. Although botanists have long used these structures (often called foveoles) as taxonomic characters (4), for nearly a century the alleged relationship between mites and leaf domatia was ignored or derided (40).

Lundström believed that mites induced acarodomatia (31); however, that belief was clearly wrong. Leaf domatia are plant morphogenetic structures (4, 40). Unlike the hair-like erinose patches, pocket galls, and other deformities produced by eriophyoid mites (25), leaf domatia are formed whether or not mites are present. Nevertheless, more than a century after Lundström's initial observations, near simultaneous publication of studies from Australasia (40) and North America (46) demonstrated a clear association between mites and leaf domatia across a broad range of plant species.

Structure, Distribution, and Occupants of Leaf Domatia

In North America, the most common type of leaf domatium is a tuft of hairs, often associated with a shallow pit or pocket-like flap of tissue, in vein axils on the underside of leaves (46, 61, 68). O'Dowd & Pemberton (39) found that more than a third of 425 Korean temperate broadleaf species surveyed had domatia and that 84% of these structures were tufts or pockets similar to those found on North American species. In deciduous forests in the central United States, 48% of the 71 species of woody trees, shrubs, and vines examined by Willson (68) had tuft or tuft-pocket leaf domatia. An additional 23% of species had raised, tomentose veins. Willson called these structures "halfway houses" in comparison to the discrete tuft domatia that occurred in vein axils. Thus, 71% of the woody plants sampled in forests in east-central Illinois had leaves with leaf domatia or halfway houses. All of these structures commonly sheltered predatory tydeid, stigmaeid, and phytoseiid mites but only rarely contained plant-parasitic mites (68).

Leaf domatia in the form of raised domes, relatively large pouches, and pits that penetrate deep within the leaf lamina are common on leaves in many warm-temperate to tropical forests (4, 37, 40–42, 48, 55). Analogous cavities are formed from rolled leaf margins (orlettes) (46). Evidence supporting the homology or analogy between the relatively simple hair tufts and the more elaborate cavities includes their structure (all enclose mite-sized spaces and are accessed through minute openings), their location, and the occurrence within the same genus of tufts on some species and more elaborate cavities on others (40). With a single exception, studies have shown that predatory mites and those that scavenge and feed on fungi and other microbes predomi-

Table 1 Tests of mite-leaf domatia mutualism

Evidence for or against mutualism	Supporting Studies
In plants with leaf domatia, the predominate arthropods are:	
Beneficial mites	15, 37, 39, 40, 46, 48, 49, 55, 60–66
Plant-parasitic mites	None
Insects	33
Experimental removal of leaf domatia causes:	
Reduced numbers of fungivorous mites	49, 65, 66
Reduced numbers of predatory mites	15, 49, 62–66
Increased numbers of plant-parasitic mites	15, 66
Mites benefit from inhabiting domatia because of:	
Domatia as mite nurseries	15, 37, 46, 62, 63
Enhanced reproduction by predatory mites	15, 49

nate in both tuft domatia and the more elaborate pits, pockets, pouches, and domes, representing over 90% of the arthropods found in these structures (see Table 1). Small insects, especially scales and thrips, are occasionally found in leaf domatia but are rarely seen in abundance (37, 40, 46, 64, 66). The exception is an extremely small species of thrips described from leaf domatia in Costa Rica; however, no serious attempt was made to find any mites that also may have been present (33). None of the mites consistently found in domatia are known to harm plants (Table 1), and many may benefit plants by reducing parasite load or mobilizing nutrients sequestered by microbes (64).

In forests, mites that graze on microbes are the most abundant inhabitants of leaf domatia (Table 1). The suggestion that plants benefit from grazing of mites on epiphyllous fungi and from the release of nutrients in mites' feces originated with Lundström (31). In a study using bitumen paint to block domatia in leaves of a dozen species of rain-forest plants, after five months, fungivorous mites were six times more common on leaves with open domatia than on adjacent leaves with the domatia painted shut (65). However, the effects of fungi on plants are likely to vary from beneficial to deleterious, and the importance of elevated numbers of fungivores on leaves with domatia is difficult to evaluate (64). A far clearer example of a benefit to plants would be elevated numbers of predatory mites known to act as bodyguards, protecting the plant from parasites (6).

The Plant-Predator Mutualism Hypothesis

Although chance could explain occasional associations of predatory mites with leaf domatia, the consistent and abundant presence of predators within domatia requires a more involved hypothesis, such as an interaction between the mites and plants. A simple yet sufficient explanation is that leaves with domatia are more attractive to these minute bodyguards than leaves without domatia (55, 63).

Mite use of leaf domatia has now been demonstrated in North America, Costa Rica, Hawaii (46, 55, 60, 68), Australia, New Guinea, New Zealand (15, 37, 40, 49, 62–66), Bangladesh (48), and Korea (39). In all of these studies, predatory mites, especially Phytoseiidae, Stigmaeidae [both containing important biological-control agents of plant-parasitic mites (19)], and Tydeidae (see 22, 29, 64 for discussions of feeding behavior in Tydeidae) are common inhabitants of leaf domatia. Plants with leaf domatia were more likely to be inhabited by a predator and had many more predators per leaf compared with sympatric plants lacking these structures (15, 48, 49, 62–66).

If plant-parasitic mites also were major inhabitants of domatia, then a functional explanation for the plant-predator relationship would be suspect. Spider mites are almost never found in leaf domatia (63, 66; DE Walter, unpublished information). Gall-forming eriophyid mites sometimes inhabit domatia in large numbers; however, eriophyid use of domatia is relatively rare, transitory, and generally unsuccessful (i.e. galls are rarely initiated in domatia) (66). False spider mites (*Brevipalpus* spp.) sometimes occur in domatia (40, 49). For example, during an outbreak of *Brevipalpus obovatus* on garden shrubs in Brisbane, Australia, 23% of the eggs and 16% of postegg stages of this pest were present within leaf domatia. However, 85% of the eggs and half of the active stages of the phytoseiid mites preying on *B. obovatus* were also found within domatia, and populations of the plant-parasite soon collapsed (DE Walter, unpublished information). Therefore, all of the data at hand indicate that plant-parasitic mites are rare and transient inhabitants of domatia but that predatory mites are consistently and abundantly present.

That mites seek out appropriately sized spaces on the plants they inhabit is not in itself surprising, but that plants produce structures that have no known function except to house mites calls for an explanation (40). When plants produce other structures that attract arthropods, e.g. flowers, extrafloral nectaries, or elaiosomes, an evolved benefit to both the plant and the arthropods is usually assumed. A similar mutually beneficial relationship may be the case in mite-domatia associations.

WHAT THE PLANT GETS   Plant-parasitic mites reduce yields (by limiting growth and reproduction) and may cause death of plants. The evidence for these effects in agricultural systems has been overwhelming, especially since the advent of synthetic insecticides elevated spider mites to egregious secondary pests. As already discussed above, plant-parasitic mites are small, use stylet-like mouthparts to puncture individual plant cells (and thus can avoid many plant defenses), and have short generation times and high reproductive outputs allowing them to overexploit their hosts (6, 18, 28, 29). However, plant-parasitic mites must run a gauntlet of predators when they attempt to colonize leaves. In agricultural systems, phytoseiid mites are extremely useful biological-control agents. Many phytoseiid species are reared commercially for inundative or inoculative release, and numerous other species naturally colonize crops and suppress plant-parasitic mites and some small insect pests (2, 9, 10, 19–21, 25, 34, 50, 51).

If plants with domatia attract phytoseiid mites to leaves before attack by plant parasites, i.e. if domatia are a type of constitutive defense against herbivores, then the removal of domatia should result in fewer mite bodyguards on leaves. Walter & O'Dowd tested this hypothesis in two ways: by using bitumen paint to close the domatia of a rain-forest tree (63) and by shaving off tuft domatia of a garden shrub (62). In both cases, phytoseiid mite numbers were strongly reduced (to 24 and 18% of controls, respectively) on leaves where mites had no access to domatia compared with mock-treated controls. Similar effects have been demonstrated for 12 additional species of trees and lianas (65).

The reverse holds true as well. In an innovative series of experiments, Rozario (49) added tufts of polyester fibers to the vein axils of grape varieties with poorly expressed domatia. In the laboratory, leaves with artificial domatia retained more of the predator *M. occidentalis*, and these mites had higher reproductive outputs than they did on the same variety without polyester tufts. In the field, results varied, but in some experiments, tydeid and phytoseiid mite populations were significantly higher when artificial domatia were present.

Given the potential of plant-parasitic mites to reduce plant fitness and the ability of predatory mites to protect plants from attack, the apparent independent derivation of domatia in many lineages of angiosperms (38–41) is easily understood. Under attack by parasitic mites, plants with leaves that attract predatory mites would experience better survival, growth, and reproduction than those less able to attract bodyguards. Definitive demonstration of the protective function of domatia in the field appears to be experimentally intractable (64), but a protective effect has been demonstrated in the laboratory (66). In this study, we cut shoots of an ornamental shrub, shaved off tuft domatia, and mounted the cuttings in small flasks of water. Then we added the predatory mite *M. occidentalis* and the spider mite *Tetranychus urticae* to each cutting (and to mock-shaved controls). After five days, cuttings with intact domatia had three times as many predatory mites as those with domatia removed. In contrast, spider-mite reproduction was more than three times higher on cuttings with domatia removed. Using a similar system, Grostal & O'Dowd (15) demonstrated that *M. occidentalis* ate more spider-mite eggs when domatia were present.

WHAT THE MITE GETS   To call an interaction between two organisms a mutualism implies that the relationship improves the fitness of both parties (3). The studies reviewed above support the hypothesis that plants benefit from having leaf domatia, but what of the mites? If mites gain no particular benefit from entering domatia, or if predators are lured away from better hunting grounds by hideaways on barren leaves, then the interaction would be better classified as a commensalism or, perhaps, a parasitism of the mites by the plants. The facultative nature of the interaction could be considered an argument against an interpretation of mutualism. Most of the predatory mites associated with leaf domatia tend to have broad host distributions (59) and prey preferences (19–21), and unlike many mutualisms (e.g. pollination, many ant plants) no nutritional reward is directly associated with domatia.

However, many nutritional mutualisms are diffuse (3), and even without a ready source of food, eggs and newly molted mites probably are better protected from predators and from desiccation inside domatia. Several studies have demonstrated that phytoseiid mites preferentially use leaf domatia for oviposition (46, 62, 63). In the field, more than three quarters of all eggs laid by phytoseiid mites were found within domatia (62, 63). In a series of laboratory experiments, Grostal & O'Dowd (15) showed that *M. occidentalis* laid fewer eggs at low than at high relative humidity. However, leaf domatia buffered the effect of low humidity, and the predator laid more than twice as many eggs in the presence than in the absence of domatia. Whenever domatia were available, 84–100% of the predator's eggs were laid within them. Predatory and fungivorous mites also preferentially use leaf domatia for molting (62).

CONCLUSIONS

The recent fascination with the chemistry of plant-mite interactions has been useful for understanding the complexity of tritrophic systems but has obscured our view of the importance of microhabitat to the success of predatory mites. The hiding places present on hairy leaves and within leaf domatia offer a simple method for maintaining populations of predatory mites in agricultural systems. This method could be implemented through companion plantings of predator-friendly plants, by enhancing domatia in crop varieties, and by developing artificial mite houses for plant varieties with leaf surfaces unfavorable to predators.

The insights from the studies of mite-plant associations also have important evolutionary implications. Intricate, mutually beneficial interactions between the largest forest trees and the smallest of arboreal arthropods appear to be common and widespread. The plant participates in these interactions both chemically (infochemicals, nectary secretions) and morphologically. Phylloplane architecture strongly affects the diversity and abundance of foliar mites, and where these animals are an important component of biodiversity (e.g. if predators protect forest relicts from invasion by polyphagous spider mites), then appropriate leaf habitats must be maintained.

SUMMARY

During their long evolutionary association, mites and terrestrial plants have developed several ways of using each other. Plant-parasitic mites are capable of overexploiting their hosts; however, numerous voracious predators, especially phytoseiid mites, can suppress populations of plant parasites. The minute herbivores often make shelters on the leaf surface (e.g. silken webs or galls) that protect them from environmental extremes, whereas predatory mites are at the mercy of wind, rain, and larger predators. However, predatory mites naturally seek out protected areas on leaves, and leaves with tomenta of erect hairs usually house more predators than leaves with smooth surfaces. Predatory mites were probably never a factor in the evolution of extensive leaf tomenta; nevertheless, many woody plants have small patches of hairs restricted to the vein axils on the undersides of leaves. These tufts of hairs, and more elaborate pits, pouches, pockets, and domes, which also occur in vein axils, are called leaf domatia. Comparative observations and experimental data support the hypothesis that plants have evolved leaf domatia as a constitutive defense against herbivores. Plants benefit from having leaf domatia by an increased probability that populations of predatory mites will inhabit their leaves before attack by plant parasites. Mites benefit from leaf domatia by securing a safe place for oviposition and molting. Like entomophilous flowers, elaiosomes, extrafloral nectaries, and formicaria, leaf domatia are the manifestation of a long-term and mutually beneficial interaction between plants and arthropods.

ACKNOWLEDGMENTS

My thanks to Dennis O'Dowd and Heather Proctor for offering, respectively, old- and new-chum perspectives on the phylloplane, and to Bruce Halliday and Matt Colloff who critically reviewed an earlier version of this manuscript. I also especially thank Shelley Rozario for permission to discuss her thesis work and Andrew Ward for skeptically collecting the data on the distribution of *B. obovatus*.

Literature Cited

1. Aina OJ, Rodríguez JG, Knavel DE. 1972. Characterizing resistance to *Tetranychus urticae* (Koch) in tomato. *J. Econ. Entomol.* 65:641–43
2. Bakker FM, Klein ME. 1992. Transtro-phic interactions in cassava. *Exp. Appl. Acarol.* 14:293–311
3. Bronstein JL. 1994. Conditional outcomes in mutualistic interactions. *Trends Ecol. Evol.* 9:214–17

4. Brouwer YM. 1983. Domatia and their occurrence in the Australian flora. *Aust. Syst. Soc. Newsl.* 34:6–9
5. Coy R, Greenslade P, Rounsevell D. 1993. A survey of invertebrates in Tasmanian rainforest. *Tasmanian NRCP Tech. Rep. No. 9.* Dep. Arts, Sport, Environ. Territories, Canberra
6. Dicke M, Sabelis M. 1988. How plants obtain predatory mites as bodyguards. *Neth. J. Zool.* 38:148–65
7. Dicke M, Sabelis MW, Takabayashi J, Bruin J, Posthumus MA. 1990. Plant strategies of manipulating predator-prey interactions through allelochemicals: prospects for application in pest control. *J. Chem. Ecol.* 16:3091–118
8. Downing RS, Moilliet TK 1967. Relative densities of predacious and phytophagous mites on three varieties of apple trees. *Can. Entomol.* 99:738–41
9. Duso C. 1992. Role of *Amblyseius aberrans* (Oud.), *Typhlodromus pyri* Scheuten and *Amblyseius andersoni* (Chant) (Acari, Phytoseiidae) in vineyards. *J. Appl. Entomol.* 114:455–62
10. Duso C. 1993. Factors affecting the potential of phytoseiid mites (Acari: Phytoseiidae) as biocontrol agents in North-Italian vineyards. *Exp. Appl. Acarol.* 17:241–58
11. Gentile AG, Webb RE, Stoner AK. 1969. *Lycopersicon* and *Solanum* spp. resistant to the carmine and the two-spotted spider mite. *J. Econ. Entomol.* 62:834–36
12. Gibson RW. 1976. Trapping of the spider mite *Tetranychus urticae* by glandular hairs on the wild tomato *Solanum berthaultii. Potato Res.* 19:179–82
13. Gibson RW, Valencia L. 1978. A survey of potato species for resistance to the mite *Polyphagotarsonemus latus,* with particular reference to the protection of *Solanum berthaultii* and *S. tarijense* by glandular hairs. *Potato Res.* 21:217–23
14. Gregory P, Avé DA, Bouthyette PY, Tingey WM. 1986. Insect-defensive chemistry of potato glandular trichomes. See Ref. 26, pp. 173–84
15. Grostal P, O'Dowd DJ. 1994. Plants, mites and mutualism: leaf domatia and the abundance and reproduction of mites on *Viburnum tinus* (Caprifoliaceae). *Oecologia* 97:308–15
16. Harvey TL, Martin TJ. 1980. Effects of wheat pubescence on infestations of wheat curl mite and incidence of wheat streak mosaic. *J. Econ. Entomol.* 73: 225–27
17. Harvey TL, Martin TJ, Seifers DL. 1990. Wheat curl mite and wheat streak mosaic virus in moderate trichome density wheat cultivars. *Crop Sci.* 30:534–36
18. Helle W, Sabelis MW, eds. 1985. *Spider Mites, Their Biology, Natural Enemies, and Control,* Vol. 1A. New York: Elsevier
19. Helle W, Sabelis MW, eds. 1985. *Spider Mites, Their Biology, Natural Enemies, and Control,* Vol. 1B. New York: Elsevier
20. Hoy MA, ed. 1982. *Recent Advances in Knowledge of the Phytoseiidae.* Publ. 3284. Div. Agric. Sci., Univ. Calif. Berkeley
21. Hoy MA, Cunningham L, Knutson L, eds. 1983. *Biological Control of Pests by Mites.* Special Publ. 3304, Div. Agric. Nat. Res., Univ. Calif. Berkeley. 185 pp.
22. Hussein N, Perring TM. 1986. Feeding habits of the Tydeidae with evidence of *Homeopronematus anconai* (Acari: Tydeidae) predation on *Aculops lycopersici* (Acari: Eriophyidae). *Int. J. Acarol.* 12:215–21
23. Huxley CR. 1986. Evolution of benevolent ant-plant relationships. See Ref. 26, pp. 257–82
24. Jeffree CE. 1986. The cuticle, epicuticular waxes and trichomes of plants, with reference to their structure, functions and evolution. See Ref. 26, pp. 23–64
25. Jeppson LR, Keifer HH, Baker EW. 1975. *Mites Injurious to Economic Plants.* Berkeley, CA: Univ. Calif. Press
26. Juniper B, Southwood R, eds. 1986. *Insects and the Plant Surface.* London: Edward Arnold. 360 pp.
27. Koptur S. 1992. Extrafloral nectary-mediated interactions between insects and plants. In *Insect-Plant Interactions,* ed. E Bernays, 4:81–129. Boca Raton: CRC
28. Krantz GW. 1978. *A Manual of Acarology.* Corvallis: Oreg. State Univ. Book Stores
29. Krantz GW, Lindquist EE. 1979. Evolution of phytophagous mites (Acari). *Annu. Rev. Entomol.* 24:121–58
30. Krivolutsky DA, Druk A Ya. 1986. Fossil oribatid mites. *Annu. Rev. Entomol.* 31:533–45
31. Lundström AN. 1887. Planzenbiologische Studien. II. Die Anpassungen der Planzen an Thiere. *Nova Acta Reg. Soc. Sci. Ups. Ser. 3.* 13:1–87
32. May RM. 1988. How many species are there on earth? *Science* 241:1441–49
33. Mound LA. 1993. The first thrips species (Insecta) inhabiting leaf domatia: *Domatiathrips cunninghamii* gen. et sp. nov. (Thysanoptera: Phlaeothripidae). *J. NY Entomol. Soc.* 101:424–30

34. Needham GR, Welbourn WC, Horn DJ, Mitchell R. 1995. *Acarology IX: Symp. Int. Congr.* Columbus: Ohio Biol. Survey. In press
35. Nihoul P. 1993. Do light intensity, temperature and photoperiod affect the entrapment of mites on glandular hairs of cultivated tomato? *Exp. Appl. Acarol.* 17:709–18
36. Norton RA, Bonamo PM, Grierson JD, Shear WA. 1988. Oribatid mite fossils from a terrestrial Devonian deposit near Gilboa, New York. *J. Paleontol.* 62:259–69
37. O'Dowd DJ. 1994. Association between mites and leaf domatia of coffee (*Coffea arabica*) in north Queensland. *Bull. Entomol. Res.* 84:361–66
38. O'Dowd DJ, Brew CR, Christophel DC, Norton RA. 1991. Mite-plant associations from the Eocene of southern Australia. *Science* 252:99–101
39. O'Dowd DJ, Pemberton RW. 1994. Leaf domatia in Korean plants: floristics, frequency, and biogeography. *Vegetatio* 114:137–49
40. O'Dowd DJ, Willson MF. 1989. Leaf domatia and mites on Australian plants: ecological and evolutionary implications. *Biol. J. Linn. Soc.* 37:191–236
41. O'Dowd DJ, Willson MF. 1991. Associations between mites and leaf domatia. *Trends Ecol. Evol.* 6:179–82
42. O'Dowd DJ, Willson MF. 1991. A pocket full of mites. *Aust. Nat. Hist.* 23:840–47
43. Overmeer WPJ, van Zorn AQ. 1984. The preference of *Amblyseius potentillae* (Acarina: Phytoseiidae) for certain plant substrates. In *Acarology VI*, ed. DA Griffiths, CE Bowman, 1:591–96. Chichester: Horwood
44. Patterson CG, Thurston R, Rodriguez JG. 1974. Two-spotted spider mite resistance in *Nicotiana* species. *J. Econ. Entomol.* 67:341–44
45. Pemberton RW. 1993. Observations of extrafloral nectar feeding by predaceous and fungivorous mites. *Proc. Entomol. Soc. Wash.* 95:642–43
46. Pemberton RW, Turner CE. 1989. Occurrence of predatory and fungivorous mites in leaf domatia. *Am. J. Bot.* 76:105–12
47. Price PW, Bouton CE, Gross P, McPheron BA, Thompson JN, Weis AE. 1980. Interactions among three trophic levels: influence of plants on the interactions between insect herbivores and natural enemies. *Annu. Rev. Ecol. Syst.* 11:41–65
48. Rosario SA. 1995. Associations between mites and leaf domatia: evidence from South Asia. *J. Trop. Ecol.* 11:99–108
49. Rosario SA. 1994. *Domatia and mites: effects of leaf morphology on beneficial mites in semi-natural and managed systems.* PhD thesis. Monash Univ., Melbourne. 187 pp.
50. Sabelis MW, Bakker FM. 1992. How predatory mites cope with the web of their tetranychid prey: a functional view on dorsal chaetotaxy in the Phytoseiidae. *Exp. Appl. Acarol.* 16:203–25
51. Smith D, Papacek DF. 1991. Studies of the predatory mite *Amblyseius victoriensis* (Acarina: Phytoseiidae) in citrus orchards in south-east Queensland: control of *Tegolophus australis* and *Phyllocoptruta oleivora* (Acarina: Eriophyidae), effect of pesticides, alternative host plants and augmentative release. *Exp. Appl. Acarol.* 12:195–217
52. Stoner AK, Frank JA, Gentile AG. 1968. The relationship of glandular hairs on tomato to spider mite resistance. *Proc. Am. Soc. Hortic. Sci.* 93:532–38
53. Sutherst RW, Jones RJ, Schnitzerling HJ. 1982. Tropical legumes of the genus *Stylosanthes* immobilize and kill cattle ticks. *Nature* 295:320–21
54. Sutherst RW, Wilson LJ. 1986. Tropical legumes and their ability to immobilize and kill cattle ticks. See Ref. 26, pp. 185–94
55. Turner CE, Pemberton RW. 1989. Leaf domatia and mites: a plant protection-mutualism hypothesis. In *The Evolutionary Ecology of Plants*, ed. JH Bock, YB Linhart, pp. 341–59. Boulder, CO: Westview
56. van Haren RJF, Steenhuis MM, Sabelis MW, De Ponti OMB. 1987. Tomato stem trichomes and dispersal success of *Phytoseiulus persimilis* relative to its prey *Tetranychus urticae*. *Exp. Appl. Acarol.* 3:115–21
57. van Lenteren JC, Ponti OMB. 1990. Plant-leaf morphology, host-plant resistance and biological control. *Symp. Biol. Hung.* 39:365–86
58. Vet LEM, Dicke M. 1992. Ecology of infochemical use by natural enemies in a tritrophic context. *Annu. Rev. Entomol.* 37:141–72
59. Walter DE. 1992. Leaf surface structure and the distribution of *Phytoseius* mites (Acarina: Phytoseiidae) in south-east Australian forests. *Aust. J. Zool.* 40:593–603
60. Walter DE. 1995. Dancing on the head of a pin: mites in the rainforest canopy. *Rec. West. Aust. Mus. Suppl.* 52:49–53
61. Walter DE, Denmark HA. 1991. Use of

leaf domatia on wild grape (*Vitis munsoniana*) by arthropods in central Florida. *Fla. Entomol.* 74:440–46

62. Walter DE, O'Dowd DJ. 1992. Leaves with domatia have more mites. *Ecology* 73:1514–18

63. Walter, DE, O'Dowd DJ. 1992. Leaf morphology and predators: effect of leaf domatia on the abundance of predatory mites (Acari: Phytoseiidae). *Environ. Entomol.* 21:478–84

64. Walter DE, O'Dowd DJ. 1995. Life on the forest phylloplane: hairs, little houses, and myriad mites. In *Forest Canopies—A Review of Research on This Biological Frontier*, ed. M Lowman, N Nadkarni, pp. 325–51. New York: Academic

65. Walter DE, O'Dowd DJ. 1995. Beneath biodiversity: factors influencing the diversity and abundance of canopy mites. *Selbyana.* 16:12–20

66. Walter DE, O'Dowd DJ. 1995. The good, the bad, and the ugly: which really inhabit leaf domatia? See Ref. 34, In press

67. Walter DE, O'Dowd DJ, Barnes V. 1994. The forgotten arthropods: foliar mites in the forest canopy. *Mem. Queensl. Mus.* 36:221–26

68. Willson MF. 1991. Foliar mites in the Eastern deciduous forest. *Am. Midl. Nat.* 126:111–17

69. Zacharda M, Krivolutsky DA. 1985. Prostigmatic mites (Acarina: Prostigmata) from the Upper Cretaceous and Paleocene amber of the USSR. *Vestn. Cesk. Spol. Zool.* 49:147–52

Annu. Rev. Entomol. 1996. 41:115–39

THE ROLE OF MACROINVERTEBRATES IN STREAM ECOSYSTEM FUNCTION

J. Bruce Wallace

Department of Entomology and Institute of Ecology, University of Georgia, Athens, Georgia 30602

Jackson R. Webster

Department of Biology, Virginia Polytechnic Institute and State University, Blacksburg, Virginia 24061

KEY WORDS: insects, rivers, grazers, shredders, collectors, filterers, predators

ABSTRACT

This review focuses on some of the roles of macroinvertebrate functional groups, i.e. grazers, shredders, gatherers, filterers, and predators, in stream-ecosystem processes. Many stream-dwelling insects exploit the physical characteristics of streams to obtain their foods. As consumers at intermediate trophic levels, macroinvertebrates are influenced by both bottom-up and top-down forces in streams and serve as the conduits by which these effects are propagated. Macroinvertebrates can have an important influence on nutrient cycles, primary productivity, decomposition, and translocation of materials. Interactions among macroinvertebrates and their food resources vary among functional groups. Macroinvertebrates constitute an important source of food for numerous fish, and unless outside energy subsidies are greater than in-stream food resources for fish, effective fisheries management must account for fish-invertebrate linkages and macroinvertebrate linkages with resources and habitats. Macroinvertebrates also serve as valuable indicators of stream degradation. The many roles performed by stream-dwelling macroinvertebrates underscore the importance of their conservation.

INTRODUCTION

Consumers maintain and modify ecosystem function in ways that often transcend simple consumption of food (131). Chew (21) suggested that consumers

115

0066-4170/96/0101-0115$08.00

benefit ecosystems as regulators rather than energy movers. Consumers' regulatory functions include regulation of rates of succession and nutrient cycling, transportation and mixing of materials, top-down influences (by predators and herbivores), and physical structuring of ecosystems [as "ecosystem engineers" sensu Jones et al (87)]. Macroinvertebrates are usually ignored in such discussions. Our purpose here is to underscore the important roles of macroinvertebrates in streams.

The macroinvertebrate assemblage of most streams is highly diverse, and many of the individual species may be redundant (98) in the sense that ecosystem functions can proceed if they are absent (191). Categorization of any stream macroinvertebrate as a keystone species would be difficult (119, 138), but as a group they perform essential functions and are critical to the maintenance of stream functional integrity (8). Even normally rare species may have a critical role that becomes evident only after a major disturbance (188).

Functional Groups

In the heterogeneous physical environment of streams, benthic invertebrates have evolved a diverse array of morphological and behavioral mechanisms for exploiting foods. Throughout this review, we follow the functional classification of Cummins, which is based on morpho-behavioral mechanisms used by invertebrates to acquire foods (32, 113). These groups include scrapers, animals adapted to graze or scrape materials (periphyton, or attached algae, and its associated microbiota) from mineral and organic substrates; shredders, organisms that comminute primarily large pieces of decomposing vascular plant tissue (>1 mm diameter) along with the associated microflora and fauna, feed directly on living vascular macrophytes, or gouge decomposing wood; gatherers (= collectors), animals that feed primarily on fine particulate organic matter (FPOM; <1 mm diameter) deposited in streams; filterers, animals with specialized anatomical structures (e.g. setae, mouth brushes, fans, etc) or silk and silk-like secretions that act as sieves to remove particulate matter from suspension (189, 206); and predators, organisms that feed primarily on animal tissue by either engulfing their prey or piercing prey and sucking body contents.

These functional feeding groups refer primarily to modes of feeding or to the food-acquisition system (sensu 31) and not to the type of food per se (e.g. as determined from gut-content analysis). For example, many filter-feeding insects of high-gradient streams are primarily carnivores (e.g. 13). Scrapers consume not only attached algae but also quantities of what must be characterized as epilithon (100). Likewise, although shredders may select those leaves that have been microbially conditioned by colonizing fungi and bacteria (e.g. 32), these shredders also ingest attached algal cells, protozoans, and various other components of the fauna during feeding (113). Some shredders apparently obtain very little of their assimilated energy directly from microbial biomass (47), although enzymes derived from microbial endosymbionts or microbes ingested with leaf tissue may be important in cellulose hydrolysis (170). Although these mechanisms used to obtain foods seem valid criteria for separating taxa, many questions remain concerning the ultimate sources of protein, carbohydrates, fats, and assimilated energy for each of these functional groups.

Physical Template

The physical environment of streams places many constraints on organisms as well as on the type and form of food that is available. Most stream reaches are characterized by many diverse microhabitats (51), which result from physical factors, such as relief, lithology, runoff, and large woody debris, that generate an array of channel forms (15). Physical heterogeneity, including the substrate and the current velocity of a stream channel, is an important factor that may influence local biotic diversity (80, 120, 182), nutrient dynamics (117, 155), algae and macrophyte distribution (155, 156), retention and distribution of organic matter (78, 79, 96, 173, 175, 198), predator-prey interactions (61, 151), presence or absence of refugia during disturbance (96), and secondary production of invertebrates (12, 78, 172). Growns & Davis (58) demonstrated linkages between functional feeding groups and several near-bed hydraulic parameters, but such linkages are not surprising. For example, within the mosaic of habitat types in a southern Appalachian stream, secondary production by most filterers occurs in high velocity, low-retention habitats, whereas secondary production by gatherers and shredders dominate low-velocity, high-retention pools (e.g. 78). The linkages between flow parameters, resource availability, respiratory and thermal requirements, and biotic interactions such as competition and predation influence the structure and function of diverse stream ecosystems. These parameters (as well as others) presumably influence an organism's energy costs and gains as well as its ultimate success in a given habitat (59).

Despite the attractiveness of a habitat-based approach to studying stream ecosystems, many problems remain in its application for several reasons: (a) Many interacting factors influence biota; (b) an array of microhabitats may exist within and among streams; (c) boundaries among patches are often indiscreet and vary in space and time; and (d) the resolution and classification of such boundaries may vary with the research or management objectives (64).

Top-Down and Bottom-Up Influences on Stream Ecosystems

Top-down regulation by consumers can have an important influence on nutrient cycles, primary productivity, decomposition, and translocation of materials.

Despite the abundant evidence from both field and laboratory studies for either strong bottom-up or top-down influence by consumers in streams, many of the studies reviewed below (as well as many others) suffer from one or more deficiencies: (a) They were conducted at specific times of the year and were often of short duration. (b) They failed to consider indirect effects that may require several generations to detect. (c) Unnatural densities of primary and secondary consumers were used. (d) Enclosures or exclosures were used that did not allow sufficient exchange with the stream environment. (e) There was no replication (primarily a problem with ecosystem-level studies). (f) The study failed to consider or incorporate abiotic forces such as hydrologic regime (i.e. account for floods and drought). Hunter & Price (77) suggested that environmental heterogeneity may have influenced many of the classic debates on whether abiotic or biotic factors determine population change and that differences in results can be partly attributed to the relative stabilities of the various environments studied. They suggested that although top-down trophic cascades are dramatic in some instances, a bottom-up perspective seems more logical because "the removal of higher trophic levels leaves lower trophic levels present (if perhaps greatly modified), whereas the removal of primary producers leaves no system at all" (77, p. 725). Despite experimental limitations, many studies have demonstrated a significant role for stream-dwelling macroinvertebrates at intermediate levels of food webs, regardless of whether the results pointed to top-down or bottom-up effects. It is through these intermediate stages, incorporating macroinvertebrate populations, that the effects are propagated either up or down. In the following section, we focus on some of the ascribed roles of specific macroinvertebrate functional groups in stream ecosystem processes.

FUNCTIONAL ROLES OF INVERTEBRATES IN STREAMS

Grazers

Invertebrate herbivores in streams use various mechanisms to feed on plants: Some feed on aquatic vascular plants by shredding (shredder-herbivores) or piercing; others graze algal films attached to rocks or other submerged objects. In most streams, algal grazing is by far the most important. Mechanistically, these grazers are scrapers (32) because they feed by scraping the epilithon (100) from mineral and organic substrates.

The algae-grazers interaction in streams is tightly coupled; that is, algal production directly affects grazers and grazer feeding directly affects algae (158). Gregory (56) and Lamberti & Moore (94) noted that this interaction had not been extensively studied. However, considerable research, perhaps stimulated by those reviews, has focused on this area in the past decade.

Gregory (56) noted that many studies have shown a correlation between algal and grazer abundance, which suggests food limitation. For example, following clear-cutting of stream-side vegetation, high grazer abundance is generally associated with increased light levels and high algal abundance (e.g. 65, 188). More compelling evidence for resource limitation emerges from studies showing the density dependence of grazer growth (73, 74, 95, 108). Moreover, other studies revealed that algal production is increased by the addition of nutrients, light, or both and that this increase is followed by enhanced grazer abundance, growth, and/or production (41, 63, 69, 71, 92, 120, 129, 146, 147, 160). Finally, light reduction has been shown to decrease grazer abundance (52) and nutrient reduction to decrease snail growth rates (128).

Gregory (56) reviewed studies demonstrating that grazers can greatly reduce algae abundance, showing that removal of grazers increased algae abundance. Yet other studies have revealed little or no grazer effect (43, 74, 130, 183). A grazing effect may not be seen when algae are light limited (43, 130) or when grazer abundance is low (85, 91). In recent studies, investigators have used raised platforms in natural streams (43, 85, 95, 108), enclosures and exclosures in natural streams (44, 45, 109, 110, 160), stream-side or in-stream channels (63, 71–74, 93, 129, 160), and laboratory streams (91, 92, 128, 176, 177) to examine grazer-algae interactions. The effects of insecticides (203, 207) or top-down effects induced by predator manipulation have also been studied (150, 151). Most of these authors reported that grazers reduce algal biomass.

Most studies have shown that algal primary production is lower when grazers are present than when they are absent (71, 85, 91, 95, 128, 160). However, Lamberti et al (91, 92) found evidence that grazing snails may increase primary production. Lamberti & Moore (94) suggested that at low densities grazers have no effect on algal production, at intermediate densities they may increase algal production by stimulating productivity (i.e. production per unit biomass, P/B), and at high densities grazers decrease algal production. Stimulation of primary production by grazers is controversial but has been demonstrated in some terrestrial ecosystems (e.g. 112) and been suggested for streams (40). Grazers might enhance algal productivity (P/B) by removing dead or senescent algal cells; shifting algal community composition to more productive species; decreasing the thickness of the algal film, which would allow light and nutrient penetration; and remobilizing nutrients (92, 94). Productivity may also be enhanced by mucus trails of grazing gastropods, which may provide a microenvironment rich in regenerated nutrients that stimulates growth of microalgae, as shown for marine habitats (141).

On the other hand, algal production increases caused by grazers may be impossible to confirm because the loss of photosynthetic tissue may exceed any enhanced algal *P/B*. Grazer enhancement of algal *P/B* has often but not universally been demonstrated. Several studies (72, 91, 95) have shown that grazers may increase productivity per unit chlorophyll a (*P*/Chla), but other studies have shown decreases (71, 125, 160) or no change (85, 128). In general, grazing appears to increase the assimilation number (chlorophyll a per unit biomass) (56, 73, 85, 91).

Stream invertebrates may affect algal abundance by methods other than direct feeding. Hart (62) found that a grazing caddisfly reduced the overstory consisting of filamentous blue-green algae, and this promoted the growth of diatoms. Within deep water habitats, crayfish reduced *Cladophora* spp. and thereby indirectly aided diatoms and diatom-feeding grazing insects (26). Case-building chironomids increased the surface area for diatom growth and perhaps protected diatoms from grazing mayflies (153). In tropical streams, freshwater shrimp may enhance algal growth by clearing surficial sediment via their feeding activity (154). By reducing periphyton biomass, grazers indirectly influence localized hydraulic characteristics, zones of stationary water (transient storage), and nutrient cycling (110, 126). Several studies have also shown that grazing snails (92, 125) and insects (37) increase amounts of downstream export of FPOM from grazed surfaces, as well as the resistance of stream algae to disturbance from floods (127).

Shredders

Upland streams, especially those draining forested catchments, receive a large portion of their energy input as coarse particulate organic matter (CPOM) from terrestrial litter inputs. For example, in the eastern United States, average litter-fall inputs to streams are about 600 g (dry mass) m^{-2} $year^{-1}$, over half of which consists of leaves (200). This detrital material represents an important source of energy for many stream detritivores (6, 32), and food limitation to CPOM-consuming detritivores has been shown (36, 157).

In both laboratory studies (33, 111, 124, 145, 167) and stream manipulations (22, 29, 192), shredders increase conversion of CPOM to FPOM and dissolved organic matter (DOM) (118). CPOM-shredding insects generally have low assimilation efficiencies (55, 81, 111). Hence, a large portion of leaf litter inputs are transformed into FPOM, which is more amenable to downstream transport (e.g. 29, 34, 192).

In addition to facilitating downstream transfer of FPOM, some burrowing animals increase FPOM transfer to sediments. For example, larvae of the European sericostomatid caddisfly *Sericostoma personatum* feed on surficial CPOM at night and burrow into the stream bed during the day. Larval defecation by *S. personatum* increased subsurface sediment organic content by

75–185% over the organic content of control sediments in controls containing no sericostomatids (185). Presumably, similar increases may result from the presence of other sericostomatids such as the western North American species *Gumaga nigricula,* which also transfers case-associated algae that is subsequently lost via abrasion during burrowing (14).

WOOD FEEDING   Invertebrate shredders also promote wood decomposition by scraping, gouging, and tunneling wood (7, 38). These activities expose additional wood to further microbial colonization and decomposition (7). Also, the wood-gouging habits of net-spinning caddisflies during retreat formation over many decades may be important; these filterers were implicated in a 53–58% reduction in a cross-sectional area of untreated timber pilings and the subsequent collapse of a highway bridge (133).

SHREDDER HERBIVORES   Shredding and consumption of living macrophytes (by shredder herbivores) is another potential pathway of FPOM production in stream ecosystems. Although macrophytes are generally assumed to enter stream foodwebs as detritus during autumn senescence, invertebrate consumption of living macrophytes may also contribute detritus to the food web (101, 135, 163). Some floating-leaf macrophytes may be heavily grazed, and decomposition associated with grazing can contribute considerable amounts of organic matter to detrital foodwebs throughout the growing season (190). However, invertebrate consumption of submerged macrophytes in streams has not been well studied. Shredder herbivores such as some trichopteran larvae may rely on macrophytes for food during late spring and summer in downstream reaches where leaf inputs are reduced and CPOM standing crops are low (82). In a Danish stream, consumption of the macrophyte *Potamogeton perfoliatus* was low, ranging from 1.3 to 1.8% of annual plant production; however, consumption was higher (4–18% of macrophyte production) early in the growing season (83). Larval feeding and growth rates of the limnephilid caddisfly *Anabolia nervosa,* the dominant herbivore in this study, indicate that *Potamogeton* tissue is probably as suitable a food as any terrestrial leaf litter (84). Other than the results of these studies, little information is available on feeding ecology or the effects of aquatic invertebrates on macrophytes in streams, or on the ecosystem consequences of macrophyte feeding (135).

SHREDDER-GENERATED FPOM   Heard (66) described a "resource chain" in which consumers specialize on a resource and consequently influence the rate of transfer as the resource passes through different conditions. The processing of allochthonus inputs to streams is an example. Cummins et al (33) demonstrated that FPOM-feeding collectors exhibit faster growth rates in the presence of leaf-shredding invertebrates. Likewise, in short-term feeding studies with

^{32}P-labeled leaves, two FPOM feeders accumulated significantly more radio-phosphorus when shedders were present to facilitate leaf breakdown (167). Transformation of organic matter by shredders is probably far more important than their ability to directly degrade organic material via metabolic respiration. Direct metabolic respiration by invertebrate fauna in Bear Brook, New Hampshire, was estimated at <1% of the annual flux of organic matter through the stream (49). However, when feeding activities, bioenergetic efficiencies, and secondary production of invertebrates were considered, the overall impact of shredders on conversion of CPOM to FPOM was 13–35% (e.g. 197).

Until recently, little direct evidence was available to quantify the importance of shredders (114). In southern Appalachian headwater streams, the application of an insecticide eliminated >90% of insect biomass and greatly reduced secondary production (104). This manipulation significantly reduced leaf litter breakdown and export of FPOM compared with adjacent, untreated reference streams (29, 186, 192). Restoration of the shredder functional group coincided with restoration of leaf-litter processing rates (22, 191) and FPOM export (186, 191). These studies demonstrated that macroinvertebrates accounted for 25–28% of annual leaf-litter processing (29) and 56% of FPOM export over a 3-year period (186). Thus, biological processes in small, high-gradient streams that exhibit high physical retention of CPOM inputs favor entrainment by processing CPOM to smaller, more easily transported particles (FPOM) (29, 186, 192).

The above studies were conducted in small first-order streams, and the extent to which these studies apply to larger streams, and/or other geographical areas where shredders may not be as abundant, has not been assessed. In addition to shredder activities, those of grazers, filterers, predators, and collectors contribute to the overall FPOM pool, as well as to detrital turnover (166). Wotton (204) reviewed many other mechanisms of FPOM generation, including mechanical breakage of CPOM, flocculation of DOM, direct inputs of bacteria from allochthonous sources, microbial degradation, breakdown of large woody debris, soil organic matter, and grazing and algal sloughing. The degree to which gatherers or filterers actually depend on FPOM generated from CPOM shredders vs FPOM derived from shredder-independent mechanisms remains unknown (67). Furthermore, how much shredders facilitate collectors' activities probably varies over time and space (67). Undoubtedly, physical forces associated with flow and deposition interact to form a much more dynamic suspended and deposited FPOM pool than commonly recognized.

Gatherers

Gatherers are adapted to feeding primarily on fine particles (<1 mm diameter) deposited on substrate surfaces or in depositional areas. Gatherers usually are the most abundant stream macroinvertebrates (12, 78, 104, 121, 173), and many gatherers, such as chironomids, are among the most frequently reported prey in guts of predaceous insects (e.g. 2). Despite their obvious importance in stream food webs, their functional role is probably among the least studied.

To date, the role of gatherers in bioturbation and resuspension of organic matter has received little attention. Although surficial FPOM represents a small portion of the total FPOM standing crop in Idaho streams, it may contribute disproportionately to metabolism as it is readily available to gatherer organisms (34). Continuous deposition and resuspension of FPOM also may cause the impact of bottom-feeding gatherers on food resources to be felt downstream rather than local food depletion (34). Feeding activities of macroinvertebrates may also affect deposition. For instance, in a montane Puerto Rican stream, atyid shrimp reduce depositional organic matter as well as the abundance of smaller collectors (chironomid larvae) (154).

In Sycamore Creek, Arizona, gatherers exhibit low assimilation efficiencies (7–15%) and very high ingestion rates (food consumption equivalent to their body weight every 4–6 h) (48). As ingestion rates of collectors in Sycamore Creek exceed primary production, coprophagy is obviously an important component of gatherer feeding. Although they performed no actual measurements, Fisher & Gray (48) suggested that bactivory associated with fecal reingestion is an important component of collector diets. Using ^{13}C sodium acetate to label bacteria in a headwater spring seep in North Carolina, Hall (60) found that several FPOM-feeding gatherer taxa (chironomids and copepods) had a higher $\delta\,^{13}C$ than their FPOM food resources, which suggests preferential assimilation of bacterial carbon relative to FPOM.

MICROBE-DETRIVORE RELATIONSHIPS   Views concerning the relative roles of detritus, decomposer microbes, and detritivorous animals differ widely. Aquatic insects can be microbial predators or competitors or can depend on microbes as a link to detritus (70); however, clarification of conditions under which microbes are detrital consumers vs competitors awaits more data (115). Many stream ecosystems depend strongly on allochthonous inputs of DOM and bacteria. In low-gradient streams of the southeastern US, metazoans such as filter-feeding black flies can directly consume large numbers of bacteria, effectively short-circuiting several trophic transfers associated with the microbial loop (39, 115, 116). Fine particle–feeding meiobenthos assimilate a much larger proportion of microbial biomass (142) found on FPOM than that consumed by insects that shred larger leaves (46). The microbial "peanut butter" and detrital "cracker" usually associated with macroinvertebrate shredders (30) may be more applicable to microdetritivores (142).

In another study in Sycamore Creek, gatherers ingested (and reingested through coprophagy) an estimated 131% of the nitrogen retained during a

20-day, postflood recovery period (57). Most of the nitrogen was returned to the particulate nitrogen pool by gatherer egestion, excretion, and mortality. Gatherers recycled between 15 and 70% of the retained nitrogen back to primary producers as excreted ammonia. However, insects, even aquatic species, can excrete nitrogen in forms other than ammonia (20). The fate of organic nitrogen and other nutrients eliminated by invertebrates and the potential availability of this material to microbes and algae merits much more study.

Filter Feeders

Filter feeders, especially filter-feeding macroinvertebrates, have evolved various mechanisms for removing particles from suspension (189, 206). Although many trichopterans may be filter feeders based on their mode of capture, they are also predators, that rely primarily on animal drift (13, 53, 143, 144). Conversely, some ephemeropterans, trichopterans such as Philopotamidae, and dipterans such as Simuliidae and some Chironomidae (189) exploit minute particles (<1–50 μm in diameter), which dominate the seston in most streams. For bivalves, the range of particles consumed is generally smaller, ranging from <1 to 10 μm (181). Some filter feeders, such as the Philopotamidae, Simuliidae, and bivalves may actually increase particle sizes by ingesting minute particles and egesting compacted fecal particles larger than those originally consumed. Thus, these animals may perform two very important functions: (a) the removal of FPOM from suspension (which would otherwise pass unused through the stream segment) and (b) the supply of larger particles via their feces to a broad spectrum of deposit-feeding detritivores.

Filter feeders may retard downstream transport of suspended particulate organic matter (POM) (193). In doing so, they would significantly decrease spiraling distances of nutrients and organic matter (134, 199). Newbold et al (134) suggested that filter feeders have their greatest effect on nutrient spiraling length when particulate transport and nutrient limitation are high. Studies have indicated low rates of seston removal by filter feeders, i.e. generally well below 1% seston removal per meter of stream length (53, 67). The highest rates of seston removal were obtained in studies that incorporated fine-particle feeders such as Simuliidae. Morin et al (122) found that simuliid larvae ingested 0.8–1.4% of the seston per meter of stream below a Quebec lake outlet. This study took place during a late spring period when flows were low and standing stock of black flies was high, whereas other studies, performed on an annual basis, indicated lower rates of seston removal.

Larger particle–feeding hydropsychids (Trichoptera) select higher-quality food items such as diatoms and animal drift (13, 143, 144). This selectivity, and generally low rates of seston removal by hydropsychids, suggests that their major impact is on the quantity and type of POM in suspension (13, 53, 144).

Experimental studies such as that of Georgian & Thorp (53) are especially relevant to ecologists studying invertebrate drift in streams. They estimated that two *Hydropsyche* species in riffles of a New York stream removed 18% of drifting invertebrate prey per meter. Their results suggest that when large net-spinning caddisfly populations are present in shallow streams, their predation may suppress stream drift (53).

Streams with limited stable substrate, sufficient current velocity, and high-quality organic seston concentrations often support massive standing stocks of filter-feeding hydropsychids and/or black flies (50, 139, 184, 205). Filterer densities that are higher than those of other functional groups are possible because filterers use the kinetic energy of the current to exploit foods produced in upstream habitats (28). As a consequence, filterers expend less energy in search of food; consequently, the stream segment in which they occur can support a higher biomass per unit area (28). In addition, some of highest secondary production values reported per unit habitat space are those of filtering invertebrates in streams. The high filterer biomass or production found below impoundments or lake outflows is especially noteworthy (50, 105, 139, 159, 184), as is that on woody debris in low-gradient streams with unstable sandy bottoms (12, 28, 171). Thus, in habitats with a high degree of particle transport, filterers exploit the physical environment and increase particle retention. In contrast, as noted above, highest shredder densities are often found in CPOM-retentive reaches, where they exploit retained food resources, increase conversion of CPOM to more easily transported FPOM, and decrease particle retention.

In addition to their influence on suspended organic matter, filterers may modify local benthic community structure. For example, hydropsychid predation may have an important influence on community structure occurring near their retreats in lake outlets (42) and natural streams (35, 68). In a sandy-bottomed Australian stream, hydropsychid larvae appear to facilitate colonization by grazing *Baetis* spp. mayflies by increasing retention and abundance of food resources of *Baetis* species (algae and detritus) on the silken hydropsychid retreats (137).

Predators

As in other types of ecosystems, predators in streams have top-down effects on their prey through direct consumption and reduction of prey populations. During the past 15 years, numerous studies have examined various aspects of predator-prey interactions in streams. Many of these studies have been previously reviewed (3, 25, 169, 201, 205). For the purpose of this review, we focus on the impact of predation on benthic communities and specific processes.

Earlier reviews by Allan (2) and Thorp (180) suggested a lack of strong evidence that predators significantly influence lotic community structure. More recent studies have yielded mixed results regarding the impact of predators on prey populations: Many studies show significant effects of predators on prey, whereas others have shown little or no impact of predation on prey populations (25, 169). Results of a meta-analysis of 20 studies showed that, on average, predators deplete prey density by ~0.4 standard deviations from prey densities found in predator-free areas, which is a small-to-moderate, but significant, impact (204).

Cooper et al (25) suggested that the magnitude of prey exchange (i.e. immigration and emigration) among substrate patches has an overwhelming influence on the perceived effects of predators on prey populations in enclosure and exclosure studies conducted in streams. They suggest that this exchange may be a reason that lentic studies, or studies in isolated stream pools, show a greater proportion of significant predator impacts than stream studies. Sih & Wooster (169) extended the analysis by addressing predator impacts in patches surrounded by background environments lacking predators and having a constant or decreasing prey density. They then examined situations in which the per capita emigration rates of prey are altered by the presence of predators. Outcomes were also influenced by the presence or absence of predators and the degree of prey recruitment in the background environment, as well as by the ability of prey species to hide in refuges (169). Sih et al (168) found that approximately 25% of all prey populations showed negative predator impacts in studies involving experimental manipulations of predators. Invertebrate predators appear to have a greater impact on benthic prey than do fish predators, apparently because of different behavioral responses of prey (see references in 169). Vertebrate predators such as fish often cause invertebrate prey to reduce their movement rate and seek refuge in the substrate, whereas invertebrate predators increase prey movement and their propensity to drift (see references in 169, 204). Studies that examine the impact of both vertebrate and invertebrate predators simultaneously are difficult and require detailed knowledge of behavioral interactions between predators as well as between predators and prey. For example, stonefly and fish predation either interfered with or facilitated the other predator depending upon whether *Baetis* or *Ephemerella* species were the prey (174).

NONLETHAL EFFECTS OF INVERTEBRATE PREDATORS   Predators may also influence growth and reproduction of prey populations. For example, in the absence of predatory crayfish, snails (*Physella* spp.) reproduced earlier and grew to a terminal body size less than half of that of snails found in the presence of crayfish (27). By shunting more assimilated energy into rapid growth and delaying the onset of reproduction, snails achieve a larger terminal body size and thereby decrease mortality resulting from size-specific predation, which is much greater for smaller than larger snails (27). Peckarsky et al (140) found that *Baetis* mayflies raised in the presence of stonefly predators matured at significantly smaller sizes, showed little or no growth, and had lower egg biomass per female than *Baetis bicaudatus* reared in the absence of plecopteran predators. Scrimgeour & Culp (165) reared *Baetis tricaudatus* in the laboratory under safe (no predation threat) and risky (model predator present) conditions. *B. tricaudatus* reared in safe environments matured earlier, reached a larger terminal size, and exhibited both greater fecundity and larger egg size than those reared under predation threat at each food level. Although these results suggest that predators can influence prey fitness under laboratory conditions, such effects under field conditions have not been demonstrated (165).

Evidence from one study shows that predaceous plecopterans and caddisflies can significantly decrease the rate of leaf-litter processing by reducing shredder populations in leaf packs (136). However, predator densities in this experiment were almost 10 times that of background. Malmqvist (106) used more realistic densities of predators and tested the effects of a predatory stonefly, *Diura nanseni,* confined in cages with and without predators on decomposition of leaf litter. He found that less leaf material was processed in cages with predators, even though no reduction in prey densities could be demonstrated. Furthermore, in laboratory feeding experiments, two of three shredder species produced less FPOM when exposed to predators (106).

MACROINVERTEBRATE-FISH INTERACTIONS   Hynes (80) stated that invertebrates are the most widespread and important food of running-water fish and that very few groups of fish do not feed on invertebrates. However, actually demonstrating and quantifying the importance of this energy flow is difficult. Perhaps the best evidence for the importance of macroinvertebrates to fish comes from studies showing higher fish production in response to nutrient or carbon addition. Richardson (158) reviewed several studies suggesting that moderate nutrient or organic enrichment enhances fish production. In a classic study, Warren et al (194) added sucrose to an Oregon stream and observed increased growth, biomass, and production of the bacterium *Sphaerotilus natans,* aquatic invertebrates, and trout. Peterson et al (146) found a strong bottom-up effect resulting from phosphorus fertilization of a tundra river. Stable isotope analyses allowed the enrichment to be traced through the food web from algae to insects and fish (146). Studies showing higher fish abundance in streams draining clear-cut watersheds can be interpreted similarly (e.g. 202). Clear-cutting increases sunlight to streams, resulting in higher autochthonous production; higher, or at least modified, invertebrate production; and higher fish abundance.

Evidence for the importance of the macroinvertebrate-fish trophic linkage also comes from top-down studies. Strong top-down fish effects have been demonstrated in lakes (e.g. 18) and in some streams with piscivorous and algivorous fish (152). However, the results of studies of lotic invertebrate-feeding fish using enclosures or exclosures, fish removal, or fish addition have varied (3, 169). Except in a few studies, effects were seen only on some species or only on some substrates (e.g. 150, 151, 164), and most studies did not show dramatic effects. Several authors have discussed how these variations might be attributed to problems associated with enclosure and exclosure studies (e.g. 3, 25, 54, 169). Also, some macroinvertebrate-fish studies were done with drift-feeding salmonids, which may have little impact on benthos (19, 196).

Allen (4, 5) studied trout and invertebrates in a small New Zealand trout stream, the Horokiwi, and noted, "We find therefore, that the quantity of bottom fauna which the trout eat in a year, 14 tons, is seventy times as great as the average amount of fauna present at one time" (5, p. 34). Allen's classic study became widely known as the Allen paradox (80). However, more recent studies have shown that the ratio of fish ingestion to macroinvertebrate standing crop may not greatly exceed possible turnover ratios of the macroinvertebrates (195, 196).

SECONDARY PRODUCTION   The data required to document the linkage between benthic macroinvertebrate prey and their invertebrate and vertebrate predators are difficult to obtain. Measurements of macroinvertebrate abundance and biomass are not sufficient to estimate the quantity of food available to predators. Benke (10) argued that production is the most comprehensive measure of success of a population because it includes a composite of several features: abundance, biomass, growth, reproduction, survivorship, and generation time. Unfortunately, while secondary-production measurements of numerous taxa exist, in Benke's (10) extensive review, he found total invertebrate production for <50 streams worldwide. Moreover, few studies have estimated production of macroinvertebrate prey and their predators in the same stream. In some fishless first- and second-order streams at Coweeta, North Carolina, invertebrate predators are responsible for 25–36% of total benthic production (78, 104). Likewise, Smock et al (171) found that invertebrate predators represented 30% of macroinvertebrate production in a low-gradient South Carolina stream.

APPLIED ASPECTS OF MACROINVERTEBRATE FUNCTION IN STREAMS

Exotic Species

In streams, noninsects usually constitute the most notorious exotic invertebrate invaders. Some well-known examples include the Asiatic clam, *Corbicula fluminea;* the zebra mussel, *Dreissena polymorpha;* and crayfish, e.g. *Orconectes rusticus. C. fluminea* were estimated to filter the entire water column of a reach of the Potomac River in 3–4 days (24) and that of the Chowan River, North Carolina, every 1.0–1.6 days, depending on chlorophyll concentrations (97). However, the extent to which this impact extends throughout the entire water column may depend on mixing at different flow regimes (97). In the lower Potomac, Phelps (148) implicated the invasion of the Asiatic clam as triggering a series of ecosystem changes such as decreased turbidity, which increased submerged aquatic vegetation. These changes in turn influenced alterations in biota including algae, fish, and birds.

Since its discovery in the Great Lakes in the late 1980s, the zebra mussel has spread rapidly throughout many lakes and connecting waterways, including the Mississippi River drainage (103). The spread was so rapid and documented impacts were so alarming that by 1993, at least one book was devoted entirely to the zebra mussel (132). Although the potential impacts on streams and rivers of North America are uncertain, in the Rhine, population densities of 30,000–40,000 per m^2 young zebra mussels have been observed (132). Its extended planktonic larval stage (lasting several days to weeks) has tremendous dispersal potential in running waters (132). The most obvious concerns are the mussel's high filtration rates and its reputation as a notorious fouler of various water works (103). However, some positive environmental impacts may include removal of nutrients and seston from the water column; use as sentinal organisms for various pollutants and trace metals (86); and conversion of various toxic wastes to consumable nutrients for other benthos and phytoplankton. Nevertheless, the degree to which potential benefits offset negative effects is, at best, uncertain. Negative impacts have been projected for some native biota, phytoplankton, and fisheries, although some benthos may benefit from the deposition of nutrients and organic matter.

Numerous species of crayfish have been introduced to lakes and streams worldwide (76). One native midwestern species, *O. rusticus,* has a grossly disjunct range, apparently the result of numerous multiple introductions (76); this species may have displaced native crayfish in some streams and lakes (17, 76, 107). Mechanisms of displacement may vary from locality to locality. Mather & Stein (107) suggested that slow displacement of *Orconectes sanborni* by *O. rusticus* in an Ohio stream is mediated in part by lower fish predation on the larger *O. rusticus.* Thus, indirect and direct effects of predation as influenced in part by body size seem to be important in the displacement process.

The introduced stream-inhabiting insects that have received the most attention are those used to control noxious aquatic plants (e.g. 16). Undoubtedly, stream insects have been much more successful invaders than commonly recognized. They are often readily assimilated into the local fauna, and the extent to which they have altered food webs is unclear. A relatively recent introduction of two alien caddisfly species into Hawaiian Island streams has resulted in some long-term dietary shifts of an endemic goby away from the relatively few native aquatic insect species to these exotic species (90).

Fisheries

An important question for commercial and sports fisheries is, what limits productivity of aquatic ecosystems? Richardson (158) reviewed several lines of evidence and reached the conclusion that fish production, at least for salmonids, is limited by benthic production. Evidence also indicates that some habitats may contribute more than others to productivity of higher trophic levels such as fish. In a low-gradient, warm-water river lacking salmonids, Benke et al (11) found that woody-debris habitats represented only 4% of total benthic habitat but contributed 60% to total invertebrate biomass and 16% of total invertebrate production in a study reach of the Satilla River, Georgia. Four of the eight major fish species in the Satilla obtained at least 60% of their diet from snag-inhabiting invertebrates, and significant portions of the diets of piscivorous species relied on prey that used snag-inhabiting invertebrates in their diets.

A disproportionate contribution of specific habitats to invertebrate production and/or drift, which is subsequently available at higher trophic levels, can have far-reaching consequences for fisheries management. However, such contributions are rarely assessed. In some situations, analysis of various habitats will require a tremendous effort. For example, Baker et al (9) delineated about 13 different freshwater habitats along a reach of the lower Mississippi River. Obviously, biotic inventories in this large riverine system require tremendous effort, and assessing productivity and processes in such diverse systems demands even more effort and resources. Numerous direct and indirect linkages between habitats add another layer of complexity. As pointed out by Richardson (158), most models of stream ecosystem function have addressed flow of energy and materials without incorporating feedback mechanisms that regulate population and trophic interactions. Most of these studies have failed to identify feedback loops that may be strongly regulating in stream ecosystems. The task of identifying such feedback loops will be formidable. Clearly, unless outside energy subsidies are greater than in-stream food resources for fish, effective fisheries management must account for fish-invertebrate linkages and macroinvertebrate linkages with resources and habitats.

Pollutants

Many of the processes relating to translocation of nutrients and food resources in streams also influence translocation of industrial pollutants in streams. For example, both grazers and shredders enhance downstream movement of radio-labeled organically bound toxicants from periphyton and leaf litter, respectively (162, 178). Furthermore, downstream populations of filter-feeding hydropsychids accumulated significantly greater amounts of a radiolabeled PCB in periphyton-dominated channels in the presence of grazing invertebrates, and the presence of shredders significantly increased release of radiolabeled PCB from leaf litter to downstream hydropsychid populations (162). Sallenave et al (162) suggested that modeling the fate and transport of lipophilic, organically bound compounds in streams will require an understanding not only of the physiochemical properties of the system, but of biotic processing as well. The potential role of biota in translocation and retention of contaminants in stream ecosystems deserves more attention than it has received to date.

Macroinvertebrates as Biological Monitors and Indicators

Macroinvertebrates have been used to monitor accumulation of heavy metals (86) and insecticides (123) in streams. Aquatic insects, as well as other components of the aquatic biota, have been used extensively to evaluate the degree of anthropogenic disturbance to both lotic and lentic ecosystems. In recent years, interest in this area has grown tremendously, as evidenced by several books devoted entirely to the subject (e.g. 1, 102, 149, 161). Invertebrates have been used in numerous biological-monitoring methods (e.g. 161). The most widely used are based on tolerance values for specific taxa, which normally range from 0 (very intolerant) to 10 (very tolerant) according to the ability of a taxon to inhabit streams differing in water quality (e.g. 23, 75, 99, 149). In view of the many roles performed by macroinvertebrates in streams, indices that incorporate concepts such as biological diversity and integrity are an important and economical means of assessing ecosystem health (88, 89), although other views have been expressed (179). Experimental manipulation of headwater streams has established a link between degraded biotic indexes for an insecticide-induced disturbance and ecosystem processes such as secondary production, detritus processing, and FPOM export (187). Large increases in algal standing crops after an insecticide reduced macroinvertebrate populations were also noted in a Japanese stream (207). However, the extent to which biotic indices and modified community structure indicate altered ecosystem-level processes for other types of anthropogenic disturbances (e.g. organic pollution, heavy metals, sediments, acid-mine drainage, and forest clear-cutting) remains unknown. This area of investigation deserves a much greater blend of basic and applied ecology than it has received to date and should be a rewarding and important area of future research.

SUMMARY

As consumers at the intermediate levels of lotic food webs, macroinvertebrates are influenced by both bottom-up and top-down forces in streams and serve as conduits by which these effects are propagated. Although the bottom-up role of macroinvertebrates is clear, the top-down impacts have been less well documented. Shredder detritivores can exercise strong top-down effects by depleting their food resources, but they do not influence CPOM renewal. Similarly, gatherers have little top-down effect on renewal of their foods, whereas feeding and fecal production by other groups influence FPOM availability.

The extent to which feeding by gatherers qualitatively modifies FPOM resources depends on nutrient-microbial-detrital-animal linkages. Despite their abundance and importance to higher trophic levels, the functional role of gatherers is poorly known. In contrast, algal-grazer interactions are tightly coupled, as grazers influence both standing crop and rate of renewal of their algal resource. However, filterers, with the exception of microfilterers in some localities, have minimal quantitative influence on their resources and on resource renewal, but they exert their strongest effect on seston quality. Filterers also link suspended particles and FPOM supply to gatherers.

Available evidence suggests the impact of invertebrate predators on their prey is probably at least as great, if not greater than, that of vertebrate predators in many streams. In addition to direct mortality, their impact includes nonlethal effects on prey feeding activities, growth rates, fecundity, and behavior. The long coevolution of invertebrate and vertebrate predators and their prey, coupled with the complex mosaic of stream habitats, demands that we use complex and innovative approaches to understand predator impact on benthic communities.

The many roles performed by macroinvertebrates in streams underscores the importance of their conservation. Macroinvertebrates have served as valuable indicators of degradation of streams, and as increasing demands are placed on our water resources, their value in assessments of these impacts will increase.

Literature Cited

1. Abel PD. 1989. *Water Pollution Biology*. Chichester: Ellis Horwood
2. Allan JD. 1983. Predator-prey relationships in streams. See Ref. 9a, pp. 191–229
3. Allan JD. 1995. *Stream Ecology: Struc-* *ture and Function of Running Waters*. London: Chapman & Hall
4. Allen KR. 1951. The Horokiwi stream. A study of a trout population. *NZ Dep. Fish. Bull.* 10:1–231
5. Allen KR. 1952. A New Zealand trout

stream: some facts and figures. *NZ Dep. Fish. Bull.* 10A:1–70

6. Anderson NH, Sedell JR. 1979. Detritus processing by macroinvertebrates in stream ecosystems. *Annu. Rev. Entomol.* 24:351–77

7. Anderson NH, Steedman RJ, Dudley T. 1984. Patterns of exploitation by stream invertebrates of wood debris (xylophagy). *Verh. Int. Ver. Limnol.* 22:1847–52

8. Angermeier PL, Karr JR. 1994. Biological integrity versus biological diversity as policy directives. *BioScience* 44:690–97

9. Baker JA, Killgore KJ, Kasul RL. 1991. Aquatic habitats and fish communities in the lower Mississippi River. *Rev. Aquat. Sci.* 3:313–56

9a. Barnes JR, Minshall GW, eds. 1983. *Stream Ecology: Application and Testing of General Ecological Theory.* New York: Plenum

10. Benke AC. 1993. Concepts and patterns of invertebrate production in running waters. *Verh. Int. Ver. Limnol.* 25:15–38

11. Benke AC, Henry RL, Gillespie DM, Hunter RJ. 1985. Importance of snag habitat for animal production in Southeastern streams. *Fisheries* 10:8–13

12. Benke AC, Van Arsdall TC, Gillespie DM, Parrish FK. 1984. Invertebrate productivity in a subtropical blackwater river: the importance of habitat and life history. *Ecol. Monogr.* 54:25–63

13. Benke AC, Wallace JB. 1980. Trophic basis of production among net-spinning caddisflies in a southern Appalachian stream. *Ecology* 61:108–18

14. Bergey EA, Resh VH. 1994. Effects of burrowing by a stream caddisfly on case-associated algae. *J. North Am. Benthol. Soc.* 13:379–90

15. Brussock PP, Brown AV, Dixon JC. 1985. Channel form and stream ecosystem models. *Wat. Res. Bull.* 21:859–66

16. Buckingham GR. 1984. Biological control of weeds by insects. *J. Ga. Entomol. Soc. 2nd Suppl.* 19:62–78

17. Butler MJ, Stein RA. 1985. An analysis of the mechanisms governing species replacements in crayfish. *Oecologia* 66:168–77

17a. Calow P, Petts GE, eds. 1992. *The Rivers Handbook: Hydrological and Ecological Principles.* Vol. 1. Oxford: Blackwell

18. Carpenter SR, Kitchell JF, Hodgson JR. 1985. Cascading trophic interactions and lake productivity. *BioScience* 35:634–39

19. Chapman DW. 1966. Food and space as regulators of salmonid populations in streams. *Am. Nat.* 100:345–57

20. Chapman RF. 1982. *The Insects: Structure and Function.* Cambridge: Harvard Univ. Press

21. Chew RM. 1974. Consumers as regulators of ecosystems: an alternative to energetics. *Ohio J. Sci.* 74:359–70

22. Chung K, Wallace JB, Grubaugh JW. 1993. The impact of insecticide treatment on abundance, biomass, and production of litterbag fauna in a headwater stream: a study of pretreatment, treatment, and recovery. *Limnologica* 28:93–106

23. Chutter FM. 1972. An empirical biotic index of the quality of water in South African streams and rivers. *Water Res.* 6:19–30

24. Cohen RRH, Dresler PV, Phillips EJP, Cory RL. 1984. The effect of the Asiatic clam, *Corbicula fluminea*, on phytoplankton of the Potomac River, Maryland. *Limnol. Oceanogr.* 29:170–80

25. Cooper SD, Walde SJ, Peckarsky BL. 1990. Prey exchange rates and the impact of predators on prey populations in streams. *Ecology* 71:1503–14

26. Creed RP. 1994. Direct and indirect effects of crayfish grazing in a stream community. *Ecology* 75:2091–103

27. Crowl TA, Covich AP. 1990. Predator-induced life history shifts in a freshwater snail. *Science* 247:949–51

28. Cudney MD, Wallace JB. 1980. Life cycles, microdistribution and production dynamics of six species of net-spinning caddisflies in a large southeastern (U.S.A.) river. *Holarct. Ecol.* 3:169–82

29. Cuffney TF, Wallace JB, Lugthart GJ. 1990. Experimental evidence quantifying the role of benthic invertebrates in organic matter dynamics of headwater streams. *Freshw. Biol.* 23:281–99

30. Cummins KW. 1974. Structure and function of stream ecosystems. *BioScience* 24:631–41

31. Cummins KW. 1986. The functional role of black flies in stream ecosystems. In *Black Flies: Ecology, Population Management, and Annotated World List,* ed. KC Kim, RW Merritt, pp. 1–10. University Park, PA: Penn. State Univ. Press

32. Cummins KW, Klug MJ. 1979. Feeding ecology of stream invertebrates. *Annu. Rev. Ecol. Syst.* 10:147–72

33. Cummins KW, Petersen RC, Howard FO, Wuycheck JC, Holt VI. 1973. The utilization of leaf litter by stream detritivores. *Ecology* 54:336–45

34. Cushing CE, Minshall GW, Newbold JD. 1993. Transport dynamics of fine

particulate organic matter in two Idaho streams. *Limnol. Oceanogr.* 38:1101–15

35. Diamond JM. 1986. Effects of larval retreats of the caddisfly *Cheumatopsyche* on macroinvertebrate colonization in Piedmont, USA streams. *Oikos* 47: 13–18

36. Dobson M, Hildrew AG. 1992. A test of resource limitation among shredding detritivores in low order streams in southern England. *J. Anim. Ecol.* 61:69–77

37. Dudley T. 1992. Beneficial effects of herbivores on stream macroalgae via epiphyte removal. *Oikos* 65:121–27

38. Dudley T, Anderson NH. 1982. A survey of invertebrates associated with wood debris in aquatic habitats. *Melanderia* 39:1–21

39. Edwards RT, Meyer JL. 1987. Bacteria as a food source for black fly larvae in a blackwater river. *J. North Am. Benthol. Soc.* 6:241–50

40. Elwood JW, Nelson DJ. 1972. Periphyton production and grazing rates in a stream measured with ^{32}P material balance method. *Oikos* 23:295–303

41. Elwood JW, Newbold JD, Trimble AF, Stark RW. 1981. The limiting role of phosphorus in a woodland stream ecosystem: effects of P enrichment on leaf decomposition and primary producers. *Ecology* 62:146–58

42. Englund G. 1993. Interactions in a lake outlet stream community: direct and indirect effects of net-spinning caddis larvae. *Oikos* 66:431–38

43. Feminella JW, Power ME, Resh VH. 1989. Periphyton response to invertebrate grazing and riparian canopy in three northern California coastal streams. *Freshw. Biol.* 22:445–57

44. Feminella JW, Resh VH. 1990. Hydrologic influences, disturbance, and intraspecific competition in a stream caddisfly population. *Ecology* 7:2083–94

45. Feminella JW, Resh VH. 1991. Herbivorous caddisflies, macroalgae, and epilithic microalgae: dynamic interactions in a stream grazing system. *Oecologia* 87:247–56

46. Findlay S, Meyer JL, Smith PJ. 1986. Contribution of fungal biomass to the diet of a freshwater isopod (*Lirceus* sp.). *Freshw. Biol.* 16:377–85

47. Findlay SEG, Meyer JL, Smith PJ. 1984. Significance of bacterial biomass in the nutrition of a freshwater isopod (*Lirceus*). *Oecologia* 63:38–42

48. Fisher SG, Gray LG. 1983. Secondary production and organic matter processing by collector macroinvertebrates in a desert stream. *Ecology* 64:1217–24

49. Fisher SG, Likens GE. 1973. Energy flow in Bear Brook, New Hampshire: an integrative approach to stream ecosystem metabolism. *Ecol. Monogr.* 43:421–39

50. Fremling CR. 1960. Biology and possible control of nuisance caddisflies of the upper Mississippi River. *Res. Bull. Iowa Agric. Exp. Stn.* 348:856–79

51. Frissell CA, Liss WJ, Warren CE, Hurley MD. 1986. A hierarchical framework for stream habitat classification: viewing streams in a watershed context. *Environ. Manage.* 10:199–214

52. Fuller RL, Roelofs JR, Fry TJ. 1986. The importance of algae to stream invertebrates. *J. North Am. Benthol. Soc.* 5:290–94

53. Georgian T, Thorp JH. 1992. Effects of microhabitat selection on feeding rates of net-spinning caddisfly larvae. *Ecology* 73:229–40

54. Gerking SD. 1994. *Feeding Ecology of Fish.* San Diego: Academic

55. Golladay SW, Webster JR, Benfield EF. 1983. Factors affecting food utilization by a leaf shredding aquatic insect: leaf species and conditioning time. *Holarct. Ecol.* 6:157–62

56. Gregory SV. 1983. Plant-herbivore interactions in stream systems. See Ref. 9a, pp. 157–89

57. Grimm NB. 1988. Role of macroinvertebrates in nitrogen dynamics of a desert stream. *Ecology* 69:1884–93

58. Growns IO, Davis JA. 1994. Longitudinal changes in near-bed flows and macroinvertebrate communities in a western Australian stream. *J. North Am. Benthol. Soc.* 13:417–38

59. Hall CAS, Stanford JA, Hauer FR. 1992. The distribution and abundance of organisms as a consequence of energy balances along multiple environmental gradients. *Oikos* 65:377–90

60. Hall RO. 1995. The use of a stable carbon isotope addition to trace bacterial carbon through a stream food web. *J. North Am. Benthol. Soc.* 14:269–77

61. Hansen RA, Hart DD, Merz RR. 1991. Flow mediates predator-prey interactions between triclad flatworms and larval black flies. *Oikos* 60:187–96

62. Hart DD. 1985. Grazing insects mediate algal interactions in a stream benthic community. *Oikos* 44:40–46

63. Hart DD, Robinson CT. 1990. Resource limitation in a stream community: phosphorus enrichment effects of periphyton and grazers. *Ecology* 71:1494–502

64. Hawkins CP, Kershner JL, Bisson PA, Bryant MD, Decker LM, et al. 1993. A hierarchical approach to classifying

stream habitat features. *Fisheries* 18:3–12

65. Hawkins CP, Murphy ML, Anderson NH. 1982. Effects of canopy, substrate composition, and gradients on structure of macroinvertebrate communities in Cascade Range streams of Oregon. *Ecology* 62:387–97

66. Heard SD. 1994. Processing chain ecology: resource condition and interspecific interactions. *J. Anim. Ecol.* 63:451–64

67. Heard SD, Richardson JS. 1995. Shredder-collector facilitation in stream detrital food webs: Is there enough evidence? *Oikos* 72:359–66

68. Hemphill N. 1991. Disturbance and variation in competition between two stream insects. *Ecology* 72:864–72

69. Hershey AE, Hiltner AL, Hullar MAJ, Miller MC, Vestal JR, et al. 1988. Nutrient influence on a stream grazer: *Orthocladius* microcommunities repond to nutrient input. *Ecology* 69:1383–92

70. Hildrew AG. 1992. Food webs and species interactions. See Ref. 17a, pp. 309–30

71. Hill WR, Boston HL, Steinman AD. 1992. Grazers and nutrients simultaneously limit lotic primary productivity. *Can. J. Fish. Aquat. Sci.* 49:504–12

72. Hill WR, Harvey BC. 1990. Periphyton responses to higher trophic levels and light in a shaded stream. *Can. J. Fish. Aquat. Sci.* 47:2307–14

73. Hill WR, Knight AW. 1987. Experimental analysis of the grazing interaction between a mayfly and stream algae. *Ecology* 68:1955–65

74. Hill WR, Knight AW. 1988. Concurrent grazing effects of two stream insects on periphyton. *Limnol. Oceanogr.* 33:15–26

75. Hilsenhoff WL. 1987. An improved biotic index of organic stream pollution. *Great Lakes Entomol.* 20:31–39

76. Hobbs HH, Jass JP, Huner JV. 1989. A review of global crayfish introductions with particular emphasis on two North American species (Decapoda: Cambaridae). *Crustaceana* 56:299–316

77. Hunter MD, Price PW. 1992. Playing chutes and ladders: herterogeneity and the relative roles of bottom-up and top-down forces in natural communities. *Ecology* 73:724–32

78. Huryn AD, Wallace JB. 1987. Local geomorphology as a determinant of macrofaunal production in a mountain stream. *Ecology* 68:1932–42

79. Huryn AD, Wallace JB. 1988. Community structure of Trichoptera in a mountain stream: spatial patterns of production and functional organization. *Freshw. Biol.* 20:141–55

80. Hynes HBN. 1970. *The Ecology of Running Waters.* Toronto: Univ. Toronto Press

81. Iversen TM. 1979. Laboratory energetics of larvae of *Sericostoma personatum* (Trichoptera). *Holarct. Ecol.* 2:1–5

82. Jacobsen D. 1993. Trichopteran larvae as consumers of submerged angiosperms in running waters. *Oikos* 67:379–83

83. Jacobsen D, Sand-Jensen K. 1994. Invertebrate herbivory on the submerged macrophyte *Potamogeton perfoliatus* in a Danish stream. *Freshw. Biol.* 31:43–52

84. Jacobsen D, Sand-Jensen K. 1994. Growth and energetics of a trichopteran larva feeding on fresh submerged and terrestrial plants. *Oecologia* 97:412–18

85. Jacoby JM. 1987. Alterations in periphyton characteristics due to grazing in a Cascade foothill stream. *Freshw. Biol.* 18:495–508

86. Johnson RK, Wiederholm T, Rosenberg DM. 1993. Freshwater biomonitoring using individual organisms, populations, and species assemblages of benthic macroinvertebrates. See Ref. 161, pp. 40–158

87. Jones CG, Lawton JH, Shachak M. 1994. Organisms as ecosystem engineers. *Oikos* 69:373–86

88. Karr JR. 1991. Biological intergrity: a long-neglected aspect of water resource management. *Ecol. Appl.* 1:66–84

89. Karr JR. 1993. Defining and assessing ecological integrity: beyond water quality. *Environ. Toxicol. Chem.* 12:1521–31

90. Kido MH, Ha P, Kinzie RA. 1993. Insect introductions and diet changes in an endemic Hawaiian amphidromous Goby, *Awaous stamineus* (Pisces: Gobiidae). *Pac. Sci.* 47:43–50

91. Lamberti GA, Ashkenas LR, Gregory SV, Steinman AD. 1987. Effects of three herbivores on periphyton communities in laboratory streams. *J. North Am. Benthol. Soc.* 6:92–104

92. Lamberti GA, Gregory SV, Ashkenas LR, Steinman AD, McIntire CD. 1989. Productive capacity of periphyton as a determinant of plant-animal interactions in streams. *Ecology* 70:1840–56

93. Lamberti GA, Gregory SV, Hawkins CP, Wildman RC, Ashkenas LR, Denicola DM. 1992. Plant-herbivore interactions in streams near Mount St Helens. *Freshw. Biol.* 27:237–47

94. Lamberti GA, Moore JW. 1984. Aquatic insects as primary consumers. See Ref. 155a, pp. 164–95

136   WALLACE & WEBSTER

95. Lamberti GA, Resh VH. 1983. Stream periphyton and insect herbivores: an experimental study of grazing by a caddisfly population. *Ecology* 64:1124–35
96. Lancaster J, Hildrew AG. 1993. Flow refugia and the microdistribution of lotic macroinvertebrates. *J. North Am. Benthol. Soc.* 12:385–93
97. Lauritsen DD. 1986. Filter-feeding in *Corbicula fluminea* and its effect on seston removal. *J. North Am. Benthol. Soc.* 5:165–72
98. Lawton J. 1991. Are species useful? *Oikos* 62:3–4
99. Lenat DR. 1993. A biotic index for the southeastern United States: derivation and list of tolerance values, with criteria for assigning water-quality ratings. *J. North Am. Benthol. Soc.* 12:279–90
100. Lock MA. 1981. River epilithon—a light and energy transducer. In *Perspectives in Running Water Ecology*, ed. MA Lock, DD Williams, pp. 3–40. New York: Plenum
101. Lodge DM. 1991. Herbivory on freshwater macrophytes. *Aquat. Bot.* 41:195–224
102. Loeb SL, Spacie A, eds. 1994. *Biological Monitoring of Aquatic Systems.* Boca Raton: Lewis
103. Ludyanskiy ML, McDonald D, MacNeill D. 1993. Impact of the zebra mussel, a bivalve invader. *BioScience* 43:533–44
104. Lugthart GJ, Wallace JB. 1992. Effects of disturbance on benthic functional structure and production in mountain streams. *J. North Am. Benthol. Soc.* 11:138–64
105. Mackay RJ, Waters TF. 1986. Effects of small impoundments on hydropsychid caddisfly production in Valley Creek, Minnesota. *Ecology* 67:1680–86
106. Malmqvist B. 1993. Interactions in stream leaf packs: effects of a stonefly predator on detritivores and organic matter processing. *Oikos* 66:454–62
107. Mather ME, Stein RA. 1993. Direct and indirect effects of fish predation on the replacement of a native crayfish by an invading congener. *Can. J. Fish. Aquat. Sci.* 50:1279–88
108. McAuliffe JR. 1983. Competition, colonization patterns, and disturbance in stream benthic communities. See Ref. 9a, pp. 137–56
109. McCormick PV, Stevenson JV. 1989. Effects of snail grazing on benthic community structure in different nutrient environments. *J. North Am. Benthol. Soc.* 8:162–72
110. McCormick PV, Stevenson RJ. 1991.

Grazer control of nutrient availability in the periphyton. *Oecologia* 86:287–91
111. McDiffett WF. 1970. The transformation of energy by a stream detritivore, *Pteronarcys scotti* (Plecoptera). *Ecology* 51:975–88
112. McNaughton SJ. 1986. On plants and herbivores. *Am. Nat.* 128:765–70
113. Merritt RW, Cummins KW, eds. 1984. *An Introduction to the Aquatic Insects of North America.* Dubuque: Kendall/Hunt
114. Merritt RW, Cummins KW, Burton TM. 1984. The role of aquatic insects in the processing and cycling of nutrients. See Ref. 155a, pp. 134–63
115. Meyer JL. 1990. A blackwater perspective on riverine ecosystems. *BioScience* 40:643–51
116. Meyer JL. 1994. The microbial loop in flowing waters. *Microb. Ecol.* 28:195–99
117. Meyer JL, McDowell WH, Bott TL, Elwood JW, Ishizaki C, et al. 1988. Elemental dynamics in streams. *J. North Am. Benthol. Soc.* 7:410–32
118. Meyer JL, O'Hop J. 1983. Leaf-shredding insects as a source of dissolved organic carbon in headwater streams. *Am. Midl. Nat.* 109: 175–83
119. Mills LS, Soule ME, Doak DF. 1993. The keystone-species concept in ecology and conservation. *BioScience* 43: 219–24
120. Minshall GW. 1984. Aquatic insect-substratum relationships. See Ref. 155a, pp. 358–400
121. Minshall GW, Petersen RC, Cummins KW, Bott TL, Sedell JR, et al. 1983. Interbiome comparison of stream ecosystem dynamics. *Ecol. Monogr.* 53:1–25
122. Morin A, Back C, Chalifour A, Boisvert J, Peters RH. 1988. Effect of black fly ingestion and assimilation on seston transport in a Quebec lake outlet. *Can. J. Fish. Aquat. Sci.* 45:705–14
123. Muirhead-Thomson RC. 1987. *Pesticide Impact on Stream Fauna: With Special Reference to Macroinvertebrates.* Cambridge: Cambridge Univ. Press
124. Mulholland PJ, Elwood JW, Newbold JD, Ferren LA. 1985. Effect of a leaf-shredding invertebrate on organic matter dynamics and phosphorus spiralling in heterotrophic laboratory streams. *Oecologia* 66:199–206
125. Mulholland PJ, Newbold JD, Elwood JW, Hom CL. 1983. The effect of grazing intensity on phosphorus spiralling in autotrophic streams. *Oecologia* 58: 358–66
126. Mulholland PJ, Steinman AD, Marzolf

ER, Hart DR, DeAngelis DL. 1994. Effect of periphyton biomass on hydraulic characteristics and nutrient cycling in streams. *Oecologia* 98:40–47

127. Mulholland PJ, Steinman AD, Palumbo AV, DeAngelis DL. 1991. Influence of nutrients and grazing on the response of stream periphyton communities to a scour disturbance. *J. North Am. Benthol. Soc.* 10:127–42

128. Mulholland PJ, Steinman AD, Palumbo AV, Elwood JW, Kirschtel DB. 1991. Role of nutrient cycling and herbivory in regulating periphyton communities in laboratory streams. *Ecology* 72:966–82

129. Mundie JH, Simpson KS, Perrin CJ. 1991. Responses of stream periphyton and benthic insects to increases in dissolved inorganic phosphorus in a mesocosm. *Can. J. Fish. Aquat. Sci.* 48:2061–72

130. Murphy ML. 1984. Primary production and grazing in freshwater and intertidal reaches of a coastal stream, southeast Alaska. *Limnol. Oceanogr.* 29:805–15

131. Naiman RJ. 1988. Animal influences on ecosystem dynamics. *BioScience* 38:750–52

132. Nalepa TF, Schloesser DW, eds. 1993. *Zebra Mussels: Biology, Impacts, and Control.* Boca Raton: Lewis

133. National Transportation Safety Board. 1989. *Collapse of the S.R. 675 Bridge Spans over the Pocomoke River near Pocomoke City, Maryland. PB89-916205.* Washington, DC: Natl. Transp. Safety Board/HAR-89/04.1–80

134. Newbold JD, O'Neill RV, Elwood JW, Van Winkle W. 1982. Nutrient spiralling in streams: implications for nutrient limitation and invertebrate activity. *Am. Nat.* 120:628–52

135. Newman RM. 1991. Herbivory and detritivory on freshwater macrophytes by invertebrates: a review. *J. North Am. Benthol. Soc.* 10:89–114

136. Oberndorfer RY, McArthur JV, Barnes JR. 1984. The effect of invertebrate predators on leaf litter processing in an alpine stream. *Ecology* 65:1325–31

137. O'Connor NA. 1993. Resource enhancement of grazing mayfly nymphs by retreat-building caddisfly larvae in a sandbed stream. *Aust. J. Mar. Freshwat. Res.* 44:353–62

138. Paine RT. 1966. Food web complexity and species diversity. *Am. Nat.* 100:65–75

139. Parker CR, Voshell JR. 1983. Production of filter-feeding Trichoptera in an impounded and free flowing river. *Can. J. Zool.* 61:70–87

140. Peckarsky BL, Cowan CA, Penton MA,

Anderson C. 1993. Sublethal consequences of stream-dwelling predatory stoneflies on mayfly growth and fecundity. *Ecology* 74:1836–46

141. Peduzzi P, Herndl GJ. 1991. Mucus trails in the rocky intertidal: a highly active microenvironment. *Mar. Ecol. Prog. Ser.* 75:267–74

142. Perlmutter DG, Meyer JL. 1991. The impact of a stream-dwelling harpacticoid copepod upon detritally associated bacteria. *Ecology* 72:2170–80

143. Petersen LB-M. 1985. Food preferences in three species of *Hydropsyche* (Trichoptera). *Verh. Int. Ver. Limnol.* 22:3270–74

144. Petersen LB-M. 1989. Resource utilization of coexisting species of Hydropsychidae (Trichoptera). *Arch. Hydrobiol. Suppl.* 83:83–119

145. Petersen RC, Cummins KW. 1974. Leaf processing in a woodland stream. *Freshw. Biol.* 4:343–68

146. Peterson BJ, Deegan L, Helfrich J, Hobbie JE, Hullar M, et al. 1993. Biological responses of a tundra river to fertilization. *Ecology* 74:653–72

147. Peterson BJ, Hobbie JE, Hershey A, Lock M, Ford T, et al. 1985. Transformation of a tundra river from heterotrophy to autotrophy by addition of phosphorus. *Science* 229:1383–86

148. Phelps HL. 1994. The Asiatic clam (*Corbicula fluminea*) invasion and system-level ecological changes in the Potomac River estuary near Washington, DC. *Estuaries* 17:614–21

149. Plafkin JL, Barbour MT, Porter KD, Gross SK, Hughes RM. 1989. *Rapid Bioassessment Protocols for Use in Streams and Rivers: Benthic Macroinvertebrates and Fish. EPA/444/4-89-001.* Washington: US Environ. Prot. Agency

150. Power ME. 1990. Effects of fish in river food webs. *Science* 250:811–14

151. Power ME. 1992. Habitat heterogeneity and the functional significance of fish in river food webs. *Ecology* 73:1675–88

152. Power ME, Mathews WJ, Stewart AJ. 1985. Grazing minnows, piscivorous bass, and stream algae: dynamics of a strong interaction. *Ecology* 66:1448–56

153. Pringle CM. 1985. Effects of chironomid (Insecta: Diptera) tube-building activities on stream diatom communities. *J. Phycol.* 21:185–94

154. Pringle CM, Blake GA, Covich AP, Buzby KM, Finley A. 1993. Effects of omnivorous shrimp in a montane tropical stream: sediment removal, disturbance of sessile invertebrates and en-

hancement of understory algal biomass. *Oecologia* 93:1–11

155. Pringle CM, Naiman RJ, Bretschko G, Karr JR, Oswood MW, et al. 1988. Patch dynamics in lotic ecosystems: the stream as a mosaic. *J. North Am. Benthol. Soc.* 7:503–24

155a. Resh VH, Rosenberg DM, eds. 1984. *The Ecology of Aquatic Insects.* New York: Praeger

156. Reynolds CS. 1992. Algae. See Ref. 17a, pp. 195–215

157. Richardson JS. 1991. Seasonal food limitation of detritivores in a montane stream: an experimental test. *Ecology* 72:873–87

158. Richardson JS. 1993. Limits to productivity in streams: evidence from studies of macroinvertebrates. *Can. Spec. Publ. Fish. Aquat. Sci.* 118:9–15

159. Richardson JS, Mackay RJ. 1991. Lake outlets and the distribution of filter feeders: an assessment of hypotheses. *Oikos* 62:370–80

160. Rosemond AD, Mulholland PJ, Elwood JW. 1993. Top-down and bottom-up control of stream periphyton: effects of nutrients and herbivores. *Ecology* 74:1264–80

161. Rosenberg DM, Resh VH, eds. 1993. *Freshwater Biomonitoring and Benthic Macroinvertebrates.* New York: Chapman & Hall

162. Sallenave RM, Day KE, Kreutzweiser DP. 1994. The role of grazers and shredders in the retention and downstream transport of a PCB in lotic environments. *Environ. Toxicol. Chem.* 13:1843–847

163. Sand-Jensen K, Madsen TV. 1989. Invertebrates graze submerged rooted macrophytes in streams. *Oikos* 55:420–23

164. Schofield K, Townsend CR, Hildrew AG. 1988. Predation and the prey community of a headwater stream. *Freshw. Biol.* 20:85–95

165. Scrimgeour GJ, Culp JM. 1994. Feeding while evading predators by a lotic mayfly: linking short-term foraging behaviors to long-term fitness consequences. *Oecologia* 100:128–34

166. Shepard RB, Minshall GW. 1984. Role of benthic insect feces in a Rocky Mountain stream: fecal production and support of consumer growth. *Holarct. Ecol.* 7:119–27

167. Short RA, Maslin PE. 1977. Processing of leaf litter by a stream detritivore: effect on nutrient availability to collectors. *Ecology* 58:935–38

168. Sih A, Crowley P, McPeek M, Petranka J, Strohmeier K. 1985. Predation, com-

petition, and prey communities: a review of field experiments. *Annu. Rev. Ecol. Syst.* 16:269–311

169. Sih A, Wooster D. 1994. Prey behavior, prey dispersal, and predator impacts on stream prey. *Ecology* 75:1199–207

170. Sinsabaugh RL, Linkins AE, Benfield EF. 1985. Cellulose digestion and assimilation by three leaf-shredding insects. *Ecology* 66:1464–71

171. Smock LA, Gilinsky E, Stoneburner DL. 1985. Macroinvertebrate production in a southeastern United States blackwater stream. *Ecology* 66:1491–503

172. Smock LA, Gladden JE, Riekenberg JL, Smith LC, Black CR. 1992. Lotic macroinvertebrate production in three dimensions: channel surface, hyporheic, and floodplain environments. *Ecology* 73:876–88

173. Smock LA, Metzler GM, Gladden JE. 1989. The role of organic debris dams in the structuring and functioning of low-gradient headwater streams. *Ecology* 70:764–75

174. Soluk DA. 1993. Multiple predator effects: predicting combined functional response of stream fish and invertebrate predators. *Ecology* 74:219–25

175. Speaker RW, Moore KW, Gregory SV. 1984. Analysis of the process of retention of organic matter in stream ecosystems. *Verh. Int. Ver. Limnol.* 22:1835–41

176. Steinman AD, Kirschtel D, Mulholland PJ. 1991. Interactive effects of nutrient reduction and herbivory on biomass, taxonomic composition, and phosphorus uptake in lotic periphyton communities. *Can. J. Fish. Aquat. Sci.* 48:1951–59

177. Steinman AD, McIntire CD, Gregory SV, Lamberti GA, Ashkenas LR. 1987. Effects of herbivore type and density on taxonomic structure and physiognomy of algal assemblages in laboratory streams. *J. North Am. Benthol. Soc.* 6:175–88

178. Stewart AJ, Hill WR, Boston HL. 1993. Grazers, periphyton and toxicant movement in streams. *Environ. Toxicol. Chem.* 12:955–57

179. Suter GW. 1993. A critique of ecosystem health concepts and indexes. *Environ. Toxicol. Chem.* 12:1533–39

180. Thorp JH. 1986. Two distinct roles for predators in freshwater assemblages. *Oikos* 47:75–82

181. Thorp JH, Covich AP, eds. 1991. *Ecology and Classification of North American Freshwater Invertebrates.* San Diego: Academic

182. Vannote RL, Minshall GW, Cummins

KW, Sedell JR, Cushing CE. 1980. The river continuum concept. *Can. J. Fish. Aquat. Sci.* 37:130–37

183. Vaughn CC, Gelwick FP, Matthews WJ. 1993. Effects of algivorous minnows on production of grazing stream invertebrates. *Oikos* 66:119–28

184. Voshell JR. 1985. *Trophic Basis of Production for Macroinvertebrates in the New River Below Bluestone Dam.* Blacksburg: Va. Polytechnic Inst. State Univ.

185. Wagner R. 1991. The influence of the diel activity pattern of the larvae of *Sericostoma personatum* (Kirby and Spence) (Trichoptera) on organic matter distribution in stream sediments: a laboratory study. *Hydrobiologia* 224: 65–70

186. Wallace JB, Cuffney TF, Webster JR, Lugthart GJ, Chung K, Goldowitz BS. 1991. A five-year study of export of fine organic particles from headwater streams: effects of season, extreme discharges, and invertebrate manipulation. *Limnol. Oceanog.* 36:670–82

187. Wallace JB, Grubaugh JW, Whiles MR. 1995. Biotic indices and stream ecosystem processes: results from an experimental study. *Ecol. Appl.* In press

188. Wallace JB, Gurtz ME. 1986. Response of *Baetis* mayflies (Ephemeroptera) to catchment logging. *Am. Midl. Nat.* 115: 25–41

189. Wallace JB, Merritt RW. 1980. Filter-feeding ecology of aquatic insects. *Annu. Rev. Entomol.* 25:103–32

190. Wallace JB, O'Hop J. 1985. Life on a fast pad: water-lily leaf beetle impact on water-lilies. *Ecology* 66:1534–44

191. Wallace JB, Vogel DS, Cuffney TF. 1986. Recovery of a headwater stream from an insecticide-induced community disturbance. *J. North Am. Benthol. Soc.* 5:115–26

192. Wallace JB, Webster JR, Cuffney TF. 1982. Stream detritus dynamics: regulation by invertebrate consumers. *Oecologia* 53:197–200

193. Wallace JB, Webster JR, Woodall WR. 1977. The role of filter-feeders in flowing waters. *Arch. Hydrobiol.* 79: 506–32

194. Warren CE, Wales JH, Davis GE,

Doudoroff P. 1964. Trout production in an experimental stream enriched with sucrose. *J. Wildl. Manage.* 28:617–60

195. Waters TF. 1988. Fish production-benthos relationships in trout streams. *Pol. Arch. Hydrobiol.* 35:545–61

196. Waters TF. 1993. Dynamics in stream ecology. *Can. Spec. Publ. Fish. Aquat. Sci.* 118:1–8

197. Webster JR. 1983. The role of benthic macroinvertebrates in detritus dynamics of streams: a computer simulation. *Ecol. Monogr.* 53:383–404

198. Webster JR, Covich AP, Tank JL, Crocket TV. 1994. Retention of coarse organic particles in streams in the southern Appalachian Mountains. *J. North Am. Benthol. Soc.* 13:140–50

199. Webster JR, Patten BC. 1979. Effects of watershed perturbation on stream potassium and calcium dynamics. *Ecol. Monogr.* 49:51–72

200. Webster JR, Wallace JB, Benfield EF. 1995. Organic processes in streams of the eastern United States. In *River and Stream Ecosystems,* ed. CE Cushing, GW Minshall, KW Cummins, pp. 117–87. Amsterdam: Elsevier

201. Williams DD, Feltmate BW. 1992. *Aquatic Insects.* Wallingford: CAB Int.

202. Wilzbach MA, Cummins KW, Hall JD. 1986. Influence of habitat manipulations on interactions between cutthroat trout and invertebrate drift. *Ecology* 67:898–911

203. Winterbourn MJ. 1990. Interactions among nutrients, algae and invertebrates in a New Zealand mountain stream. *Freshw. Biol.* 23:463–74

204. Wooster D. 1994. Predator impacts on stream benthic prey. *Oecologia* 99:7–15

205. Wotton RS. 1987. Lake outlet blackflies—the dynamics of filter feeders at very high population densities. *Holarct. Ecol.* 10:65–72

206. Wotton RS. 1994. *The Biology of Particles in Aquatic Systems.* Boca Raton: Lewis

207. Yasuno M, Fukushima S, Hasegawa J, Shioyama F, Hatakeyama S. 1982. Changes in the benthic fauna and flora after application of temephos to a stream on Mt. Tsukuba. *Hydrobiologia* 89:205–14

Annu. Rev. Entomol. 1996. 41:141–61
Copyright © 1996 by Annual Reviews Inc. All rights reserved

EVOLUTION OF TICKS

J. S. H. Klompen and W. C. Black IV
Department of Microbiology, Colorado State University, Fort Collins, Colorado 80523-1677

J. E. Keirans and J. H. Oliver, Jr.
Institute of Arthropodology and Parasitology, Georgia Southern University, Landrum Box 8056, Statesboro, Georgia 30460-8056

KEY WORDS: Ixodida, molecular systematics, ecological specificity, host specificity, biogeography

ABSTRACT

Evolutionary patterns in ticks have traditionally been cast in terms of host associations. Largely untested assumptions of cospeciation and observations of current host associations are used to estimate the age of different taxa. Several recent phylogenetic studies of supraspecific relationships in ticks, based on both morphological and DNA-sequence data, allow the first rigorous testing of these assumptions. Reanalysis of patterns of tick-host associations suggests that the perception of host specificity in ticks may be an artifact of incomplete sampling. An analysis of tick-host and -habitat associations and biogeographical patterns, in the context of the newly derived phylogenies, suggests that much of the existing host-association patterns may be explained as artifacts of biogeography and ecological specificity.

PERSPECTIVES AND OVERVIEW

Ticks are obligate, nonpermanent ectoparasites of terrestrial vertebrates. All species are exclusively hematophagous in all feeding stages. They have considerable medical-veterinary importance (58, 59) because of their ability to harm their hosts through direct action (tick paralysis, exsanguination) or by vectoring disease organisms. As a result, many aspects of tick biology, including tick ecology, physiology, and reproduction, have been studied in considerable detail. In fact, more data are available for ticks than for any other group of Acari. Excellent and relatively recent reviews include studies by Oliver (49) and Sonenshine (58, 59). This review focuses on tick systematics and evolu-

141

tion, an area that, until recently, had not undergone rigorous and testable analyses. The traditional approach to tick evolution has been to stress the importance of the hosts, often suggesting host specificity as the main factor driving tick evolution. This hypothesis proposes that ticks associated with primitive hosts must be primitive themselves and that each tick lineage arose when the oldest taxon of its current host arose. In general, it follows the tenets of the old rules of parasitology, in which parasites are depicted as dependent on their hosts in nearly all aspects of their biology.

During the past decades, the approach to the study of parasite-host associations has changed drastically because of the development of more rigorous analytical methods. The most important of these methods is the use of phylogenetic systematics to produce independently derived parasite phylogenies (6, 7, 42). Such phylogenies allow rigorous tests of adaptive hypotheses (13, 14) as well as hypotheses of cospeciation or vicariance (7, 8, 50, 51). The new picture of parasites no longer depicts them as degenerate forms, largely dependent on their hosts in nearly all aspects of their evolution, but as independently evolving lineages showing various patterns of association with their hosts and their off-host environment (for nonpermanent parasites). This new view challenges workers in this field to question the relative influence of ecological factors on the evolution of such parasites as ticks.

Most published studies using arthropod parasites deal with permanent parasites, such as lice (21, 22, 39, 40) or psoroptid mites (48). However, the considerable diversity of nonpermanent arthropod parasites raises the question as to what role the different components of parasite ecology, the on- and off-host phases, play in the evolution of such arthropods. One might expect different selection regimes in the two phases of these parasites' life cycles, selection pressures that are not necessarily compatible. Ticks are quite interesting in this respect because, unlike other groups such as fleas, mosquitoes, and chiggers, each active life stage experiences the dual pressures of survival on and off a host.

Since the last review of tick biology and evolution in an *Annual Review* series (49), two major developments have provided new data that allow us to revisit the problems of tick systematics and evolution. First, the introduction of molecular techniques has not only provided new data sets for phylogenetic analyses but, through the use of novel techniques such as single-strand conformation polymorphism (23), is also allowing studies of population genetics (45). Second, the use of phylogenetic methods in tick systematics (5, 31, 32, 34, 54; JSH Klompen, JH Oliver Jr, JE Keirans & PJ Homsher, in preparation) has, for the first time, generated independently derived and testable hypotheses on tick relationships. These phylogenies are being used to reexamine the current hypotheses of tick evolution.

LIFE HISTORY

Ticks comprise the suborder Ixodida of the order Parasitiformes, one of the two orders of mites (Acari) (47). They are unique among Acari in having a large body size (2–30 mm), specialized mouthparts (hypostome), and a peculiar highly specialized aggregation of sensory structures on tarsus I (Haller's organ) (2, 36). The group includes about 820 species, divided into three families: the Argasidae (soft ticks), the Ixodidae (hard ticks), and the Nuttalliellidae (28, 49). The family Nuttalliellidae is monotypic and shares characters of both Argasidae and Ixodidae, in addition to having many derived features (30).

Nymphal and adult Argasidae are characterized by a ventral position of the capitulum, a highly sculptured integument, and the absence of a dorsal shield or scutum. Their developmental stages consist of an egg, a hexapod larva, a variable number of octopod nymphal instars, and the adult. The variable number of nymphal instars in Argasidae is unique among known Acari. With few exceptions, blood meals are required for the development of immature stages and for adult reproduction. Larvae may need several days to feed, but most nymphs and adults become replete in a few hours or less (known as rapid feeding). Given a life span that may last one or more years, the time spent on the host is relatively quite short. Females may feed repeatedly, depositing a small batch of eggs after each feeding session.

Most Argasidae are nest, burrow, or roost parasites, and mating usually takes places off the host, in the habitat. Because nest habitats are often used by only one individual host, or by several individuals of the same species, many argasids will feed on the same host individual or species throughout their life. Even so, the overall host range of the family spans widely diverse terrestrial vertebrates, including turtles, squamates (i.e. lizards, snakes), birds, and mammals. A few species may also feed on amphibians. Their geographic range includes the tropical and temperate parts of all major biogeographical regions.

Members of the family Ixodidae differ from the soft ticks in that they retain an anterior capitulum; a simple, striate integument; and a scutum. They also have only one nymphal instar. In most species, all stages feed once, after which they drop off the host. Compared with the Argasidae, Ixodidae of all stages feed to repletion slowly (several days). Despite this so-called slow feeding behavior, ixodid ticks may still spend 94–97% of their life off-host (44). Replete adult females produce a single, large clutch of eggs and then die.

Although some lineages are associated with nests, burrows, and so on, most taxa are not confined to such sites. As a result, the hosts of the different stages may be the same individual, different individuals of the same species, or members of different species. The last is frequently the case, as the different stages of many ixodid species commonly appear on hosts of different sizes.

The overall host range is similar to that of the Argasidae, whereas the geographic range is even more extensive, including the colder temperate and arctic regions (e.g. some sea bird ticks occur on islands of the coast of Antarctica) (26).

EVOLUTION, TRADITIONAL VIEWS

Host Specificity and the Origin of Ticks

HOOGSTRAAL HYPOTHESES   Most of the studies on tick evolution done since the 1950s have been heavily influenced by the work of Harry Hoogstraal. He published over 400 papers in nearly four decades of work, and much of our understanding of tick ecology, habitat preferences, and host associations comes from the tremendous collections and observations he made. Because his views have been so influential in shaping tick research in much of the western world (Soviet workers followed different paradigms), we outline Hoogstraal's ideas on tick evolution below.

Hoogstraal and earlier workers suggested that various structural modifications of the mouthparts and coxae were associated with specialization for particular hosts (27). They noted that changes in these characters in different stages appeared to be correlated with the different host species parasitized by each stage and concluded that adaptation to the hosts played a major role in tick evolution. Moreover, this adaptation was assumed to lead to host specificity and eventually to parallel evolution between ticks and their hosts. This view became the prevailing theme in all of Hoogstraal's writing on tick evolution. He developed a classification system that placed each tick taxon into one of six categories of host specificity, ranging from strict-total, in which both adults and immatures are specific for the same limited host group, to nonparticular, in which adults and immatures are catholic in host acceptability (26). Several of his papers involved extensive categorization of species from various taxonomic groups into these six divisions (25–27). In the strict-total category, he included all Argasidae; all species in the ixodid genera *Aponomma, Boophilus,* and *Margaropus;* and many species in the remaining ixodid genera. The nonparticular category contained only ixodid ticks but included representatives of all major ixodid genera. Although Hoogstraal recognized that some ticks have a broad host range, he felt that these situations were artificial, arising through human intervention in the host-parasite relationship that had predominated during tick speciation. In Hoogstraal's view, ticks clearly are highly host specific, which constituted prima facie evidence of cospeciation. These conclusions regarding the importance of host specificity in tick evolution are frequently quoted in medical entomology textbooks and in tick taxonomic literature.

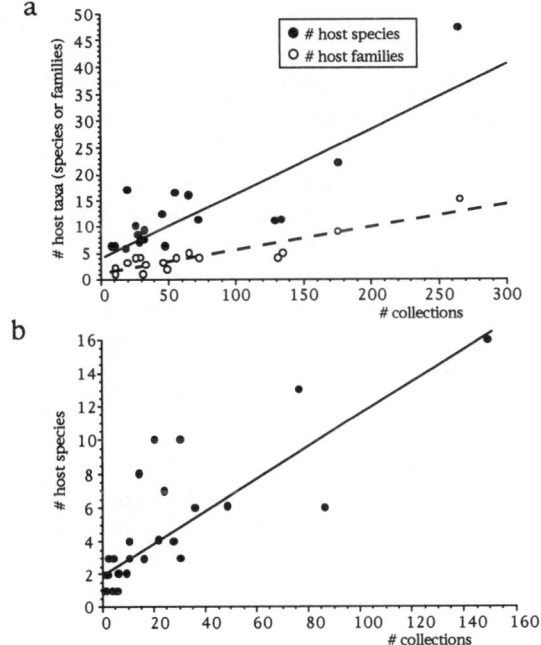

Figure 1 Regression analysis of host specificity in ticks. (*a*) *Ixodes* species. Number of host species = 0.120 (number of collections) + 4.52 (R^2 = 0.70, P < 0.0001). Number of host families = 0.044 (number of collections) + 1.12 (R^2 = 0.83, P < 0.0001). (*b*) *Carios* species. Number of host families = 0.099 (number of collections) + 1.95 (R^2 = 0.68, P < 0.0001).

Hoogstraal's views of the long-term evolution of ticks (as summarized in Figure 2) combine a scenario of host specificity and broad cospeciation with the assumption that ticks are a group of very ancient derivation (24–27). In particular he suggested that the ancestral ticks, resembling the present-day Argasidae, arose in the late Paleozoic or early Mesozoic [approximately 225 million years ago (mya)], in warm humid climates. These ticks had a three-host life cycle, and hosts in these climates were available throughout the year. These ancestral hosts were assumed to be "large, smooth-skinned reptiles" (24, p. 3). Ticks subsequently radiated along two pathways, the Argasidae and the Ixodidae. The Prostriata was among the earliest lineages to differentiate from those ancestral forms. Within the Metastriata, Hoogstraal proposed that the Amblyomminae originated on reptiles in the late Permian and radiated on those during the Triassic and Jurassic. The Haemaphysalinae and Hyalomminae also arose on reptiles in, respectively, the Triassic and in the late Cretaceous. The Rhipicephalinae did not appear until the Tertiary and evolved on mammals.

The above model of tick long-term evolution has two assumptions in common with many other scenarios: First, the most primitive groups of ticks are associated with the most primitive groups of hosts, and second, these groups of ticks must have arisen approximately when their particular host group arose. For example, Pomerantsev (52) rejected reptiles as ancestral hosts for at least the Ixodidae (he preferred mammals or birds) and used the observation that the endemic Australian subgenera of *Ixodes* were associated with the most primitive mammals—monotremes and marsupials—to support his conclusion that these ticks were the most primitive Ixodinae. Morel (43) suggested that the most primitive Metastriata were *Amblyomma* spp. associated with turtles. Because turtles were assumed to be the ancestral hosts, he suggested that these ticks originated at the same time as turtles, in this case the Jurassic. This is not the only study using host associations to back up hypotheses on ixodid origins. Hoogstraal (25) correlated the origin of ticks with the appearance of large reptiles in the late Paleozoic, and Filippova (17) assumed an origin of the Ixodidae somewhere in the Cretaceous based on an assumed ancestral association with monotremes and marsupials. All of these models depend on the notions that ticks are generally host specific and that their evolution is largely characterized by cospeciation with those hosts. Neither hypothesis has ever been adequately tested.

FOSSIL RECORD   Tests of the above hypotheses on the time of origin of ticks, and the hypotheses of cospeciation underlying them, cannot be conducted using fossil evidence. Fossils of Parasitiformes, in particular of ticks, are quite rare and relatively recent. Only four descriptions of fossil ticks have been published: a male indistinguishable from *Dermacentor reticulatus* found in the exterior auditory canal of a woolly rhinoceros, *Tichorhinus antiquitatis,* from the Pliocene (56, 60); *Ixodes succineus,* found in Baltic amber (35–40 mya) (60); *Ixodes tertiarius,* found in Oligocene deposits (ca. 30 mya) in Wyoming (57); and a male *Amblyomma,* closely resembling *Amblyomma testudinis,* found in Dominican amber dating between 30 and 40 mya (37). Interestingly, fossil argasids have not been found, and all of these specimens were either identical to extant species or highly similar to them, suggesting a low rate of divergence (3).

This fossil record for ticks contrasts sharply with that for the other main group of Acari, the Acariformes. Fossils of that group date from the Devonian (46). If mites are monophyletic, and if the Acariformes and Parasitiformes are sister groups (38), then these two groups must be of similar age. Based on this notion, Oliver (49) suggested that ticks might have originated as early as the Devonian (350–400 million years ago). Although this estimate might be questioned on the basis of its assumption that ticks and the Parasitiformes as a whole share a similar time of origin (not to mention the assumption of acarine monophyly), it is the first published estimate of the age of the Ixodida that does not depend on hosts. With regard to biology, Oliver (49) suggested that the ancestral tick stock was probably predaceous, a characteristic of all orders of Arachnida. He proposed that when the ticks abandoned the predatory habit and became ectoparasites, their hosts were probably amphibians and not "reptiles."

TESTS OF HOST SPECIFICITY   In the absence of fossil evidence, the above hypotheses on the time of origin of ticks can be tested indirectly by examining their most basic assumption, the perception that most tick species are strictly host specific. To test host specificity, one of us (JSH Klompen) examined the frequency of host records for species of hard ticks in the genus *Ixodes* and bat-parasitizing soft ticks in the genus *Carios*. Both groups have broad geographic distributions, covering several biogeographic regions. In addition, all *Carios* species parasitizing bats were categorized as strict-total by Hoogstraal and thus seemed ideal to test Hoogstraal's hypotheses of tick-host specificity.

All data were extracted from the literature. In response to Hoogstraal's admonition that "the patterns of limited host specificity may be altered when physiologically acceptable domestic or feral mammals intrude into the primary host-parasite associations" (27, pp. 516–17), all records of associations with humans or domestic animals were excluded from these analyses. For this analysis, the number of host species associated with a given tick species was plotted against the number of independent collections of that species. Under conditions of strict host specificity, the number of hosts and the number of collections should share no correlation: Regardless of the number of collections, the number of hosts should remain very low. Alternatively, if host specificity is an artifact of inadequate sampling, then the number of hosts will increase with sampling effort. Both analyses showed a strong positive correlation between the total number of collections and the number of known host species (Figures 1*a,b*). Some broad level of host specificity might remain if parasite speciation was slow relative to host speciation. Using the same preliminary data, we tested this idea for *Ixodes* spp. by plotting the number of host families parasitized against the total number of collections (Figure 1*a*). Although some species appear to be completely or mostly restricted to hosts in a single family or even genus [e.g. *Ixodes* (*Ixodiopsis*) *soricis* was restricted to *Sorex* spp.], the overall trend is a gradual increase in the number of families with increased numbers of collections.

These results raise the possibility that much of the distinction between strict-total and less-specific categories might be the difference between rarely and frequently collected species. This conclusion in turn implies that the assertion of host specificity should be reevaluated. The reader should note that although this result weakens support for a model in which host specificity leads to cospeciation, it does not exclude the existence of cospeciation: Parasite and host phylogenies must be compared to rule out cospeciation.

METHODOLOGICAL PROBLEMS   Hoogstraal's scenario also raises an important methodological problem. In none of his writings did he differentiate clearly between the two processes of adaptation and speciation, which suggests that he believed them to be causally related: Adaptation and host specificity implied cospeciation. The idea that the degree of host specificity of a group of parasites provides direct clues to speciation patterns of those parasites is a view embedded in the Rules of Parasitology (see 7). However, recent analyses of parasite-host associations using independently derived phylogenies of both parasites and hosts have disproved this scenario, showing host specificity to be a poor predictor of speciation patterns. Although host specificity is a property of parasite adaptation, parasite speciation, the characteristic determining cospeciation, is not (9). In other words, although speciation and adaptation are always phylogenetically correlated, neither is causally dependent upon the other (9). Supporting hypotheses of cospeciation with perceived patterns of host specificity is therefore a methodologically flawed approach. Moreover, although the presence of host adaptations may lead to host specificity, observed host specificity is not necessarily an indicator of host adaptation. Host specificity may also arise because of no opportunity to transfer to alternative hosts (which secondarily may result in cospeciation!) or as a secondary effect of adaptation to off-host habitat (see below).

Ecological Specificity

The alternative to host specificity, habitat or ecological specificity, has not received nearly as much attention. In this view, tick evolution may be determined largely by adaptation to a particular habitat type, not by adaptation to a particular host taxon (3). Support for this hypothesis comes from the observation that many tick species parasitize phylogenetically distant hosts with similar nesting or perching habitats. This observational evidence has never been fully integrated and therefore remains largely anecdotal. One of the most convincing examples might be the argasid tick *Ornithodoros turicata*. This species is associated with the burrows of gopher tortoises in the southeastern United States but will feed on gopher frogs, gopher tortoises, snakes, small mammals, and burrowing owls, all of which may inhabit gopher-tortoise burrows (41). Moreover, these ticks develop at least as well on gopher frogs as on gopher tortoises (41), thus casting doubt on the universality of the hypothesis stating that many hosts are fed upon, but few are suitable for tick development (26). In another example, both *Argas cooleyi* and *Carios concanensis* feed mostly on cliff swallows (*Hirundo pyrrhonota*). They use the same aggregation pheromone (20), but these species rarely aggregrate together, possibly because of habitat partitioning based on differences in humidity, temperature, etc (20a). Moreover, these ticks appear to exhibit little actual host specificity. For example, *C. concanensis* feeds quite readily on bats inhabiting the cliffs in which the swallows nest. Overall, this association pattern seems more specific for cliffs with cliff-swallow nests than for cliff swallows. Given that many related host species will use the same or related habitat types (e.g. bats of different species in a cave), the ecological specificity hypothesis still allows for a certain degree of broad host specificity, but only as a secondary result of ecological specificity (33).

SYSTEMATICS

Ixodida

To assess the roles of ecological and host specificity in tick evolution, we must examine the various attributes of tick species in the context of the evolutionary histories of those species. This requires objective hypotheses of phylogenetic relationships among species, subgenera, and genera.

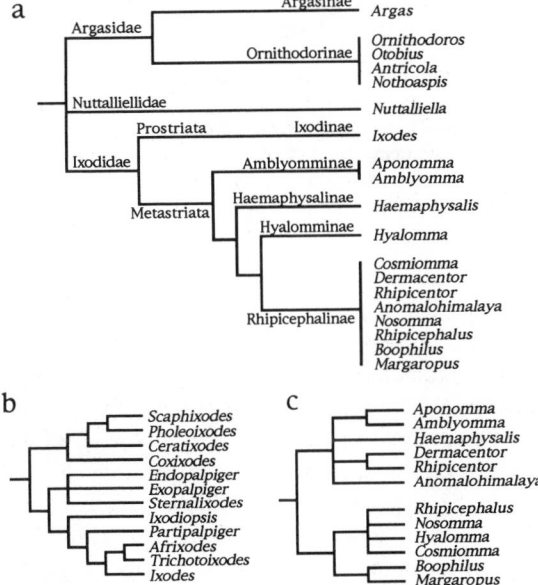

Figure 2.   Hypotheses of relationships derived from traditional classifications. (*a*) Hoogstraal tree for all Ixodida (based on Reference 26). (*b*) *Ixodes* subgenera (based on Reference 17). (*c*) Metastriata (based on Reference 18).

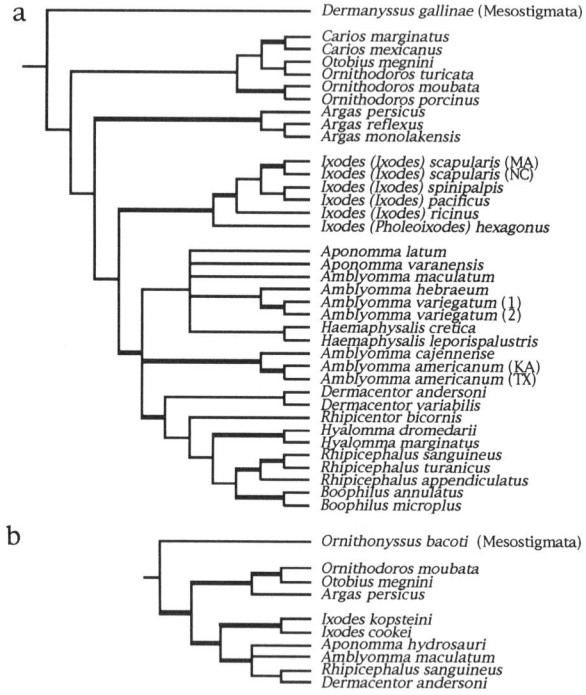

a

Dermanyssus gallinae (Mesostigmata)
Carios marginatus
Carios mexicanus
Otobius megnini
Ornithodoros turicata
Ornithodoros moubata
Ornithodoros porcinus
Argas persicus
Argas reflexus
Argas monolakensis
Ixodes (Ixodes) scapularis (MA)
Ixodes (Ixodes) scapularis (NC)
Ixodes (Ixodes) spinipalpis
Ixodes (Ixodes) pacificus
Ixodes (Ixodes) ricinus
Ixodes (Pholeoixodes) hexagonus
Aponomma latum
Aponomma varanensis
Amblyomma maculatum
Amblyomma hebraeum
Amblyomma variegatum (1)
Amblyomma variegatum (2)
Haemaphysalis cretica
Haemaphysalis leporispalustris
Amblyomma cajennense
Amblyomma americanum (KA)
Amblyomma americanum (TX)
Dermacentor andersoni
Dermacentor variabilis
Rhipicentor bicornis
Hyalomma dromedarii
Hyalomma marginatus
Rhipicephalus sanguineus
Rhipicephalus turanicus
Rhipicephalus appendiculatus
Boophilus annulatus
Boophilus microplus

b

Ornithonyssus bacoti (Mesostigmata)
Ornithodoros moubata
Otobius megnini
Argas persicus
Ixodes kopsteini
Ixodes cookei
Aponomma hydrosauri
Amblyomma maculatum
Rhipicephalus sanguineus
Dermacentor andersoni

Figure 3. Molecular-based phylogenies of the Ixodida. Both trees are based on 1000 bootstrap replications. Bold lines indicate branches with bootstrap support over 90%. The outgroup for both analyses was the mesostigmatid mites. (*a*) Based on 16S mitochondrial ribosomal DNA sequences (5); (*b*) based on 18S nuclear ribosomal DNA sequences (WC Black, in preparation). Total of 1124 base pairs (aligned), with 147 informative positions, tree length 582, consistency index (CI) = 0.676 (excluding uninformative characters), retention index (RI) = 0.635.

As noted above, the suborder Ixodida includes three generally recognized families, Argasidae, Ixodidae, and Nuttaliellidae. Relationships among the three families are largely unresolved, and most classifications depict them as a trichotomy (see Figure 2*a*). Until recently, the monophyly of these groupings had not been questioned, but a study based on 16S mitochondrial ribosomal DNA sequence data (5) suggested the possibility of paraphyly of the Argasidae (Figure 3*a*). It showed considerable bootstrap support (71–89%, depending on the type of analysis) for an arrangement in which the Argasinae were more closely related to the Ixodidae than to the Ornithodorinae. However, this unexpected result was not duplicated in a follow-up analysis based on the more conserved 18S nuclear ribosomal gene (Figure 3*b*).

Figure 4. Host associations and geographical distributions of the Argasidae. Morphology-based phylogeny as in Klompen & Oliver (34). Relationships within *Carios* are modified based on more recent data (JSH Klompen & JH Oliver, in preparation). Traditional subgeneric designations are added in square brackets where appropriate. Associations with bats are listed separately from other placentals based on their highly distinct ecology, not on their phylogeny. Abbreviation and symbols: *S., Secretargas;* +, common association; o, rare association. The asterisk denotes that geographical records exclude recent introductions into other biogeographical regions.

Argasidae

Historically, the classification of the Argasidae has been the subject of considerable disagreement, mostly between Russian and other Eastern European workers on the one hand and western systematists on the other (34, 53). Russian workers recognized two subfamilies: Argasinae, with a single genus, *Argas,* and the Ornithodorinae, with two tribes, Otobiini and Ornithodorini. Each tribe has two genera: *Otobius* and *Alveonasus* and *Ornithodoros* and *Antricola,* respectively (16, 53). American workers (11, 25) maintain the division between the two subfamilies, but reject the tribes. They also assigned the subgenus *Ogadenus* to the Argasinae (in Russian classifications, this subgenus belongs in the Ornithodorinae, tribe Otobiini, genus *Alveonasus*). Moreover, they added the monotypic genus *Nothoaspis* (29).

The first cladistic analysis of subgeneric relationships in the Argasidae (34) led to a revised classification differing quite dramatically from both the Russian and the American views. The basic division into two subfamilies, Argasinae

(57 species in the genus *Argas*) and Ornithodorinae (about 114 species in the genera *Ornithodoros, Otobius,* and *Carios*), is maintained, but their composition has changed [species numbers following those of Keirans (28)]. The number of genera is reduced: *Antricola, Nothoaspis,* the bat-associated *Argas,* and part of the former *Ornithodoros* are combined into a new ornithodorine genus, *Carios,* and *Alveonasus* is now considered a subgenus of *Argas* (Figure 4). With the exception of the relationship of Argasinae with the Ornithodorinae (see above), these results are consistent with results from a molecular analysis of tick relationships (5).

Ixodidae

Unlike argasid classifications, basal relationships in the Ixodidae are generally agreed upon. Prostriata, comprising about 240 species in a single genus, *Ixodes,* is separate from Metastriata, which includes all other genera (26). This division has held up in recent cladistic analyses that used either molecular (5) or morphological (JSH Klompen, JH Oliver Jr, JE Keirans & PJ Homsher, in preparation) character sets. In both studies, support for the monophyly of Metastriata is considerably stronger than for monophyly of Prostriata.

Ixodes, Prostriata's single genus, contains at least 14 subgenera (12). Filippova (15, 17) proposed a basal trichotomy within the genus (Figure 2*b*). The first lineage includes *Sternalixodes, Exopalpiger,* and *Endopalpiger;* the second *Coixodes, Ceratixodes, Pholeoixodes,* and *Scaphixodes;* and the third *Ixodiopsis, Partipalpiger, Afrixodes, Trichotoixodes,* and *Ixodes* sensu strictu (s.s.). Camicas & Morel proposed an alternative hypothesis (10) featuring two groupings, the Eschatocephalinae with five genera, *Ceratixodes, Lepidixodes, Eschatocephalus, Scaphixodes* (including *Multidentatus* and *Trichotoixodes*), and *Pholeoixodes* (including *Ixodiopsis*), and the Ixodinae with a single genus incorporating the remaining taxa. The results of a morphology-based phylogenetic analysis of subgeneric relationships in *Ixodes* sensu lato (s.l.) (32) confirm Filippova's grouping of *Sternalixodes, Endopalpiger,* and some *Exopalpiger,* but indicate that the remaining subgenera (including many *Exopalpiger*) belong in a single lineage (Figure 5). Support for most lineages is relatively weak. However, the basal position of the Australian taxa (represented by *Coixodes, Ceratixodes, Sternalixodes,* and the Australian *Exopalpiger*) was confirmed in analyses based on 16S mitochondrial ribosomal DNA sequences (JSH Klompen & WC Black, in preparation).

The classification for the Metastriata most widely used in the western literature is summarized in the Hoogstraal tree (Figure 2*a*). Largely on the basis of the size and shape of the palps, Hoogstraal divided the Metastriata into 4 subfamilies (26) on a scale from primitive to derived: Amblyomminae (125 species in 2 genera), Haemaphysalinae (147 species in 1 genus), Hyalom-

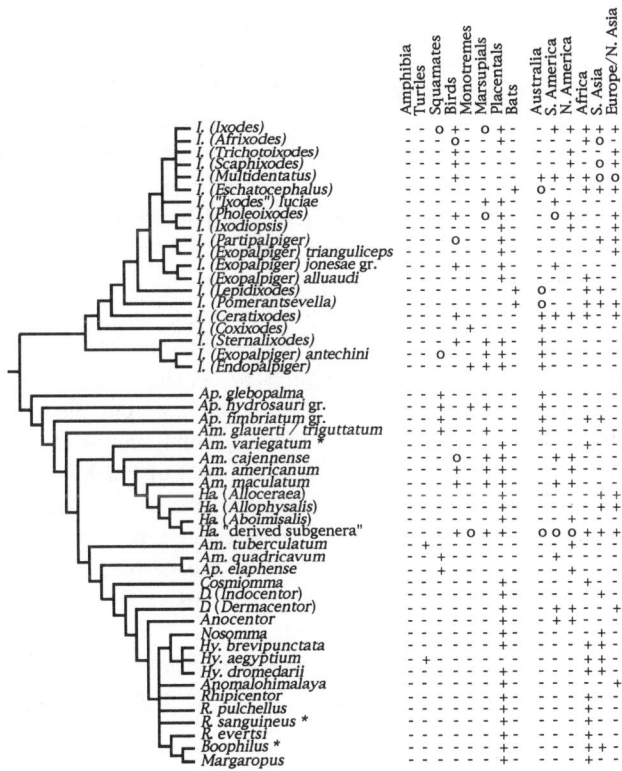

Figure 5. Host associations and geographical distributions of the Ixodidae. Morphology-based phylogeny for Prostriata (32) and Metastriata (JSH Klompen, JH Oliver Jr, JE Keirans & PJ Homsher, in preparation). Abbreviations: *Ap., Aponomma; Am., Amblyomma; D., Dermacentor; Ha., Haemaphysalis; Hy., Hyalomma; I., Ixodes; R., Rhipicephalus;* +, common association; o, rare association. The asterisk denotes that geographical records exclude recent introductions into other biogeographical regions.

minae (22 species in 1 genus), and Rhipicephalinae (119 species in 8 genera). The main alternative, developed by Russian workers (18, 19, 52), features groupings equivalent to Hoogstraal's Amblyomminae and Haemaphysalinae, but these groupings are arranged quite differently; several others are not recognized by Hoogstraal at all. In the most recent refinement of this classification (Figure 2c), Filippova (19) proposes two tribes, the Amblyommini, which has four subtribes (Amblyommini s.s., Haemaphysalini, Dermacentorini, and Anomalohimalaini), and the Rhipicephalini, which has two subtribes (Rhipicephalini s.s. and Margaropini). The above classifications were primarily intended as aids for identification, not necessarily as hypotheses of relationships. In a proposed phylogeny for Metastriata, Morel (43) suggested a basal position of the turtle-associated *Amblyomma*, with *Aponomma, Haemaphysalis,* and the derived Metastriata related to different lineages within *Amblyomma.* Within the derived Metastriata, he proposed close affinities shared by *Nosomma* and *Hyalomma; Boophilus, Margaropus,* and *Rhipicephalus (Digineus)*; and *Dermacentor, Anocentor,* and *Rhipicentor.*

Two recent phylogenetic analyses (5; JSH Klompen, JH Oliver Jr, JE Keirans & PJ Homsher, in preparation) show good support for monophyly of some of the specific groupings proposed in the above studies. These groupings include the Haemaphysalinae/Haemaphysalini (genus *Haemaphysalis*), the Hyalomminae (genus *Hyalomma*), and the assemblage of Hyalomminae and Rhipicephalinae (Figures 3a, 5). However, neither analysis supports monophyly of Hoogstraal's Rhipicephalinae (paraphyletic because of the exclusion of *Hyalomma*) or Filippova's Amblyomminae (an apparent grade taxon, paraphyletic because of the exclusion of her Rhipicephalinae). All studies (traditional and phylogenetic) agree that the genera *Aponomma* and *Amblyomma* are basal within the Metastriata, a position supported by the observation of unusually long (relative to the derived lineages) branch lengths for those taxa (5). However, these genera may be paraphyletic relative to *Haemaphysalis* and the derived Metastriata (the *Cosmiomma-Margaropus* lineage in Figure 5). Such a result, which is consistent with Morel's phylogeny, has important consequences for assessments of the basal evolution of the family.

EVOLUTION, REVISED VIEWS

The revised views on tick systematics, as outlined above, require some changes in our thinking about tick evolution, including the evolution of host associations. In assessing the evolution of host-association patterns, three different factors must be considered: hosts, off-host habitats, and biogeography. Biogeography should have considerable influence on host-selection patterns. After all, hosts not occurring in a given area cannot be used by the resident tick populations. Unfortunately, although several studies have focused on lesser taxa (e.g. 54), only a few have dealt specifically with biogeographical patterns at the familial or subfamilial level. Morel (43) proposed a scenario that relied almost exclusively on long-distance dispersal, whereas Balashov (4) explicitly included ecological data and continental-drift patterns. On the other hand, Morel considered different biogeographical patterns for different groups within *Aponomma* and *Amblyomma* (an approach consistent with the results of the systematic analyses), while Balashov's scenario was based on the less likely assumption that all current genera are monophyletic.

Argasidae

ARGASINAE   Within the genus *Argas* two subgenera, *A.* (*Argas*) and *A.* (*Proknekalia*), are exclusively associated with birds while another, *A.* (*Alveonasus*), is exclusively associated with mammals. The monotypic *A.* (*Ogadenus*) is associated with various desert vertebrates (mammals, lizards, and birds), whereas *A.* (*Secretargas*) includes one species, *A. echinops,* on a Madagascan tenrec; another, *A. hoogstraali,* on a Madagascan lizard; and a third, *A. transgariepinus,* mostly on bats, but which occasionally feeds on lizards occurring near the bat roosts (26).

The above host association pattern (Figure 4) superficially seems to indicate that at least some lineages are specific for birds, but it is also consistent with ecological specificity. For example, the eight species associated with swallow nests are unlikely to encounter any other host species in that habitat. Notably, *A.* (*Argas*) *brevipes* parasitizes hosts in several orders of birds. It is also one of the few *Argas* species associated with tree holes, a nest type shared by several bird taxa. In addition, even if we acknowledge some broad specificity for birds, we can find little evidence of cospeciation. If a species is found on hosts from more than one order, these orders are rarely phylogenetically, but nearly always ecologically, related (e.g. cliff-nesting sea birds or tree-hole breeders). The one generalization that seems to be valid for the habitat of most Argasinae is an association with low humidity and rocks. The latter includes nests under rocks, on rock cliffs, etc. Exceptions include *A.* (*Argas*) *macrostigmatus,* which is found in humid habitats, and most members of the *monachus* and *persicus* groups of *A.* (*Argas*), which are associated with arboreal stick nests.

Most argasine species associated with nonvolant hosts are restricted to the Old World. Only one, the poorly known *A.* (*Alveonasus*) *cooleyi,* occurs in North America, and none are reported from South America or Australia. Not unexpectedly, members of the bird-associated subgenus *A.* (*Argas*) have attained a worldwide distribution. Although the evidence is currently inconclusive, we suggest that the subfamily most likely originated in the Old World.

ORNITHODORINAE   Host-association patterns in the Ornithodorinae vary. The three members of the genus *Otobius* are associated with North American mammals, i.e. small rodents, lagomorphs, or artiodactyls (deer, pronghorn, cattle). Habitat associations vary from open fields, to deer and jackrabbit beds, to the nests of deer mice (*Peromyscus* spp.) and wood rats (*Neotoma* spp.).

Various nonvolant vertebrates serve as hosts for the basal taxa of *Carios;* three species (*C. cyclurae, C. darwini, C. galapagensis*) occur on lizards, and eight are largely restricted to mammals. Within the latter group, at least three species (*C. chironectes, C. echimys, C. marmosae*) parasitize South American marsupials (often in addition to placental mammals), and at least two (*C. rudis, C. talaje*) will feed readily on birds. Two lineages appear more specific. The *capensis* group is largely associated with wide-ranging sea birds, primarily shore birds (Charadriiformes), but also pelicans (Pelicaniformes) and penguins (Sphenisciformes), whereas another, very large lineage is associated exclusively with bats (Chiroptera).

Once again, this pattern of apparent host specificity is consistent with specificity for particular types of habitats: nests of sea birds for the *capensis* group, bat roosts in trees and caves for the bat associates. For example, several species (e.g. *C. marginatus, C. mexicanus, C. silvai, C. viguerasi*) are largely restricted to bat hosts in the family Mormoopidae and the distantly related phyllostomid tribe Glossophagini (genera *Phyllonycteris, Erophylla, Brachyphylla*). This host-association pattern seems more compatible with a preference for very dark parts of caves, the typical habitat of these bat taxa, than with host specificity. Similarly, members of the *batuensis* species group associated with Old World fruit bats (Pteropodidae) occasionally exhibit overlapping host use but could not be shown to have overlapping habitat use (33).

Carios, like *Otobius*, its putative sister group, appears to be a New World genus. Most taxa occur in the area between the southern USA and the northern half of South America, including the Caribbean, which harbors considerable diversity. Unlike *Otobius* lineages, which did not disperse out of the New World before human commerce began there (transport of horses and cattle), at least two lineages of *Carios* have dispersed naturally to the Old World and Australia. We are assuming dispersal in this case because the basal lineages are New World endemics (JSH Klompen & JH Oliver, in preparation). The first dispersing lineage, the *capensis* group, has several species that are cosmopolitan or restricted to the Old World. Dispersal may have resulted from the association with wide-ranging sea birds (such associations appear to have produced similar near-cosmopolitan distributions for the *Ixodes* subgenera *Ceratixodes* and *Multidentatus*). The other dispersing lineage(s) include(s) bat associates. How this association might have led to dispersal is less obvious, because few bat taxa are known to disperse across broad stretches of ocean. Published phylogenies of *Carios* (34) suggest two dispersal events (one leading to the *batuensis* group, the other to the *C. vespertilionis-C. boueti* lineage), but a recent analysis of all species in the genus (JSH Klompen & JH Oliver, in preparation) suggests that these two lineages are sister groups and that a single dispersal event occurred (Figure 4).

Members of the genus *Ornithodoros* show very little host specificity. Individual species are often associated with hosts in one or more classes, including anuran amphibians, turtles, squamates (both lizards and snakes), and many bird and mammal taxa (including monotremes and marsupials). A few species appear to be more host specific. Among these are the only two argasids restricted to tortoises, the African *Ornithodoros compactus* and *Ornithodoros transversus,* an associate of giant Galapagos tortoises. *O. transversus* is the only permanent parasite among ticks; it even deposits its eggs on its host (26).

Most *Ornithodoros* species are associated with burrows in soil, but *O. macmillani* is found in bird nests in tree holes, whereas *O. gurneyi,* associated with kangaroo beds, and *O. coriaceus,* associated with deer beds, are found in open fields. Overall, ecological specificity may be far more important in this genus than host specificity.

The geographical range of *Ornithodoros* species includes Australia, the Old World, and the New World. However, conclusions regarding biogeography cannot yet be made because relationships within the genus are so poorly resolved. A few oddities have appeared in phylogenetic analyses (34): a potential sister-group relationship between *O. gurneyi* (Australia) and *O. coriaceus* (western North America), and a close relationship between the species in the *moubata* group (Africa) and *O. transversus* (Galapagos islands).

Ixodidae

HOSTS AND HABITAT   In the interest of brevity, the discussion of ixodid host-association patterns cannot be comprehensive but instead focuses on the few questions most affected by revised ideas on tick relationships.

Most Prostriata are associated with mammals (monotremes, marsupials, and placentals), but all members of the subgenera *Scaphixodes, Multidentatus, Trichotoixodes, Ceratixodes,* and *Xiphixodes* feed almost exclusively on birds. Other subgenera (*Sternalixodes, Pholeoixodes, Afrixodes, Ixodes*) have individual species that are either largely restricted to birds [e.g. *I. (Pholeoixodes) baergi*], or that occasionally use birds [e.g. *I. (Ixodes) scapularis*]. Although several species may use lizards and snakes as hosts for the immatures [e.g. *I. (Ixodes) ricinus*], only one seems restricted to lizards [*I. (Ixodes) asanumai*], and none are associated with turtles or amphibians. The genus *Ixodes* does include all four ixodid species associated with bats. Neither a morphology-based (32) nor a molecular (JSH Klompen & WC Black, in preparation) phylogenetic analysis support a monophyletic lineage that includes all or even most taxa associated with birds. However, combined morphological and molecular data did suggest the possibility of a monophyletic lineage that includes all bat associates. This result may, once again, reflect specificity for the unusual ecology of bats, rather than specificity for the bats themselves. This notion is supported by the observation that *I. (Lepidixodes) kopsteini* uses hosts in both suborders of bats. Numerous prostriate species are burrow associates, e.g. *I. (Pholeoixodes)* and *I. (Ixodiopsis)*, but many others are nonburrow species.

The total host range of the Metastriata is even broader than that of the Prostriata. Most taxa, especially the derived Metastriata (the *Cosmiomma-Margaropus* lineage), are almost exclusively associated with mammals, but several other species are associated with birds (*Amblyomma* and *Haemaphysalis* species), squamates (*Aponomma* and *Amblyomma* species), turtles [*Amblyomma* species and *Hyalomma* (*Hyalomma*) *aegyptium*], and amphibians (*Amblyomma* spp.). Immatures often have wider host ranges than adults.

BIOGEOGRAPHY   Biogeographical correlations appear to be very important at all taxonomic levels among Ixodidae. For example, at low taxonomic levels, recent molecular systematic analyses based on mitochondrial 16S and 12S ribosomal DNA sequences (45) suggest strong support for a western United States clade (*I. pacificus, I. jellisoni, I. neotomae, I. spinipalpis*) within the subgenus *Ixodes,* excluding related eastern (*I. scapularis, I. affinis*) or Old World species (*I. ricinus, I. persulcatus*).

At the subfamily level (Figure 5), the most prominent pattern involves the exclusively or partly Australian distribution of the basal lineages in both Prostriata (32) and Metastriata (JSH Klompen, JH Oliver Jr, JE Keirans & PJ Homsher, in preparation). All other native Australian ixodid ticks (those not obviously introduced by humans) either are associated with wide-ranging sea birds [e.g. *I. (Multidentatus)*] or are part of a lineage of derived *Haemaphysalis* species that includes many Southeast Asian species. This pattern suggests an origin of the family Ixodidae in Australia, or at least a basal division between Australian and non-Australian species in both lineages (the choice between the two alternatives will depend on further resolution of the phylogenies). If the geographical origin turns out to be Australia, we can roughly estimate the temporal origin of the family, which would presumably be after the breakup of Gondwanaland was under way and the Australian continent was relatively isolated. If correct, this hypothesis sets the upper age limit for the origins of the family in the middle Cretaceous, about 120 mya (1). This estimate is quite different from Hoogstraal's Permian (24), Balashov's Triassic (3), or Oliver's (49) Devonian estimates but fits fairly well with Filippova's (17) estimate of late Cretaceous for the genus *Ixodes.*

Biogeography occasionally provides alternatives to host-based scenarios. Pomerantsev (52) suggested that Australian *Ixodes* species occur there because they are adapted to primitive mammals (monotremes and marsupials). We suggest that they may be associated with such hosts because that is all that is available in Australia. This reinterpretation is supported by the observation that many of these ticks readily feed on birds as well as both endemic and recently introduced placental mammals (55). Similar reasoning can be used to explain the associations of basal Metastriata (*Aponomma glebopalma, Aponomma hydrosauri* group, so-called typical *Aponomma,* and Australian *Amblyomma* species) with varanid lizards, monotremes, and marsupials. However, availability does not explain the entire host-association pattern. The observation that the various tick taxa do not use all available hosts suggests some level of specificity. Nonetheless, the match is consistent, and the most interesting question might not be why these taxa feed on a certain host group, but why basal Prostriata do not feed on varanid lizards and basal Metastriata do not feed on birds.

CONCLUSION

Reanalysis of patterns of tick-host associations does not support the generally accepted view that tick evolution arose through host adaptation, host specificity, or cospeciation. The fit of such hypotheses with available data is weak at best. Support for hypotheses predicated on host specificity is especially poor. In contrast, considerable portions of these patterns seem to fit a hypothesis in which restrictions imposed by biogeography and ecological specificity, rather than host specificity, lead to the observed host-association patterns. We therefore suggest that biogeography and ecological specificity should be accorded a far more prominent role in tick evolution.

ACKNOWLEDGMENTS

This research was supported by National Institute of Allergy and Infectious Diseases Grant No. AI 30026 (JSHK, JEK, JHO), a Colorado State University College Research Council Grant (WCB), a contract from the Centers for Disease Control (WCB), and National Science Foundation Grant No. DEB-9420658 (JSHK, WCB, JEK).

Literature Cited

1. Audley-Charles MG, Hurley AM, Smith AG. 1981. Continental movements in the Mesozoic and Cenozoic. In *Wallace's Line and Plate Tectonics*, ed. TC Whitmore, pp. 9–23. Oxford: Clarendon
2. Balashov YS. 1972. Bloodsucking ticks (Ixodoidea): vectors of disease of man and animals. *Misc. Publ. Entomol. Soc. Am.* 8:160–376
3. Balashov YS. 1989. Coevolution of ixodid ticks and terrestrial vertebrates. *Parazitologiya* 23:427–67 (In Russian)
4. Balashov YS. 1994. Importance of continental drift in the distribution and evolution of ixodid ticks. *Entomol. Rev.* 73:42–50
5. Black WC IV, Piesman J. 1994. Phylogeny of hard- and soft-tick taxa (Acari: Ixodida) based on mitochondrial 16S rDNA sequences. *Proc. Natl. Acad. Sci. USA* 91:10034–38
6. Brooks DR. 1979. Testing the context and extent of host-parasite coevolution. *Syst. Zool.* 28:299–307

160   KLOMPEN ET AL

7. Brooks DR. 1985. Historical ecology: a new approach to studying the evolution of ecological associations. *Ann. Miss. Bot. Garden* 72:660–80

8. Brooks DR, McLennan DA. 1991. *Phylogeny, Ecology, and Behavior.* Chicago: Univ. Chicago Press. 434 pp.

9. Brooks DR, McLennan DA. 1993. *Parascript. Parasites and the Language of Evolution.* Washington: Smithsonian Inst. 429 pp.

10. Camicas JL, Morel PC. 1977. Position systematique et classification des tiques (Acarida: Ixodida). *Acarologia* 18:410–20

11. Clifford CM, Kohls GM, Sonenshine DE. 1964. The systematics of the subfamily Ornithodorinae (Acarina: Argasidae). I. The genera and subgenera. *Ann. Entomol. Soc. Am.* 57:429–37

12. Clifford CM, Sonenshine DE, Keirans JE, Kohls GM. 1973. Systematics of the subfamily Ixodinae (Acarina: Ixodidae). I. The subgenera of *Ixodes. Ann. Entomol. Soc. Am.* 66:489–500

13. Coddington JA. 1988. Cladistic tests of adaptational hypotheses. *Cladistics* 4:3–22

14. Farrell BD, Mitter C, Futuyma DJ. 1992. Diversification at the insect-plant interface. Insight from phylogenetics. *BioScience* 42:34–42

15. Filippova NA. 1958. A contribution to the morphology and systematics of the immature phases of the ticks (Ixodinae). *Parasitol. Sb.* 18:10–77 (In Russian)

16. Filippova NA. 1966. *Argasid Ticks (Argasidae).* Moscow/Leningrad: Izd. Nauka, 255 pp. (In Russian)

17. Filippova NA. 1977. *Ixodid Ticks of the Subfamily Ixodinae.* Leningrad: Izd. Nauka, 393 pp. (In Russian)

18. Filippova NA. 1984. Taxonomy of ticks of the family Ixodidae (Acarina, Parasitiformes) in the fauna of the USSR and plans for studying it. *Parazitol. Sb.* 32:61–78; Transl. from Russian in T1796, NAMRU–3

19. Filippova NA. 1994. Classification of the subfamily Amblyomminae (Ixodidae) in connection with reinvestigation of chaetotaxy of the anal valve. *Parazitologiya* 28:3–12 (In Russian)

20. George JE. 1981. The influence of aggregation pheromones on the behavior of *Argas cooleyi* and *Ornithodoros concanensis* (Acari: Ixodoidea: Argasidae). *J. Med. Entomol.* 18:129–33

20a. George JE, Cook B. 1979. Exokinetic repsonse of *Argas cooleyi* and *Ornithodoros concanensis* (Acari: Ixodoidea: Argasidae) to various environmental temperatures. *J. Med. Entomol.* 16:275–85

20b. Griffiths DA, Bowman CE, eds. 1984. *Acarology VI*, Vol. 1. Chichester, UK: Ellis Horwood

21. Hafner MS, Nadler SA. 1990. Cospeciation in host-parasite assemblages: comparative analysis of rates of evolution and timing of cospeciation events. *Syst. Zool.* 39:192–204

22. Hafner MS, Sudman PD, Villablanca FX, Spradling TA, Demastes JW, Nadler SA. 1994. Disparate rates of molecular evolution in cospeciating hosts and parasites. *Science* 265:1087–90

23. Hiss RH, Norris DE, Dietrich CH, Whitcomb RF, Bosio CF, et al. 1994. Molecular taxonomy using single-strand conformation polymorphism (SSCP) analysis of mitochondrial ribosomal DNA genes. *Insect Mol. Biol.* 3:171–82

24. Hoogstraal H. 1978. Biology of ticks. In *Tick-Borne Diseases and Their Vectors. Proc. Int. Conf.*, ed. JKH Wilde, pp. 3–14. Edinburgh: Univ. Edinburgh

25. Hoogstraal H. 1985. Argasid and nuttalliellid ticks as parasites and vectors. *Adv. Parasitol.* 24:135–238

26. Hoogstraal H, Aeschlimann A. 1982. Tick-host specificity. *Bull. Soc. Entomol. Suisse* 55:5–32

27. Hoogstraal H, Kim KC. 1985. Tick and mammal coevolution, with emphasis on *Haemaphysalis.* In *Coevolution of Parasitic Arthropods and Mammals*, ed. KC Kim, pp. 505–68. New York: Wiley & Sons

28. Keirans JE. 1992. Systematics of the Ixodida (Argasidae, Ixodidae, Nuttalliellidae): an overview and some problems. In *Tick Vector Biology. Medical and Veterinary Aspects*, ed. BH Fivaz, TN Petney, IG Horak, pp. 1–21. Berlin: Springer-Verlag

29. Keirans JE, Clifford CM. 1975. *Nothoaspis reddelli*, new genus and new species (Ixodoidea: Argasidae), from a bat cave in Mexico. *Ann. Entomol. Soc. Am.* 68:81–85

30. Keirans JE, Clifford CM, Hoogstraal H, Easton ER. 1976. Discovery of *Nuttalliella namaqua* Bedford (Acarina: Ixodoidea: Nuttalliellidae) in Tanzania and redescription of the female based on scanning electron microscopy. *Ann. Entomol. Soc. Am.* 69:926–32

31. Klompen JSH. 1992. Comparative morphology of argasid larvae (Acari: Ixodida: Argasidae), with notes on phylogenetic relationships. *Ann. Entomol. Soc. Am.* 85:541–60

32. Klompen JSH. 1995. Phylogenetic re-

lationships in the family Ixodidae with emphasis on the genus *Ixodes* (Parasitiformes: Ixodida). In *Acarology IX. Proc. Int. Congr. Acarol., 17–22 July 1994*, ed. GR Needham, DJ Horn, WC Welbourn. Columbus: Ohio State Univ. In press

33. Klompen JSH, Keirans JE, Durden LA. 1995. Three new species of ticks (Ixodida: Argasidae: *Carios*) on fruit bats (Chiroptera: Pteropodidae) in the Australasian region, with notes on host associations. *Acarologia* 36:25–40

34. Klompen JSH, Oliver JH Jr. 1993. Systematic relationships in the soft ticks (Acari: Ixodida: Argasidae). *Syst. Entomol.* 18:313–31

35. Deleted in proof

36. Krantz GW. 1978. *A Manual of Acarology*. Corvallis, OR: Oreg. State Univ. Book Stores. 509 pp. 2nd ed.

37. Lane RS, Poinar GO Jr. 1986. First fossil tick (Acari: Ixodidae) in New World amber. *Int. J. Acarol.* 12:75–78

38. Lindquist EE. 1984. Current theories on the evolution of major groups of Acari and on their relationship with other groups of Arachnida, with consequent implications for their classification. See Ref. 20b, 28–62

39. Lyal CHC. 1986. Coevolutionary relationships of lice and their hosts: a test of Fahrenholz's Rule. In *Coevolution and Systematics*, ed. AR Stone, DL Hawksworth, pp. 77–91. Oxford: Clarendon/Syst. Assoc.

40. Lyal CHC. 1987. Co-evolution of trichodectid lice (Insecta: Phthiraptera) and their mammalian hosts. *J. Nat. Hist.* 21:1–28

41. Milstrey EG. 1987. *Bionomics and ecology of Ornithodoros (P.) turicata americanus (Marx) (Ixodoidea: Argasidae) and other commensal invertebrates present in the burrows of the gopher tortoise, Gopherus polyphemus Daudin.* PhD thesis. Univ. Fla., Gainesville. 295 pp.

42. Mitter C, Brooks DR. 1983. Phylogenetic aspect of coevolution. In *Coevolution*, ed. DJ Futuyma, M Slatkin, pp. 65–98. Sunderland, Massachusetts: Sinauer

43. Morel PC. 1969. *Contribution a la connaissance de la distribution des tiques (Acariens, Ixodidae et Amblyommidae) an Afrique Ethiopienne continentale.* PhD thesis. Univ. Paris, Paris. 388 pp.

44. Needham GR, Teel PD. 1991. Off-host physiological ecology of ixodid ticks. *Annu. Rev. Entomol.* 36:659–81

45. Norris DE, Klompen JSH, Keirans JE, Black WC IV. 1995. Population genetics of *Ixodes scapularis* (Acari: Ixodidae) based on mitochondrial 16S and 12S genes. *J. Med. Entomol.* In press

46. Norton RA, Bonamo PM, Grierson JD, Shear WM. 1988. Oribatid mite fossils from a Devonian deposit near Gilboa, New York State. *J. Paleontol.* 62:259–69

47. Norton RA, Kethley JB, Johnston DE, OConnor BM. 1993. Phylogenetic perspectives on genetic systems and reproductive modes of mites. In *Evolution and Diversity of Sex Ratio in Insects and Mites*, ed. DL Wrench, MA Ebbert, pp. 8–99. New York: Chapman & Hall

48. OConnor BM. 1984. Co-evolutionary patterns between astigmatid mites and primates. See Ref. 20b, pp. 186–95

49. Oliver JH Jr. 1989. Biology and systematics of ticks (Acari: Ixodida). *Annu. Rev. Ecol. Syst.* 20:397–430

50. Page RDM. 1990. Temporal congruence and cladistic analysis of biogeography and cospeciation. *Syst. Zool.* 39:205–26

51. Page RDM. 1991. Clocks, clades, and cospeciation: comparing rates of evolution and timing of cospeciation events in host-parasite assemblages. *Syst. Zool.* 40:188–98

52. Pomerantsev BI. 1947. Basic directions of evolution of Ixodoidea. *Parazitol. Sb.* 10:5–18; Transl. from Russian in T55, NAMRU–3

53. Pospelova-Shtrom MV. 1969. On the system of classification of ticks of the family Argasidae Can., 1890. *Acarologia* 11:1–22

54. Robbins RG, Keirans JE. 1992. Systematics and ecology of the subgenus *Ixodiopsis* (Acari: Ixodidae: *Ixodes*). *Thomas Say Found. Monogr.* 14:1–159

55. Roberts RFS. 1970. *Australian Ticks.* Melbourne: CSIRO. 267 pp.

56. Schille F. 1916. Entomologie aus der Mammut- und Rhinoceros-Zeit Galiziens. *Entomol. Z.* 30:42–43

57. Scudder SH. 1885. A contribution to our knowledge of Paleozoic Arachnides. *Proc. Am. Acad. Sci.* 2:12

58. Sonenshine DE. 1991. *Biology of Ticks*, Vol. 1. New York: Oxford Univ. Press. 447 pp.

59. Sonenshine DE. 1994. *Biology of Ticks*, Vol. 2. New York: Oxford Univ. Press. 465 pp.

60. Weidner H. 1964. Eine Zecke, *Ixodes succineus* sp. n., im baltischen Bernstein. *Veröff. Überseemus. Bremen* 3:143–51

Annu. Rev. Entomol. 1996. 41:163–90

ION CHANNELS AS TARGETS FOR INSECTICIDES

Jeffrey R. Bloomquist

Department of Entomology, Virginia Polytechnic Institute and State University, Blacksburg, Virginia 24061-0319

KEY WORDS: neurotoxicity, toxin, baculovirus, convulsants, ligand-gated receptor, mode of action

ABSTRACT

Ion channels are the primary target sites for several classes of natural and synthetic insecticidal compounds. The voltage-sensitive sodium channel is the major target site for DDT and pyrethroids, the veratrum alkaloids, and N-alkylamides. Recently, neurotoxic proteins from arthropod venoms, some of which specifically attack insect sodium channels, have been engineered into baculoviruses to act as biopesticides. The synthetic pyrazolines also primarily affect the sodium channel, although some members of this group target neuronal calcium channels as well. The ryanoids have also found use as insecticides, and these materials induce muscle contracture by irreversible activation of the calcium-release channel of the sarcoplasmic reticulum. The arylheterocycles (e.g. endosulfan and fipronil) are potent convulsants and insecticides that block the GABA-gated chloride channel. In contrast, the avermectins activate both ligand- and voltage-gated chloride channels, which leads to paralysis. At field-use rates, a neurotoxic effect of the ecdysteroid agonist RH-5849 is observed that involves blockage of both muscle and neuronal potassium channels. The future use of ion channels as targets for chemical and genetically engineered insecticides is also discussed.

OVERVIEW AND SCOPE

This review focuses on the neurotoxic actions of insecticides, including their effects on excitable membrane physiology and the underlying interactions with ion channels. The term insecticide has been interpreted broadly, and the discussion encompasses synthetic compounds, botanicals, and arthropod toxins employed in engineered biopesticides. Sections on synthetic insecticides focus on established compounds, as well as on published information on experimental materials that have not yet been commercialized. Natural products that

163

0066-4170/96/0101-0163$08.00

attack ion channels are covered because of the use of these materials in organic farming (82), and such compounds may serve as leads for new synthetic insecticides. Finally, arthropod toxins are covered that have been introduced into baculoviruses to improve the efficacy of the virus for insect control.

The target sites of several classes of neurotoxic insecticides are ion channel proteins. This review discusses only those compounds that act upon, and are thought to bind directly to, ion-channel proteins and consequently does not cover neurotoxic compounds that act as neurotransmitter mimics (e.g. nicotine and imidacloprid) or that function as enzyme inhibitors, such as the organophosphorus anticholinesterases. Wherever possible, the focus remains on the principal ion-channel target site for a particular class of insecticide, and the review does not attempt an exhaustive comparison of all documented ion channel–related effects. Recent reviews have covered the major and minor target sites of action for the pyrethroids (9, 67, 108), as well as the role of various types of chloride channels in the actions of insecticidal polychlorocycloalkanes, arylheterocycles, and avermectins (10, 17, 22). Thus, rather than attempt to present a comprehensive treatment of the primary literature on the mode of action of these compounds, I reference comprehensive reviews and major papers on particular topics, with emphasis on new findings.

SODIUM CHANNELS

Pyrethroids and DDT

The chemistry (Figure 1) and symptomology of pyrethroids as they relate to the established type 1 and type 2 classification system have been extensively reviewed (9, 48, 94). Type 1 compounds include DDT, its analogues, and all pyrethroids containing descyano-3-phenoxybenzyl or other alcohols. At doses toxic to mammals, these compounds cause a whole-body tremor, or T syndrome (106). The type 2 compounds specifically contain an α-cyano-3-phenoxybenzyl alcohol and produce choreoathetosis (sinuous writhing) and profuse salivation, designated the CS syndrome (106). Differential poisoning syndromes caused by type 1 and 2 pyrethroids also occur in insects, but they are less distinct than those observed in mammals (9, 94). Dermal exposure to type 1 and 2 pyrethroids can cause paresthesia, a tingling or burning sensation of the skin; the effect is more intense for type 2 compounds (108). The type 1 and type 2 classifications are not absolute in either insects or mammals, and certain compounds show effects intermediate between the two classes (94).

Pyrethroid intoxication results from their potent actions on the central and peripheral nervous systems. In mammals, pyrethroids augment electrical activity in the brain, spinal column, and peripheral neurons, and these actions, in their various forms, are thought to underlie pyrethroid-induced tremors,

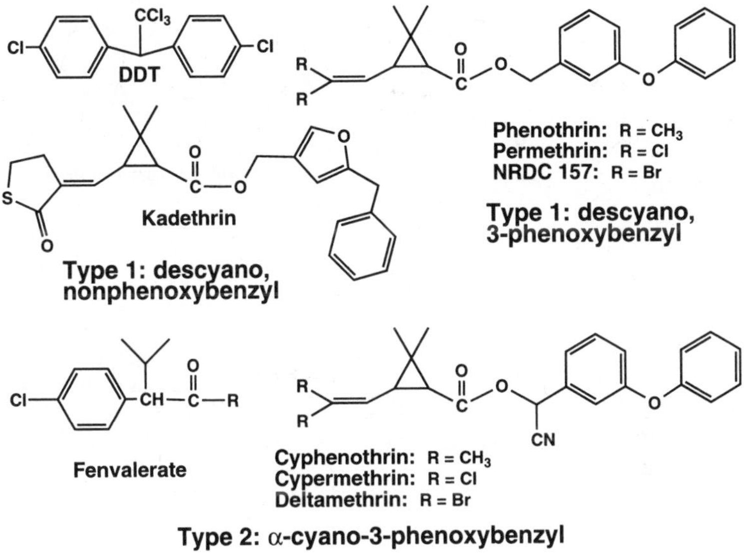

DDT

Kadethrin

Type 1: descyano, nonphenoxybenzyl

Phenothrin: R = CH₃
Permethrin: R = Cl
NRDC 157: R = Br

Type 1: descyano, 3-phenoxybenzyl

Fenvalerate

Cyphenothrin: R = CH₃
Cypermethrin: R = Cl
Deltamethrin: R = Br

Type 2: α-cyano-3-phenoxybenzyl

Figure 1   Representative chemical structures of type 1 and type 2 pyrethroids.

convulsions, and paresthesia (9). Type I compounds induce multiple-spike discharges in insect, crayfish, and frog axons (9, 94). In contrast, type 2 pyrethroids produce a stimulus-dependent depolarization of the axon membrane potential, reduction in the amplitude of the action potential, and loss of electrical excitability (9, 94). At increased concentrations, type 1 pyrethroids produce effects on axons similar to those of type 2 compounds (9, 60). The type 1 and 2 effects are also expressed at insect motor nerve terminals, where type 1 pyrethroids cause presynaptic repetitive discharge, and type 2 compounds cause a tonic release of transmitter indicative of membrane depolarization (90, 91). Moreover, type 2 compounds were more potent than type 1 for depolarizing motor nerve terminals, and this effect was correlated with their acute toxicity (91). Pyrethroids also depolarize mammalian nerve terminals (36) and thereby potentiate the release of various transmitters (32, 37, 72). In some sensory structures, increased electrical activity could be induced by both types of pyrethroids, with generally more potent and intense responses induced by the type 2 compounds (21, 23, 104). This hyperexcitatory effect of type 2 pyrethroids contrasts with their blocking effect on axons and results from their strong membrane-depolarizing action, which causes certain types of sensory structures to increase their firing rate (9, 60).

Pyrethroids also possess some effects on muscle that may be relatively more important in mammals. Type 2 pyrethroids induce repetitive activity in mam-

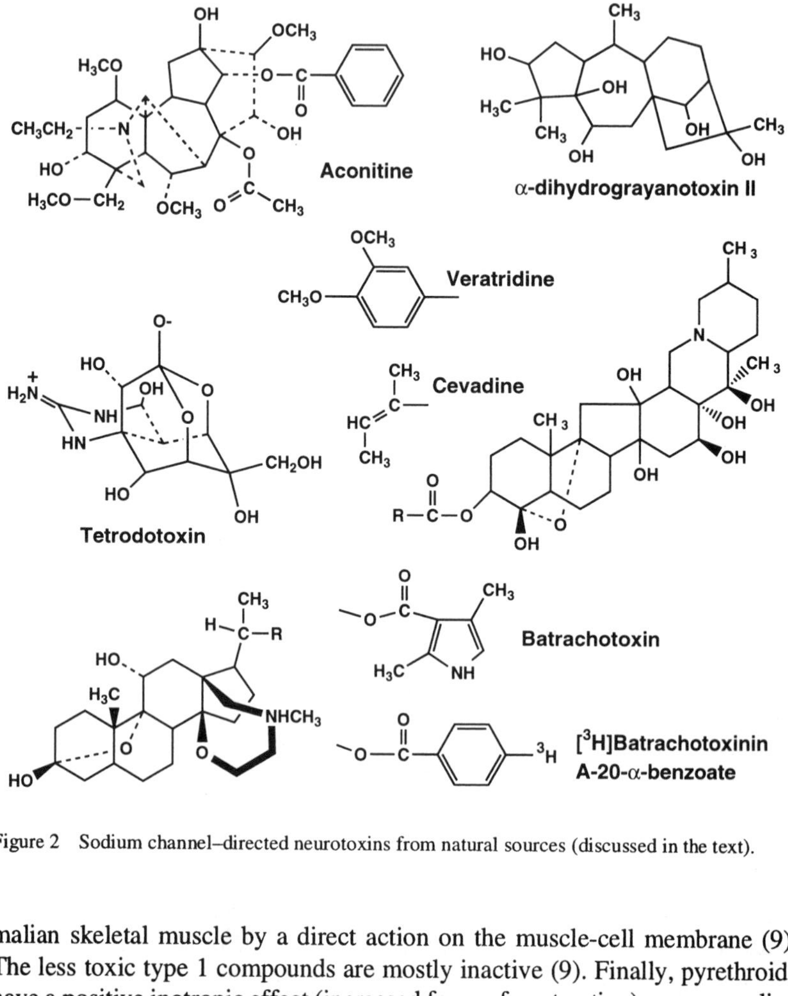

Figure 2   Sodium channel–directed neurotoxins from natural sources (discussed in the text).

malian skeletal muscle by a direct action on the muscle-cell membrane (9). The less toxic type 1 compounds are mostly inactive (9). Finally, pyrethroids have a positive inotropic effect (increased force of contraction) on mammalian heart, and type 2 compounds are more potent than type 1 (6, 25, 42). This effect is probably mediated by an augmented release of catecholamines from sympathetic nerve terminals, as well as by a direct action on the cardiac musculature (6).

Several lines of evidence suggest that the neurophysiological effects of pyrethroids result from an action on the voltage-sensitive sodium channel, which plays a critical role in generating the nerve membrane action potential (19). In several studies, the repetitive activity, membrane depolarization, and secretion of neurotransmitter induced by pyrethroids was reversed or prevented by tetrodotoxin (Figure 2), a specific blocker of the voltage-sensitive sodium channel (9, 94). Voltage clamp experiments, which can demonstrate effects on specific ion channels in the nerve membrane, showed that the repetitive firing and depolarization caused by the pyrethroids resulted from a prolongation of sodium-channel current (9, 67, 94, 108). Normally, the membrane sodium current activates (channels open) and then inactivates (channels shut) within a few milliseconds in response to a step depolarization of the membrane potential (67, 108). Pyrethroids slowed or delayed the inactivation of the sodium channel in voltage-clamped axons, thus prolonging the open state (9, 67, 94, 108). Pyrethroids also produced a residual, slowly decaying sodium current (tail current) that continues to flow after the membrane is repolarized, indicating that the activation gate of the channel opens and closes more slowly (67, 108). The delayed shutting of pyrethroid-modified sodium channels allows a persistent inward current to flow after an action potential, resulting in repetitive firing and depolarization of the nerve membrane (9, 67, 94, 108).

Patch-clamp techniques allow more detailed analyses of pyrethroid modification of sodium currents at the single-channel level. Analysis of single sodium-channel currents in the presence of pyrethroids revealed evidence for a population of channels with slowed kinetic transitions between channel states (9, 67). Channels modified by pyrethroids display normal single-channel conductance but exhibit a prolonged open state that displays altered activation kinetics (9, 67). Finally, single-channel and tail-current analyses indicate that pyrethroids modify the closed state of the sodium channel but have a higher affinity for the open state if it is present (67). An increased rate of open channel modification would, at least in part, explain the stimulus-dependent effects of the pyrethroids (67).

Sodium current prolongation, acute lethality, and signs of intoxication are closely correlated in the action of pyrethroids. The time constant for the decay rate of the tail current is a measure of the prolongation of the sodium current and can be used to evaluate the efficacy of a pyrethroid (9). The tail current values calculated for a range of pyrethroids match structure-activity relationships for the time course of the negative afterpotential, repetitive firing, and depolarization of the nerve membrane (60, 107). In general, the tail currents of type 1 compounds decay relatively rapidly, which correlates with induction of repetitive firing and a depolarizing afterpotential that subsides within a few tens or hundreds of milliseconds (60, 107). The tail current of the type 2 compounds is more persistent and decays over several seconds or minutes, if at all (60, 107). This effect is correlated with the slow depolarization of the membrane potential and block of the action potential caused by type 2 compounds (9, 60). The more rapid decay of modified sodium currents by type 1 pyrethroids is consistent with their lower toxicity and less intense effects on nerve firing (9). More detailed analysis of time-constant values (60, 107) showed that they fell along a continuum in which the DDT analogues <

nonphenoxybenzyl pyrethroids (e.g. allethrin and tetramethrin) < descyano,3-phenoxybenzyl pyrethroids (e.g. phenothrin and permethrin) < α-cyano-3-phenoxybenzyl pyrethroids (e.g. cypermethrin and deltamethrin). Regression analysis showed that the sodium current time constant, signs of intoxication, and acute lethality were closely correlated over a broad range of values, with a correlation coefficient, r, of 0.913 (9).

Pharmacological interactions of pyrethroids with the voltage-dependent sodium channel have been investigated using tracer ion flux methodologies (9). Incubation of pyrethroids with cultured mouse neuroblastoma cells (51) or mouse brain synaptosomes (14, 45, 95) showed that these compounds did not stimulate ^{22}Na uptake alone but enhanced uptake caused by other sodium-channel toxins. When incubated in the presence of batrachotoxin, veratridine, and dihydrograyanotoxin II (Figure 2), which are known to activate the sodium channel, both deltamethrin and kadethrin enhanced toxin-dependent ^{22}Na uptake into neuroblastoma cells (51). A similar allosteric enhancement occurred (51) when pyrethroids were incubated with scorpion (*Leiurus quinquestriatus*) or sea anemone (*Anemonia sulcata*) polypeptide neurotoxins. Pyrethroids also enhanced batrachotoxin- and veratridine-dependent sodium uptake in mouse brain synaptosomes, generally causing a small (two- to threefold) increase in affinity and, in the case of veratridine, a large increase in maximal uptake (14). Unlike neuroblastoma cells, in synaptosomes the sea anemone toxin ATX II does not activate sodium influx, and its effects are not enhanced by pyrethroids (14). In a structure-activity study (45) of deltamethrin and its descyano analogue NRDC 157, only the toxic isomers of these compounds enhanced veratridine-dependent sodium uptake. This effect was attained at half-maximal concentrations of 25 and 220 nM, respectively, for deltamethrin and NRDC 157. This 8.8-fold difference in potency was closely correlated with the 10-fold difference in the respective toxicities of deltamethrin and NRDC 157 to mice after intracerebral injection (45). Thus, these studies confirm at the biochemical level many of the structure-activity relationships observed in neurophysiological studies.

Radioligand binding studies have demonstrated that pyrethroids bind to a unique site on the sodium channel. The high lipophilicity of the pyrethroids has prevented their use as radiolabeled probes for direct measurement of channel-binding interactions (9). However, the allosteric interactions between pyrethroids and alkaloid activators evident in flux assays were investigated in binding studies with [^{3}H]batrachotoxin A-20-α-benzoate ([^{3}H]BTX-B), a batrachotoxin analogue (Figure 2) that labels the activator site on the sodium channel (15, 79). DDT, deltamethrin, and the neurotoxic isomers of cypermethrin increased the affinity of rat brain synaptosomal sodium channels for BTX-B approximately threefold without affecting maximal binding (15). Moreover, five pyrethroids and DDT exhibited potencies for enhancing BTX-B binding ($r = 0.99$) closely correlated with their ability to enhance sodium influx in neuroblastoma cells (54). In other experiments, pyrethroids did not affect the binding of [^{3}H]tetrodotoxin, or labeled sea anemone or scorpion toxins, indicating that these binding sites are distinct (54).

These studies identified unique toxin-binding sites on the sodium channel, a subset of which are important sites for agents utilized for insect control. The following classification of these toxin-binding sites is a generally accepted numbering system that facilitates discussion. Site 1 binds the specific channel antagonists tetrodotoxin and saxitoxin. Site 2 recognizes the diverse group of lipophilic activators batrachotoxin, veratridine, aconitine, and the grayanotoxins. Site 3 binds the α-scorpion toxins and sea anemone toxins, which slow sodium-channel inactivation and also enhance sodium uptake by site 2 compounds. Finally, pyrethroids bind to a site distinct from sites 1–3 that has been designated either site 5 (101) or site 6 (54). The exact nature of the pyrethroid-binding site is unknown, but because of the lipophilic nature of pyrethroids, it probably resides within the membrane-bilayer portion of the sodium channel (76).

Knockdown resistance to pyrethroids is associated with a reduced sensitivity of the nervous system and involves modification of sodium-channel pharmacology. There is ample evidence for reduced sodium-channel density and altered sodium-channel structure as mechanisms of knockdown resistance in particular strains and species of insect (9). The pharmacological profile of knockdown resistance is best characterized in the house fly, in which the major knockdown resistance alleles, designated *kdr* and *super-kdr,* provide broad cross resistance to all pyrethroids and virtually all DDT analogues (9). Several studies have confirmed that the alteration in the sensitivity of the pyrethroid site in these strains is not accompanied by any significant changes in sites 1 or 3, which recognize tetrodotoxin and saxitoxin and the α-scorpion toxins, respectively (9). However, the picture is not as clear with respect to toxins acting at site 2, where the expression of resistance is not uniform. Thus, resistance to only a subset of site 2 compounds often occurs; the extent of resistance varies from strain to strain; and the magnitude of resistance to site 2 toxins in *kdr* and *super-kdr* strains is not correlated with that of the pyrethroids (9).

Veratrum Alkaloids

The veratrum alkaloids used in insect control are usually extracted from the seeds of sabadilla (*Schoenocaulon* spp., Liliaceae). The compounds from this plant with the greatest insecticidal activity are the lipophilic alkaloids veratridine and cevadine (112). Sabadilla extracts strongly irritate mucous membranes and cause muscle rigor in mammals, as well as paralysis in insects

(112). The primary toxic compound, veratridine, causes an increase in the negative action potential in the axon membrane, repetitive firing, and a depolarization of the nerve membrane potential (75). Veratridine-dependent depolarization requires the presence of extracellular sodium ions (75) and can stimulate the release of various neurotransmitters (58). In structure-activity studies, veratridine was the most active compound, followed by cevadine, and then the protoveratrines (75). Under voltage-clamp conditions, veratridine slows sodium-channel inactivation in frog myelinated nerve and induces a tail current after step repolarization (83). In addition, veratridine causes a large hyperpolarizing shift (up to 93 mV) in the voltage dependence of channel activation (57). At the single-channel level, this effect is expressed as a large increase in the probability of channel opening, specifically mediated by an increase in the channel-opening rate constant (43). This action accounts for veratridine's ability to stimulate ^{22}Na flux into synaptosomes of mammals (14) and insects (35). Competition studies confirmed that veratridine and cevadine act at the same site on the sodium channel (50). They are thought to gain access to this site by moving through the lipid bilayer in their free-base form, becoming protonated and then interacting with negatively charged and hydrophobic residues on the internal face of the sodium channel (50). As mentioned previously, veratridine acts at the same site (site 2) as batrachotoxin, aconitine, and the grayanotoxins (51, 98). Evidence supporting this conclusion comes from ^{22}Na flux studies, which show that veratridine and aconitine are partial agonists and are competitive antagonists of batrachotoxin-dependent channel activation (98). Veratridine and aconitine also block the binding of [^{3}H]TBX-B with affinities that closely match those for activating sodium influx (20). Veratridine cross resistance in *kdr* strains is generally mild (two- to fivefold), although one report described 100-fold resistance in a house fly sensory nerve assay (9). Thus, knockdown-resistant strains might be controlled with particular compounds acting at site 2.

N-Alkylamides

The N-alkylamides (Figure 3) represent a chemical class of insecticide modeled on a group of structurally related unsaturated lipid amides that are similar to the natural products pellitorine and affinin (38). These compounds have been isolated from the roots, stems, or fruit of several plants, especially those belonging to the genus *Piper* (97). The synthetic isobutylamide compound (2E,6Z,8E)-N-isobutyl-9-phenylnona-2,6,8-trienamide (IPT) produced neurotoxic signs of intoxication when topically applied to houseflies, and in whole-insect preparations caused increased discharge and block of conduction in motor neurons (8). When applied to the cockroach nerve cord in vitro, the action potentials developed a large depolarizing afterpotential, which was

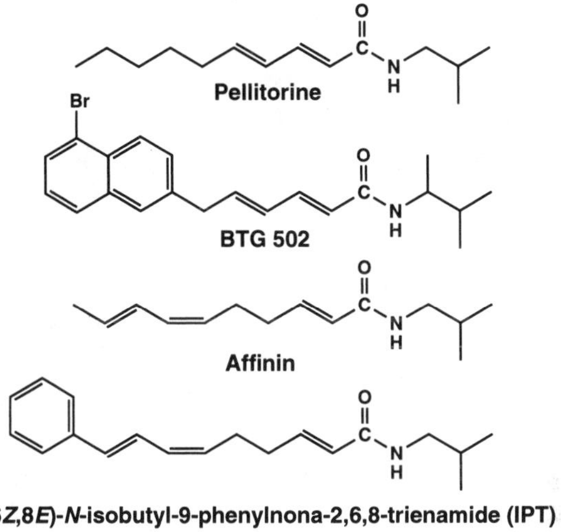

Figure 3   Natural and synthetic *N*-alkylamides.

sometimes associated with trains of spikes and would always eventually block impulse conduction (8, 55). Intracellular recordings from dissociated locust neuronal somata showed that 5 μM IPT caused repetitive firing, prolonged action potentials, and membrane depolarization that was blocked by application of tetrodotoxin (55). Under whole-cell voltage clamp conditions, this compound caused large amplitude, slowly decaying tail currents, a suppression of the peak sodium current, and a 5- to 15-mV negative shift in the peak sodium conductance curve; these effects were similar to those observed for pyrethroids (55).

Studies on the sodium-channel pharmacology of the *N*-alkylamides were investigated in transmitter release, ion flux, and radioligand-binding assays. IPT was tested for its ability to release preloaded [³H]choline from cockroach synaptosomes (69). In this assay, it had no effect alone but inhibited the ability of veratridine to stimulate release via its interaction with sodium channels (69). Ottea et al (77) undertook a more detailed analysis of the actions of *N*-alkylamides on the voltage-sensitive sodium channel using the compound BTG 502. This compound had little effect on ²²Na uptake into synaptosomes when tested alone, but when co-incubated with *L. quinquestriatus* scorpion venom, it exhibited marked synergism of uptake. Half-maximal uptake in the presence of scorpion venom occurred at a concentration of 1.7 μm BTG 502, and it blocked batrachotoxin-dependent sodium influx by 50% at a concentration of 1.5 μM. BTG 502 was a competitive inhibitor of [³H]BTX-B binding, produc-

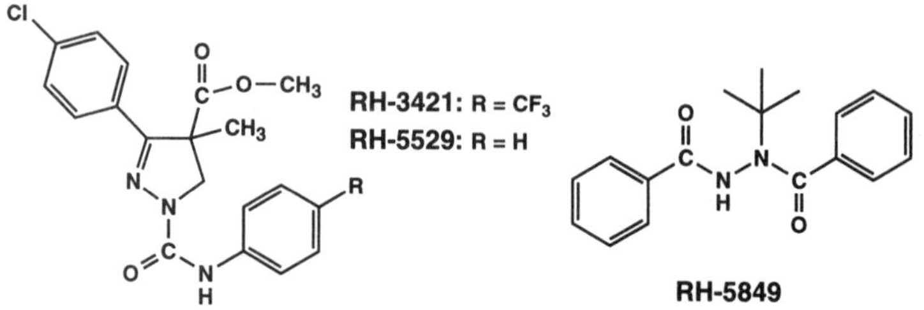

Figure 4   Experimental insecticides discovered by Rhom and Haas Company with unique blocking actions on ion channels. The pyrazolines RH-3421 and RH-5529 are shown (*left*), along with the diacylhydrazine RH-5849 (*right*).

ing half-maximal inhibition at 2 μM. Taken together, these results suggested that the *N*-alkylamides were exerting their effects at the activator site (site 2) on the voltage-sensitive sodium channel (77). Like veratridine, the *N*-alkylamides have greater activity against knockdown-resistant strains of insects than do the pyrethroids (46).

Structure-activity studies with a series of *N*-alkylamides revealed that the ability to activate sodium influx was better correlated with insecticidal activity than the ability to displace [^{3}H]BTX-B binding (78). Three noninsecticidal *N*-alkylamides were identified that displaced [^{3}H]BTX-B binding but apparently could not significantly affect channel function, as measured by stimulation of sodium uptake. Concentration-response curves showed that some *N*-alkylamides can stimulate sodium influx through a second, high-affinity site on the sodium channel distinct from site 2, although this effect was not correlated with insecticidal activity. These studies also established that insecticidal activity in this chemical series was optimal with 2,4-dieneamides and was also affected by the total alkyl chain length (78).

Pyrazolines and Dihydropyrazoles

Another class of insecticide acting upon the voltage-sensitive sodium channel consists of the 1-(phenylcarbamoyl)-2-pyrazolines, also known as dihydropyrazoles (Figure 4). These materials were first synthesized some years ago, and toxicity tests with larval *Aedes aegypti*, *Pieris brassicae*, and *Leptinotarsa decemlineata* showed excellent insecticidal activity (105, 114). Recent molecular shape analyses revealed that the most bioactive conformation was one that corresponded to a near all-planar structure (84). This consideration was much more critical than overall lipophilicity for dictating insecticidal activity.

A major drawback to the pyrazolines is their unacceptable mammalian toxicity, which includes a delayed-onset neurological syndrome (71).

Salgado (86) found that the pyrazolines induce unique symptomology in poisoned insects through an idiosyncratic blocking action on nerves. Adult *Periplaneta americana* or *Manduca sexta* larvae treated with the pyrazoline RH-3421 became uncoordinated and lost postural control. The insects then entered a quiescent state but exhibited convulsions when disturbed. Death occurred over a period of several days. Recordings from symptomatic cockroaches showed that RH-3421 blocked spontaneous activity in sensory, motor, and central neurons. However, tactile stimulation could still activate sensory-to-motor reflex arcs. A similar blockade of spontaneous nerve activity occurred in *M. sexta* larvae. Sensory nerve preparations from both insects were especially sensitive to the action of RH-3421, and the global loss of sensory input seemed to be the primary effect of intoxication.

Intracellular recordings from crayfish giant axons showed that nerve block by RH-3421 was enhanced by membrane depolarization and relieved by hyperpolarization. From these studies, Salgado (86) concluded that the pyrazolines exerted a voltage-dependent block of sodium channels. Voltage-clamp studies (87) confirmed that pyrazoline blockage of the sodium current was specific and voltage dependent. The current was virtually unaffected at a holding potential of -100 mV but was blocked rapidly at a holding potential of -82 mV. The kinetics of pyrazoline blockage were slow, and attainment of steady state required over 20 min. The voltage dependence and slow time course of current blockage suggested that pyrazolines bound to inactivated states of the sodium channel and that binding was especially dependent on the process of slow inactivation (87).

Biochemical studies revealed synaptic effects of pyrazolines and investigated their pharmacological interactions with the sodium channel. RH-3421 inhibited the spontaneous release of [^{3}H]γ-aminobutyric acid (GABA) from preloaded mouse brain synaptosomes with an IC_{50} value of 40 nM (71). Moreover, RH-3421 also blocked the release of label by deltamethrin and α-scorpion toxin (from *L. quinquestriatus*) (71). Pyrazolines (RH-3421 and RH-5529) also antagonized the veratridine-dependent depolarization of synaptosomes, as measured with the voltage-sensitive fluorescent dye rhodamine 6G (68). The lack of effect of either compound on potassium-dependent depolarization confirmed a specific effect on the sodium channel (68). Other studies found that RH-3421 blocked ^{22}Na uptake stimulated by veratridine, batrachotoxin, α-scorpion venom, and pumiliotoxin B, confirming that the pyrazolines were general antagonists of sodium-channel activation (28). In addition, interactions between pyrazolines and the alkaloid activators veratridine and batrachotoxin displayed a reciprocal competitive inhibition (28). Although this fact might suggest that the pyrazolines were acting at site 2 of the sodium channel, binding studies with BTX-B clearly showed that RH-3421 noncompetitively displaced BTX-B through an increase in its dissociation rate (27). These actions are similar to those of local anesthetics, anticonvulsants, and antiarrhythmics, suggesting that the pyrazolines bind to the same site as these compounds, a site that differs from sites 1–3 and the pyrethroid site (27).

Dong & Scott (33) reported that a *kdr*-like strain of *Blattella germanica* showed fourfold cross resistance to RH-23421, as well as to pyrethroids, batrachotoxin, and aconitine. This finding suggests a rather extensive structural change in the sodium channel of this strain, which deserves further study.

POLYPEPTIDE TOXINS

Much interest surrounds the development of genetically engineered biopesticides with enhanced pesticidal and pestistatic properties by introducing genes for neurotoxic polypeptides into the baculovirus genome. For example, a toxin from the straw itch mite (*Pyemotes tritici*) that has paralytic activity specific for insects was cloned into a baculovirus, and this gene was expressed in cultured cells and in caterpillars (100). However, the site of action and pharmacology of this toxin is unknown. Other studies (26, 96) reported the expression of a virus containing a gene for the insect-specific neurotoxin from the scorpion *Androctonus australis*. This *A. australis* insect toxin, AaIT, is a protein of 70 amino acids with 4 disulfide bridges (119). AaIT causes a rapid-contraction paralysis when injected into blowfly larvae and, in neurophysiological studies, prolonged trains of action potentials in presynaptic motor nerve terminals (119). Similar repetitive activity is observed in cockroach axons, and the decay of the sodium current is prolonged by AaIT under voltage-clamp conditions (119). [^{125}I]AaIT binds to nervous tissue with a K_d of 1.19 nM and a maximal binding capacity of 1.37 pmol mg^{-1} membrane protein (120). This binding capacity was similar to that of [^{3}H]saxitoxin, 1.43 pmol/ mg^{-1} membrane protein (120). These data provide compelling evidence that this toxin interacts with the voltage-sensitive sodium channel of nerve membrane (119, 120). No information is available on the interactions of AaIT with alkaloid activators or pyrethroids on insect sodium channels. Experimental evidence that they bind to a separate site on the channel comes from studies of knockdown-resistant strains of insects, which exhibit no cross resistance to any of the scorpion toxins (9). Thus, the performance of these toxins in the field should remain unaffected by the presence of pyrethroid resistance.

The continued search for toxins suitable for use in bioengineered pesticides resulted in the recent discovery of a new toxin, NPS-901, from the spider

Diguetia canities. When injected, this toxin showed paralytic activity that was specific for insects, and cloning into a baculovirus enhanced the virus' insecticidal activity (53). Mode-of-action studies (13) showed that both native and recombinant NPS-901 toxin elevated spike discharges in peripheral sensory nerves and induced trains of excitatory junction potentials in skeletal muscle after a single stimulus to the motor nerve. Tetrodotoxin both prevented and reversed these effects. Similarly, NPS-901 caused a depolarization of the membrane potential of cockroach giant axon preparations that was also reversed by tetrodotoxin. The action of NPS-901 was different from the potassium channel blocker 4-aminopyridine, which caused a predictable broadening of the cockroach axon spike without depolarizing the membrane potential (13). These studies suggested that NPS-901 acts upon the voltage-sensitive sodium channel. In an effort to identify the binding domain for NPS-901, it was screened in a binding assay with [^{125}I]AaIT. NPS-901 proved to be a poor inhibitor of [^{125}I]AaIT binding to cell membranes isolated from house fly heads, which indicates that it may act at a different site on the voltage-sensitive sodium channel (13).

CHLORIDE CHANNELS

Channel-Blocking Convulsants

The channel-blocking convulsants represent a structurally diverse group of compounds that includes commercial and experimental insecticides, as well as the botanical toxin picrotoxinin (Figure 5). The oldest group of commercial compounds are the polychlorocycloalkane insecticides that include the environmentally stable cyclodienes and lindane (17). Another chemical family of potent convulsants are the trioxabicyclooctanes, and the first compound synthesized in this group was *t*-butylbicyclophosphorothionate (TBPS) (17). Chemical synthesis around the trioxabicyclooctane moiety resulted in the discovery of the bicycloorthobenzoates, as epitomized by various *t*-butylbicycloorthobenzoate (TBOB) analogues (17). Later studies showed that the trioxabicyclooctane moiety could be replaced with silatrane, dithiane, spirosultam (LY 219048), pyrazole (fipronil, JKU 0422), or triazole (SN 606011) ring systems (10).

In both insects and mammals, the chloride channel blockers are highly toxic convulsants that cause hyperexcitation of the nervous system through antagonism of the inhibitory neurotransmitter GABA (10, 17). A series of studies on motor neuron discharge in larval central nervous system (CNS) preparations from *Drosophila melanogaster* showed that 1 mM GABA inhibited neuron discharge and that inhibition was reversed by treatment with dieldrin (10), picrotoxinin (10), LY 219048 (12), JKU 0422 (11), and TBPS (11). Similarly,

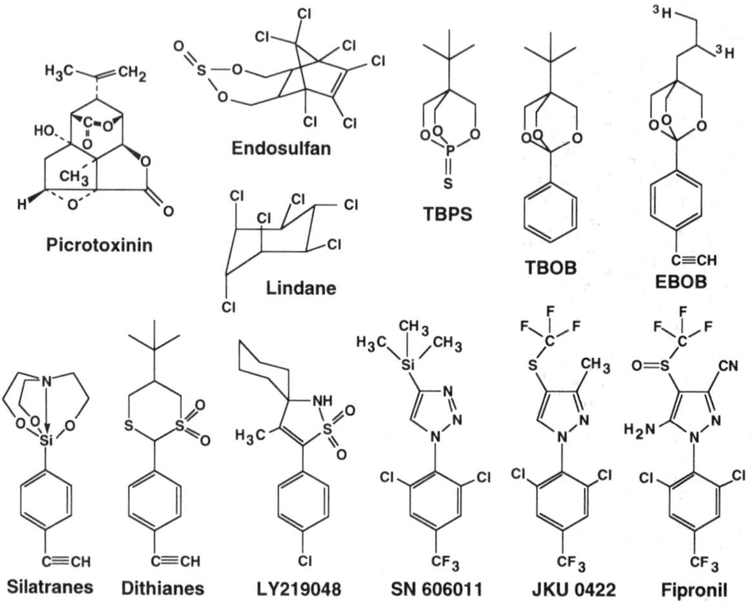

Figure 5 Chemical structures of compounds that block the GABA-gated chloride channel. TBOB epitomizes the structure of various *t*- butylbicycloorthobenzoates.

a high concentration of GABA was shown to delay the onset of hyperexcitation by 10 μM lindane and picrotoxinin in the cockroach CNS (99). Ionophoretic application of GABA to the cockroach D_f neuron hyperpolarized the membrane potential, and this effect was antagonized in a noncompetitive manner by 1 μM endrin or 10 μM lindane (110). A similar antagonism of GABA-induced currents in voltage-clamped locust neurons was observed with SN 606011 (109). The noncompetitive kinetics of current blockage by polychlorocycloalkanes and the similarity of their action to that of picrotoxinin are consistent with allosteric GABA antagonism via a blocking action on the chloride channel (44). In rat dorsal root ganglion cells, GABA induced both transient and sustained currents, and the transient inward current was antagonized noncompetitively by lindane (74) and dieldrin (66). Similar effects on transient GABA-stimulated currents were observed in voltage-clamped *Xenopus laevis* oocytes expressing rat brain GABA receptors (10). Noise analysis experiments on cultured locust and cockroach neurons found that picrotoxinin, dieldrin, and lindane reduced the frequency of GABA-stimulated channel opening with no change in channel open time (7), which is consistent with a stabilization of a closed-liganded form of the channel (10). Recent patch clamp studies in *D. melanogaster* neurons led to the conclusion that chloride-channel blockers such as picrotoxinin, lindane, and TBPS preferentially bind to a nonconducting, desensitized state of the GABA receptor (118).

GABA-gated ^{36}Cl flux experiments confirmed the GABA antagonism observed in neurophysiological studies and facilitated analysis of structure-activity relationships. When any of the compounds described above were tested in flux assays, they proved to be potent inhibitors of GABA-stimulated chloride uptake (10). Kinetic studies demonstrated that the inhibition was noncompetitive, which confirmed the effects observed in neurophysiological assays (110). Across several studies, inhibitory potencies for polychlorocycloalkanes consistently grouped together: the IC_{50} values for 12-ketoendrin, isobenzan, endrin, and endosulfan I < dieldrin and heptachlor epoxide < aldrin, heptachlor, and lindane (10). The observed potency profile for inhibiting GABA-dependent chloride flux showed good correlation ($r = 0.78–0.92$) with the acute toxicity of cyclodienes (10). These quantitative analyses strengthened the link between blockage of the GABA-gated chloride channel and the neuroexcitation and acute lethality of the polychlorocycloalkanes.

Characterization of convulsant binding at the chloride channel was originally attempted with [^{3}H]α-dihydropicrotoxinin, but incorporation of radioisotope to give [^{35}S]TBPS or [^{3}H]TBOB resulted in improved ligands with higher affinities and a higher percentage of specific binding (17). Essentially all the channel-blocking convulsants are competitive inhibitors of [^{35}S]TBPS or [^{3}H]TBOB binding, indicating that they act at a similar or identical site (10, 17). The phenylpyrazoles (e.g. fipronil) are exceptional in that they are noncompetitive antagonists, either because their binding is irreversible or because they bind to a coupled but distinct site (24). In several studies, the displacement of binding by polychlorocycloalkanes showed excellent correlation with a block of GABA-stimulated chloride uptake in mammalian brain vesicles, suggesting that occupation of the binding site coincides with blockage of the chloride ion channel (10). Detailed kinetic studies of binding inhibition showed that the receptor affinity for cyclodiene binding was controlled by the dissociation rate, whereas the receptor affinities for the TBOB analogues and dithianes depended on the association rate (49). This latter finding suggests an induced-fit model in which binding of these ligands initiates a conformational change into a blocked state (49). The potency of cyclodienes for displacing [^{35}S]TBPS binding is also correlated with their acute toxicity, but in most cases the data from functional chloride flux assays are better predictors of the acute toxicity of convulsant channel blockers (10).

Casida (17) reviewed evidence that chloride channel blockers fell into two groups (types A and B) with respect to correlations of chloride flux inhibition vs acute lethality. Type A comprised larger or more extended molecules with high insecticidal activity (e.g. polychlorocycloalkanes, picrotoxinin, and TBOB analogues), whereas type B included smaller, more compact compounds of reduced potency and insecticidal activity (e.g. TBPS)

Although studies have demonstrated that [^{35}S]TBPS and [^{3}H]TBOB are extremely useful as radioligands in mammals, binding studies with these radioligands in insects have often produced contradictory and unpredictable results (10). Currently, the best ligand for the insect GABA receptor is 1-(4-ethynylphenyl)-4-[2,3-^{3}H$_2$]propyl-2,6,7-trioxabicyclo[2.2.2]octane, or [^{3}H]EBOB (10). Membranes prepared from house fly heads have high-affinity binding sites for this compound (K_d = 1.4 nM), and displacement of [^{3}H]EBOB binding by lindane displays competitive kinetics (30). In addition, displacement of [^{3}H]EBOB binding and acute toxicity to house flies were closely correlated for a wide range of channel-blocking convulsants, including TBOB analogues, cyclodiene insecticides, picrotoxinin, and substituted dithianes and silatranes (30). Using house fly head membranes, Deng et al (30, 31) compared several structurally diverse chloride channel–directed toxicants in terms of inhibition of [^{35}S]TBPS and [^{3}H]EBOB binding, acute lethality, and temperature coefficient of toxicity. From these studies, they expanded their original concept of types A and B to include types C and D as well. Types A and B remained essentially the same as described above, with the addition of some trithiabicyclooctanes and dithianes as mixed type A/B. Type C comprised the phenylpyrazoles, and type D the avermectins (31).

Resistance to cyclodiene insecticides is mediated by an alteration in the structure of a GABA receptor subunit and provides cross resistance to a broad range of channel-blocking convulsants. ffrench-Constant and colleagues (39, 40) have documented that cyclodiene resistance occurs through a single amino acid substitution (Ala to Ser or Gly) in the M2 region of a GABA-receptor subunit. Numerous studies of cyclodiene-resistant strains, using both toxicity determinations and neurophysiological responses, have reported cross resistance to polychlorocycloalkane insecticides, TBPS, a TBOB analogue, picrotoxinin, LY219048, and phenylpyrazoles (3, 10, 11, 24). The findings of Deng and coworkers (31) are generally in accord with these results, except that these authors observed low levels of cross resistance to compounds designated type A/B or type B in dieldrin-resistant house flies.

Channel-Activating Avermectins

The avermectins (Figure 6) are a group of related macrocyclic lactones isolated from *Streptomyces avermitilis* (22). The basic structural motif is found in the natural product avermectin B$_{1a}$, which constitutes the major component in the insecticide abamectin (22). Chemical modification of avermectin B$_{1a}$ has yielded a number of semisynthetic materials. Reduction of a single unsaturation

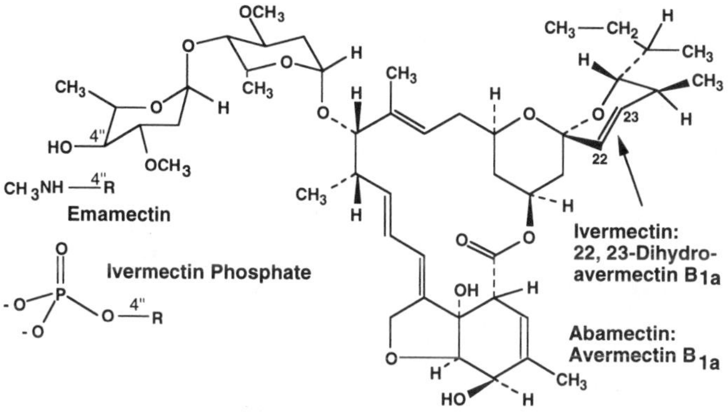

Figure 6   Natural and semi-synthetic avermectins. The natural insecticide abamectin contains at least 80% avermectin B_{1a} (10).

in avermectin B_{1a} gives ivermectin, a commercial anthelmintic. A related compound with insecticidal activity is MK-244 or emamectin (4″-epimethylamino-4″-deoxyavermectin B_{1a}) (22).

Mammals poisoned with avermectins exhibit hyperexcitability, uncoordination, and tremor, which typically give way to ataxia and coma-like sedation. In insects and nematodes poisoned by avermectins, ataxia and paralysis predominate, and overt signs of hyperexcitation are largely absent (10, 22).

The avermectins block electrical activity in a variety of vertebrate and invertebrate nerve and muscle preparations by increasing the membrane conductance to chloride ions (10, 22). In tissues containing GABA receptors, this conductance increase is often accompanied by a loss of sensitivity to exogenously applied GABA (10, 22). At the single-channel level, patches of *Ascaris suum* muscle bathed in ivermectin showed activation of multiple chloride channels that were sensitive to changes in chloride-ion concentration (61). Moreover, when GABA was applied in the presence of 20 nM ivermectin, both the frequency of channel opening and the single-channel amplitude were reduced (61).

The avermectins also affect a variety of other ligand-gated chloride channels (10, 22), including one gated by glutamate in insect and nematode skeletal muscle that may be primarily responsible for avermectin-induced muscle paralysis (4, 5, 34). This glutamate-gated chloride is activated by several avermectins, including the water-soluble compound ivermectin phosphate (4, 5). Two subunits of this receptor (α and β) were recently cloned from the nematode *Caenorhabditis elegans* (24a), which should facilitate the molecular analysis of avermectin action in invertebrates.

Effects of avermectins are also evident in chloride flux assays; in these studies, they block GABA-stimulated uptake (93) and cause release of pre-loaded ^{36}Cl from brain vesicles by activating voltage-dependent chloride channels (1, 80). Avermectins also cause a chloride channel–dependent release of neurotransmitters in mammals (111) and insects (70). This effect, in conjunction with GABA antagonism, probably contributes to excitatory signs of intoxication.

In binding studies, [^{3}H]abamectin and [^{3}H]ivermectin possess high affinity for mammalian brain-membrane preparations, with K_d values of approximately 2 nM and 22 nM, respectively (10). High affinity binding of [^{3}H]ivermectin ($K_d = 0.26$ nM) was also observed in membranes from the nematode C. elegans, and the greater affinity of the nematode receptor is probably responsible for the wide therapeutic index of this compound for controlling parasitic nematodes (92). Structure-activity studies of eight avermectin analogues showed that their potency for inhibiting [^{3}H]ivermectin binding was closely correlated ($r = 0.923$) with their potency for paralyzing C. elegans (92). Kinetic analysis of [^{3}H]ivermectin binding to C. elegans membranes demonstrated a rapid, reversible complex formation that was followed by essentially irreversible binding (92). More recent studies analyzed the interactions of a series of 10 avermectins with [^{3}H]avermectin and [^{3}H]EBOB binding to membrane preparations from house fly heads (29). Binding displacement of both ligands was correlated with the toxicity of these compounds. However, displacement of [^{3}H]EBOB binding was noncompetitive and incomplete, indicating that the avermectins probably act at a different site on the GABA-gated chloride channel (29). Unique interactions of avermectins with chloride channels are supported by the lack of cross resistance observed in cyclodiene-resistant strains (10, 22).

CALCIUM CHANNELS

Ryanodine

Ryania has been used as an insecticide since 1945 and consists of the powdered stem of the tropical shrub Ryania speciosa (112). The extract contains several structurally related ryanoids (52), but most research on the mode of action of these materials has examined ryanodine. Ryanodine augments contracture of insect, crustacean, and vertebrate skeletal muscles without depolarizing the muscle membrane (47). Under voltage clamp conditions, 1 μM ryanodine increased the peak calcium current that flowed in response to a step depolarization and shifted the activation of this current to more hyperpolarized potentials (47). Fill & Coronado (41) reviewed numerous studies demonstrating that ryanodine primarily attacks the calcium-release channel of sarcoplasmic reticulum (SR). In heavy vesicles isolated from SR, ryanodine altered the permeability of the vesicles to ^{45}Ca (62). In patch-clamp studies, nanomolar concentrations of ryanodine irreversibly activated the calcium-release channels, but with a reduced amplitude of single-channel current (41). Higher concentrations tended to block the channel (41). Irreversible activation of the SR calcium channel is consistent with the tetanic contracture observed in ryanodine poisoning of skeletal muscle (103).

Interaction of ryanoids with the SR release channel was evident in radioligand-binding assays. [^{3}H]Ryanodine displayed specific, high-affinity binding to SR membranes from various mammalian sources (K_d values of 20–40 nM) that was sensitive to the calcium-release blocker ruthenium red (81). Similar high-affinity, ruthenium red–sensitive binding was observed in cockroach and house fly muscle preparations, from which the K_d value obtained was 4–6 nM (56). However, insect preparations do differ in some aspects of their ryanodine pharmacology. In calcium-free media, [^{3}H]ryanodine binding is stimulated by caffeine in cockroach and mouse muscle preparations but not in preparations from house fly thorax (56). Molecular modeling studies showed that the dominant electrostatic interaction for ryanodine binding is localized to the hydroxyl group (position O34) contained in a hydrophobic area of the molecule (113). The pyrrole and isopropyl groups have the greatest steric effect on binding and are thought to be buried within a cleft of the channel protein (113). In contrast, the other end of the molecule appears to be oriented near the entrance to the binding pocket, as bulky groups placed in the 9 position or a fluorescent adduct substituted in the 21 position have little effect on binding (113).

Inhibition of [^{3}H]ryanodine binding and the toxicity of ryanoids to mammals and insects are qualitatively correlated. For a series of seven ryanoids, inhibition of [^{3}H]ryanodine binding to mammalian muscle preparations was correlated with toxicity to mice (81). The most potent and toxic compounds (Figure 7) were ryanodine and 9,21-dehydroryanodine (81), which had IC_{50} values for inhibiting [^{3}H]ryanodine binding of 27 and 23 nM, respectively, in mammalian skeletal muscle. Both compounds had LD_{50} values to mice of 0.1 mg kg^{-1}. At the other end of the activity spectrum were anhydroryanodine and ryanodol, which were essentially inactive and had IC_{50} values > 10,000 nM and LD_{50} values > 20 mg kg^{-1} (81). Ryanodine and 9,21-dehydroryanodine are the most abundant ryanoids in stem extract and account for virtually all of the insecticidal activity (52). Unfortunately, they possess high mammalian toxicity. Greater selectivity was available from other stem constituents, such as 10-(O-methyl)-ryanodine (52). This compound was twice as active as ryanodine at the LD_{50} level in flies, but was about 28-fold less toxic to mice (52). Similarly, ryanodol is of low toxicity to mice (81) but has 26–50% of the toxic activity of ryanodine in cockroaches and flies (56). Despite its high toxicity to insects, ryanodol is a relatively poor inhibitor of [^{3}H]ryanodine binding and therefore may act at a distinct binding site (56).

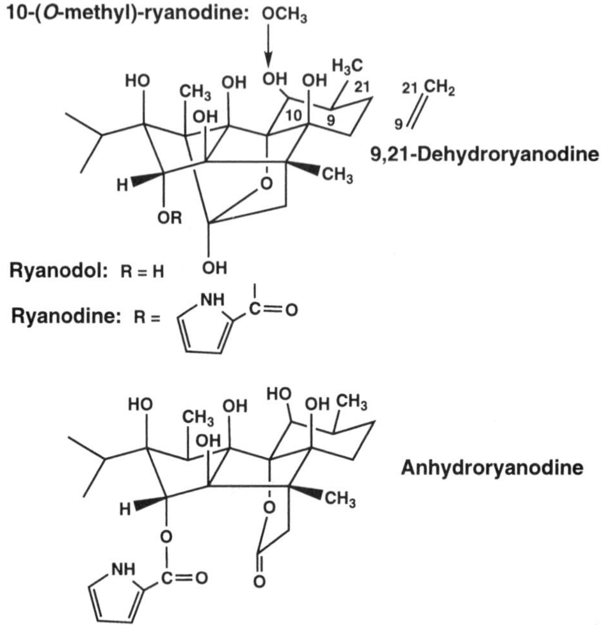

Figure 7 Chemical structures of ryanodine analogues that illustrate significant aspects of their structure-activity relationships or insect-selective toxicity.

Pyrazolines and Dihydropyrazoles

In addition to their blocking action on sodium channels, at least some pyrazolines appear to significantly affect calcium channels as well. A single study (117) has compared the effects of the pyrazoline RH-5529 (Figure 4) on both sodium and calcium channels by monitoring synaptosomal influx of calcium with the fluorescent indicator, fura-2. Actions on sodium channels were studied by measuring the ability of RH-5529 to block veratridine-dependent uptake of calcium ions. This calcium uptake results from veratridine-stimulated influx of sodium, which depolarizes the synaptosomes and thereby activates presynaptic calcium channels. The IC_{50} for blocking veratridine-dependent action on sodium channels in this assay was approximately 3 μM. Similarly, direct activation of presynaptic calcium channels by depolarization with 60 mM K$^+$ was blocked by RH-5529 with an IC_{50} value of 3.5 μM. Thus, RH-5529 blocks sodium and calcium channels of mouse brain synaptosomes with near equal affinity (117). Previous studies on RH-5529 (71) showed that it could block K$^+$-induced release of neurotransmitter, an action that was not shared by the related pyrazoline RH-3421. This effect, which was specific to RH-5529, is consistent with the blockage of calcium channels reported by Zhang & Nicholson (117). Finally, the structural dependence of pyrazoline interactions with the calcium channel can be traced to a single difference in phenyl substituent present in these two compounds. It may be that the electron withdrawing character of the trifluoromethyl group increases effective binding to, or interaction with, the sodium channel, while the presence of hydrogen allows for a lower-affinity interaction with either channel (71).

POTASSIUM CHANNELS

Diacylhydrazines

The 1,2-dibenzoyl-1-*tert*-butylhydrazines (e.g. RH-5849; Figure 4) were introduced as insecticides on the basis of their activity as ecdysone agonists (115). Subsequent investigation of these compounds revealed that they had some neurotoxic actions in Coleoptera, and moreover that these effects occurred at field-use rates (64). Salgado (88) observed that injections of 50 μg g^{-1} RH-5849 in *P. americana* caused hyperactivity within 3 min and prostration within 7 min. Nerve ablation experiments indicated that these signs were caused by an action on the CNS. This finding was confirmed by sucrose-gap recordings of synaptic transmission across the cockroach cercal nerve-giant fiber synapse, where 170 μM RH-5849 prolonged the excitatory postsynaptic potential and caused repetitive firing in the axon (88). RH-5849 (100 μM) also greatly prolonged the neurally evoked responses in ventral longitudinal muscles of house fly larvae, but without causing repetitive firing. The slowing of action potential repolarization suggested an action on potassium channels. Voltage-clamp experiments (88) showed that RH-5849 preferentially blocked the late sustained component (I_K) of the potassium current with little effect on the initial transient component (I_A). In more detailed voltage-clamp studies on larval house fly muscle, Salgado (89) estimated the IC_{50} value of RH-5849 for blocking I_K to be 59 μM, with a Hill coefficient of 1.5. This latter finding suggested that the binding of more than one molecule of RH-5849 was required to block the channel. However, because of the physical properties of RH-5849 and its behavior in aqueous media in vitro, these quantitative findings are somewhat ambiguous (89).

PERSPECTIVES AND CONCLUSIONS

The voltage-sensitive sodium channel remains an important insecticide target site that continues to be exploited with new compounds and technologies. This ion channel is well established as the primary molecular target site for all pyrethroids and virtually all DDT analogues (9). Other natural toxins targeting the sodium channel have been used in insect control and have provided a starting point for chemical synthesis efforts. For example, compounds binding to site 2, such as veratridine and the N-alkylamides, are natural insecticides found in plants, and these chemicals have piqued the interest of chemists. These compounds also show some potential for circumventing knockdown resistance to pyrethroids. The potential for utilizing site 3 toxins for insect control has been demonstrated by inserting genes into baculoviruses that code for sodium channel-directed polypeptides. It may even be possible to design a pseudopeptide or nonpeptide compound that would act at site 3, via techniques similar to those used in the development of synthetic mimics and antagonists of insect neuropeptides (65). The pyrazolines apparently bind to a unique site on the sodium channel and possess a highly insecticidal blocking action that is unlike that of other compounds. However, whether pyrazolines will become commercially viable remains to be seen, given what appears to be an undesirable mammalian toxicity. The continued molecular analysis of mammalian and insect sodium channels promises to illuminate the structure of the various toxin-binding sites and provide insight into the molecular basis of knockdown resistance. This knowledge will be indispensable in the future development of new synthetic compounds or toxins aimed at this membrane ion channel.

The development of chloride channel–directed insecticides was given a boost by the discovery and commercialization of the avermectins, as well as the discovery of the highly potent bicycloorthobenzoates. Drawing on this knowledge, chemists have synthesized several classes of arylheterocycles with high potency as GABA antagonists. One commercial compound, fipronil, has been introduced, and more will probably follow. Fipronil has somewhat better selectivity than the bicycloorthobenzoates by virtue of its reduced mammalian toxicity (24). A potential problem for compounds with a mode of action similar to the cyclodienes is the high level of cross resistance present in cyclodiene-resistant strains (10). The prevalence of this type of target-site resistance suggests that new chloride channel-blocking insecticides should be used judiciously to manage the selection pressure on pest populations. As stated previously, cyclodiene resistance does not affect the toxicity of the avermectins, but metabolic and penetration resistance to these compounds has appeared in field populations and after laboratory selection (22).

The ryanoids are the only insecticides to have a calcium channel as their primary target site. The recent identification of more insect-selective ryanodine analogues may facilitate the use of second-generation ryanoids in the future. The nonselective action of RH-5529 on both sodium and calcium channels, as well as the much greater effect of RH-3421 on sodium channels, indicates that the pyrazolines are primarily sodium channel–directed toxicants. It seems a bit surprising that voltage-sensitive calcium channels have not been exploited by a greater variety of commercial insecticidal compounds. Voltage-sensitive calcium channels have been cloned (16), are ubiquitous in nerve and muscle, and play pivotal roles in cellular excitability and release of neurotransmitters (63, 102). In addition, disruption of calcium ion fluxes and homeostasis is cytotoxic (73). Numerous synthetic chemical (63, 102) and toxin (2) probes for calcium channels are also available. Thus, this ion channel would appear to be an excellent candidate for insecticide development.

The diacylhydrazines represent one of the few cases where potassium channels constitute a target site for an insecticide. Even in this case, however, the main insecticidal action of these compounds is probably their ability to act as ecdysteroid agonists. For many years, neuronal and muscle potassium channels lacked the toxin probes available for the sodium channel. The only available probes were channel-blocking compounds of relatively low potency, such as tetraethylammonium and the aminopyridines (116). The past few years have seen the discovery of several potent natural toxins, including apamin, charybdotoxin, noxiustoxin, and dendrotoxin, that block various potassium-channel subtypes (18). These toxins might also prove to be useful in engineered baculoviruses. Synergistic activity might be achieved by cloning into a virus a cocktail of toxins that have appropriate actions at different ion channels. This mixture might include toxins to enhance sodium influx, while others might synergize these effects by blocking potassium channels and thereby prevent repolarization of the nerve membrane potential.

ACKNOWLEDGMENTS

I thank the Editors of *Annual Reviews* for their kind invitation to contribute this chapter. Other thanks are extended to my collaborators at FMC and NPS Pharmaceuticals for permission to report results on the *Diguetia canites* toxin. Ms. Debbie Price is thanked for her secretarial assistance.

Literature Cited

1. Abalis IM, Eldefrawi AT, Eldefrawi ME. 1986. Actions of avermectin B_{1a} on the γ-aminobutyric acid$_A$ receptor and chloride channels in rat brain. *J. Biochem. Toxicol.* 1:69–82

2. Adams ME, Bindokas VP, Hasegawa L, Venema VJ. 1990. Ω-Agatoxins: novel calcium channel antagonists of two subtypes from funnel web spider (*Agelenopsis aperta*) venom. *J. Biol. Chem.* 265:861–67

3. Anthony NM, Benner EA, Rauh JJ, Sattelle DB. 1991. GABA receptors of insects susceptible and resistant to cyclodiene insecticides. *Pestic. Sci.* 33: 223–30

4. Arena JP, Liu KK, Paress PS, Cully DF. 1991. Avermectin-sensitive chloride currents induced by *Caenorhabditis elegans* RNA in *Xenopus* oocytes. *Mol. Pharmacol.* 40:369–74

5. Arena JP, Liu KK, Paress PS, Schaeffer

JM, Cully DF. 1992. Expression of a glutamate-activated chloride current in *Xenopus* oocytes injected with *Caenorhabditis elegans* RNA: evidence for modulation by avermectin. *Mol. Brain Res.* 15:339–48

6. Berlin JR, Akera T, Brody TM, Matsumura F. 1984. The inotropic effects of a synthetic pyrethroid decamethrin on isolated guinea pig atrial muscle. *Eur. J. Pharmacol.* 98:313–22

7. Bermudez I, Hawkins CA, Taylor AM, Beadle DJ. 1991. Actions of insecticides on the insect GABA receptor complex. *J. Receptor Res.* 11:221–32

8. Blade RJ, Burt PE, Hart RJ, Moss MDV. 1985. The action of insecticidal isobutylamide compounds on the insect nervous system. *Pestic. Sci.* 16:554

9. Bloomquist JR. 1993. Neuroreceptor mechanisms in pyrethroid mode of action and resistance. *Rev. Pestic. Toxicol.* 2:185–230

10. Bloomquist JR. 1993. Toxicology, mode of action and target site–mediated resistance to insecticides acting on chloride channels. *Comp. Biochem. Physiol. Ser. C* 106:301–14

11. Bloomquist JR. 1994. Cyclodiene resistance at the insect GABA receptor/chloride channel complex confers broad cross resistance to convulsants and experimental phenylpyrazole insecticides. *Arch. Insect Biochem. Physiol.* 26:69–79

12. Bloomquist JR, Jackson JL, Karr LL, Ferguson HJ, Gajewski RP. 1993. Spirosultam LY219048: a new chemical class of insecticide acting upon the GABA receptor/chloride ionophore complex. *Pestic. Sci.* 39:185–92

13. Bloomquist JR, Kinne L, Deutsch V, Simpson S. 1994. *Insecticidal peptide toxin from Diguetia canites: mode of action studies.* Presented at Annu. Meet. Am. Chem. Soc., 208th, Washington DC

14. Bloomquist JR, Soderlund DM. 1988. Pyrethroid insecticides and DDT modify alkaloid-dependent sodium channel activation and its enhancement by sea anemone toxin. *Mol. Pharmacol.* 33:543–50

15. Brown GB, Gaupp JE, Olsen RW. 1988. Pyrethroid insecticides: stereospecific allosteric interaction with the batrachotoxinin-A benzoate binding site of mammalian voltage-sensitive sodium channels. *Mol. Pharmacol.* 34:54–59

16. Campbell KP, Leung AT, Sharp AH. 1988. The biochemistry and molecular biology of the dihydropyridine-sensitive

calcium channel. *Trends Neurosci.* 11:425–30

17. Casida JE. 1993. Insecticide action at the GABA-gated chloride channel: recognition, progress, and prospects. *Arch. Insect Biochem. Physiol.* 22:13–23

18. Castle NA, Haylett DG, Jenkinson DH. 1989. Toxins in the characterization of potassium channels. *Trends Neurosci.* 12:59–65

19. Catterall WA. 1984. The molecular basis of neuronal excitability. *Science* 223:653–61

20. Catterall WA, Morrow CS, Daly JW, Brown GB. 1981. Binding of batrachotoxinin A 20-α-benzoate to a receptor site associated with sodium channels in synaptic nerve ending particles. *J. Biol. Chem.* 256:8922–27

21. Chalmers AE, Osborne MP. 1986. The crayfish stretch receptor organ: a useful model system for investigating the effects of neuroactive substances. *Pestic. Biochem. Physiol.* 26:128–38

22. Clark JM, Campos F, Scott JG, Bloomquist JR. 1995. Resistance to avermectins: extent, mechanisms, and management implications. *Annu. Rev. Entomol.* 40:1–30

23. Clements AN, May TE. 1977 The actions of pyrethroids upon the peripheral nervous system and associated organs in the locust. *Pestic. Sci.* 8:661–80

24. Cole LM, Nicholson RA, Casida JE. 1993. Action of phenylpyrazole insecticide at the GABA-gated chloride channel. *Pestic. Biochem. Physiol.* 46:47–54

24a. Cully DF, Vassilatis DK, Liu KK, Paress PS, van der Ploeg LHT, et al. 1994. Cloning of an avermectin-sensitive glutamate-gated chloride channel from *Caenorhabditis elegans. Nature* 371:707–11

25. Daly JW, McNeal ET, Gusovsky F. 1987. Cardiotonic activities of pumiliotoxin B, pyrethroids and a phorbol ester and their relationships with phosphatidylinositol turnover. *Biochim. Biophys. Acta* 930:470–74

26. Dee A, Belagaje M, Ward K, Chio E, Lai MT. 1990. Expression and secretion of a functional scorpion insecticidal toxin in cultured mouse cells. *Bio/Technology* 8:339–42

27. Deecher DC, Payne GT, Soderlund DM. 1991. Inhibition of [³H]batrachotoxinin A–20-α-benzoate binding to mouse brain sodium channels by the dihydropyrazole insecticide RH 3421. *Pestic. Biochem. Physiol.* 41:265–73

28. Deecher DC, Soderlund DM. 1991. RH 3421, an insecticidal dihydropyrazole, inhibits sodium channel–dependent so-

dium uptake in mouse brain preparations. *Pestic. Biochem. Physiol.* 39:130–37

29. Deng Y, Casida JE. 1992. House fly head GABA-gated chloride channel: toxicologically relevant binding site for avermectins coupled to site for ethynylbicycloorthobenzoate. *Pestic. Biochem. Physiol.* 43:116–22

30. Deng Y, Palmer CJ, Casida JE. 1991. House fly brain γ-aminobutyric acid-gated chloride channel: target for multiple classes of insecticides. *Pestic. Biochem. Physiol.* 41:60–65

31. Deng Y, Palmer CJ, Casida JE. 1993. House fly head GABA-gated chloride channel: four putative insecticide binding sites differentiated by [³H]EBOB and [³⁵S]TBPS. *Pestic. Biochem. Physiol.* 47:98–112

32. Doherty JD, Lauter CJ, Salem N. 1986. Synaptic effects of the synthetic pyrethroid resmethrin in rat brain in vitro. *Comp. Biochem. Physiol. Ser. C* 84:373–79

33. Dong K, Scott JG. 1991. Neuropharmacology and genetics of *kdr*-type resistance in the German cockroach, *Blattella germanica* L. *Pestic. Biochem. Physiol.* 41:159–69

34. Scott RH, Duce IR. 1985. Effects of 21,23-dihydroavermectin B₁ₐ on locust *(Schistocerea gregaria)* muscles may involve several sites of action. *Pestic. Sci.* 16:599–604

35. Dwivedy AK. 1988. Alkaloid neurotoxin–dependent sodium transport in insect synaptic nerve-ending particles. *Comp. Biochem. Physiol. Ser. C* 91:349–54

36. Eels JT, Bandettini PA, Holman PA, Propp JM. 1992. Pyrethroid insecticide–induced alterations in mammalian synaptic membrane potential. *J. Pharmacol. Exp. Ther.* 262:1173–81

37. Eels JT, Dubocovich ML. 1988. Pyrethroid insecticides evoke neurotransmitter release from rabbit striatal slices. *J. Pharmacol. Exp. Ther.* 246:514–21

38. Elliott M, Farnham AW, Janes NF, Johnson DM, Pulman DA. 1987. Synthesis and insecticidal activity of lipophilic amides. Part 1. Introductory survey, and discovery of an active synthetic compound. *Pestic. Sci.* 18:191–201

39. ffrench-Constant RH, Rocheleau TA, Steichen JC, Chalmers AE. 1993. A point mutation in a *Drosophila* GABA receptor confers insecticide resistance. *Nature* 363:449–51

40. ffrench-Constant RH, Steichen JC, Rocheleau TA, Aronstein K, Roush RT.

1993. A single-amino acid substitution in a γ-aminobutyric acid subtype A receptor locus is associated with cyclodiene insecticide resistance in *Drosophila* populations. *Proc. Natl. Acad. Sci. USA* 90:1957–61

41. Fill M, Coronado R. 1988. Ryanodine receptor channel of sarcoplasmic reticulum. *Trends Neurosci.* 11:453–57

41a. Ford M, Lunt G, Reay C, Usherwood P, eds. 1986. *Neuropharmacology and Pesticide Action.* Chichester, UK: Horwood

42. Forshaw PJ, Bradbury JE. 1983. Pharmacological effects of pyrethroids on the cardiovascular system of the rat. *Eur. J. Pharmacol.* 91:207–13

43. Garber SS, Miller C. 1987. Single Na⁺ channels activated by veratridine and batrachotoxin. *J. Gen. Physiol.* 89:459–80

44. Ghiasuddin SM, Matsumura F. 1982. Inhibition of gamma-aminobutyric acid (GABA)-induced chloride uptake by gamma-BHC and heptachlor epoxide. *Comp. Biochem. Physiol. Ser. C* 73:141–44

45. Ghiasuddin SM, Soderlund DM. 1985. Pyrethroid insecticides: potent, stereospecific enhancers of mouse brain sodium channel activation. *Pestic. Biochem. Physiol.* 24:200–6

46. Gibson AJ, Osborne MP, Ross HF. 1990. An electrophysiological study of susceptible (Cooper) and resistant (*kdr*; *super-kdr*) strains of the adult housefly (*Musca domestica* L.) using an isolated mesothoracic leg preparation. *Pestic. Sci.* 30:379–96

47. Goblet C, Mounier Y. 1981. Effects of ryanodine on the ionic currents and the calcium conductance in crab muscle fibers. *J. Pharmacol. Exp. Ther.* 219:526–33

48. Gray AJ, Soderlund DM. 1985. Mammalian toxicology of pyrethroids. In *Progress in Pesticide Biochemistry and Toxicology,* ed. D Hutson, T Roberts, 5:193–248. New York: Wiley

49. Hawkinson JE, Casida JE. 1992. Binding kinetics of γ-aminobutyric acidₐ receptor noncompetitive antagonists: trioxabicyclooctane, dithiane, and cyclodiene insecticide–induced slow transition to blocked chloride channel conformation. *J. Pharmacol. Exp. Ther.* 42:1069–76

50. Honerjager P, Dugas M, Zong X-G. 1992. Mutually exclusive action of cationic veratridine and cevadine at an intracellular site of the cardiac sodium channel. *J. Gen. Physiol.* 90:699–720

51. Jacques Y, Romey G, Cavey M, Kar-

talovski B, Lazdunski M. 1980. Interaction of pyrethroids with the Na⁺ channel in mammalian neuronal cells in
culture. *Biochim. Biophys. Acta* 600:
882–97

52. Jefferies PA, Toia RF, Brannigan B,
Pessah I, Casida JE. 1992. Ryania insecticide: analysis and biological activity of 10 natural ryanoids. *J. Agric. Food
Chem.* 40:142–46

53. Krapcho K, Kral R, Johnson J, Van
Wagenen B, DelMar E, et al. 1994.
Insecticidal peptide toxin from Diguetia
canites: *chemistry, toxicology, molecular biology, and application in a recombinant baculovirus.* Presented at Annu.
Meet. Am. Chem. Soc., 208th, Washington DC

54. Lazdunski M, Lombet A, Mourre C.
1988. Specific binding sites for pyrethroids on the voltage-dependent sodium
channel. See Ref. 60a, pp. 289–300

55. Lees G, Burt PE. 1988. Neurotoxic
actions of a lipid amide on the cockroach nerve cord and on locust somata
maintained in short term culture: a
novel preparation for the study of Na⁺
channel pharmacology. *Pestic. Sci.* 24:
189–91

56. Lehmberg E, Casida JE. 1994. Similarity of insect and mammalian ryanodine
binding sites. *Pestic. Biochem. Physiol.*
48:145–52

57. Leibowitz MD, Sutro JB, Hille B. 1986.
Voltage-dependent gating of veratrid-
ine-modified Na channels. *J. Gen.
Physiol.* 87:25–46

58. Levi G, Gallo V, Raiteri M. 1980. A
reevaluation of veratridine as a tool for
studying the depolarization-induced release of neurotransmitters from nerve
endings. *Neurochem. Res.* 5:281–95

59. Deleted in proof

60. Lund AE, Narahashi T. 1983. Kinetics
of sodium channel modification as the
basis for the variation in the nerve membrane effects of pyrethroids and DDT
analogs, *Pestic. Biochem. Physiol.* 20:
203–16

60a. Lunt G, ed. 1988. *Neurotox '88: Molecular Basis of Drug and Pesticide
Action.* Amsterdam: Excerpta Medica

61. Martin RJ, Pennington AJ. 1989. A
patch-clamp study of effects of dihy-
droavermectin on *Ascaris* muscle. *Br.
J. Pharmacol.* 98:747–56

62. Meissner G. 1986. Ryanodine activation
and inhibition of the Ca²⁺ release channel of sarcoplasmic reticulum. *J. Biol.
Chem.* 14:6300–6

63. Miller RJ. 1987. Multiple calcium channels and neuronal function. *Science* 235:
46–52

63a. Miyamoto J, Kearney PC, eds. 1983.
*Pesticide Chemistry: Human Welfare
and the Environment.* Oxford, UK: Pergamon

64. Monthean C, Potter DA. 1992. Effects
of RH 5849, a novel insect growth
regulator, on Japanese beetle (Coleoptera: Scarabeidae) and fall armyworm
(Lepidoptera: Noctuidae) in turfgrass. *J.
Econ. Entomol.* 85:507–13

65. Nachman RJ, Holman GM, Haddon
WF. 1993. Leads for insect neuropeptide
mimetic development. *Arch. Insect Biochem. Physiol.* 22:181–97

66. Nagata K, Narahashi T. 1993. Suppression of neuronal GABA-activated Cl⁻
current by the insecticide dieldrin. *Toxi-
cologist* 13:126

67. Narahashi T. 1992. Nerve membrane
Na⁺ channels as targets of insecticides.
Trends Pharmacol. Sci. 13:236–41

68. Nicholson RA. 1992. Insecticidal dihydropyrazoles antagonize the depolarizing action of veratridine in mammalian
synaptosomes as measured with a voltage-sensitive dye. *Pestic. Biochem.
Physiol.* 42:197–202

69. Nicholson RA, Botham R, Blade RJ.
1985. The interaction of sodium channel
directed neurotoxicants and a novel insecticidal isobutylamide with central
nerve terminals prepared from the cockroach (*Periplaneta americana*). *Pestic.
Sci.* 24:554–55

70. Nicholson RA, Kumi CO. 1991. The
effects of pesticides, brevetoxin B, and
the cardiotonic drug DPI 201 106 on
release of acetylcholine from insect central nerve terminals. *Pestic. Biochem.
Physiol.* 40:86–97

71. Nicholson RA, Merletti EL. 1990. The
effect of dihydropyrazoles on release of
[³H]GABA from nerve terminals isolated from mammalian cerebral cortex.
Pestic. Biochem. Physiol. 37:30–40

72. Nicholson RA, Wilson R, Potter C,
Black M. 1983. Pyrethroid- and DDT-
evoked release of GABA from the nervous system in vitro. See Ref. 63a, 3:
75–78

73. Nicotera P, Bellomo G, Orrenius S.
1990. The role of Ca²⁺ in cell killing.
Chem. Res. Toxicol. 3:484–94

74. Ogata N, Vogel SM, Narahashi T. 1988.
Lindane but not deltamethrin blocks a
component of GABA-activated chloride
channels. *FASEB J.* 2:2895–900

75. Ohta M, Narahashi T, Keeler RF. 1973.
Effects of veratrum alkaloids on membrane potential and conductance of
squid and crayfish giant axons. *J. Pharmacol. Exp. Ther.* 184:143–54

76. Osborne MP, Smallcombe A. 1983. Site

of action of pyrethroid insecticides in neuronal membranes as revealed by the kdr resistance factor. See Ref. 63a, 3: 103–7

77. Ottea JA, Payne GT, Bloomquist JR, Soderlund DM. 1989. Activation of sodium channels and inhibition of [³H]batrachotoxinin A–20-α-benzoate binding by an N-alkylamide neurotoxin. Mol. Pharmacol. 36:280–84

78. Ottea JA, Payne GT, Soderlund DM. 1990. Action of insecticidal N-alkylamides at site 2 of the voltage-sensitive sodium channel. J. Agric. Food Chem. 38:1724–28

79. Payne GT, Soderlund DM. 1989. Allosteric enhancement by DDT of the binding of [³H]batrachotoxinin A–20-α-benzoate to sodium channels. Pestic. Biochem. Physiol. 33:276–82

80. Payne GT, Soderlund DM. 1992. Activation of γ-aminobutyric acid insensitive chloride channels in mouse brain synaptic vesicles by avermectin B₁ₐ. J. Biochem. Toxicol. 6:283–92

81. Pessah IN, Waterhouse AL, Casida JE. 1985. The calcium-ryanodine receptor complex of skeletal and cardiac muscle. Biochem. Biophys. Res. Commun. 128: 449–56

82. Pleasant B. 1991. The return of an old insect killer. Org. Garden. 38:52–53

83. Rando T. 1989. Rapid and slow gating of veratridine-modified sodium channels in frog myelinated nerve. J. Gen. Physiol. 93:43–65

84. Rowberg KA, Even M, Martin E, Hopfinger AJ. 1994. QSAR and molecular shape analyses of three series of 1-(phenylcarbamoyl)-2-pyrazoline insecticides. J. Agric. Food Chem. 42:374–80

85. Deleted in proof

86. Salgado VL. 1990. Mode of action of insecticidal dihydropyrazoles: selective block of impulse generation in sensory nerves. Pestic. Sci. 28:389–411

87. Salgado VL. 1992. Slow voltage-dependent block of sodium channels in crayfish nerve by dihydropyrazole insecticides. Mol. Pharmacol. 41:120–26

88. Salgado VL. 1992. The neurotoxic insecticidal mechanism of the nonsteroidal ecdysone agonist RH-5849:K⁺ channel block in nerve and muscle. Pestic. Biochem. Physiol. 43:1–13

89. Salgado VL. 1992. Block of voltage-dependent K⁺ channels in insect muscle by the diacylhydrazine insecticide RH-5849, 4-aminopyridine, and quinidine. Arch. Insect Biochem. Physiol. 21:239–52

90. Salgado VL, Irving SN, Miller TA. 1983. Depolarization of motor nerve terminals by pyrethroids in susceptible and kdr-resistant house flies. Pestic. Biochem. Physiol. 20:100–14

91. Salgado VL, Irving SN, Miller TA. 1983. The importance of nerve terminal depolarization in pyrethroid of insects. Pestic. Biochem. Physiol. 20:169–82

92. Schaeffer JM, Haines HW. 1989. Avermectin binding in Caenorhabditis elegans: a two-state model for the avermectin binding site. Biochem. Pharmac. 38:2329–38

93. Soderlund DM, Adams PM, Bloomquist JR. 1987. Differences in the action of avermectin B₁ₐ on the GABAₐ receptor complex of mouse and rat. Biochem. Biophys. Res. Commun. 146:692–98

94. Soderlund DM, Bloomquist JR. 1989. Neurotoxic actions of pyrethroid insecticides. Annu. Rev. Entomol. 34:77–96

95. Soderlund DM, Bloomquist JR, Ghiasuddin S, Stuart A. 1987. Enhancement of veratridine-dependent sodium channel activation by pyrethroids and DDT analogs. In Sites of Action for Neurotoxic Pesticides. ACS Symp. Ser., ed. R Hollingworth, M Green, 356:251–61. Washington DC: Am. Chem. Soc.

96. Stewart LMD, Hirst M, Ferber ML, Merryweather AT, Cayley PJ, Possee RD. 1991. Construction of an improved baculovirus insecticide containing an insect-specific toxin gene. Nature 352:85–88

97. Su HCF. 1985. N-isobutylamides. In Comprehensive Insect Physiology, Biochemistry, and Pharmacology, ed. GA Kerkut, LI Gilbert, 12:273–89. Oxford: Pergamon

98. Tamkun MM, Catterall WA. 1981. Ion flux studies of voltage-sensitive sodium channels in synaptic nerve-ending particles. Mol. Pharmacol. 19:78–86

99. Tanaka K, Scott JG, Matsumura F. 1984. Picrotoxinin receptor in the central nervous system of the American cockroach: its role in the action of cyclodiene-type insecticides. Pestic. Biochem. Physiol. 22:117–27

100. Tomalski MD, Miller LK. 1991. Insect paralysis by baculovirus-mediated expression of a mite neurotoxin gene. Nature 352:82–85

101. Trainer VL, Moreau E, Guedin D, Baden DG, Catterall WA. 1993. Neurotoxin binding and allosteric modulation at receptor sites 2 and 5 on purified and reconstituted rat brain sodium channels. J. Biol. Chem. 268:17114–19

102. Tsien RW, Lipscombe D, Madison DV, Bley KR, Fox AP. 1988. Multiple types of neuronal calcium channels and their

selective modulation. *Trends Neurosci.* 10:431–38

103. Usherwood PNR. 1962. The action of the alkaloid ryanodine on insect skeletal muscle. *Comp. Biochem. Physiol.* 6: 181–99

104. van den Bercken J. 1979. Effects of insecticides on the sensory nervous system. In *Neurotoxicology of Insecticides and Pheromones*, ed. T Narahashi, pp. 183–210. New York: Plenum

105. van Hes R, Wellinga K, Grosscurt AC. 1978. 1-Phenylcarbamoyl-2 pyrazolines: a new class of insecticides. 2. Synthesis and insecticidal properties of 3,5-diphenyl-1-phenylcarbamoyl-2-py razolines. *J. Agric. Food Chem.* 26:915–18

106. Vershoyle RD, Aldridge WN. 1980. Structure-activity relationships of some pyrethroids in rats. *Arch. Toxicol.* 454: 325–29

107. Vijverberg HPM, de Weille JR, Ruigt GSF, van den Bercken J. 1986. The effect of pyrethroid structure on the interaction with the sodium channel in the nerve membrane. See Ref. 41a, pp. 267–87

108. Vijverberg HPM, van den Bercken J. 1990. Neurotoxicological effects and the mode of action of pyrethroid insecticides. *CRC Crit. Rev. Toxicol.* 21:106–26

109. von Keyserlingk HC, Willis RJ. 1992. The GABA activated Cl⁻ channel in insects as target for insecticide action: a physiological study. In *Neurotox '91: Molecular Basis of Drug and Pesticide Action*, ed. IA Duce, pp. 79–104. London: Elsevier Applied Science

110. Wafford KA, Sattelle DB, Gant DB, Eldefrawi AT, Eldefrawi ME. 1989. Noncompetitive inhibition of GABA receptors in insect and vertebrate CNS by endrin and lindane. *Pestic. Biochem. Physiol.* 33:213–19

111. Wang CC, Pong SS. 1982. Actions of avermectin B_{1A} on GABA nerves. In *Membranes and Genetic Diseases*, pp. 373–95. New York: Liss

112. Ware GW. 1982. *Fundamentals of Pesticides—a Self-Instruction Guide*, pp. 78–79. Fresno, CA: Thompson

113. Welch W, Ahmad S, Airey JA, Gerzon K, Humerickhouse RA, et al. 1994. Structural determinants of high-affinity binding of ryanoids to the vertebrate skeletal muscle ryanodine receptor: a comparative molecular field analysis. *Biochemistry* 33:6074–85

114. Wellinga K, Grosscurt AC, van Hes R. 1977. 1-Phenylcarbamoyl-2-pyrazolines: a new class of insecticides. 1. Synthesis and insecticidal properties of 3- phenyl-1-phenylcarbamoyl-2-pyrazolines.   *J. Agric. Food Chem.* 25:987–92

115. Wing K, Slawecki R, Carlson G. 1988. RH-5849, a nonsteroidal ecdysone agonist: effects on larval Lepidoptera. *Science* 241:470–72

116. Yeh JZ, Oxford GS, Wu CH, Narahashi T. 1976. Dynamics of aminopyridine block of potassium channels in squid axon membrane. *J. Gen. Physiol.* 68: 519–35

117. Zhang A, Nicholson RA. 1993. The dihydropyrazole RH-5529 blocks voltage-sensitive calcium channels in mammalian synaptosomes. *Pestic. Biochem. Physiol.* 45:242–47

118. Zhang H-G, ffrench-Constant RH, Jackson MB. 1994. A unique amino acid of the *Drosophila* GABA receptor with influence on drug sensitivity by two mechanisms. *J. Physiol.* 479:65–75

119. Zlotkin E. 1986. The interaction of insect-selective neurotoxins from scorpion venoms with insect neuronal membranes. See Ref. 41a, pp. 384–413

120. Zlotkin E. 1988. On the site of action of the insect selective neurotoxin AaIT. See Ref. 60a, pp. 35–48

Annu. Rev. Entomol. 1996. 41:191–210

DEVELOPMENT OF RECOMBINANT BACULOVIRUSES FOR INSECT CONTROL

Bryony C. Bonning

Department of Entomology, Iowa State University, Ames, Iowa 50010

Bruce D. Hammock

Departments of Entomology and Environmental Toxicology, University of California, Davis, California 95616

KEY WORDS: insecticides, genetic engineering, nuclear polyhedrosis virus

---

ABSTRACT

In this review, we provide an overview of the current status of recombinant baculoviruses, describe the development of genetically engineered baculoviruses for use as rapid-action biological insecticides, and provide more detailed information on one particular set of recombinant viruses. The advantages and disadvantages of recombinant baculovirus insecticides, and the importance of risk-assessment studies of these genetically modified organisms, are reviewed. Finally the importance of sensible regulatory strategies to the success and future prospects of this technology is discussed.

PERSPECTIVES AND OVERVIEW

With the advent of recombinant DNA technology and the recent development of rapid-action recombinant baculoviruses, interest in the field potential of these viruses for insect control has increased dramatically. Several major companies and various academic and government laboratories worldwide are currently contributing to this area.

Wild-type baculoviruses are an integral component of the natural biological control of many pest species, and application of wild-type viruses has been very effective for pest management in several cases (31). However, in most cases the wild-type viruses have failed to compete with classical insecticides

191

0066-4170/96/0101-0191$08.00

because of several factors, application technology and low field persistence among them. The limitation discussed here is the slow kill by viruses, which not only makes them poorly competitive, but also limits industrial investment in application and formulation technologies for enhanced efficacy. In the past several years, the speed of kill of the viruses has been markedly improved through the use of recombinant DNA technology. These successes have led to industrial investment in related technologies that should make both the wild-type and recombinant viruses more attractive for use in insect pest control.

Several recently published reviews on baculoviruses have covered their biology (1, 3, 39), pathogenesis (99), diversity and molecular biology (7, 59), and application as protein expression vectors (63, 65, 78), as well as the use of wild-type (31, 32) or genetically engineered baculoviruses (8, 42, 67, 79, 103) for insect control. Other articles have focused on specific aspects of recombinant DNA technology and their influence on the efficacy of baculovirus insecticides (10, 74).

INTRODUCTION

Baculovirus Diversity

Baculoviruses are double-stranded DNA viruses that are pathogenic to arthropods—predominantly holometabolous insects. The Baculoviridae contains two genera, as determined by structural properties (80a): the *Nucleopolyhedrovirus* nucleopolyhedroviruses and the *Granulovirus*. The nuclear polyhedrosis viruses (NPV) have virions embedded in a crystalline matrix of the protein polyhedrin. The virions vary in their configuration in that the nucleocapsids may be enveloped singly or in multiples. The occluded viruses are referred to as polyhedra. The granuloviruses (GV) have one, or rarely two, virions embedded in a crystalline matrix of granulin. The nucleocapsid is always enveloped singly.

Efforts to genetically engineer baculoviruses for control of insect pests have centered on the nucleopolyhedroviruses for control of lepidopteran larvae. This review focuses on progress made in genetic engineering of this group for increased use in insect pest control. Systems for engineering these viruses are well developed (57, 63, 78, 86, 95). The technology discussed here can be applied to other viruses or other insect pathogens and commensals as cloning systems are developed.

Baculovirus Life Cycle

Following ingestion of the nucleopolyhedrovirus by the larva, the protective polyhedrin coat dissolves in the midgut (Figure 1). The virions, released from the matrix of the polyhedron, set up sites of primary infection in the cells

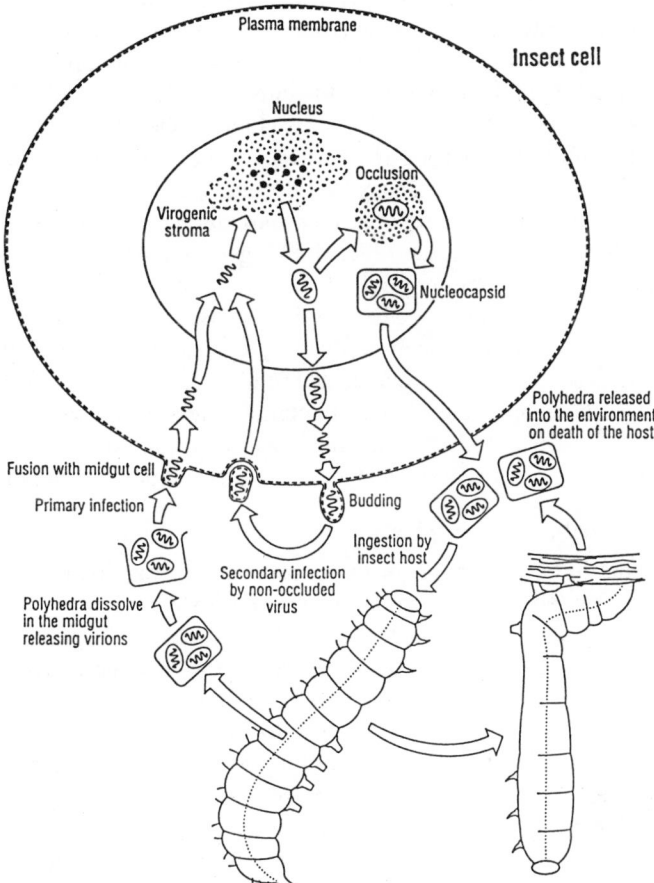

Figure 1   Life cycle of the nucleopolyhedrovirus. Following ingestion of the virus, the polyhedra dissolve in the midgut, releasing nucleocapsids that fuse with the midgut cells. The viral DNA replicates in the nucleus and progeny viruses bud through the cytoplasmic membrane to disseminate the infection. After the initial round of replication, nucleocapsids produced in the nucleus are enclosed within polyhedrin. When the insect dies, the polyhedra are released from the lysed cells. [Courtesy of Intercept Ltd. (8).]

comprising the midgut. Initial rounds of viral replication within the nucleus of the infected cell produce a second viral phenotype, the budded virus or virus particle, which spreads the infection to other tissues. The budded viruses move through the cell membrane and become coated with a viral protein–modified basal plasma membrane. Infection of different larval tissues occurs in a sequential manner, and the virus is hypothesized to use the tracheal system of the insect as a conduit (30). However, there is some debate about whether this is the primary route of viral transmission within the insect, or whether the hemocytes play this role. At later stages of virus infection, progeny viruses become occluded within the nuclei of the infected cells. Prior to death, larvae infected with some wild-type baculoviruses climb up the plant. They succumb to the virus infection and hang from an elevated position, which facilitates dissemination of the virus as the cadaver decomposes.

Following infection of an insect cell, baculovirus gene expression occurs in a temporally regulated cascade (37). Unlike gene transcription in the three other temporal phases, transcription of the immediate early (IE) genes does not depend on production of other viral proteins because these genes are transcribed by host factors. Their products upregulate the delayed early (DE) genes. Late-gene expression occurs concurrently with the onset of DNA replication, and these genes encode the structural proteins of the virus particles. The very late genes encode proteins involved in the final stages of infection and polyhedron morphogenesis, including the p10 and polyhedrin proteins.

Virus Control of Insect Pests

The potential of baculoviruses for insect pest control was first recognized in the early 1900s (93). Since then, numerous baculoviruses have been used to control, among others, hymenopteran, lepidopteran, and coleopteran pests of crops such as coconuts, cotton, and cabbages, as well as pests of beehives. The control of pests on soybean crops in Brazil and on palm and coconut in the South Pacific by *Anticarsia gemmatalis* NPV and *Oryctes rhinoceros* baculovirus, respectively, is particularly noteworthy (31). The time taken by the virus to kill the host insect from the point of infection ranges from days to weeks, depending on temperature, viral dose, insect age, and the particular host and virus species. During this time, the insect continues to feed. Hence, many of the pests currently controlled by baculoviruses are pests of crops that can sustain some damage without significant economic loss. In addition, baculoviruses have been used successfully when the insects are small and it is possible to apply virus. For example, in Brazil the soybean crops are monitored intensively for larvae for optimal timing of virus application before significant damage has occurred. Genetic engineering of the baculovirus to reduce the time taken by the virus to kill the host insect will not preclude the use of the wild-type viruses in biocontrol but promises to yield viruses more competitive with classical insecticides. This development will significantly increase the potential of baculoviruses for pest control, particularly in row crop agriculture.

Advantages of Recombinant Baculovirus Insecticides

Insect viruses are important components of natural biological control, but this article concentrates on their use as biological insecticides with reduced capacity to recycle in the field. The pressure to find novel means of pest control to reduce reliance on the synthetic chemical insecticides is increasing. The number of compounds currently available to the grower is decreasing as a result of pest resistance (89) and detrimental effects of insecticides on both human health and the environment (19). In this respect, several advantages are associated with the use of baculoviruses as insect-control agents. One of the most important attributes of baculoviruses for pest control is their host specificity. Many baculoviruses infect only a few species, often within the same family. This attribute makes them ideal for incorporation into integrated pest management (IPM) programs because one of the main concepts of IPM is to target pest species that exceed the economic-damage threshold, while leaving the rest of the fauna undisturbed (34). Thus, the whole potential of beneficial organisms can be exploited.

Viral pesticides can be applied using conventional techniques and do not create the problems associated with residues. They do not show cross resistance with chemical compounds. Although insects have shown resistance to baculoviruses in some cases (38), resistance ratios are generally low, and in many cases resistance is unstable in the absence of selection pressure. Apparent negatively correlated resistance to chemical insecticides has even been noted (38a, 72). The use of baculovirus insecticides in IPM programs with other biological-control agents or with classical chemical insecticides may reduce the likelihood of resistance developing to the baculovirus. In some pest-management situations, the recombinant viruses will be used to augment the action of classical insecticides. In fact, some recombinant viruses can synergize and be synergized by classical insecticides, and this complementarity makes the viruses even more attractive as tools for IPM (72).

For the production of baculovirus insecticides, industry is pursuing several strategies, including both in vitro and in vivo technologies. However, wild-type and recombinant viruses can be produced by cottage industries. With carefully produced innocula, local laboratories in developing countries could produce these genetically sophisticated products with minimal technological input.

Disadvantages of Recombinant Baculovirus Insecticides

The enhanced speed of kill of recombinant viruses limits their propensity to recycle in the environment as fewer polyhedra are produced compared with the wild-type virus. Apart from this limitation, the advantages and disadvantages of recombinant and wild-type baculoviruses are the same. In many cases, the traits that are considered advantageous actually limit the utility of the recombinant viruses. For example, the limited host range allows a pest-management specialist to reduce one pest population with precision without disruption of biological-control agents and without the deleterious effects on nontarget organisms associated with use of classical pesticides and even *Bacillus thuringiensis*. However, this trait also limits the market size for the baculovirus and thus the development costs that can be invested in the technology.

Cloning systems that allow the recombinant technology to be applied to different viruses is lacking even within the Baculoviridae. Hence, the application of genetically engineered baculoviruses for insect control is currently limited to a few pest species. The rate of development of these systems will, one hopes, accelerate. However, of all the limitations facing the use of genetically modified and wild-type viruses, the problems associated with correct application and low persistence are the most serious. The many advantages of the recombinant viruses have attracted industrial investment in research to solve these problems. These technologies will be applicable to wild-type viruses and other biological-control agents intended for augmentative release in pest management, in addition to recombinant viruses. Despite the great potential of recombinant baculoviruses, they do not represent a panacea or even a stand-alone technology for insect control. Rather, they will be used as additional tools for application in IPM programs.

DEVELOPMENT OF RECOMBINANT BACULOVIRUSES

Genetic Engineering of Baculoviruses

The aim of genetic engineering of baculoviruses for use as insecticides is to combine the pathogenicity of the virus with the insecticidal action of a toxin, hormone, or enzyme. Upon infection of the insect larva with the recombinant baculovirus, the foreign protein is expressed. If this protein is toxic to the insect, the insect will die rapidly from this effect, rather than from the viral infection itself. The recombinant approach will probably also be used to improve production, modify host range, and enhance the utility of various insect viruses as biopesticides. However, the goal of the research reviewed here is to reduce the time from infection with the recombinant virus to death of the insect such that feeding damage is below the economic threshold. This goal necessitates an approximate lethal-time ratio (lethal time of test virus divided by lethal time of wild-type virus) (9) of 0.4–0.5 for control of insect pests on many crops. Reduction of the lethal time may also enhance farmer or user acceptance of baculovirus insecticides.

Two major baculovirus-expression systems have been developed for the production of recombinant proteins for research and clinical use. These are based on the nucleopolyhedrovirus derived from the alfalfa looper, *Autographa californica* (AcNPV), and a similar virus from the silkworm, *Bombyx mori* (BmNPV). The sequences of the entire genomes of both AcNPV and BmNPV have now been determined (2, 69a). Early engineering work was carried out with BmNPV for high levels of protein production in larvae of *B. mori.* As *B. mori* is the only known host for BmNPV, this approach also provided biological containment for the virus (69). Most recent work in developing the virus for insect control has concentrated on AcNPV. A variety of recently developed techniques and transfer vectors greatly facilitate the engineering process (5, 27, 61). Protein-expression systems have also been established in other baculoviruses, such as *Helicoverpa zea* NPV (21) and *Lymantria dispar* NPV (107). This research provides the basis for engineering of these viruses for use as insect pest–control agents in the future.

The circular genome of AcNPV is approximately 134 kilobase pairs (kb) (2). Because of the difficulty of direct manipulation of such a large piece of DNA, engineering of a baculovirus is usually carried out in two steps. First, the foreign gene is incorporated into a baculovirus-transfer vector. Most transfer vectors used are bacterial plasmid, University of California (pUC), derivatives, which encode an origin of replication for propagation in *Escherichia coli* and an ampicillin-resistance gene. The pUC fragment is ligated to a small segment of DNA taken from the viral genome. The foreign gene sequence is incorporated into a cloning site downstream of the promoter selected to drive expression. For the second step, the transfer vector is mixed with DNA from the parental virus. The engineered DNA is incorporated into the virus via homologous recombination events within the nucleus of cultured insect cells. Unlike genetic engineering in plants, which results in a rather random incorporation of new DNA into the genome, the baculovirus system allows the precise insertion of foreign DNA without disruption of other genes. No drug-resistance markers are included in the final clone, which eliminates some of the major objections raised to recombinant organisms (35). Commercial kits and reagents are available for this work, as well as several excellent manuals (57, 86). A number of recently developed alternative approaches for genetic engineering of baculoviruses have been reviewed elsewhere (27).

Early research involved engineering of these viruses for use as protein-expression vectors rather than for insect control (90). The approach involved replacing the gene encoding polyhedrin with the foreign gene of interest. Expression of the foreign gene was driven by the polyhedrin promoter in a polyhedrin-negative virus. Although these viruses can be manipulated successfully in cell culture for production of high levels of foreign protein, they lose any advantage conferred by the polyhedrin coat that protects the virus from inactivation by desiccation and ultraviolet light under field conditions. Replacement of the viral gene encoding the p10 protein (98), which is involved in calyx attachment and nuclear lysis, also resulted in reduced viral fitness. The stability of polyhedra produced by p10-negative viruses is greatly reduced (101).

An alternative approach to replacement of a viral gene with a foreign gene sequence is to duplicate a viral promoter. In this instance, none of the viral genes are lost, and promoters of essential viral genes can be used for expression of foreign proteins. The level and timing of expression of a particular protein by a recombinant baculovirus is determined in part by the promoter chosen to drive transcription of the foreign gene sequence. Currently, the polyhedrin and p10 promoters are used most frequently for expression of recombinant proteins. However, expression under the basic protein promoter was higher in several instances (13, 60a, 91). In the future, the use of early promoters, hybrid promoters, and promoters from other species will increase, particularly with the identification of peptides and proteins that disrupt insect biology at lower expression levels.

Recombinant Baculoviruses Developed for Insect Control

Any gene coding for a protein that disrupts normal larval development or behavior and reduces feeding damage caused by the insect is a candidate for expression by a recombinant baculovirus for insect-control purposes. To establish the suitability of a gene for baculovirus expression, one must consider the mechanism of action and the behavior of the virus after infection to assess whether the gene product is likely to reach its target site. As a general principle, a product that acts on the whole insect systemically is preferable to products that only show toxicity to the infected cells. Although the baculovirus system offers several levels of specificity, agents that exhibit insect-selective toxicity are preferable from a safety standpoint to those that might have a more universal effect. Suitable genes for expression in the baculovirus include insect enzymes and hormones that may disrupt development or homeostasis and insect-selective neurotoxins. The recombinant baculoviruses developed to date have been reviewed more extensively elsewhere (8, 47, 65, 103), but an outline of current developments is presented below.

Recombinant baculoviruses are being developed as biological insecticides for repeated application rather than as biological control agents that establish and recycle in the field. Because the recombinant virus expresses genes for rapid kill of the host, it is subject to strong negative selection pressure and consequently cannot compete with its wild-type counterpart. Should regulators require redundant systems, techniques such as gene deletion can be adopted to further reduce the fitness of the virus. An alternative approach would involve applying preoccluded viruses (POVs), i.e. the nucleocapsids of polyhedrin-negative viruses (106).

Recombinant Baculoviruses Expressing Toxins

Arthropod venoms offer a rich source of insect-selective toxins (50). Indeed, expression of insect-selective toxins in the baculovirus-expression system has proved to be highly successful for increasing virus efficacy in insect-pest control. The first toxin to be engineered into a baculovirus was the insect-selective toxin derived from the scorpion *Buthus eupeus* (BeIT) (16). However, for unknown reasons, this toxin did not increase the efficacy of the recombinant virus. Two of the toxins produced by the bacterium *Bacillus thuringiensis* (Bt) have been engineered into AcNPV. Of the four classes of Bt isolates, CryI produces crystals that are toxic to lepidopteran larvae (51). These toxins are thought to generate pores in cell membranes, leading to disruption of osmotic balance and then cell death. Recombinant baculoviruses were constructed to express *B. thuringiensis* subsp. *kurstaki* HD-73 δ-endotoxin (76) and *B. thuringiensis* subsp. *aizawai* 7.21 CryIA(b) toxin (70). As expected given the mechanism of action of the recombinant proteins, neither of the recombinant viruses constructed showed promise in terms of increased insecticidal efficacy.

Baculoviruses expressing the insect-specific toxin AaIT derived from the North African scorpion *Androctonus australis* are among the most promising recombinant baculoviruses developed to date (69, 73, 94). This toxin acts on the neuronal sodium channel, causing presynaptic excitatory effects. Larvae infected with these recombinant baculoviruses exhibit dorsal arching and increased irritability and cease feeding. Lethal times are reduced by 25–40% compared with those of the wild-type virus, and feeding damage by larvae infected with recombinant virus is reduced by 50% on cabbage compared with damage caused by larvae infected with wild-type viruses (25, 94). The speed of virus kill further increases in the presence of low doses of pyrethroid insecticides (72). Pyrethroid-resistant insects are also killed more quickly by this combination (72). These observations illustrate that the recombinant viruses could be used initially in conjunction with classical insecticides in resistance-management programs.

TxP1, a toxin derived from the straw itch mite, *Pyemotes tritici,* was also expressed in AcNPV (96, 97). The mite uses this toxin, which causes muscle contraction and paralysis, to immobilize insects up to 150,000 times its size. The exact mechanism of action is not known. Recombinant baculoviruses expressing TxP1 had a lethal time 30–40% lower than that of the wild-type virus. These successes with toxins of relatively limited activity on lepidopterous larvae suggest that insect-selective peptides of far greater activity will prove useful with currently used expression systems as well as with systems under the control of earlier promoters. Moreover, the expression of combinations of genes in multiple expression vectors (5), rather than expression of individual genes, should also increase the speed of kill.

A maize mitochondrial protein involved in cytoplasmic male sterility and disease susceptibility has been expressed in AcNPV (60). This protein, URF13, is hydrophobic and binds tightly to the membranes of cells. Injection of larvae of the cabbage looper, *Trichoplusia ni,* with the polyhedrin-negative recombinant virus resulted in a 40% decrease in lethal time compared with that of the wild-type virus. URF13 decreased the ability of AcNPV to produce polyhedra when polyhedrin-positive constructs were made. Further research should shed light on the exact mechanism of toxicity of URF13 to the insect host, allowing us to determine whether this protein is suitable for pest-control purposes.

Recombinant Baculoviruses Expressing Insect Hormones

An alternative approach to the expression of insect-selective toxins by a baculovirus is expression of a component of the insect endocrine system (56, 71). The rationale behind this approach is that if a single component is overproduced by the recombinant baculovirus, homeostasis will be disrupted, causing a deleterious effect on the insect. The first recombinant baculovirus to be developed that had enhanced insecticidal activity expressed the diuretic hormone from the tobacco hornworm, *Manduca sexta,* thought to be involved in water balance (64). Larvae of *B. mori* injected with the recombinant BmNPV, which expressed the diuretic hormone, died 20% more quickly than larvae injected with the wild-type virus. This resulted from an alteration in metabolism of larval fluid.

Eclosion hormone, which is involved in ecdysis and shedding of the cuticle during molting, was expressed in AcNPV (28). The recombinant hormone was active, but the efficacy of the recombinant virus was not enhanced relative to the wild-type virus (29). The prothoracicotropic hormone (PTTH) from *B. mori,* which stimulates production of ecdysone, has also been expressed in AcNPV (82). The recombinant viruses produced large amounts of PTTH. No reduction in lethal time was seen upon infection of larvae of the fall armyworm, *Spodoptera frugiperda,* with the recombinant virus, but expression of PTTH markedly inhibited the pathogenicity of AcNPV (82). Other insect hormones have been expressed with little or no success. Considering the complex events involved in homeostasis, the failure of single neurohormones to dramatically enhance the efficacy of recombinant baculovirus insecticides is perhaps not surprising. Once we have a greater understanding of endocrine systems, an improved ability to achieve high-level expression of small peptides, and an understanding of the pharmacokinetics and interactions of neuromodulators, expression of neurohormones may be a fruitful approach in the future.

Ecdysteroid UDP-Glucosyltransferase

The baculovirus AcNPV produces ecdysteroid UDP-glucosyltransferase (egt) (81, 83–85). In vitro studies have shown that this enzyme transfers glucose from UDP-glucose to ecdysteroids involved in molting. Expression of this protein prevents the host insect from molting and effectively keeps the insect in the feeding stage (85). However, the exact role of egt in vivo has not been demonstrated, and it may have other primary and secondary effects (80b). Deletion of the egt gene from the viral genome results in a 10–20% reduction in lethal time, a 40% reduction in feeding damage, and a 30% reduction in the number of progeny viruses produced. Regulatory agencies view the egt-negative gene-deletion construct with less scrutiny than they do gene-addition constructs, such as those expressing toxins or juvenile hormone esterase. Thus, the egt-negative virus is being used for some early releases to test its effectiveness and thereby pave regulatory pathways for the recombinant baculovirus technology. The egt-negative AcNPV is likely to be the first recombinant baculovirus approved for commercial use as a pesticide (80b). This example demonstrates the utility of gene deletion for improving the speed of kill of a virus. Gene deletion may also prove useful for modification of other traits, such as host range, or for production purposes.

Recombinant Baculoviruses Expressing Juvenile Hormone Esterase

Two epithelial hormones are involved in lepidopteran larval development: 20-hydroxyecdysone, which initiates the molt, and juvenile hormone (JH), which controls the nature of the molt. If the JH titer is high, the molt is isometric to a larger larval stadium. If the JH titer is low, an anisometric molt to the pupa will occur. The reduction in JH titer is a key event in insect development that leads ultimately to termination of feeding and metamorphosis. This decrease in titer occurs through a decrease in biosynthesis caused at least in part by neurohormone(s) and neuromodulators such as allatostatin and allatohibins. Juvenile hormone esterase (JHE) increases as JH titer decreases (40). JHE catalyzes the hydrolysis of the highly stable conjugated methyl ester of the JHs to the corresponding biologically inactive JH acid. JH is intrinsically involved in regulation of gene expression in both larval and adult insects (87).

The agricultural-chemical industry has shown considerable interest in anti-JH agents, because induction of precocious development in a crop pest would reduce feeding damage (92). The expression of JHE in a baculovirus vector is a continuation of this line of research into anti-JH agents.

The coding sequence for JHE from the tobacco budworm, *Heliothis virescens* (46), was expressed in various baculovirus constructs (10). JHE has been expressed in AcNPV under control of the polyhedrin (43, 88), p10 (11, 88), and basic protein (p6.9) promoters (13), as well as a hybrid promoter based on the polyhedrin gene (29). Despite high levels of expression of JHE both in vitro (43) and in vivo (11, 29), the significant improvement of the insecticidal efficacy observed for neonate larvae of *T. ni* (43) was not seen for other stages or insect species. Compensation for the increased titer of JHE by the regulatory and feedback mechanisms intrinsic to the insect's JH system may explain the limited insecticidal activity of this recombinant virus.

Glycosylation of the molting hormone may reduce or prevent any deleterious effect caused by JHE. Eldridge et al addressed the question of whether the effects of JHE could be masked by the action of egt by expressing JHE in an egt-negative virus (29). In this study, infections of larvae with either egt-positive or egt-negative recombinant viruses expressing JHE did not differ in terms of lethal times, although differences might have appeared upon infection of larvae in earlier stadia.

JHE is an extraordinarily stable protein that does not degrade on incubation in vitro with hemolymph. However, pharmacokinetic analyses showed that the half-life of JHE injected into the hemolymph of larvae of *M. sexta* was only approximately 1 h (53). Further analysis indicated that JHE was being removed by the pericardial cells (15, 54). Electron microscopy of immunogold-labeled sections of these cells showed concentration of the JHE within lysosome-like granules, where it is presumably degraded (15). JHE is removed by a saturable, first-order process, which suggests that the mechanism of uptake may be receptor-mediated endocytosis (53).

To prevent degradation of JHE in the pericardial cells, sequences likely to be involved in the degradation process were altered using site-directed mutagenesis. Specifically, lysine residues at positions 29 and 522 within the peptide sequence of JHE were replaced with arginine residues (100). Baculoviruses were then constructed to express the modified JHEs with the single (AcJHE-29, AcJHE-522), or with both (AcJHE-KK), mutations. Neither the kinetic parameters nor the expression levels of the modified JHEs in vivo or in vitro were significantly different from those of the wild-type JHE. However, infection of *H. virescens* larvae with AcJHE-KK significantly reduced lethal times and feeding damage compared with those observed upon infection with the wild-type, nonengineered virus. Feeding damage was reduced by 50% on lettuce and 36% on cotton leaves. We examined the effects of AcJHE-KK on larvae of *H. virescens* in conjunction with topically applied JH analogues, JHE inhibitors, or anti-JH agents. The lethal time of AcJHE-KK increased when used with JH analogues or JHE inhibitors. These data suggest that catalytic activity is required for this insecticidal effect. When AcJHE-KK was bioassayed in conjunction with anti-JH agents, its lethal time decreased. This result suggests that the insecticidal efficacy of AcJHE-KK results at least in part from an anti-JH activity.

Biochemical analyses of events within the pericardial cell complex after injection of JHE into larvae of *M. sexta* suggest that JHE is bound by a heat-shock protein (Hsp) prior to degradation (BC Bonning, VK Ward, TF Booth, RD Possee & BD Hammock, unpublished data). The Hsp is hypothesized to bind to JHE at a putative lysosome-targeting sequence (17) and to facilitate transport to the lysosome (18). The mutation at position 522 is within the putative lysosome-targeting sequence of JHE. Immunoprecipitation of the wild-type and modified JHEs with anti-Hsp antiserum showed that the modifications significantly reduced binding of the Hsp to JHE. Although the exact mechanism of action of Hsps in chaperoning proteins to the lysosomes is unclear in both this and mammalian systems (49), we propose that the insecticidal activity seen upon infection with AcJHE-KK results from impaired degradation of JHE. Accumulation of JHE-KK within the pericardial cells may reduce the local titer of JH and impair the secretory functions of this organ (33). Variation in efficacy, which sometimes occurs with infection of larvae with AcJHE-KK, may relate to the precise timing of expression of JHE-KK in relation to other physiological events associated with JH action during larval development.

These data suggest that expression of other regulatory proteins and enzymes could be useful in accelerating the kill of recombinant baculoviruses. They also indicate that genetic changes made to enhance the stability or change the distribution of an expressed peptide or protein can be used to improve the performance of recombinant viruses.

Unexpected Results

During the course of research on AcJHE-KK, a simultaneous study to investigate the catalytic mechanism of JHE was under way. To establish the relative contributions of specific amino acids to the catalytic activity of this enzyme, a series of modifications were made to JHE (100). Several of these modifications completely removed the catalytic activity of JHE. For example, the active site Ser201 was replaced to produce JHE-SG, and His446 was replaced to produce JHE-HK. The viruses AcJHE-SG and AcJHE-HK, which express the inactive forms of JHE, were used as control viruses for bioassay of AcJHE-KK. Against expectations, JHE-SG turned out to be highly insecticidal (12). Lethal times were reduced by 20–30% compared with those of the wild-type virus, with a 66% reduction in feeding damage on lettuce and a 55% reduction on cotton. When bioassayed with anti-JH agents, the lethal time of AcJHE-SG was increased, suggesting that the mechanism of action is not related to the degradation or sequestration of JH. About 25% of the larvae infected with AcJHE-SG died during molting, being unable or only partially able to escape from the old exuvium. We propose that the mechanism of action of JHE-SG is related to events at the molt.

A virus was constructed to express a modified form of JHE with mutations at Lys29, Lys522, and Ser201 (MMM van Meer & BC Bonning, unpublished data). This virus, AcJHE-KSK, has comparable efficacy to AcJHE-KK and

AcJHE-SG in terms of feeding damage and lethal times. Similar to results with AcJHE-SG, the lethal time of AcJHE-KSK increased upon application of anti-JH agents during bioassay of the virus. Moreover, the symptoms were similar to those seen upon infection with AcJHE-SG.

The insecticidal efficacies of AcJHE-KK, AcJHE-SG, and AcJHE-KSK do not differ significantly from the efficacy of AcAaIT (12), which is among the best of the toxin-expressing baculoviruses. The recombinant baculoviruses expressing JHE have a possible advantage over toxin-expressing baculoviruses in terms of the public's response to these genetically engineered organisms. In addition, rapid radiochemical (45) and colorimetric assays (75) for JHE activity may greatly facilitate monitoring of the virus under field conditions.

RISK ASSESSMENT OF RECOMBINANT BACULOVIRUS INSECTICIDES

To date, evaluation of the recombinant baculoviruses indicates that they offer great benefit to agriculture with minimal risk. No members of the Baculoviridae infect plants or nonarthropod animals. Selective viruses are most attractive for IPM, but if regulatory barriers are high, industry will be forced to develop viral pesticides that control a broad spectrum of pests. Fortunately, the initial work has focused on viruses that have a relatively narrow host range compared with even highly selective materials such as *B. thuringiensis* and the insect growth regulators. Having developed prototype viruses with enhanced speed of kill, investigators have been conducting extensive tests to ensure that the recombinant viruses do not represent a hazard to nontarget organisms or to the environment (23, 24). The goal of ongoing laboratory-based tests is to ascertain that the genetic manipulation process has not altered the host range of the virus (4, 6). An understanding of the molecular determinants of host range should speed such evaluations (26, 55, 58, 68, 80).

For host-range analyses, large numbers of nontarget organisms are screened for possible infection with the recombinant virus. The potential effects of the expressed protein on predators and parasites of the targeted insect pest are also determined in the laboratory (48). In addition, the genetic stability of the viruses and the likelihood of genetic exchange between viruses infecting the same host insect will be established. Based on these data, field trials have been conducted using various constructs, including AcAaIT (25, 42), and more trials are planned for the near future.

A series of contained field trials carried out with genetically modified baculoviruses in England have addressed questions of virus stability and efficacy (4, 6). In 1989, the Boyce Thompson Institute for Plant Research at Cornell University conducted the first uncontained field trials of genetically engineered baculoviruses (104, 105). The recombinant baculovirus most recently tested for efficacy under field conditions is AcAaIT (25). Data on feeding damage caused by infected larvae of *T. ni* were promising but insufficient to evaluate the commercial potential of the virus.

Because research with recombinant baculoviruses involves recombinant DNA technology, it now requires special scrutiny (77). One hopes that the regulatory authorities will define clear processes for the evaluation and regulation of biological organisms for use in agriculture, using performance-based standards regardless of whether they are generated by classical or recombinant means. Fortunately, the scientists involved in the development of recombinant baculoviruses have themselves set high standards for environmental safety and have designed risk-assessment experiments to test worst-case scenarios. When evaluating the results and the risks and benefits of recombinant-baculovirus technology, we should consider the relative attributes and limitations of competing technologies. Opponents of the technology (20, 102) fail to recognize that all of the rapid-kill viruses are designed as nonrecycling biological pesticides that are at a selective disadvantage compared with the wild-type organisms (41). Thus, there is no evolutionary driver for the virus to replicate in the environment or colonize new hosts. The very nature of the rapid-kill response leads to a dramatic reduction in progeny viruses and a strong negative selection pressure. Thus, these recombinant viruses are more attractive ecologically than organisms provided with traits that give a positive selective advantage.

FUTURE PROSPECTS

New, more potent, insect-selective toxins should be isolated from various organisms within the foreseeable future. Agents with greater activity allow the use of earlier and weaker promoters and may also result in insect death in earlier viral replication cycles (67). Further development of recombinant baculovirus insecticides may proceed along multiple fronts (42, 44). Synergism between insect-selective toxins expressed in the same baculovirus construct represents a new avenue for investigation. Modification of proteins and peptides for enhanced stability in vivo may improve all of the recombinant baculoviruses, especially those that show minimal increases in speed of kill. Apart from the recombinant protein expressed by the virus, formulation components can also be used very effectively to enhance the speed of kill of genetically engineered baculoviruses (22).

Recombinant baculoviruses represent a valuable technology that has great potential for effective integration into pest-management systems. Both the recombinant and wild-type viruses are more environmentally attractive than classical pesticides and may represent a major step toward a more sustainable agriculture. However, the full potential of recombinant baculoviruses will only be realized if the regulation of their testing and commercial use is rationalized through a risk-based approach (77).

ACKNOWLEDGMENTS

The authors thank Drs. James Fuxa, Ivan Gard, Carlo Ignoffo, Lois Miller, Henry Miller, David O'Reilly, Just Vlak, and Loy Volkman for comments on the manuscript. This work was supported by DCB-91-19332 from NSF, 91-37302-6186 from USDA, and 23-696 from the USDA Forest Service. BCB was supported by a NATO collaborative research grant (CRG 900955).

> Any *Annual Review* chapter, as well as any article cited in an *Annual Review* chapter, may be purchased from the Annual Reviews Preprints and Reprints service.
> 1-800-347-8007; 415-259-5017; email: arpr@class.org

Literature Cited

1. Adams JR, McClintock JT. 1991. Nuclear polyhedrosis viruses of insects. In *Atlas of Invertebrate Viruses,* ed. JR Adams, JR Bonami, pp. 87–204. Boca Raton, FL: CRC
2. Ayres MD, Howard SC, Kuzio J, Lopez-Ferber M, Possee RD. 1994. The complete sequence of *Autographa californica* nuclear polyhedrosis virus. *Virology* 202:586–605
2a. Beckage NE, Thompson SN, Federici BA, eds. 1993. *Parasites and Pathogens of Insects.* New York: Academic
3. Bilimoria SL. 1991. The biology of nuclear polyhedrosis viruses. In *Viruses of Invertebrates,* ed. E Kurstak, pp. 1–72. New York: Dekker
4. Bishop DHL. 1989. Genetically engineered viral insecticides—a progress report 1986–1989. *Pestic. Sci.* 27:173–89
5. Bishop DHL. 1992. Baculovirus expression vectors. *Sem. Virol.* 3:253–64
6. Bishop DHL, Entwistle PF, Cameron IR, Allen CJ, Possee RD. 1988. Field trials of genetically engineered baculovirus insecticides. In *Release of Genetically Engineered Micro-organisms,* ed. M Sussman, CH Collins, FA Skinner, DE Stewart-Tull, pp. 143–79. New York: Academic
7. Blissard GW, Rohrmann GF. 1990. Baculovirus diversity and molecular biology. *Annu. Rev. Entomol.* 35:127–55
8. Bonning BC, Hammock BD. 1992. Development and potential of genetically engineered viral insecticides. *Biotech. Genet. Eng. Rev.* 10:453–87
9. Bonning BC, Hammock BD. 1993. Lethal ratios: an optimized strategy for

presentation of bioassay data generated from genetically engineered baculoviruses. *J. Invert. Pathol.* 62:196–97
10. Bonning BC, Hammock BD. 1994. Insect control by use of recombinant baculoviruses expressing juvenile hormone esterase. See Ref. 47a, pp. 368–83
11. Bonning BC, Hirst M, Possee RD, Hammock BD. 1992. Further development of a recombinant baculovirus insecticide expressing the enzyme juvenile hormone esterse from *Heliothis virescens.* *Insect Biochem. Mol. Biol.* 22:453–58
12. Bonning BC, Hoover K, Booth TF, Duffey S, Hammock BD. 1995. Development of a recombinant baculovirus expressing a modified juvenile hormone esterase with potential for insect control. *Arch. Biochem. Physiol.* In press
13. Bonning BC, Roelvink PW, Vlak JM, Possee RD, Hammock BD. 1994. Superior expression of juvenile hormone esterase and β-galactosidease from the basic protein promoter of *Autographa californica* nuclear polyhedrosis virus compared to the p10 protein and polyhedrin promoters. *J. Gen. Virol.* 75: 1551–56
14. Deleted in proof
15. Booth TF, Bonning BC, Hammock BD. 1992. Localization of juvenile hormone esterase during development in normal and in recombinant baculovirus-infected larvae of the moth *Trichoplusia ni. Tiss. Cell* 24:267–82
16. Carbonell LF, Hodge MR, Tomalski MD, Miller LK. 1988. Synthesis of a gene coding for an insect-specific scorpion neurotoxin and attempts to express

it using baculovirus vectors. *Gene* 73: 409–18

17. Chiang H, Dice JF. 1988. Peptide sequences that target proteins for enhanced degradation during serum withdrawal. *J. Biol. Chem.* 263:6797–805

18. Chiang H, Terlecky SR, Plant CP, Dice JF. 1989. A role for a 70-kilodalton heat shock protein in lysosomal degradation of intracellular proteins. *Science* 246: 382–85

19. Coats JR. 1994. Risks from natural versus synthetic insecticides. *Annu. Rev. Entomol.* 39:489–515

20. Coghlan A. 1994. Will the scorpion gene run wild? *New Sci.* June 25: 14–15

21. Corsaro BG, DiRenzo J, Fraser MJ. 1989. Transfection of cloned *Heliothis zea* cell lines with the DNA genome of the *Heliothis zea* nuclear polyhedrosis virus. *J. Virol. Methods* 25:283–91

22. Corsaro BG, Gijzen M, Wang P, Granados RR. 1993. Baculovirus enhancing proteins as determinants of viral pathogenesis. See Ref. 2a, pp. 127–45

23. Cory JS. 1991. Release of genetically modified viruses. *Rev. Med. Virol.* 1:79–88

24. Cory JS, Entwistle PF. 1990. Assessing the risks of genetically manipulated baculoviruses. *Aspects Appl. Biol.* 24: 187–94

25. Cory JS, Hirst ML, Williams T, Hails RS, Goulson D, et al. 1994. Field trial of a genetically improved baculovirus insecticide. *Nature* 370:138–40

26. Crozier G, Crozier L, Argaud O, Poudevigne D. 1994. Extension of *Autographa californica* nuclear polyhedrosis virus host range by interspecific replacement of a short DNA sequence in the p143 helicase gene. *Proc. Natl. Acad. Sci. USA* 91:48–52

27. Davies AH. 1994. Current methods for manipulating baculoviruses. *Biotechnology* 12:47–50

28. Eldridge R, Horodyski FM, Morton DB, O'Reilly DR, Truman JW, et al. 1991. Expression of an eclosion hormone gene in insect cells using baculovirus vectors. *Insect Biochem.* 21:341–51

29. Eldridge R, O'Reilly DR, Hammock BD, Miller LK. 1992. Insecticidal properties of genetically engineered baculoviruses expressing an insect juvenile hormone esterase gene. *Appl. Environ. Microbiol.* 58:1583–91

30. Engelhard EK, Kam-Morgan LNW, Washburn JO, Volkman LE. 1994. The insect tracheal system: a conduit for the systemic spread of *Autographa californica* M nuclear polyhedrosis virus. *Proc. Natl. Acad. Sci. USA* 91:3224–27

31. Entwistle PF, Evans HF. 1985. Viral control. See Ref. 38b, pp. 347–412

32. Federici BA. 1993. Viral pathogenicity in relation to insect control. See Ref. 2a, pp. 81–101

33. Fife HG, Palli SR, Locke M. 1987. A function for pericardial cells in an insect. *Insect Biochem.* 17:829–40

34. Food and Agriculture Organization of the United Nations. 1968. Rep. 1st session of the FAO panel of experts on integrated pest control, Rome. *Meet Rep. PL/1967/M/7 FAO*

35. Fox JL. 1995. EPA's first commercial release is still pending. *Biotechnology* 13:114–15

36. Deleted in proof

37. Friesen PD, Miller LK. 1986. The regulation of baculovirus gene expression. *Curr. Top. Microbiol. Immunol.* 131:31–49

38. Fuxa JR. 1993. Insect resistance to viruses. See Ref. 2a, pp. 197–209

38a. Fuxa JR, Richter AR. 1990. Response of nuclear polyhedrosis virus–resistant *Spodoptera frugiperda* larvae to other pathogens and to chemical insecticides. *J. Invert. Pathol.* 55:272–77

38b. Gilbert LI, Kerkut GA, eds. 1985. *Comprehensive Insect Physiology, Biochemistry and Pharmacology*. Oxford: Pergamon

39. Granados RR, Federici BA. 1986. *The Biology of Baculoviruses. Practical Application in Insect Control*. Boca Raton, FL: CRC. 276 pp.

40. Hammock BD. 1985. Regulation of juvenile hormone titer: degradation. See Ref. 38b, pp. 431–72

41. Hammock BD. 1992. Virus release evaluation. *Nature* 355:119

42. Hammock BD. 1993. Recombinant baculoviruses as biological insecticides. In *Pest Management: Biologically Based Technologies*, ed. RD Lumsden, JL Vaughn, pp. 313–25. Washington, DC: Am. Chem. Soc.

43. Hammock BD, Bonning B, Possee RD, Hanzlik TN, Maeda S. 1990. Expression and effects of the juvenile hormone esterase in a baculovirus vector. *Nature* 344:458–61

44. Hammock BD, McCutchen BF, Beetham J, Choudary P, Fowler E, et al. 1993. Development of recombinant viral insecticides by expression of an insect specific toxin and insect specific enzyme in nuclear polyhedrosis viruses. *Arch. Insect Biochem. Physiol.* 22:315–44

45. Hammock BD, Sparks TC. 1977. A rapid assay for insect juvenile hormone esterase activity. *Anal. Biochem.* 82: 573–79

46. Hanzlik TN, Abdel-Aal YAI, Harshman LG, Hammock BD. 1989. Isolation and sequencing of cDNA clones coding for juvenile hormone esterase from *Heliothis virescens:* evidence for a charge relay network of the serine esterases different from the serine proteases. *J. Biol. Chem.* 264:12419–25

47. Hawtin RE, Possee RD. 1993. Genetic manipulation of the baculovirus genome for insect pest control. See Ref. 2a, pp. 179–95

47a. Hedin P, Menn JJ, Hollingworth R, eds. 1994. *Natural and Derived Pest Management Agents.* Washington, DC: Am. Chem. Soc.

48. Heinz KM, McCutchen BF, Herrmann R, Parrella MP, Hammock BD. 1995. Direct effects of recombinant nuclear polyhedrosis viruses on selected nontarget organisms. *J. Econ. Entomol.* 88(2): 259–64

49. Hendrick JP, Hartl FU. 1993. Molecular chaperone functions of heat-shock proteins. *Annu. Rev. Biochem.* 62:349–84

50. Herrmann R, Moskowitz H, Zlotkin E, Hammock BD. 1995. Positive cooperativity among insecticidal scorpion neurotoxins. *Toxicon.* In press

51. Hofte H, Whiteley HR. 1989. Insecticidal crystal proteins of *Bacillus thuringiensis. Microbiol. Rev.* 53:242–55

52. Hoover K, Schultz CM, Lane SS, Bonning BC, Duffey SS, et al. 1995. Reduction in damage to cotton plants by a recombinant baculovirus that causes moribund larvae of *Heliothis virescens* to fall off the plant. *Biol. Control.* 5: In press

53. Ichinose R, Kamita SG, Maeda S, Hammock BD. 1992. Pharmacokinetic studies of the recombinant juvenile hormone esterase in *Manduca sexta. Pestic. Biochem. Physiol.* 42:13–23

54. Ichinose R, Nakamura A, Yamoto T, Booth TF, Maeda S, et al. 1992. Uptake of juvenile hormone esterase by pericardial cells of *Manduca sexta. Insect Biochem. Mol. Biol.* 22:893–904

55. Kamita SG, Maeda S. 1993. Inhibition of *Bombyx mori* nuclear polyhedrosis virus (NPV) replication by the putative DNA helicase gene of *Autographa californica* NPV. *J. Virol.* 67:6239–45

56. Keeley LL, Hayes TK. 1987. Speculations on biotechnology applications for insect neuroendocrine research. *Insect Biochem.* 17:639–51

57. King LA, Possee RD. 1992. *The Baculovirus Expression System.* London: Chapman & Hall. 229 pp.

58. Kondo A, Maeda S. 1991. Host range expansion by recombination of the baculoviruses *Bombyx mori* nuclear polyhedrosis virus and *Autographa californica* nuclear polyhedrosis virus. *J. Virol.* 65: 3625–32

59. Kool M, Vlak JM. 1993. The structural organization of the *Autographa californica* nuclear polyhedrosis virus genome. *Arch. Virol.* 130:1–16

60. Korth KL, Levings CS. 1993. Baculovirus expression of the maize mitochondrial protein URF13 confers insecticidal activity in cell cultures and larvae. *Proc. Natl. Acad. Sci. USA* 90: 3388–92

60a. Lawrie AM, King LA. 1993. *Baculovirus expression of urokinase-type plasminogen activator: comparison of late and very late promoters.* Presented at Annu. Meet. Am. Soc. Virol. 12th, Davis, CA

61. Luckow VA. 1994. Insect cell expression technology. In *Principles and Practice of Protein Engineering,* ed. JL Cleland, CS Craik, pp. 1–27. New York: Wiley & Sons

62. Deleted in proof

63. Maeda S. 1989. Expression of foreign genes in insects using baculovirus vectors. *Annu. Rev. Entomol.* 34:351–72

64. Maeda S. 1989. Increased insecticidal effect by a recombinant baculovirus carrying a synthetic diuretic hormone gene. *Biochem. Biophys. Res. Commun.* 165: 1177–83

65. Maeda S. 1994. Expression of foreign genes in insect cells using baculovirus vectors. In *Insect Cell Biotechnology,* ed. K Maramorosch, AH McIntosh, pp. 1–31. Boca Raton, FL: CRC

66. Deleted in proof

67. Maeda S, Hammock BD. 1993. Recombinant baculoviruses expressing foreign genes for improved insect control. In *Pest Control with Enhanced Environmental Safety,* ed. SO Duke, JJ Menn, JR Plimmer, pp. 281–97. Washington, DC: Am. Chem. Soc.

68. Maeda S, Kamita SG, Kondo A. 1993. Host range expansion of *Autographa californica* nuclear polyhedrosis virus (NPV) following recombination of a 0.6-kilobase-pair DNA fragment originating from *Bombyx mori* NPV. *J. Virol.* 67:6234–38

69. Maeda S, Volrath SL, Hanzlik TN, Harper SA, Maddox DW, et al. 1991. Insecticidal effects of an insect-specific neurotoxin expressed by a recombinant baculovirus. *Virology* 184:777–80

69a. Majima K, Gomi SJ, Chua JW, Maeda S. 1994. *Nucleotide sequence and gene-deletion analysis of the* Bombyx mori *nuclear polyhedrosis (BmNPV) genome.*

Annu. Meet. Am. Soc. Virol. 13th, Madison, WI

70. Martens JWM, Honee G, Zuidema D, van Lent JWM, Visser B, et al. 1990. Insecticidal activity of a bacterial crystal protein expressed by a recombinant baculovirus in insect cells. *Appl. Environ. Microbiol.* 56:2764–70

71. Masler EP, Kelly TJ, Menn JJ. 1991. Biologically active insect peptides; prospects for applied and fundamental knowledge. In *Insect Neuropeptides: Chemistry, Biology and Action,* ed. JJ Menn, TJ Kelly, EP Masler, pp. 6–18. Washington, DC: Am. Chem. Soc.

72. McCutchen BF, Betana MD, Herrmann R, Hammock BD. 1995. Interactions of recombinant and wild-type baculoviruses with classical insecticides and pyrethroid-resistant tobacco budworm (Lepidoptera: Noctuidae). *J. Econ. Entomol.* In press

73. McCutchen BF, Choudary PV, Crenshaw R, Maddox D, Kamita SG, et al. 1991. Development of a recombinant baculovirus expressing an insect-selective neurotoxin: potential for pest control. *Biotechnology* 9:848–52

74. McCutchen BF, Hammock BD. 1994. A recombinant baculovirus expressing an insect-selective neurotoxin: characterization, strategies for improvement and risk assessment. See Ref. 47a, pp. 348–67

75. McCutchen BF, Uematsu T, Székács A, Huang TL, Shiotsuki T, et al. 1993. Development of surrogate substrates for juvenile hormone esterase. *Arch. Biochem. Biophys.* 307:231–41

76. Merryweather AT, Weyer U, Harris MPG, Hirst M, Booth TF, et al. 1990. Construction of genetically engineered baculovirus insecticides containing the *Bacillus thuringiensis* subsp. *kurstaki* HD-73 delta endotoxin. *J. Gen. Virol.* 71:1535–44

77. Miller HI. 1994. A need to reinvent biotechnology regulation at the EPA. *Science* 266:1815–18

78. Miller LK. 1988. Baculoviruses as gene expression vectors. *Annu. Rev. Microbiol.* 42:177–99

79. Miller LK. 1993. Baculoviruses: high-level expression in insect cells. *Curr. Opin. Genet. Dev.* 3:97

80. Mori H, Nakazawa H, Shirai N, Shibata N, Sumida M, et al. 1992. Foreign gene expression by a baculovirus vector with an expanded host range. *J. Gen. Virol.* 73:1877–80

80a. Murphy FA, Fauquet CM, Bishop DHLB, Ghabrial SA, Jarvis AW, et al, eds. 1995. *Virus Taxonomy. Sixth Rep.*

Int. Comm. Taxonomy of Viruses. New York: Springer-Verlag

80b. O'Reilly DR. 1995. Baculovirus-encoded ecdysteroid UDP-glucosyltransferases. *Insect Biochem. Mol. Biol.* 25(5): 541–50

81. O'Reilly DR, Brown MR, Miller LK. 1992. Alteration of ecdysteroid metabolism due to baculovirus infection of the fall armyworm *Spodoptera frugiperda:* host ecdysteroids are conjugated with galactose. *Insect Biochem. Mol. Biol.* 22:313–20

82. O'Reilly DR, Kelly TJ, Masler EP, Thyagaraja BS, Robson RM, et al. 1993. Overexpression of *Bombyx mori* prothoracicotrophic hormone using baculovirus vectors. *Insect Biochem. Mol. Biol.* 25:475–85

83. O'Reilly DR, Miller LK. 1989. A baculovirus blocks insect molting by producing ecdysteroid UDP-glucosyl transferase. *Science* 245:1110–12

84. O'Reilly DR, Miller LK. 1990. Regulation of expression of a baculovirus ecdysteroid UDP-glucosyl transferase gene. *J. Virol.* 64:1321–28

85. O'Reilly DR, Miller LK. 1991. Improvement of a baculovirus pesticide by deletion of the EGT gene. *Biotechnology* 9:1086–89

86. O'Reilly DR, Miller LK, Luckow VA. 1992. *Baculovirus Expression Vectors—a Laboratory Manual.* New York: Freeman. 347 pp.

87. Riddiford LM. 1994. Cellular and molecular actions of juvenile hormone. I. General considerations and premetamorphic actions. *Adv. Insect Physiol.* 24:213–74

88. Roelvink PW, van Meer MMM, de Kort CAD, Possee RD, Hammock BD, et al. 1992. Temporal expression of *Autographa californica* nuclear polyhedrosis virus polyhedrin and p10 gene. *J. Gen. Virol.* 73:1481–89

89. Roush RT, Tabashnik BE. 1990. *Pesticide Resistance in Arthropods.* New York: Chapman & Hall

90. Smith GE, Summers MD, Fraser MJ. 1983. Production of human β-interferon in insect cells infected with a baculovirus expression vector. *Mol. Cell. Biol.* 3:2156–65

91. Sridhar P, Panda AK, Pal R, Talwar GP, Hasnain SE. 1993. Temporal nature of the promoter and not the relative strength determines the expression of an extensively processed protein in a baculovirus system. *FEBS Lett.* 315: 282–86

92. Staal GB. 1986. Anti juvenile hormone agents. *Annu. Rev. Entomol.* 31:391–429

93. Steinhaus EA. 1956. Microbical control—the emergence of an idea. *Hilgardia* 26:107–60
94. Stewart LMD, Hirst M, Ferber ML, Merryweather AT, Cayley PJ, et al. 1991. Construction of an improved baculovirus insecticide containing an insect-specific toxin gene. *Nature* 352:85–88
95. Summers MD, Smith GE. 1987. A manual of methods for baculovirus vectors and insect cell culture procedures. *Tex. Agric. Exp. Stn. Bull.* 1555:1–57
96. Tomalski MD, Miller LK. 1991. Insect paralysis by baculovirus-mediated expression of a mite neurotoxin gene. *Nature* 352:82–85
97. Tomalski MD, Miller LK. 1992. Expression of a paralytic neurotoxin gene to improve insect baculoviruses as biopesticides. *Biotechnology* 10:545–49
98. Vlak JM, Schouten A, Usmany M, Belsham GJ, Klinge-Roode EC, et al. 1990. Expression of a cauliflower mosaic virus gene I using a baculovirus vector based on the p10 gene and a novel selection method. *Virology* 178:312–20
99. Volkman LE, Keddie BA. 1990. Nuclear polyhedrosis virus pathogenesis. *Sem. Virol.* 1:249–56
100. Ward VK, Bonning BC, Huang T, Shiotsuki T, Griffith VN, et al. 1992. Analysis of the catalytic mechanism of juvenile hormone esterase by site-directed mutagenesis. *Int. J. Biol. Chem.* 24:1933–41
101. Williams GV, Rohel DZ, Kuzio J, Faulkner P. 1989. A cytopathological investigation of *Autographa californica* nuclear polyhedrosis virus p10 gene function using insertion/deletion mutants. *J. Gen. Virol.* 70:187–202
102. Williamson M. 1991. Biocontrol risks. *Nature* 353:394
103. Wood HA, Granados RR. 1991. Genetically engineered baculoviruses as agents for pest control. *Annu. Rev. Microbiol.* 45:69–87
104. Wood HA, Hughes PR, Shelton A. 1994. Field studies of the co-occlusion strategy with a genetically altered isolate of the *Autographa californica* nuclear polyhedrosis virus. *Environ. Entomol.* 23:211–19
105. Wood HA, Hughes PR, van Beek N, Hamblin M. 1990. An ecologically acceptable strategy for the use of genetically engineered baculovirus pesticides. In *Insect Neurochemistry and Neurophysiology 1989,* ed. AB Borkovec, EP Masler, pp. 285–88. Clifton, NJ: Humana
106. Wood HA, Trotter KM, Davis TR, Hughes PR. 1993. Per os infectivity of preoccluded virions from polyhedrin-minus recombinant baculoviruses. *J. Invert. Pathol.* 62:64–67
107. Yu Z, Podgwaite JD, Wood HA. 1992. Genetic engineering of a *Lymantria dispar* nuclear polyhedrosis virus for expression of foreign genes. *J. Gen. Virol.* 73:1509–14

Annu. Rev. Entomol. 1996. 41:211–29

SEXUAL SELECTION IN RELATION TO PEST-MANAGEMENT STRATEGIES

Christine R. B. Boake

Department of Ecology and Evolutionary Biology, The University of Tennessee, Knoxville, Tennessee 37996

Todd E. Shelly and Kenneth Y. Kaneshiro

Hawaiian Evolutionary Biology Program, University of Hawaii, Honolulu, Hawaii 96822

KEY WORDS: mate choice, male competition, pheromone, SIT

ABSTRACT

The application of principles derived from the sexual selection literature can assist attempts to subvert the normal mating behavior of pests. Sexual selection encompasses both intermale competition for access to females and female choice of mates. It can operate during long-range attraction and short-range courtship, as well as after copulation. We review the major aspects of sexual selection and illustrate their application to pheromonal and SIT pest-management programs. Pheromones are important both in long-range attraction and in close-range mate choice; parapheromones may be very useful in pest management because of their influence on male mating success. Sexual selection theory provides a scheme for studying the normal mating behavior of a pest species and thus determining which attributes of the mass-reared sterile males are critical to their success with wild females. We hope that our review will suggest novel ways of attacking pests as well as encourage behavioral ecologists to study pest species.

INTRODUCTION

Insect pest-management programs that incorporate the use of pheromones or sterile insect techniques can be designed to interfere with normal reproduction in the targeted species. Pheromones are used to monitor pest-population levels, to trap and kill pests, and to disrupt sexual communication through misdirection

211

or jamming of the signal (20, 72, 112). The sterile insect technique (SIT) consists of mass rearing a pest species, irradiating it to cause sterilization, and releasing millions of sterile insects to mate with the pest in the wild either before an outbreak becomes significant, or once an outbreak is reduced with other methods (64). The success of both of these approaches depends on a knowledge of the natural mating behavior of the pest species (18, 21, 64). Detailed assessments of the factors that contribute to mating success in natural populations can now be made because the study of animal mating behavior, under the name sexual selection, has been a major focus of research in behavioral ecology for more than two decades.

While much of the research on sexual selection is focused on the evolutionary reasons for particular patterns of mating behavior and on the evolutionary consequences of such behavior, the large body of literature regarding the mechanisms of sexual selection contains information that is very likely to be applicable to insect-management programs. Some of the data come from pest species, although much come from nonpest species. In this review, we outline the salient features of sexual selection theory and relate them to pest management explicitly, taking as many illustrations as possible from the pest literature. The two ways that sexual behavior is used to subvert pest species, pheromones and SIT, are not conceptually equivalent, because each pheromone is a particular signal used in one stage of the mating process, and SIT involves all stages. Thus, some of the material we cover is relevant to SIT alone, and other topics are relevant to both SIT and pheromones. Most of our examples for SIT come from studies of tephritid pests, which have been a major target of SIT. In the final part of the review, we consider evolutionary issues that we believe relevant to pest management.

CONCEPTS IN SEXUAL SELECTION

Darwin (31) used the phrase sexual selection to explain the evolution of elaborate male characters that are apparently detrimental under natural selection, possibly because they attract predators or have high energetic costs. These traits appear to be maintained because they confer a mating advantage on their bearers. The exact nature and even existence of the balance between natural and sexual selection is a controversial topic (62), but certain characters clearly do influence male mating success. Such traits have been identified through the use of both correlational and experimental studies. Moreover, the traits that influence male mating success vary widely between species. In a general sense, some classes of sexual signals may be common within certain taxonomic groups; for example, crickets, grasshoppers, and katydids all rely to some extent on acoustic sexual signals (84), but the particular characters that are important for a given species can only be determined by studying the breeding biology of that species.

We use the convention that males are the more strongly sexually selected sex and have the more elaborate displays. The elaborations can be in any signal system, including visual, acoustic, and chemical. The statement that males are more strongly sexually selected means that some males may mate numerous times whereas others mate rarely or not at all; this pattern is common in species without parental care, although relevant data may be hard to collect for insects because of their small size. Many of the Odonata are large enough to mark individually, and in the damselfly *Enallagma hageni,* Fincke (39) showed that about 43% of 132 males in a natural population fertilized no eggs, whereas every one of 76 females in the study produced several hundred eggs. In a field study of the dragonfly *Libellula luctuosa,* Moore (79) found that 4% of the males were responsible for 70% of the copulations and that 91% of the territorial males never copulated.

The sex bias in sexual selection has its evolutionary origins in anisogamy, the difference in gamete size between males and females. Females produce a few large gametes, whereas males produce huge numbers of very small gametes. Consequently, all females are likely to reproduce whereas the sperm from one or a few males is sufficient to fertilize all the eggs in a population. There is no sexual disparity in the average number of offspring per individual (in a population that is not growing, each individual will on average leave one descendant), but there is the potential for an enormous disparity in the variances in reproductive success between the two sexes. Males, which have a high variance in reproductive success, are thus under stronger sexual selection: The males with the most attractive displays, the best territories, etc, will mate more frequently and leave more offspring. The proper way to quantify the relative reproductive success of an individual in a population has been a topic of debate, thoroughly reviewed elsewhere (26).

Anisogamy results in differential parental investment and consequently different degrees of mate selectivity in each sex. The major prediction is that the sex that invests more, generally females, shall be less sexually selected, and more selective in its choice of mates (108). The tests of this prediction involve identifying species in which males invest substantially in reproduction and then determining whether the males are more discriminating than those in species without such a heavy male investment. Work in this field shows quite clearly, for a variety of Orthoptera, that when males make a nutritional contribution during mating (for example by producing a large spermatophylax that the female consumes), they discriminate among potential mates to a greater extent than do males in less heavily investing species (47, 49, 50). These classic examples of so-called sex-role reversal thus strengthen the case for the assumption that in most species, males are the more strongly sexually selected sex. However, the assumption may not be true for species in which males defend territories with resources or provide nuptial gifts to females, as many insects do (107).

In this review, we use the terminological shorthand common in the literature. When we state that a female "chooses" among potential mates, "assesses" males, "discriminates" among them, or exhibits a "preference," we are not necessarily implying that a female has consciousness or volition. Rather, we are abbreviating a statement that the female's behavior has influenced a male's probability of mating and that the probability of mating is correlated with a phenotypic character of that male, such as the quantity or quality of his calling. We do not necessarily mean that a female is consistent in her preferences or that a male is consistent in his signal production; instead, over the population at large, certain kinds of male signals are significantly correlated with a high probability of mating.

In the early parts of this review, "mating success" is equated with "reproductive success." Mating success is not the only contributor to reproductive success, and later we shall consider postcopulatory processes, such as sperm precedence, that can have a strong influence on male reproductive success.

CONTRIBUTORS TO MATING SUCCESS

The process of mate acquisition can be broken into several stages, which we describe briefly and then review in detail with respect to pest management. Sexual selection is often dichotomized into the processes of intrasexual (or male-male) competition and intersexual (or female) choice. These form a continuum, and the mating behavior of any one species will be most likely to incorporate aspects of both. In the first stage, a male may need to attract a female. This process can involve first acquiring a territory or high dominance rank, and later producing a display that has a large active space. The active space is the region around the producer in which the signal is above the threshold for a behavioral response from conspecifics (11). For example, the calling song of male crickets has a large active space; it has been implicated in influencing spacing between males and, in laboratory experiments and in the field, has also been demonstrated to attract flying and walking females (38). Cricket courtship song, which is only produced after a male and female have encountered each other, is not loud and has a much smaller active space.

The stage after attraction is often called courtship. In some species, such as butterflies, it is elaborate, involving chemicals, ultraviolet signals, and motion (100). In other species, such as some tephritids, screwworms, and flour beetles, it seems perfunctory if it exists at all (14, 75, 110). In many species, courtship during copulation may have a strong influence on a male's ability to fertilize eggs (36); this type of courtship has received very little attention. Generally, males produce a specific display that can only be detected at short distances and that affects a female's probability of copulation. After copulation, male accessory-gland secretions may influence female receptivity to additional males (92) and oviposition rate (10). Finally, patterns of sperm precedence can influence male reproductive success (103).

What Is the Mating System?

The term mating system has two definitions, one used by behavioral ecologists and the other by geneticists. Behavioral ecologists describe the mating system as the pattern of mating behavior shown by a population (32). For example, in a lek mating system, males defend territories that contain no resources used by females, males do not provide parental care, and females visit the lek territories solely to assess males and to copulate. In genetics, the term mating system describes the relatedness of males and females that mate, ranging from inbred to outbred, and the variances in mating success of each sex (87). Behavioral ecologists are concerned with the behavior involved in mating, whereas geneticists are concerned with the genetic consequences of patterns of mating for the population at large. Both of these perspectives are valuable for applications to pest management.

The general form of the behavioral mating system can be predicted from the species' ecology (37), particularly the distribution in space and time of resources that are needed for reproduction. When resources are discrete and can be monopolized, male territoriality may evolve, and the possible mating systems include monogamy and harem polygyny. When resources are dispersed or otherwise not monopolizable, males may need to search for females or compete to attract them, which can result in polygynous mating and leks. This reasoning has been successfully applied to an analysis of the mating systems of tephritid fly species (14, 102): Females of monophagous species can be encountered by males at oviposition sites, but males of polyphagous species cannot rely on such predictable encounters and thus cannot monopolize the locations of females. In the tephritids, polyphagy has been shown to be associated with an increased tendency to have a lek mating system (14, 102).

Lek mating systems occur in several pest species, particularly tephritid fruit flies (2, 13, 14, 67, 90, 93, 97, 101, 116). Here, we examine leks to illustrate the implications of a mating system for pest management. Males in leks defend territories that females visit for the sole purpose of mating. Because the territories are not used as oviposition or feeding sites, leks are a type of nonresource-based mating system. Furthermore, they are characterized by a very high variance in male mating success; a single male in a lek has the potential for mating many times, whereas females may need to mate only once. For management purposes, one would wish to know what proportion of males are responsible for copulation, what notable behavior these males exhibit, and whether particular males are highly successful from day to day or only on one or a few days. In some insect species for which data on individually marked males are available, males may appear at the mating site on only one or a few days, although the mating site is active for weeks or months (19, 39); presumably the males are feeding when they are away. In some species, particular territories within the lek site seem more attractive to females, and the male on such a territory appears to be successful at mating regardless of his identity, but this has not been demonstrated in insects (98).

Both pheromonal and SIT control methods require an understanding of the type of mating system. In the use of pheromones for mating disruption, some knowledge of the mating system should help determine where and when a pheromone is most likely to be successful (21). Some of the difficulties with interpreting pheromone-trap data in monitoring systems may be clarified with a more detailed knowledge of the mating system (112). For example, attempts to define the relationship between trap catches, population density, and damage might benefit from an understanding of the reproductive stage at which pests are most responsive to pheromones, and the reasons why they are responsive. The use of pheromones for attraction and annihilation is most effective when females can be killed; thus, information about female responsiveness to pheromones in relation to mating status would be valuable (72).

In SIT methods, knowledge of the behavioral mating system is also important. The ability of irradiated male tephritids to compete with wild males for matings has most often been tested in field cages (e.g. 66, 99). Although cage studies are more tractable than studies of completely free flies, they may not reflect the natural situation. For example, in both the melon fly, *Bactrocera cucurbitae*, and the Mediterranean fruit fly, *Ceratitis capitata*, caged sterile males are often not at a disadvantage compared with wild males, but those released into a natural population are less likely to mate than expected (60, 99). The reduction in mating success may result from the unintentional selection during mass rearing for different mating behaviors (18) that are only detrimental when the full range of behavior in the field is necessary for success.

An understanding of the genetic mating system is also relevant to pest-management strategies. The kinds of questions stimulated by a genetic perspective include the description of the variation in male mating success and the number of males that inseminate each female, both of which have been modeled in relation to the eradication of lek mating species (59).

Intermale Competition

Males may sometimes need to win aggressive encounters with other males to gain access to mates. The males might be competing for access to a lek or other territory that is most attractive to females (19, 39) or for the top position in a dominance hierarchy (9). In the case of the cockroach *Nauphoeta cinerea,* both male and female conspecifics can distinguish the odor of dominant males, which is preferred by females, from that of subordinates (80). Males apparently begin to produce this odor when they become dominant.

Males that are unsuccessful in acquiring territories may become satellites, waiting near territorial males and intercepting females that approach that male. Sometimes, satellite males take over a territory when the original owner is removed. The satellite strategy results in mating success in several insect species (17, 78, 79). In such species, sterile males that cannot win territories can still mate if they become satellites.

Observations regarding intermale aggression are needed to learn whether sterile males can effectively compete with wild males to gain access to wild females. Although radiation doses are calibrated to ensure that captive-bred males are not behaviorally weakened, mass rearing can result in selection against male aggressive tendencies or against the appropriate signals that are used in aggressive encounters. Although breeding for highly aggressive males is tempting, anecdotal evidence from *Drosophila silvestris* (C Boake, unpublished data) suggests that highly aggressive males may be aggressive toward females as well as toward other males, thus reducing their mating success.

Female Preferences

The reasons why female mating preferences evolve are controversial (12) and are the topic of considerable research. Theoreticians and empiricists generally agree that female preferences arise in the process of discrimination within a species, rather than as a way to discriminate against males of other species (68–70); discrimination against heterospecific males is an extension of behavioral patterns that evolved for intraspecific encounters. The evolution of female preferences may be a result of genetic correlations between behavior in the two sexes, or preferences may have resulted from the choice of features that indicate higher genetic quality, such as lack of parasites, bilateral symmetry, or vigorous displays (for recent reviews, see 63, 76, 113). Most of the debates about the evolution of female preferences do not directly bear on issues in pest management, which require us to measure the strength of female preferences and to determine how these preferences affect male mating success. The consequence of a female preference is selection for certain male phenotypes; below we discuss the stages in mating behavior (mate attraction, courtship) at which female preferences could be shown, and how these could influence pest management.

Are female preferences directional or stabilizing? Female preferences, which could be expressed at either the mate-attraction or the courtship stage, are not necessarily for the most conspicuous male. In some cases, an extreme signal could elicit a lesser female response, as demonstrated for certain components of frog calls (43). The experiments to determine the limits of female preference are extremely time-consuming and may not be appropriate in many studies of pest species. If mass rearing is accompanied by directional selection for some male traits, the limits of female preferences may influence the success of selected males in the field.

MATE ATTRACTION   Males and females may encounter each other at or near resources vital to females, at traditional mating sites, or as a result of a long-distance signal produced by one sex that attracts the other. When the site for sexual encounters is on a resource, the location of males may be nonrandom within the habitat. For example, the most successful males defending territories on fruit trees may be those on leaves rather than those on the fruit resource. This is the case for the Mediterranean fruit fly (14). Finding the sites at which males can mate successfully may be crucial in efforts to study and influence mating behavior. For example, honeybees (119) and some Diptera (34) use traditional mating sites. In studies of mating disruption, the researcher may need to find these sites to verify that the sterile males have found them.

Long-range signals that cause conspecifics to approach are acoustic or chemical more often than visual. A major advantage of acoustic and chemical signals in long-range communication is that they can travel around barriers and are not restricted to the line of sight. Species that use long-range acoustic signals tend to be those with enough muscle power to produce a high-intensity signal or to enhance their environment for its effective radiation (40). Long-distance mate attraction by acoustic cues is well-illustrated by cricket song. In several cricket subfamilies, solitary males produce a loud repetitive calling song; phonotaxis tests have demonstrated that the song attracts both walking and flying females (38). In most of the studies, females have been given a choice between a signal from their own and from another species (38). However, Hedrick (51) has demonstrated that naturally occurring variation in the degree of interruption of trills is a cue used by females of *Gryllus integer* in discriminating among potential mates: Females prefer calls that are more continuous. Thus, long-range signals serve not only to attract a female to the general vicinity of a male, but also to influence mate choice.

The risk of predation can influence a female's response to an attractant. In the cricket *G. integer,* the female's preference for more continuous song (51) is reduced by the perceived risk of predation. Females prefer a less continuous song if the pathway to the (audio) speaker playing the intermittent call has more protective cover than the pathway to the speaker broadcasting the otherwise more preferred call (52).

Parasitoid flies are attracted to the calling songs of some cricket species (17,

111). They attack males more commonly than females, but sometimes search in the vicinity of a calling male to find females and satellite males. The use of acoustically orienting parasitoids may become a tool for pest management (118).

Chemical signals are often persistent and rarely used by predators (as far as is known), which has probably contributed to their evolution as sexual attractants. Many lepidopteran females use pheromones to attract males over long distances; this is clearly a component of mating behavior, but it has received relatively little investigation from the perspective of sexual selection. Chemical signals may be produced endogenously, or they may require dietary precursors, such as those used by euglossine bees (33). At least one parapheromone, methyl eugenol, may be used in the synthesis of the male sex pheromone (83), and males exposed to methyl eugenol attract more females to their calling site than males that are not exposed to the lure (96). Mediterranean fruit fly males do not actually feed on trimedlure but merely rest nearby, but exposure to the lure still appears to increase male calling activity and mating success (TS Whittier, TE Shelly & KY Kaneshiro, unpublished data). Landolt & Heath (71) report a similar case for the cabbage looper moth (*Trichoplusia ni*): Exposure to host plant volatiles increased the production of certain pheromonal components by males, which in turn conferred a mating advantage over nonexposed males.

Pheromones are the long-range attractant signals most often used by investigators to monitor a population size or to trap and kill pests (72, 112). Acoustic signals can be used to attract pests (111), but may be far less practical than pheromones for monitoring or killing pests because of the complexity and cost of the relevant technology. A clever use of acoustic traps is to attract mole crickets (*Scapteriscus* spp.) to traps that contain infective-phase nematodes, then release the crickets about a day later. This procedure resulted in the establishment of nematodes in 9 of 21 cases, and considerably fewer nematodes needing to be reared than would have been necessary if they had been spread on the desired areas (85).

SHORT-RANGE INTERACTIONS   The precopulatory interaction between a male and a female that are close enough to touch each other is generally called courtship. It can involve the repetition of one or several male and female acts, and it can occur after a male has mounted the female (36). In some species, the females give a signal when they are ready to copulate, and in others, the male insect seems unable to tell when a female is receptive, based on observations of failed attempts to copulate. Complex courtships can be described with transition diagrams (e.g. 23, 57) in which arrows between acts are coded to show the probabilities of different transitions. Comparisons of these sequences for successful and unsuccessful courtships may give information as to the stages of courtship that are critical for success. Quantitative information about normal courtship could be used to determine whether an irradiated male was courting effectively. For example, in *Anastrepha suspensa,* an increasing pulse rate of the song is correlated with increased mating success (15), but as radiation dosage increases, the pulse rate of the song drops (114). Correlational data might also be used to identify stages of courtship that are amenable to human intervention.

Correlational descriptions of courtship need to be supplemented with experiments, as has been done with many species. The experiments have been designed to learn whether certain aspects of courtship are critical to male mating success and to determine which type of male is likely to be most successful. For example, males that cannot produce acoustic signals often experience reduced or no success (30, 56, 88), but in the cricket *Amphiacusta maya,* female preferences do not affect the probability that experimentally silenced males will mate (9). As another example, experimental removal of organs that may dispense pheromones has shown that lepidopteran scales and hairpencils are critical for male copulatory orientation and for female acceptance of a male (44, 86). An exciting new experimental technique for studying visual signals in arthropod courtship is to present computer-generated video images to females (25).

When experimental studies of courtship are conducted with pest species, the results can be used to determine which characteristics of a captive population are most important to monitor during the production of sterile males. A more ambitious approach would be to breed males that are particularly successful at the critical stages of courtship, as long as such breeding was not associated with a decrement in some other stage. Furthermore, although interfering on a large scale with some stage of courtship (for example by jamming a pheromonal signal) might be tempting, we must remember that courtship is a short-range interaction that takes place between individuals that are physiologically prepared. Successful attempts at pheromonal mating disruption (21) appear to result from the failure of the two sexes to encounter each other rather than the failure of courtship after they meet. The courtship success of irradiated males might be enhanced via selective breeding or, in appropriate species, by providing captive males with a parapheromone (96).

Copulation and Beyond: Sperm Precedence and Reproductive Success

The above discussions assumed that copulation is equivalent to reproductive success. But in many insect species, females can mate more than once, and thus sperm precedence can influence male reproductive success (103). The patterns of sperm precedence vary between and often within species. In many species, the sperm from the most recent male to copulate are more likely to fertilize the female's eggs, although this probability is rarely 100% (46, 48). The patterns can depend on the types of males being tested, on the protocol of the experiment, and on the interval between matings (82, 94, 110). In several insect species (45, 53, 110), the pattern of precedence depends on the nature of the mate: When a female is mated with a conspecific male and with a male from another species, the sperm from the conspecific male fertilize the eggs regardless of whether he was the first or second mate. In odonates and in *Tenebrio molitor,* males possess an organ that removes the sperm of previous males from the female's storage organs (42, 109), thus ensuring last-male paternity.

Patterns of sperm transfer and sperm usage may also influence the outcome of a sterile male program. The duration of copulation is roughly correlated with the number of sperm transferred or with the probability of a spermato-phore being transferred. However, in many species, copulation lasts longer than the time necessary for sperm transfer (73, 74, 81, 120). This extra time may allow the transfer of substances that render the female unreceptive or unattractive to other males (1, 24, 92). The question of whether sterile males are as successful as normal males at influencing subsequent female receptivity has been addressed for several species. In experiments with *B. cucurbitae, Helicoverpa zea,* and *Heliothis virescens,* sterile males were as effective as normal males in inhibiting female remating (22, 65, 91). On the other hand, females of *Ceratitis capitata* that mated with sterile males (either ones that had been irradiated or that had mated once previously) were reportedly more likely to remate than females that had mated with wild males (5).

Consequently, a sterile male's reproductive success can depend on many factors: whether territories are involved in mating, a male's ability to locate and acquire such a territory, his ability to attract females to his vicinity, his ability to court them successfully, his ability to transfer sperm, and the female's use of the sterile sperm. A survey of the natural history of any given species will probably be adequate to determine which of the above processes deserve detailed attention.

BROADER ISSUES

Here we supplement our survey with a discussion of various issues common to many of the steps. First, we consider the criteria necessary to demonstrate that a putative signal is in fact a signal. Then we discuss several genetic issues, including the difficulties involved in laboratory culture, the possibility of correlations between different aspects of the male phenotype, the nature of variation between populations, and the point that killing pests is a form of artificial selection, which may have evolutionary implications.

In studies of animal communication, something that an animal produces is called a signal if it affects the probability of behavior of another individual (77). The long controversy about the waggle dance of honeybees (35) is a good example: Most researchers agree that the dance is produced by returning foragers, but debate arose over the information contained in the dance, the sufficiency of the dance as a signal, and whether hive-mates use the dance to find a food source. In this debate, a physical characterization of the dance was not sufficient evidence to demonstrate that it is a signal. In an example involving sexual behavior, males of the cricket *Amphiacusta maya* produce a chirp during courtship, but this chirp does not influence a female's receptivity to the male. Rather, it serves as an aggressive signal to other males and reduces the chance that the courting male will be interrupted (9). These examples illustrate the principle that both the response to and the production of a putative signal must be examined before it can be called a signal and before its function can be elucidated.

The genetic issues range from those at the individual to those at the population level. The most pressing is the possibility that insects reared in captivity will not be as successful as wild ones (18). The laboratory environment is unlikely to put a premium on male access to mating territories or on a male's ability to attract a female over a long distance. In captivity or in a field cage, a male strategy of being persistent might result in many matings, but in the wild, where females are free to leave, males may have a limited time in which to court, and vigor rather than persistence may be favored. Most tests of the success of captive-reared insects have been conducted with field cages, yet such tests may give overly optimistic results, because they do not include an assessment of whether males can locate appropriate mating sites and mate with unconstrained females. The captive population may have diverged from the natural population to such an extent that sterile flies mate preferentially with each other (assortative mating) (41, 99, 104).

Another genetic issue regards possible correlations among various factors that contribute to a male's mating success. We need to know whether these correlations are positive, negative, or zero. For example, male size and aggressive success may be positively correlated, but males that are larger and/or better fighters are also successful in courtship in only some (15, 89, 121) but not all (6, 93) species. When male size and courtship success are correlated, the underlying cause is rarely if ever known. Do females detect male size directly, or do larger males produce more intense signals? If the cues by which females assess size were known, sterile males might be supplemented with the appropriate signal, much as males have been supplied with parapheromones (see above).

Traits may be correlated at both the phenotypic and genetic levels. Phenotypic correlations could be caused almost entirely by the environment. For example, when larval food availability affects adult body size, the repercussions will show in other size-dependent traits such as aggressive success. Genetic correlations, evidenced by phenotypic correlations running in families, result from the influence of the same genes on several traits. The study of these correlations in relation to mating behavior has only just begun (3, 115). Genetic correlations could have a substantial impact in captive-rearing programs. If two traits, such as body size and aggressive success, were positively correlated, increasing one during captive breeding might also increase the other. Alternatively, for two traits with a negative genetic correlation, say aggressive success and courtship success, an increase in one during captive breeding could lead to a decrease in the other. However, genetic correlations are difficult to assess: In *Drosophila melanogaster,* larger males have an advantage in territorial battles, but selection for increased territoriality did not produce larger flies (55). The experiments necessary to reveal genetic correlations between behavioral characters are time-consuming and labor intensive (8) and may not be appropriate as a regular part of a mass-rearing program. A less difficult approach would be to monitor selected males for traits that are relevant to male mating success, not just those that were the focus of artificial selection.

A more complex genetic correlation is that between the senders and receivers of a sexual signal (7, 16). In this case, the female relatives of males that produce a particular type of signal, such as a certain pheromone blend or pitch of a sound, prefer that signal type. This kind of correlation can arise easily through assortative mating, in which males producing a certain type of signal mate with females that prefer that type. Behavioral ecologists have reported such correlations in some cases (4, 117), but frequently the studies are based on small sample sizes and the statistical power is weak.

In pest management, a technique that relied on intervention in the sexual signaling system (such as pheromone traps) could select for a change in that system, much as other pest-management efforts have resulted in resistance to insecticides and, recently, to *Bacillus thuringiensis* (95, 105). The strength of this possibility depends on several factors, including the magnitude of any such genetic correlation, the proportion of the population that is killed in the traps, the population size, and gene flow with other populations. Artificial selection of *Pectinophora gossypiella* has succeeded in changing the ratio of pheromone components, the quantity of pheromone, and the response of females to male pheromone (27–29). In both *P. gossypiella* and *Argyrotaenia velutinana,* male preferences for an extreme proportion of one component of the pheromone blend appear to show a genetic correlation with female production of a similar proportion (27, 94a). Similar genetic analyses with other species, although arduous (8), would be valuable.

Although all individuals within a population may have similar behavior, substantial differences can exist between populations. Variation within and between populations is well known in pest species, particularly in chemical-communication systems (61, 106). The effects of such variation on pest-management strategies will depend on its magnitude and on which traits are affected.

When numerous members of a natural population fail to reproduce, either because individuals cannot distinguish between pheromones from a live and a synthetic source or because females have mated with sterile males, strong selection in favor of more accurate discrimination results. The use of pheromones for pest management, as mentioned above, can cause both pheromone preference and production to shift, perhaps to a different blend of the same compounds. In the case of SIT, females capable of discriminating against sterile males would have a major reproductive advantage, and any genes underlying their abilities would be likely to spread. This possibility is no longer in the realm of speculation. Discrimination against captive-reared sterile males has evolved in wild females in two reported cases: the melon fly on Okinawa (54) and a population of Mediterranean fruit fly on the island of Kauai, Hawaii (D McInnis, personal communication).

The genetic problem of the adaptation of a wild population to captive rearing is very similar whether the species being reared is a pest or a pest-control agent. Genetic factors in the mass rearing of pest-control agents were reviewed recently (58). The authors' recommendations that the culture time be kept short (less than 5 generations) and that laboratory conditions be similar to those in the field are appropriate for the mass-rearing of species for SIT. Keeping mating conditions similar to those in the field may be very labor intensive. Consequently, the value of restoring near-natural gene frequencies to the captive stock by frequent replacement becomes very important (but see 18).

CONCLUDING REMARKS

One SIT issue recognized since the beginning (64) is whether sterile males can successfully attract and mate with females. The considerations raised above demonstrate that a careful study of the mating behavior of the pest species is an important component of management procedures. The difficulties with SIT may be reduced in some cases if the factors that contribute to male mating success are better known (18).

Pheromones need to be released at times of day and year when they are likely to be effective and in locations where they will be useful. The major considerations in pheromone release appear to result from the psychophysics of pheromone detection and the behavioral response (20, 21). The use of pheromones to interfere with courtship rather than with attraction may not be practical because courtship pheromones may be released at a particular stage of courtship rather than be broadcast.

During captive rearing of a pest species, various genetic factors, such as genetic drift and selection for specific mating behaviors favorable under laboratory conditions, could modify the general mating behavior of the stock. Irradiation itself could affect a male's mating behavior. Many of these possible difficulties in a mass-rearing or pheromone program have been recognized for a long time (64), but now very specific ways of monitoring the success of the program can be developed. These methods depend on using sexual selection theory to provide a framework for a systematic study of the mating behavior of natural populations of the pest. The most critical stages in the process and the characteristics of successful males can be identified for each species. This information will provide a basis on which to monitor the captive stock, and it may provide new insights into ways to subvert the natural mating system.

We believe that funds for research into the sexual behavior of pest species will be well spent, particularly where irradiated males have been less successful than wild males. We hope that more of our colleagues who study sexual selection will consider working on insect pests. Many pest species have mating systems that are suitable for tests of current evolutionary debates, have large population sizes, and are suitable for both laboratory and field work.

ACKNOWLEDGMENTS

We thank X Meng for help with finding references. While writing the review, TES was supported by the USDA (grant No. 59-5320-1-189) and by the California Department of Food and Agriculture (grant No. 93-0481), and CRBB received support from the USDA (grant No. 92-37302-7793).

Literature Cited

1. Alcock J. 1994. Postinsemination associations between males and females in insects: the mate-guarding hypothesis. *Annu. Rev. Entomol.* 39:1–21
2. Arita LH, Kaneshiro KY. 1989. Sexual selection and lek behavior in the Mediterranean fruit fly, *Ceratitis capitata* (Diptera: Tephritidae). *Pacific Sci.* 43: 135–43
3. Aspi J, Hoikkala A. 1993. Laboratory and natural heritabilities of male courtship song characters in *Drosophila montana* and *D. littoralis. Heredity* 70: 400–6
4. Bakker TCM. 1993. Positive genetic correlation between female preference and preferred male ornament in sticklebacks. *Nature* 363:255–57
5. Bloem K, Bloem S, Rizzo N, Chambers D. 1993. Female medfly refractory period: effect of male reproductive status. In *Fruit Flies: Biology and Management,* ed. M Aluja, P Liedo, pp. 189–90. New York: Springer-Verlag
6. Boake CRB. 1989. Correlations between courtship success, aggressive success, and body size in a picture-winged fly, *Drosophila silvestris. Ethology* 80:318–29
7. Boake CRB. 1991. Coevolution of senders and receivers of sexual signals: genetic coupling and genetic correlations. *Trends Ecol. Evol.* 6:225–27
8. Boake CRB, ed. 1994. *Quantitative Genetic Studies of Behavioral Evolution.* Chicago: Univ. Chicago Press

9. Boake CRB, Capranica RR. 1982. Aggressive signal in the "courtship" chirp of a gregarious cricket. *Science* 21:580–82

10. Boggs CL. 1995. Male nuptial gifts: phenotypic consequences and evolutionary implications. In *Insect Reproduction*, ed. SR Leather, J Hardie, pp. 213–40. Boca Raton, FL: CRC

11. Bossert WH, Wilson EO. 1963. The analysis of olfactory communication among animals. *J. Theor. Biol.* 5:443–69

12. Bradbury JW, Andersson MB, ed. 1987. *Sexual Selection: Testing the Alternatives.* Chichester, UK: Wiley & Sons

13. Burk T. 1983. Behavioral ecology of mating in the Caribbean fruit fly, *Anastrepha suspensa* (Loew) (Diptera: Tephritidae). *Fla. Entomol.* 66:330–44

14. Burk T, Calkins CO. 1983. Medfly mating behavior and control strategies. *Fla. Entomol.* 66:3–16

15. Burk T, Webb JC. 1983. Effect of male size on calling propensity, song parameters, and mating success in Caribbean fruit flies, *Anastrepha suspensa* (Loew) (Diptera: Tephritidae). *Ann. Entomol. Soc. Am.* 76:678–82

16. Butlin RK, Ritchie MG. 1989. Genetic coupling in mate recognition systems: what is the evidence? *Biol. J. Linn. Soc.* 37:237–46

17. Cade WH. 1979. The evolution of alternative male reproductive strategies in field crickets. In *Sexual Selection and Reproductive Competition in Insects,* ed. M Blum, NA Blum, pp. 343–79. London: Academic

18. Calkins CO. 1989. Quality control. See Ref. 93a, pp. 153–65

19. Campanella PJ, Wolf LL. 1974. Temporal leks as a mating system in a temperate zone dragonfly (Odonata: Anisoptera). I. *Plathemis lydia* (Drury). *Behaviour* 51:49–87

20. Cardé RT. 1990. Principles of mating disruption. See Ref. 91a, pp. 47–71

21. Cardé RT, Minks AK. 1995. Control of moth pests by mating disruption: successes and constraints. *Annu. Rev. Entomol.* 40:559–85

22. Carpenter JE, Sparks AN, Cromroy HL. 1987. Corn earworm (Lepidoptera: Noctuidae): influence of irradiation and mating history on the mating propensity of females. *J. Econ. Entomol.* 80:1233–37

23. Charlton RE, Cardé RT. 1990. Behavioral interactions in the courtship of *Lymantria dispar* (Lepidoptera: Lymantriidae). *Ann. Entomol. Soc. Am.* 83:89–96

24. Chen PS, Stumm-Zollinger E, Aigaki T, Balmer J, Bienz M, Bohlen P. 1988. A male accessory gland peptide that regulates reproductive behavior of female *D. melanogaster.* *Cell* 54:291–98

25. Clark DL, Uetz GW. 1990. Video image recognition by the jumping spider, *Maevia inclemens* (Araneae: Salticidae). *Anim. Behav.* 40:884–90

26. Clutton-Brock TH, ed. 1988. *Reproductive Success: Studies of Individual Variation in Contrasting Breeding Systems.* Chicago: Univ. Chicago Press

27. Collins RD, Cardé RT. 1989. Selection for altered pheromone-component ratios in the pink bollworm moth, *Pectinophora gossypiella* (Lepidoptera: Gelechiidae). *J. Insect Behav.* 2:609–21

28. Collins RD, Cardé RT. 1990. Selection for increased pheromone response in the male pink bollworm, *Pectinophora gossypiella* (Lepidoptera: Gelechiidae). *Behav. Genet.* 20:325–31

29. Collins RD, Rosenblum SL, Cardé RT. 1990. Selection for increased pheromone titre in the pink bollworm moth, *Pectinophora gossypiella* (Lepidoptera: Gelechiidae). *Physiol. Entomol.* 15:141–47

30. Colwell AE, Shorey HH. 1976. Courtship stimuli affecting female receptivity in the house fly, *Musca domestica.* *Ann. Entomol. Soc. Am.* 69:80–84

31. Darwin C. 1871. *The Descent of Man and Selection in Relation to Sex.* London: Murray

32. Davies NB. 1991. Mating systems. In *Behavioural Ecology, an Evolutionary Approach,* ed. JR Krebs, NB Davies, pp. 263–94. Oxford: Blackwell

33. Dodson CH, Dressler RL, Hills HG, Adams RM, Williams NH. 1969. Biologically active compounds in orchid fragrances. *Science* 164:1243–49

34. Downes JA. 1969. The swarming and mating flight of Diptera. *Annu. Rev. Entomol.* 14:271–98

35. Dyer FC, Gould JL. 1983. Honey bee navigation. *Am. Sci.* 71:587–97

36. Eberhard WG. 1994. Evidence for widespread courtship during copulation in 131 species of insects and spiders, and implications for cryptic female choice. *Evolution* 48:711–33

37. Emlen ST, Oring LT. 1977. Ecology, sexual selection, and the evolution of mating systems. *Science* 197:215–23

38. Ewing A. 1989. *Arthropod bioacoustics: Neurobiology and Behaviour.* Ithaca, NY: Cornell Univ. Press

39. Fincke OM. 1988. Sources of variation in lifetime reproductive success in a nonterritorial damselfly (Odonata: Coenagrionidae). See Ref. 26, pp. 24–43

40. Forrest TG. 1982. Acoustic communi-

cation and baffling behaviors of crickets. *Fla. Entomol.* 65:33–44

41. Fye RL, LaBrecque GC. 1966. Sexual acceptability of laboratory strains of male house flies in competition with wild strains. *J. Econ. Entomol.* 59:538–40

42. Gage MJG. 1992. Removal of rival sperm during copulation in a beetle, *Tenebrio molitor. Anim. Behav.* 44:587–89

43. Gerhardt HC. 1991. Female mate choice in treefrogs: static and dynamic acoustic criteria. *Anim. Behav.* 42:615–35

44. Grant GG. 1981. Mating behavior of the whitemarked tussock moth and role of female scales in releasing male copulatory attempts. *Ann. Entomol. Soc. Am.* 74:100–5

45. Gregory PG, Howard DJ. 1994. A postinsemination barrier to fertilization isolates two closely related ground crickets. *Evolution* 48:705–10

46. Gromko MH, Gilbert DG, Richmond RC. 1984. Sperm transfer and use in the multiple mating system of *Drosophila.* See Ref. 103, pp. 371–426

47. Gwynne DT. 1981. Sexual difference theory: Mormon crickets show role reversal in mate choice. *Science* 213:779–80

48. Gwynne DT. 1984. Male mating effort, confidence of paternity, and insect sperm competition. See Ref. 103, pp. 117–49

49. Gwynne DT. 1985. Role-reversal in katydids: habitat influences reproductive behaviour (Orthoptera: Tettigoniidae, *Metaballus* sp.). *Behav. Ecol. Sociobiol.* 16:355–61

50. Gwynne DT, Simmons LW. 1990. Experimental reversal of courtship roles in an insect. *Nature* 346:172–74

51. Hedrick AV. 1986. Female preferences for male calling bout duration in a field cricket. *Behav. Ecol. Sociobiol.* 19:73–77

52. Hedrick AV, Dill LM. 1993. Mate choice by female crickets is influenced by predation risk. *Anim. Behav.* 46:139–96

53. Hewitt GM, Mason P, Nichols RA. 1989. Sperm precedence and homogamy across a hybrid zone in the alpine grasshopper *Podisma pedestris. Heredity* 62:343–53

54. Hibino Y, Iwahashi O. 1991. Appearance of wild females unreceptive to sterilized males on Okinawa Is. in the eradication program of the melon fly, *Dacus cucurbitae* Coquillett (Diptera: Tephritidae). *Appl. Entomol. Zool.* 26:265–70

55. Hoffmann AA. 1988. Heritable variation for territorial success in two *Drosophila melanogaster* populations. *Anim. Behav.* 36:1180–89

56. Hoikkala A, Aspi J. 1993. Criteria of female mate choice in *Drosophila littoralis, D. montana,* and *D. ezoana. Evolution* 47:768–77

57. Hoikkala A, Kaneshiro K. 1993. Change in the signal-response sequence responsible for asymmetric isolation between *Drosophila planitibia* and *Drosophila silvestris. Proc. Natl. Acad. Sci. USA* 90:5813–17

58. Hopper KR, Roush RT, Powell W. 1993. Management of genetics of biological-control introductions. *Annu. Rev. Entomol.* 38:27–51

59. Horng S-B, Plant RE. 1992. Impact of lek mating on the sterile insect technique: a modeling study. *Res. Popul. Ecol.* 34:57–76

60. Iwahashi O, Itô Y, Shiyomi M. 1983. A field evaluation of the sexual competitiveness of sterilized melon flies, *Dacus (Zeugodacus) cucurbitae. Ecol. Entomol.* 8:43–48

61. Kim KC, McPheron BA, eds. 1993. *Evolution of Insect Pests: Patterns of Variation.* New York: Wiley & Sons

62. Kirkpatrick M. 1987. Sexual selection by female choice in polygynous animals. *Annu. Rev. Ecol. Syst.* 18:43–70

63. Kirkpatrick M, Ryan MJ. 1991. The evolution of mating preferences and the paradox of the lek. *Nature* 350:33–38

64. Knipling EF. 1955. Possibilities of insect control or eradication through the use of sexually sterile males. *J. Econ. Entomol.* 48:459–62

65. Kuba H, Itô Y. 1993. Remating inhibition in the melon fly, *Bactrocera (= Dacus) cucurbitae* (Diptera: Tephritidae): copulation with spermless males inhibits female remating. *J. Ethol.* 11:23–28

66. Kuba H, Koyama J. 1982. Mating behavior of the melon fly, *Dacus cucurbitae* Coquillett (Diptera: Tephritidae): comparative studies of one wild and two laboratory strains. *Appl. Entomol. Zool.* 17:559–68

67. Kuba H, Koyama J, Prokopy RJ. 1984. Mating behavior of wild melon flies, *Dacus cucurbitae* Coquillett (Diptera: Tephritidae) in a field cage: distribution and behavior of flies. *Appl. Entomol. Zool.* 19:367–73

68. Lande R. 1981. Models of speciation by sexual selection on polygenic traits. *Proc. Natl. Acad. Sci. USA* 78:3721–25

69. Lande R. 1982. Rapid origin of sexual

isolation and character divergence in a cline. *Evolution* 36:213–23

70. Lande R, Kirkpatrick M. 1988. Ecological speciation by sexual selection. *J. Theor. Biol.* 133:85–98

71. Landolt PJ, Heath RR. 1990. Sexual role reversal in mate-finding strategies of the cabbage looper moth. *Science* 249:1026–28

72. Lanier GN. 1990. Principles of attraction-annihilation: Mass trapping and other means. See Ref. 91a, pp. 25–45

73. Lew AC, Ball HJ. 1980. Effect of copulation time on spermatozoan transfer of *Diabrotica virgifera* (Coleoptera: Chrysomelidae). *Ann. Entomol. Soc. Am.* 73:360–61

74. Lorch PD, Wilkinson GS, Riello PR. 1993. Copulation duration and sperm precendence in the stalk-eyed fly *Cyrtodiopsis whitei* (Diptera: Diopsidae). *Behav. Ecol. Sociobiol.* 32:303–11

75. Mangan RL. 1988. Pedigree and heritability influences on mate selectivity and mating aggressiveness in the screwworm, *Cochliomyia hominivorax* (Diptera: Calliphoridae). *Ann. Entomol. Soc. Am.* 81:649–56

76. Markow T. 1995. Evolutionary ecology and developmental instability. *Annu. Rev. Entomol.* 40:105–20

77. Marler P. 1977. The evolution of communication. See Ref. 95a, pp. 45–70

78. McVey ME. 1988. The opportunity for sexual selection in a territorial dragonfly, *Erythemis simplicicollis*. See Ref. 26, pp. 44–58

79. Moore AJ. 1989. The behavioral ecology of *Libellula luctuosa* (Burmeister) (Odonata: Libellulidae). III. Male density, OSR, and male and female mating behavior. *Ethology* 80:120–36

80. Moore AJ, Breed MD. 1986. Mate assessment in a cockroach, *Nauphoeta cinerea. Anim. Behav.* 34:1160–65

81. Murvosh CM, Fye RL, Labrecque GC. 1964. Studies on the mating behavior of the house fly, *Musca domestica* L. *Ohio J. Sci.* 64:264–71

82. Newport MEA, Gromko MH. 1984. The effect of experimental design on female receptivity to remating and its impact on reproductive success in *Drosophila melanogaster. Evolution* 38:1261–72

83. Nishida R, Tan KH, Serit M, Lajis NH, Sukari AM, et al. 1988. Accumulation of phenylpropanoids in the rectal glands of males of the oriental fruit fly, *Dacus dorsalis. Experientia* 44:534–36

84. Otte D. 1977. Communication in Orthoptera. See Ref. 95a, pp. 334–61

85. Parkman JP, Frank JH. 1993. Use of a sound trap to inoculate *Steinernema scapterisci* (Rhabditida: Steinernematidae) into pest mole cricket populations (Orthoptera: Gryllotalpidae). *Fla. Entomol.* 76:75–81

86. Pliske TE, Eisner T. 1969. Sex pheromone of the queen butterfly: biology. *Science* 164:1170–72

87. Plomin R, DeFries JC, McClearn GE. 1980. *Behavioral Genetics: A Primer.* San Francisco: Freeman

88. Poramarcom R. 1988. *Sexual communication in the Oriental fruit fly, Dacus dorsalis Hendel (Diptera: Tephritidae).* PhD thesis. Univ. Hawaii, Honolulu

89. Poramarcom R, Boake CRB. 1991. Behavioural influences on male mating success in the Oriental fruit fly, *Dacus dorsalis* Hendel. *Anim. Behav.* 42:453–60

90. Prokopy RJ, Hendrichs J. 1979. Mating behavior of *Ceratitis capitata* on a field-caged tree. *Ann. Entomol. Soc. Am.* 72:642–48

91. Raina AK, Stadelbacher EA. 1990. Pheromone titer and calling in *Heliothis virescens* (Lepidoptera: Noctuidae): effect of mating with normal and sterile backcross males. *Ann. Entomol. Soc. Am.* 83:987–90

91a. Ridgway RL, Silverstein RM, Inscoe MN, eds. 1990. *Behavior-Modifying Chemicals for Insect Management: Applications of Pheromones and Other Attractants.* New York: Marcel Dekker

92. Ringo JM. 1996. Sexual receptivity in female insects. *Annu. Rev. Entomol.* 41:473–94

93. Robacker DC, Mangan RL, Moreno DS, Tarshis Moreno AM. 1991. Mating behavior and male mating success in wild *Anastrepha ludens* (Diptera: Tephritidae) on a field-caged host tree. *J. Insect Behav.* 4:471–87

93a. Robinson AS, Hooper G, eds. 1989. *Fruit Flies: Their Biology, Natural Enemies and Control.* New York: Elsevier

94. Robinson T, Johnson NA, Wade MJ. 1994. Postcopulatory, prezygotic isolation: intraspecific and interspecific sperm precedence in *Tribolium* spp., flour beetles. *Heredity* 73:155–59

94a. Roelofs W, Du J-W, Linn C, Glover TJ, Bjostad LB. 1986. The potential for genetic manipulation of the redbanded leafroller moth sex pheromone blend. In *Evolutionary Genetics of Invertebrate Behavior*, ed. MD Huettel, pp. 263–72. New York: Pleum

95. Roush RT, McKenzie JA. 1987. Ecological genetics of insecticide and acaricide resistance. *Annu. Rev. Entomol.* 32:361–80

95a. Sebeok TA, ed. 1977. *How Animals*

Communicate. Bloomington, IN: Indiana Univ. Press

96. Shelly TE, Dewire A-LM. 1994. Chemically mediated mating success in male Oriental fruit flies (Diptera: Tephritidae). *Ann. Entomol. Soc. Am.* 87:375–82

97. Shelly TE, Kaneshiro KY. 1991. Lek behavior of the Oriental fruit fly, *Dacus dorsalis,* in Hawaii (Diptera: Tephritidae). *J. Insect Behav.* 4:235–41

98. Shelly TE, Whittier TS. 1995. Lek behavior of insects. In *Social Competition and Cooperation in Insects and Arachnids,* Vol. 1, *Evolution of Mating Systems,* ed. JC Choe, BJ Crespi, Princeton, NJ: Princeton Univ. Press. In press

99. Shelly TE, Whittier TS, Kaneshiro KY. 1994. Sterile insect release and the natural mating system of the Mediterranean fruit fly, *Ceratitis capitata* (Diptera: Tephritidae). *Ann. Entomol. Soc. Am.* 87:470–81

100. Silberglied RE. 1977. Communication in the Lepidoptera. See Ref. 95a, pp. 362–402

101. Sivinski J. 1989. Lekking and the small-scale distribution of the sexes in the Caribbean fruit fly, *Anastrepha suspensa* (Loew). *J. Insect Behav.* 2:3–13

102. Sivinski J, Burk T. 1989. Reproductive and mating behaviour. See Ref. 93a, pp. 343–51

103. Smith RL, ed. 1984. *Sperm Competition and the Evolution of Animal Mating Systems.* New York: Academic

104. Soemori H, Tsukagushi S, Nakamori H. 1980. Comparison of mating ability and mating competitiveness between mass-reared and wild strains of the melon fly, *Dacus cucurbitae* coquillett (Diptera: Tephritidae). *Appl. Entomol. Zool.* 24: 246–50

105. Tabashnik BE. 1994. Evolution of resistance to *Bacillus thuringiensis. Annu. Rev. Entomol.* 39:47–79

106. Teale SA, Hager BJ, Webster FX. 1994. Pheromone-based assortative mating in a bark beetle. *Anim. Behav.* 48:569–78

107. Thornhill R, Alcock J. 1983. *The Evolution of Insect Mating Systems.* Cambridge, MA: Harvard Univ. Press

108. Trivers RL. 1972. Parental investment and sexual selection. In *Sexual Selection and the Descent of Man, 1871–1971,* ed. B Campbell, pp. 136–79. Chicago: Aldine

109. Waage JK. 1984. Sperm competition and the evolution of odonate mating systems. See Ref. 103, pp. 251–90

110. Wade MJ, Patterson H, Chang NW, Johnson NA. 1994. Postcopulatory, prezygotic isolation in flour beetles. *Heredity* 72:163–67

111. Walker TJ. 1993. Phonotaxis in female *Ormia ochracea* (Diptera: Tephritidae), a parasitoid of field crickets. *J. Insect Behav.* 6:389–410

112. Wall C. 1990. Principles of monitoring. See Ref. 91a, pp. 9–23

113. Watson PJ, Thornhill R. 1994. Fluctuating asymmetry and sexual selection. *Trends Ecol. Evol.* 9:21–25

114. Webb JC, Sivinski J, Smittle B. 1985. Acoustical courtship signals and sexual success in irradiated caribfly *Anastrepha suspensa* (Loew) (Diptera: Tephritidae). *Fla. Entomol.* 70:103–9

115. Webb KL, Roff DA. 1992. The quantitative genetics of sound production in *Gryllus firmus. Anim. Behav.* 44:823–32

116. Whittier TS, Kaneshiro KY, Prescott LD. 1992. Mating behavior of Mediterranean fruit flies (Diptera: Tephritidae) in a natural environment. *Ann. Entomol. Soc. Am.* 85:214–18

117. Wilkinson GS, Reillo PR. 1994. Female choice response to artificial selection on an exaggerated male trait in a stalk-eyed fly. *Proc. R. Soc. London Ser. B* 255:1–6

118. Wineriter SA, Walker TJ. 1990. Rearing phonotactic parasitoid flies (Diptera: Tachinidae: *Ormia* spp.). *Entomophaga* 35:621–32

119. Winston ML. 1987. *The Biology of the Honey Bee.* Cambridge, MA: Harvard Univ. Press

120. Yamagishi M, Tsubaki Y. 1990. Copulation duration and sperm transfer in the melon fly, *Dacus cucurbitae* Coquillett (Diptera: Tephritidae). *Appl. Entomol. Zool.* 25:517–19

121. Yuval B, Wekesa JW, Washino RK. 1993. Effect of body size on swarming behavior and mating success of male *Anopheles freeborni* (Diptera: Culicidae). *J. Insect Behav.* 6:333–42

Annu. Rev. Entomol. 1996. 41:231–56

ECOLOGY AND BEHAVIOR OF GROUND BEETLES (COLEOPTERA: CARABIDAE)

Gábor L. Lövei

Horticulture and Food Research Institute of New Zealand, Batchelar Science Centre, Private Bag 11030, Palmerston North 5301, New Zealand

Keith D. Sunderland

Horticulture Research International, Wellesbourne, Warwick CV35 9EF, United Kingdom

KEY WORDS: Coleoptera, Carabidae, bionomics, populations, assemblages

ABSTRACT

The ground beetles form the speciose beetle family Carabidae and, since their emergence in the Tertiary, have populated all habitats except deserts. Our knowledge about carabids is biased toward species living in north-temperate regions. Most carabids are predatory, consume a wide range of food types, and experience food shortages in the field. Feeding on both plant and animal material and scavenging are probably more significant than currently acknowledged. The most important mortality sources are abiotic factors and predators; pathogens and parasites can be important for some developmental stages. Although competition among larvae and adults does occur, the importance of competition as a community organization is not proven. Carabids are abundant in agricultural fields all over the world and may be important natural enemies of agricultural pests.

INTRODUCTION

The family Carabidae, the ground beetles, contains more than 40,000 described species classified into some 86 tribes (66). It is the largest adephagan family and one of the most speciose of beetle families. The suborder Adephaga is a relatively large group of specialized beetles that is morphologically defined by the presence of six abdominal ventrites, pygidial defense glands in the adult, and liquid-feeding mouthparts in the larvae (112). They are well-proportioned cursorial beetles with prominent mandibles and palps, long slender legs, striate

231

0066-4170/96/0101-0231$08.00

elytra, and sets of punctures with tactile setae. Most have an antenna-cleaning organ and largely pubescent antennae. The adults are dark colored, shiny or matte. Some have bright or metallic colors, and some are pubescent. The larvae are campodeiform, have well-developed legs, antennae, and mandibles, and bear fixed urogomphi (34, 112). Major taxonomic problems remain to be solved (8), despite cladograms with new phylogenetic hypotheses that have emerged from significant comparative anatomical studies on the adult feeding apparatus (71), antenna cleaner (96), thorax and locomotory adaptations (69, 70), hind-wing structure (89, 197), pygidial glands (77), ovipositor (15, 25), and the chemistry of defensive secretions (134). Different authors divide the family into different subfamilies; except for the tiger beetles (see 149), our ecological knowledge is scant concerning subfamilies outside the Carabinae [sensu Lawrence & Britton (112)].

The abundance, species richness, and attractive coloration of many species have made carabids popular objects of study for both professional and amateur entomologists. The last attempted synthesis on carabid biology was the descriptive monograph by Thiele (182); only certain aspects of the field have been reviewed since (2, 111, 126, 142). Proceedings of triennial meetings, started in 1969 by Dutch and German carabidologists, provide a series of useful snapshots of the state of the art in this field (19, 48, 50–53, 57, 68, 173). A quarterly journal, *Carabologia,* serves as a forum for both amateur and professional carabidologists.

In this review, we intend to summarize some of the significant achievements in carabid ecology and behavior since Thiele's (182) book was published. Carabids are present worldwide, with species richness highest in the tropical regions (66). However, our knowledge mainly stems from research done in the temperate regions of the Northern Hemisphere. The resulting bias in this review is inevitable. Our examples are illustrative, not exhaustive, and are intended to support generalizations that can serve as guidelines or hypotheses for the study of carabids in other regions.

CARABID EVOLUTION AND ADAPTATIONS

Carabids emerged in the early Tertiary as wet-biotope generalists in tropical habitats, where they remain the dominant predatory invertebrate group (67). Through a series of taxon pulses, they have radiated to drier environments as well as higher latitudes and altitudes (64). By the late Permian/early Triassic, several lineages developed a cosmopolitan distribution pattern, as demonstrated by the fossil record (155). Although this group has retained an easy-to-recognize generalist body plan, their body shape and leg morphology are characteristically modified for running, digging, burrowing, climbing, and swimming (69, 70). Different parts of the morphological apparatus and physiological mechanisms can evolve at different rates. Thus, a species can remain a generalist structurally and still become a specialist physiologically in order to, for example, live at glacier edges (*Nebria* spp.) (66). Several other structural, physiological, and behavioral adaptations enabled carabids to invade all major habitats, where at least some lineages have attained dominance; the only exception is deserts, where carabids are limited to streams and oases (66). This distribution pattern suggests that humidity is a general limiting factor. The main structural patterns in carabid evolution are flightlessness and arboreal, fossorial, and troglobitic adaptations (66). Flightlessness has repeatedly evolved in many groups (35). In the tropics, >30% of species are arboreal, exhibiting special morphological and behavioral adaptations (172). A few groups are adapted to life in self-made tunnels (mainly in sand or finely textured soil in the tropics), and even fewer groups, of cosmopolitan distribution, reside in caves (66).

CARABID STUDY METHODS

The combination of cryptic lifestyles and polyphagous feeding habits means that many aspects of carabid natural history and ecology are not easy to study. Techniques used include different trapping and marking methods for collecting beetles and estimating density (176); labor-intensive dissections or sophisticated immunological methods to study feeding (180); the use of video equipment to record walking (88), searching, and feeding (30) behavior; and the use of harmonic radar to study within-habitat movements (130). The most popular method is pitfall trapping.

A pitfall trap [or Barber-trap (9)] is a container—any one of many different designs—sunk into the ground so that its opening is at surface level. Many surface-dwelling arthropods fall in and cannot escape. The trap is a passive catching device; capture results from the activity of the target organism. The quantity and composition of the catch will vary depending on the size, shape, construction material, and distribution in space and time of the trap, as well as the preservative used and all the factors governing activity and behavior.

Pitfall trapping is the most frequently used field method for studying carabids. Although this method is surrounded by controversy and several critical papers (176 and references therein) have been published, general practice has changed little because no similarly convenient method has been recommended. Pitfall trapping remains suitable for studying several population parameters and certain community measurements such as species presence. Pitfall traps should probably not be used to study community patterns such as relative species composition or diversity. After detailed methodological and behavioral studies have been completed and validation techniques developed, pitfall trapping might be reinstated as an efficient method of studying carabid adults. However, this method cannot be expected to fill the profound gap in our knowledge of larval ecology.

ONTOGENY AND LONGEVITY

Carabids are holometabolous insects that usually lay their eggs singly. Some species lay eggs in small or larger batches in crevices or in the soil after a varying degree of preparatory work by the female (126, 182). The female carefully chooses the ovipositing site, sometimes excavating a chamber for the eggs. Some Pterostichini prepare a cocoon for a batch of eggs (20). Parental care, at its most developed, consists of no more than egg guarding or caching seeds in the egg chamber for the emerging larvae (20, 97).

The typical carabid larva is free moving and campodeiform (34) and usually undergoes three stages before pupating in a specially constructed pupal chamber in the soil. Some species (for example, *Harpalus* and *Amara* spp.) have only two larval stages. Seven tribes, plus a hypothesized ten more, have specialized larvae with more larval stages that, in at least the later stages, exhibit reduced mobility. These species, which are ant or termite symbionts or specialized ectoparasites or predators (65), total 24% of all carabid tribes [in Erwin's classification (64)]. However, as not all members of these tribes exhibit these traits, these specialized larval bionomics characterize only a small minority of all species.

The larvae (second or third stage) of many species undergo diapause, either hibernation or aestivation. The weakly sclerotized and whitish pupa lays on its back, supported by dorsal setae. Sclerotization and coloration of the adult takes place after eclosion; teneral beetles can be recognized for various lengths of time, usually weeks.

In general, ground beetles develop from egg to adult in less than one year, reproduce once, and then perish. However, individual development can last up to four years under harsh climates or adverse food conditions. *Carabus glabratus*, a species with larval hibernation and autumn reproduction in central and western Europe, has a biennial life cycle with spring breeding in upland areas of northern England (99) and in Norway (158). In northern England, *Carabus problematicus* has an annual life cycle at altitudes below 800 m and a biennial one above that (26). The European *Carabus auronitens* has a flexible life-history strategy (opportunistic oviposition, asynchronous development, partial survival of the old generation, fat body reserves, and long-term dormancy), which reduces the risk of the whole population being affected by bad weather during the postecdysial ripening (200).

Adult longevity can also exceed one season. Individuals from several species have been kept in the laboratory for up to four years. Many species (from the tribes Agonini, Harpalini, Pterostichini, Carabini) have life spans over one year

(182). Individuals from field populations of several species from different parts of the world, for example, Europe (83, 99, 125, 186, 194), Japan (166), and the sub-Antarctic (36), can live up to four years and reproduce more than once. Cave-inhabiting species often live long lives; *Laemostenus schreibersi* can live for up to 6.5 years (161). Generally, long adult life span is more common in large species and species with winter larvae [also called autumn breeders (see 49)] than in ones with summer larvae (spring breeders).

Several species show plasticity of individual development, whereas others seem to have a stable life cycle. The originally botanical term *polyvariance* was suggested to describe the former (129). Obligatory univoltism is apparently rare and occurs mainly in species of short longevity. Bi- and multiannual cycles are usually found in species living in harsh environments (sub-Arctic, highland, or xeroterm habitats), and dynamic polyvariance is common.

HABITATS, HABITAT FINDING, AND MICROHABITATS

Persistence in a habitat should depend mostly on the life stage that is most vulnerable, as determined by the longest duration, narrowest tolerance limits, and most limited escape repertoire. All these factors point to the larval stage as the key to understanding occupation of a habitat by a given carabid species. The egg is superficially the most vulnerable of the life stages, but ovipositing females can deliver eggs into microhabitats where their survival can be maximized. Moreover, the egg stage is usually short, and the egg sacs contain the resources necessary for the completion of this life stage. The pupal stage is similarly sensitive. It lacks mobility and often lasts for long periods, but it is often better defended than the egg or larva. The larva has limited mobility, weak chitinization, and therefore feeble tolerance of extremes, and it must also find sufficient food to develop. Larval feeding conditions often determine adult fertility as well (139). For reasons mentioned above, larvae are notoriously difficult to study. However, because larvae usually cannot migrate long distances, they have to survive in the environment where the egg-laying female left them. Therefore, the following discussion on adult habitat choice is justified. Habitat choice is so specific that carabids are often used to characterize habitats (see below).

The directed random walk, followed by a frequently turning walk in the presence of favorable conditions, would eventually lead carabids to their preferred habitats, but several different mechanisms help beetles find or remain in suitable habitats. These mechanisms include internal clocks, sun-compass orientation (33, 182), and orientation either toward or away from silhouettes (33, 159, 182). Some riparian ground beetles find their habitat by sensing volatile chemicals emitted by blue algae living in the same habitat (72). *Agonum quadripunctatum,* a forest species in Europe and North America associated with burnt areas, is a good flyer and is probably attracted to the smell of smoke (23). Carabids continuously sample their surroundings. For example, *Carabus nemoralis* walked around in different habitats before settling in seminatural habitats in preference to set-aside to arable areas (107).

Habitat and microhabitat distribution can be influenced by several factors:

1. Temperature or humidity extremes (several examples in 182). Favorite wintering sites are well aerated, and winter minimum temperatures are relatively high (58, 183).
2. Food conditions. For example, exclusively spermophagous *Ophonus* spp. are present in open habitats where seeds of Umbelliferae are available, whereas polyphagous *Harpalus* spp. aggregate in crops (209). Marked *Poecilus cupreus* and *Pterostichus melanarius* moved from winter wheat to a weed strip within the wheat field (where feeding conditions were better) much more frequently than they moved the reverse direction (128).
3. Presence and distribution of competitors. For example, forest carabids in Finland were influenced by the distribution of *Formica* ant species (141).
4. Life history and season. *Amara plebeja,* for instance, has different hibernation (woodland) and reproduction (grassland) habitats. The beetles fly between habitats in spring and autumn. Flight muscles are temporarily autolysed between flights, then completely reconstructed for the return flight. In the autumn, they fly toward woodland silhouette shapes (190).

DENSITY AND DISPERSAL

Carabids are often numerically dominant in collections of soil-active arthropods. However, for reasons mentioned above, this result cannot equate with high density. Data, especially in the older literature (including 182), are confusing because of the frequent acceptance of pitfall trap catches as density data. Data obtained by true density measurement methods indicate that densities fluctuate in space and time from <1 (in many habitats) to >1000 individuals per square meter (at suitable overwintering sites, see Table 1).

As a group, carabids originally used fully functional wings as the primary dispersal mode. However, flight is very costly and is subject to intense selection (160). Once the benefits of flight do not match its costs, as on, for example, islands and mountain tops, it is quickly lost (35, but see 145). Flightlessness and flight dimorphism (some individuals in a given species possess wings, others do not) has repeatedly evolved in carabids. For example, of the carabid fauna of Newfoundland (157 species), 12.7% are dimorphic and 21.0% flightless, a condition reached through nine or more independent evolutionary transitions (160). Wing dimorphism seems to be inherited in a simple Mendelian fashion through a dominant gene for short-wingedness (5, 117); in some

Table 1 Maximal densities (individuals per square meter) of ground beetle adults and larvae in different habitats[a]

Species category	Arable fields		Field boundary	Forest, heath
	Annual crops	Biennial/ perennial crops		
Species size <5 mm	5.96 (0.2–77; 72)	3.61 (0.3–2.4; 7)	66.62 (0.6–923; 23)	—
Species size >5 mm	1.83 (0.02–33; 47)	4.82 (0.7–22; 7)	14.32 (0.03–87; 18)	2.54 (0.04–22.5; 19)
Adult total	31.73 (1.2–96.1; 12)	—	233.27 (14.5–1113; 9)	2[b]
Larvae, individual species	5.46 (0.07–33; 10)	6[b]	14.5 (4–42; 6)	7.8[b]
Larvae, total	29.4, 49, 77[b]	—	49, 87[b]	—

[a] Data are given as mean (minimum–maximum; N). Only data giving true density values (obtained by soil samples, soil flooding, mark-recapture, fenced pitfalls, quadrat sampling, and vaccuum sampling) were considered and include data on 71 adult plus 13 larval taxa, obtained between 1970–1994 in 14 countries in Europe and North America. From KD Sunderland & GL Lövei, unpublished manuscript. A full list of references obtainable by request.
[b] Fewer than five observations; individual values given.

species, the trait is polymorphic (56). Environmental conditions may influence expression of the dimorphism (6). Flight ability varies little between the sexes (160).

The proportion of flightless individuals in dimorphic species increases with increasing habitat persistency and time since colonization (54). The proportion of macropterous *P. melanarius* can be as low as 2% in stable habitats (e.g. old forest patches) (42) or as high as 24–45% in less stable ones (e.g. newly reclaimed polders of The Netherlands) (87). In Edmonton, Canada, this species was first reported in 1959. Currently, 60–70% of the frontier population, ~70 km from the city, has wings; in the source population in the city, only 20% of the beetles are now macropterous (143).

Flight is greatly influenced by temperature, rain, and wind (191). In some species (such as the Palearctic *A. plebeja*), the flight muscles are broken down during egg production and then resynthesized; in others, flight capability during reproduction is not impaired, and up to 80% of dispersing females carry fertilized eggs (192). This percentage characterized species from both persistent and ephemeral habitats. However, females of more species from ephemeral habitats than from persistent ones carried ripe eggs (192), which increased the probability of (re)colonizing empty habitat patches.

Many carabids have been transported intercontinentally, e.g. from Europe to North America (115, 171). Studies in Canada of the effects of invasion by *P. melanarius* showed them to be negligible (143).

ACTIVITY: DIEL AND SEASONAL

Diel Activity Cycles

More carabids are nocturnal than diurnal. For example, in the United Kingdom, 60% of species are nocturnal and 20% diurnal (124). The diurnal activity dendrogram for carabids in UK woodlands revealed groupings for diurnal, nocturnal, and crepuscular species, plus species that overlapped some of these categories (55). Overall, nocturnal species are larger than diurnal ones. Also, their coloration often differs: Night-active species are dark and dull, and diurnal species display iridescent colors. Diel periodicity can vary with habitat (forest species tend to be nocturnal whereas grassland species are diurnal) (84) and time of year (*P. melanarius* is nocturnal until August and is mainly diurnal later) (59). Changes in temperature (102), light intensity, and humidity (182) also influence activity. In hot countries, nocturnalism becomes more common; conversely, species that are nocturnal in central Europe become diurnal in the arctic (182). Specialist feeders may synchronize their activity with that of their prey (1). Desert carabids exhibit peak activity at temperature minima (62). Cave-dwelling species often are active in short bursts between periods of inactivity (182); they sometimes exhibit diel activity cycles despite constant, very low illumination and air humidity. Such circadian clocks may serve to synchronize the activity of males and females (199). Individuals within a population can undergo different activity cycles; for example, some individuals of *Carabus auratus* are diurnal, some nocturnal, and others indifferent to diel periods (182). In some species, larvae and adults undergo different cycles (106).

Sublethal dosages of insecticides can cause marked increases in carabid activity (135) and may also indirectly cause activity increases by reducing food supplies (29).

Seasonal Rhythms

Seasonal rhythms involving dormant periods during winter and/or summer (aestivation) are an integral part of the life history of temperate-region ground beetles. The activity of the two most typical groups peaks in either spring or autumn. This peak usually coincides with the reproductive period, although the connection between activity and reproductive rhythms is flexible in many species (129). Such rhythms are inseparable from individual, especially larval, development.

Facultative diapause of summer larvae can synchronize the life cycle (126). Because of the variability in activity and reproductive seasons and the growing body of evidence on adult longevity, some authors have suggested rejecting the traditional concepts of spring-reproducing vs autumn-reproducing species

and adult overwinterers vs larval overwinterers in favor of categories containing species with summer larvae vs winter larvae (49) or species with vs without
diapausing larvae (100). In extratropical regions, the cues regulating these
cycles involve temperature and photoperiod (182). Seasonal activity and reproductive rhythms in tropical species are regulated by seasonal changes in
soil moisture and flooding (146).

FEEDING

Searching Behavior

Whereas many carabids presumably find their food via random search, several
diurnal species hunt by sight (80). Other species use chemical cues from aphids
(30), springtails (55a), or snails (202) to find prey. The use of chemical
information is probably more common than the few reported cases would
suggest.

Carabids exhibit the search pattern common to invertebrate predators (140).
After the beetle encounters a prey item in a patch, its search behavior characteristically intensifies for a specified "give-up" time period. In some species
that search two-dimensionally, finding a prey triggers a three-dimensional
search behavior [e.g. *Pterostichus cupreus* climbs the plant when it senses an
aphid at the base (30)]. The general walking pattern often alternates between
frequently turning and rarely turning walking phases (132, 195), but this pattern
is not necessarily nor always connected to feeding behavior.

Once prey is located, species typically switch to a well-defined prey-catching behavior. Many morphological and behavioral adaptations are at work in
this stage of feeding, mostly in specialized species. Prey catching, studied in
fine detail for several European species that hunt springtails, has revealed a
fascinating array of adaptations involving sight, behavior, and morphology in
both adults and larvae (11–13).

Most carabid adults use their well-developed mandibles to kill and fragment
prey into pieces. Specialist species attacking snails seem to paralyze their prey
by biting (147), thus preventing the mucus production that is the slugs' defense
reaction. Many large species eject a fluid rich in digestive enzymes; subsequently, they consume the liquid portion of their partially digested prey,
sometimes with undigested prey fragments. Larvae only consume extraorally
digested food (for a more detailed discussion of preoral digestion, see 32). The
alimentary canal is tripartite. The foregut, including the crop, is the main site
of digestion (80); enzymes synthesized in the midgut are passed forward to
the foregut. The enzyme set contains proteases, carboxylases, amylases (131),
and oligo- and polysaccharidases; this composition is thought to be a primitive
character (101). Absorption takes place in the hindgut. The speed of digestion

depends on temperature and the size of a food item (164) as well as on subsequent feeding (122). Traces of a meal could be detected for up to 14 days (122, 164).

Food Choice

Early data on several species indicated varying extents of polyphagy (37, 76, 163). The accumulated results have been extensively reviewed by Thiele (182) and, for agricultural species, by Allen (2) and Luff (126).

Carabids are mostly polyphagous feeders that consume animal (live prey and carrion) and plant material; several species are phytophagous (126, 182). A worldwide survey of the literature (111) reporting on 1054 species of carabids and cicindelids showed that 775 species (73.5%) were exclusively carnivorous, 85 species (8.1%) phytophagous, and 206 species (19.5%) omnivorous. These data, although they may indicate the general feeding habit of the family, are often based on laboratory data and are heavily biased toward northern hemisphere species. On a smaller scale, another survey showed that 27% of the 362 species in Fennoscandia were predators, 13% omnivores, and 24% herbivores; at the time of study, the food of 36% of the species was not known (114). More detailed analysis of the restricted range of species (see below) also indicates that the degree of predatory habit in the family has generally been overestimated, especially as the degree of plant and carrion feeding is not well known. In general, larvae are more carnivorous and restricted in food range while adults exhibit very catholic feeding habits, with some groups (Cychrini, Notiophilini, Loricerini, Nebriini) demonstrating varying degrees of specialization. The following paragraphs summarize the feeding of adult beetles.

Catholic feeding habits, frequent nocturnal activity, and extraintestinal digestion, among other factors, present problems for the study of feeding (140, 175). Methods applied to investigate feeding in carabids include casual or regular direct observation, exclusion techniques, forced feeding in the laboratory, density manipulation of prey and predator, the use of radioactive tracers, isotope-labeled prey techniques, gut dissection, various serological techniques, and electrophoresis. These methods and their limitations have been repeatedly reviewed (e.g. 175, 180).

Dissection of several thousand individuals of 24 European species (95) revealed the remains of aphids, spiders, lepidopteran larvae and adults, fly larvae, mites, heteropterans, opilionids, beetles, and springtails. Similar studies, conducted in Belgium (154) and New Zealand (177), also found enchytraeid worms, lumbricid worms, nematodes, hymenopterans, beetle larvae, eggs, centipedes, millipedes, mollusks, spores, fungal hyphae, seeds, and pollen. Cannibalism has also been reported.

All species in Hengeveld's study (95) were polyphagous and consumed plant material in addition to the other food items. A multivariate analysis (94) identified one group with a diet containing a high proportion of springtails and a restricted variety of other arthropods (some *Notiophilus, Leistus,* and *Agonum* species). Members of another group in the study, which eat what they can swallow, were species of *Amara, Harpalus,* and *Pterostichus* (94).

While the results mentioned above show the wide range of prey taken by ground beetles, most of these studies did not consider prey availability. Where it has been considered, opportunistic feeding habits are found. For example, ten abundant grassland species in Belgium fed mainly on springtails, the most abundant prey group (154).

Food Consumption

Carabids are voracious feeders, consuming close to their own body mass of food daily (182). Food is used to build fat reserves, especially before reproduction and hibernation (182). Feeding conditions during larval development determine adult size, which is a major determinant of potential fecundity (138). Realized fecundity depends on adult feeding conditions (see below).

Although potential food consumption can be assessed straightforwardly in the laboratory, quantification of feeding rates in the field is difficult for reasons mentioned earlier. One possible solution is to monitor egg production and/or body-mass changes by regularly sampling field populations and compare these data to calibration measurements taken on beetles kept in the laboratory under known conditions. Such measurements, performed on *Carabus yaconinus* in Japan (167), indicated that field prey consumption by females allowed them to realize 59% of their possible maximum egg production in May and 45% in June. Field consumption was similarly below the potential maximum in other species in The Netherlands (139, 189) and North America (201).

Carabids, like other animals, forage for nutrients and energy, which are packaged in food items. Feeding in the context of optimality of food composition has been little studied in carabids. Nutritional requirements for carabids have not been specifically identified nor has the observation that certain species are more specific than others been addressed from a nutritional point of view. The dietary advantages of mixed food over a single food type are well known for polyphagous invertebrate herbivores (14, 162). In carabids, females often have more prey types than males (154). Moreover, Wallin et al (196) found that egg number and size were influenced by food composition. Signs of optimal digestion were found in two carabid species (122). These data suggest that food composition is not irrelevant for foraging ground beetles, and beetles may have the ability to select a diet that matches their particular needs.

The feeding studies to date have left us with some notable gaps: (*a*) Although the range of methods applied is very wide, the degree of distortion obtained is not possible to assess. (b) Adult feeding is generally overemphasized, and detailed information on larval feeding is lacking. (c) Most studies have a narrow focus; they were done in agricultural fields and/or considered a single prey group (aphids, slugs, etc). (d) The degree of true carnivory vs carrion feeding is not adequately determined. (e) The degree of mixed feeding (plant and animal material) is probably underestimated. (f) The literature has a heavy geographical bias toward the Northern Hemisphere. (g) Physiological studies are scarce, and consequently, food-choice criteria are poorly understood in terms of diet composition.

REPRODUCTION

Fecundity can range from five to ten eggs per female in species with egg-guarding behavior to several hundred per female in species that do not guard eggs (208). Eggs can be laid in one batch, several batches in one season, or over several seasons. As many as 30–60% of individuals in a population can reproduce in more than one year (168, 186, 193). The dependence of fecundity on age is not well understood. For several species, young females have a higher reproductive output than old ones (e.g. 186), whereas the reverse is true in other species (24, 36, 83, 166).

Many carabids are apparently iteroparous rather than semelparous reproducers; such behavior results in less fluctuation in numbers over the years (125). Murdoch (136) hypothesized that the stabilizing mechanism worked so that female survival was inversely related to the amount of reproduction during the previous season, reporting that observations of *Agonum* species in England confirmed this assertion. His suggestion generally did not find support (7, 113, 166, 189, 198, but see 63). Increased mortality during reproduction may result from ecological rather than physiological factors (27), such as exposure of reproducing individuals to higher levels of external hazards such as predators or disease.

In all carabid species examined, as well as in several other predators, the variable egg production is related to the amount of food. The first priority of the adults is to meet energy demands for survival and use the surplus for reproduction. This makes sense in that under conditions of limited food supply, the survive-but-not-reproduce option enables predators to survive until better food conditions allow reproduction (133, 203). Data from Europe (187, 189), Japan (166), and North America (113) indicate that carabids in the field regularly experience food shortage and rarely realize their full reproductive potential.

In searching for an explanation of carabid fecundity, Grüm (86) found that egg numbers tended to decrease as body mass increased. Autumn breeders had higher egg numbers than spring breeders, and egg-laying rates were inversely correlated with female mobility (86). These results, along with observations of low egg numbers in cave-inhabiting species (41) and of species demonstrating parental care in Europe and New Zealand (20; GL Lövei, unpublished data), conform to some predictions of the r- and K-strategies theory. Also, ground beetle species living in unstable habitats have higher egg numbers than relatives living under less variable conditions. Similar differences are observed in adult life spans and egg numbers among the Polish and Dutch populations of several species (86). However, the r-K theory is only one of the hypotheses suggested to explain life-history features. The application of alternative theories such as Grime's C-S-R model (85) is promising (73).

MORTALITY AND POPULATION DYNAMICS

Although abiotic influences on survivorship are inevitable, constituting the principal mortality factors for all life-cycle stages of ground beetles (45), other factors play an important role in carabid population dynamics.

Mortality of the Different Stages

EGG MORTALITY   The traditional assumption that egg mortality is not significant (182) is probably not correct. Eggs of *Pterostichus oblongopunctatus* suffered 83% mortality in fresh litter but only 7% in sterilized soil (91). One potential advantage of brood watching could be protection from pathogens, although females have not been observed cleaning, surface sterilizing, or even doing anything with their eggs in the egg chamber. However, when abandoned by females, eggs quickly become moldy (20).

LARVAL MORTALITY   Larval mortality is probably a key factor in overall mortality of ground beetles, but because of the lack of appropriate methodology to study larvae, evidence for the importance of larval mortality is scant. Because larvae have weak chitinization and limited mobility, they are sensitive to desiccation, starvation, parasites, and diseases. Larvae are also cannibalistic. In laboratory and field experiments with surface-active larvae of *Nebria brevicollis,* mortality varied between 25 and 97%, depending on food conditions; parasitism caused up to 25% mortality (138, 139). The results of similar experiments with larvae of *P. oblongopunctatus,* combined with computer simulations, indicated a cumulative mortality rate for larvae and pupae of 96% (22, 92). These authors concluded that events during larval life are the most important for population regulation.

Parasitism is recognized as a very important factor in host population biology, both on ecological and evolutionary time scales (81, 157). Although predators, parasites, and pathogens affect all ground beetle developmental stages (126, 182), quantitative data remain scarce.

ADULT MORTALITY   Up to 41% parasitism by nematodes and ectoparasitic fungi was found on 14 species of *Bembidion* in Norway (3). Nematode infection in insects may cause sterility (153), resulting in obvious fitness effects. The benefit of living in exposed habitats could be freedom from parasites; the cost would be higher risks of predation and/or more frequent catastrophic events, such as flooding (3).

Most observational evidence indicates that predation is an important mortality factor for adults. Hundreds of vertebrate species prey on carabids (108–110). The ecological significance of predation pressure by small mammals was demonstrated in exclosure experiments in North America (148) and England (31), where excluding small mammals resulted in an increase in both species richness and density of carabids.

Antipredator Defenses

The evolution of terrestrial faunal groups that prey on carabids, such as amphibians, reptiles, birds, and mammals, has probably constituted a major driving force of carabid evolution (66). The large suite of antipredator defenses includes morphological, biochemical, and behavioral components. For example, morphological traits in arboreal carabids include cryptic or warning coloration, mimicry, narrow body shape, dorso-ventral flattening, large eyes, and long legs (172). Inactive beetles rest at safe sites, under stones, in crevices, in the soil, or on undersides of leaves; night activity is also thought to be an antipredator defense. Attacked beetles run to safety and hide (172), take to water (4, 174), demonstrate catalepsy (17), regurgitate crop contents and/or digestive fluid (79), and bite their attacker (61). Stridulation is also a widespread and effective deterrent (78, 82). Conspicuous elytral spots, which are present in many carabid species, may deflect attack from the vital anterior body parts (103). Batesian mimicry has been reported in carabids (116), and Müllerian mimicry was reported in tiger beetles (149). The hardened cuticle and fused elytra of large species (75) also provide structural protection from predators. The most effective defense is the excretion of compounds from pygidial glands that are universally present in carabids. The anatomy, chemistry, and effectiveness of these glands and their products has been extensively studied and reviewed (18, 105, 134).

Population Dynamics

Most of the available field data on carabids come from results of pitfall-trap catches. Catches of the same species in the same habitat from different years correlate well with changes in density (43, 125), and this comparison is generally accepted as a valid method for estimating density fluctuations and effective rates of reproduction.

Population variability in carabids (125) seems to be at the lower end of values for insects (204). Although environmental fluctuations in caves are smaller than in other terrestrial habitats, population fluctuations of the cave-inhabiting *Neaphaenops tellkampfi* in Mammoth Caves, Kentucky, were between those of *Calathus melanocephalus* and *Pterostichus versicolor,* two common species living on heath in Drenthe, The Netherlands (104). Different intrinsic and extrinsic factors—life span, fecundity, reproductive patterns, and rate of development—are thought to contribute to this relative stability (125).

Population Survival and Metapopulation Dynamics

The study of carabids has contributed significantly to the appreciation of landscape-scale dynamics. Particularly important are studies started in the late 1950s in The Netherlands (43). den Boer (46) synthesized the regional population fluctuation patterns of carabids collected over 23 years in the Dutch province of Drenthe. Using a distribution of population sizes (43), he distinguished several population fluctuation types. Species with high dispersal power (e.g. *Pterostichus niger*) exhibit population fluctuation patterns different from those of species with limited dispersal ability (e.g. *Pterostichus lepidus*). Species in Drenthe show a continuum between these two extremes. Based on this pattern, the frequency of extinction and the mean survival times of populations of the different species were simulated. This technique indicated that local populations of poorly dispersing species survive, on average, for 40–50 years. If changes in the locations of suitable habitat patches are faster, the species cannot recolonize new habitat patches fast enough and become regionally extinct. For most of Europe, these changes occur faster than required by the poorly dispersing species.

de Vries & den Boer (59a) compared the regional distribution of *Agonum ericeti,* a species found in moist heath, in 1959–1962 with its distribution in 1988–1989. This species cannot travel more than 200 meters between habitat fragments and showed an average survival time of 7–44 years in different-sized, small habitat fragments. In larger fragments, population fluctuation is asynchronous and the multipartite population can survive longer. These authors concluded that *A. ericeti* needs a habitat fragment of 50–70 ha for continuous population survival.

With the intensification of agriculture, fragmentation of natural habitats has occurred worldwide during the twentieth century. Turin & den Boer (184) and Turin & Peters (185) have examined the effects of these changes in The Netherlands since 1850. Poorly dispersing species (for example *Abax paral-*

lelepipedus, Calathus erratus, and *P. oblongopunctatus*) generally decreased; well-dispersing species (*Amara lunicollis, Dicheirotrichus gustavi, Stenolophus mixtus*) were stable or increasing; and species tolerating agricultural habitats (*C. melanocephalus, Dyschirius globosus, P. melanarius*) increased during this period. Whether these changes were caused by habitat fragmentation or habitat destruction was not clear.

ASSEMBLAGES AND COMMUNITIES

Patterns in Carabid Assemblages

Carabid assemblages are moderately species rich. Usually, no more than 10–40 species are active in a habitat in the same season; regional assemblages are correspondingly richer (98, 126, 182, 188). Generalizations are difficult as the extension of an assemblage in space or time is usually not defined; the number also depends on the method and intensity of the sampling. With the advent of more accessible and more powerful data handling, the regional and continental distribution patterns can be described and evaluated (144, 151). Future evaluations of the nestedness of ground beetle faunas is another promising endeavor.

Southwood (169) described the species packing of ground beetles at Silwood Park in southern England, a site containing 28 species. The report made no mention of the presence, abundance, or species richness of potential competitor groups (ants, spiders). More than 50 pair-wise interactions were considered significant. During the growing season, from March to November, the activity periods of the most common species filled the available time; for large species, this species packing was tighter during the summer than in spring or autumn. Species body sizes were regularly arranged between 5 and 25 mm, with an obvious gap between 12 and 14 mm, bordered by the two most common species. Habitat specialization occurred but complete lack of spatial overlap was found in only 8 of the potential 57 species-pair interactions.

The mean body size of carabid assemblages in woodlands, moors, and grasslands in northeastern England was related to several environmental factors (16). The outstanding factor was the level of disturbance that eliminated large species from the assemblage. Species body-size distribution within carabid assemblages was similarly displaced toward smaller values as disturbance from urbanization increased (179).

Coexistence and Competition

The occurrence and importance of competition among carabid beetles has been long debated. Generally, the evidence for interspecific competition as a regulatory force in populations is inconclusive, because of methodological limitations, unrealistic densities, noncomparable habitats, the methods used (examples in 142), and a general lack of experimental tests (142). Significant interspecific competition exists between adults of the North American *Carabus limbatus* and *Carabus sylvosus* (113). However, another study showed that most species do not compete in a western European beech forest (119). Similar conclusions emerge from evaluations of resource-partitioning descriptions; competition cannot be proven except in a few cases (44). In the sub-Antarctic, *Amblystogenium pacificum* and *Amblystogenium minimum* showed character displacement expected to result from competition (36).

These studies focused on the adult stage, but larvae have more restricted tolerance limits because of more restricted food range, mobility, and weaker chitinization, and are less adapted to evade resource shortages. Consequently, the importance of competition among larvae can be greater than that among adults (22, 92).

At the assemblage level, resource-partitioning patterns have been described in several studies (reviewed in 142), which have often invoked competition, present or past, as an explanation for the observed patterns of size distribution, food range, and seasonal or daily activity. These conclusions have generated lively but inconclusive debate (e.g. 44, 47 vs 118). Currently, there is no convincing evidence that competition has an important role in causing the observed patterns in carabid assemblages. A recent study on the invasion of a European carabid beetle into a Canadian forest (143) also showed a lack of competitive effects on the resident carabids.

The very concept of carabid communities is fallacious. This concept is based on a taxonomic affiliation, and carabids cannot even be considered to constitute an ecological guild. Although many carabid species can be classified as generalist predators, others that coexist with them clearly belong to different guilds. Carabids share the generalist, surface-active predator guild with at least some spiders and ants. For example, significant competition seems to take place between ants and ground beetles (205), so neglecting ants in "carabid community studies" leads to misleading conclusions.

ECONOMIC IMPORTANCE OF GROUND BEETLES

Occurrence in Agricultural Fields

Carabids are common in agricultural fields in the Northern Hemisphere. Since an early publication by Forbes (76), they have generally been considered beneficial natural enemies of agricultural pests, although a few species are pests themselves (126, 182). Thiele has synthesized the information on their biology, with special reference to their role as natural enemies (182); Allen (2) and Luff (126) have provided limited updates.

The carabid fauna of agricultural fields originates in riparian (182) or steppe

(121) habitats. Data are few outside those obtained in Europe and North America. In Canada, many species in cultivated land are either introduced European species (2, 171) or North American representatives of genera common in European agricultural fields, such as *Pterostichus, Harpalus,* or *Agonum.* In Japan, the fauna is similar to that of the European cultivated habitats at the generic level (126), although species of *Chlaenius* and *Carabus* can be abundant (207). In arid areas, Tenebrionidae are more prevalent than Carabidae (74). In New Zealand, carabids can be significant predators (10), but they are not as prevalent there as they are in northern cultivated fields (120).

Agriculture profoundly influences the composition, abundance, and spatial distribution of ground beetles through the use of agrochemicals, changes in habitat structure from cultivation methods and crop type, etc (57, 173; see reviews in 126, 182).

The Effectiveness of Carabids as Natural Enemies

Predator-prey studies have traditionally focused on interactions between specialist predators and their prey (90). Although *Calosoma sycophantha,* one of the first insects introduced for biological control (24), is such a specialist, most carabids do not fall into this category. The exploration of conditions under which generalist predators can limit prey has revealed that such predators are self-damping and highly polyphagous and that their life cycles are not in synchrony with their prey (38, 137). The ground beetles fit these criteria; they are self-damping during their larval stage (21), are polyphagous feeders (95), and having a long life cycle, are not normally tightly coupled to their prey. They can suppress pest outbreaks, but in general, their major beneficial role is to prolong the period between pest outbreaks, i.e. when the pest abundance is in the so-called natural enemy ravine (170). To increase carabids' effectiveness, biological control practitioners should consider the general habitat favorability that will keep carabids near their required site of action. A successful application of this technique could use habitat islands to serve as refuges and recolonization foci (128, 183).

The effectiveness of a natural enemy can be established through several sequential steps (123, 175, 206): 1. evaluating dynamics and correlating predator and pest density, 2. obtaining direct evidence of a trophic link between the prey and the predator, 3. experimentally manipulating predator density and its effect on pest numbers, 4. integrating the above information to quantify the effect of predator on prey.

Most studies of carabids and their prey are of the first and second type; fewer authors have considered steps 3 and 4. Well-founded evidence (gathered by means of all four steps above) for the significance of carabids as natural enemies comes from studies of polyphagous predators (carabids, spiders, staphylinids) in cereals in England (28, 60, 156, 164, 165, 178) showing that they can significantly decrease the peak density of aphids. Early-season predation, when aphid density is low, is the most significant. The relative importance of these predators varies among years and sites; often the effect cannot be attributed to one particular predator group. In some years, carabids are the most significant predators.

Carabids as Environmental Indicators

Carabids can and have been used as indicator organisms for assessments of environmental pollution (93), habitat classification for nature protection (127, 152), or characterization of soil-nutrient status in forestry (181). They might also serve as biodiversity indicators (N Stork, personal communication). However, most of the groups that are candidates for these purposes have not been subjected to a critical assessment using set criteria (150). Once we develop these criteria, we can realistically assess the suitability of ground beetles as indicator organisms.

ACKNOWLEDGMENTS

For discussions, comments, reprints, and access to unpublished information, we thank MEG Evans, PJ Johns, A Larochelle, M Sárospataki, NE Stork, and WOC Symondson.

Literature Cited

1. Alderweireldt M, Desender K. 1990. Variation of carabid diel activity patterns in pastures and cultivated fields. See Ref. 173, pp. 335–38
2. Allen RT. 1979. The occurrence and importance of ground beetles in agricultural and surrounding habitats. See Ref. 68, pp. 485–507
3. Andersen J, Skorping A. 1991. Parasites of carabid beetles: prevalance depends on habitat selection of the host. *Can. J. Zool.* 69:1216–20
4. Arens W, Bauer T. 1987. Diving behaviour and respiration in *Blethisa multipunctata* in comparison with two other ground beetles. *Physiol. Entomol.* 12: 255–61
5. Aukema B. 1990. Wing-length determination in two wing-dimorphic *Calathus*

species (Coleoptera: Carabidae). *Hereditas* 113:189–202
6. Aukema B. 1991. Fecundity in relation to wing-morph of three closely related species of the *melanocephalus* group of the genus *Calathus* (Coleoptera: Carabidae). *Oecologia* 87:118–26
7. Baars MA, van Dijk TS. 1984. Population dynamics of two carabid species in a Dutch heathland. II. Egg production and survival in relation to density. *J. Anim. Ecol.* 53:389–400
8. Ball GE. 1979. Conspectus of carabid classification: history, holomorphology and higher taxa. See Ref. 68, pp. 63–111
9. Barber H. 1931. Traps for cave-inhabiting insects. *J. Elisa Mitchell Sci. Soc.* 46:259–66
10. Barker GM. 1991. Biology of slugs

(Agrolimacidae and Arionidae: Mollusca) in New Zealand hill country pastures. *Oecologia* 85:581–95

11. Bauer T. 1979. The behavioural strategy used by imago and larva of *Notiophilus biguttatus* F. (Coleoptera, Carabidae) in hunting Collembola. See Ref. 53, pp. 133–42

12. Bauer T. 1986. How to capture springtails on the soil surface. The method of *Loricera pilicornis* F. See Ref. 50, pp. 43–48

13. Bauer T, Kredler M. 1993. Morphology of the compound eyes as an indicator of life-style in carabid beetles. *Can. J. Zool.* 71:799–810

14. Bernays EA, Bright KL, Gonzalez N, Angel J. 1994. Dietary mixing in a generalist herbivore: tests of two hypotheses. *Ecology* 75:1997–2006

15. Bils W. 1976. Das Abdomenende weiblicher, terrestrisch lebender Adephaga (Coleoptera) und seine Bedeutung fuer die Phylogenie. *Zoomorphologie* 84:113–93

16. Blake S, Foster GN, Eyre MD, Luff ML. 1994. Effects of habitat type and grassland management practices on the body size distribution of carabid beetles. *Pedobiologia* 38:502–12

17. Bleich OE. 1928. Thanatose und Hypnose bei Coleopteren. *Z. Wiss. Biol. Abt. A* 10:1–61

18. Blum MS. 1996. Semiochemical parsimony in the Arthropoda. *Annu. Rev. Entomol.* 41:291–312

19. Brandmayr P, den Boer PJ, Weber F, eds. 1983. *Ecology of Carabids: The Synthesis of Field Study and Laboratory Experiment.* Wageningen: Centre Agric. Publ. Doc. 196 pp.

20. Brandmayr P, Zetto-Brandmayr T. 1979. The evolution of parental care phenomena in Pterostichine ground beetles with special reference to the genera *Abax* and *Molops* (Col. Carabidae). See Ref. 53, pp. 35–49

21. Brunsting A, Heessen HJL. 1984. Density regulation in the carabid beetle *Pterostichus oblongopunctatus. J. Anim. Ecol.* 53:751–60

22. Brunsting A, Siepel H, van Schaik Zillesen PG. 1986. The role of larvae in the population ecology of Carabidae. See Ref. 52, pp. 399–411

23. Burakowski B. 1986. The life cycle and food preference of *Agonum quadripunctatum* (De Geer). See Ref. 50, pp. 35–39

24. Burgess AF. 1911. *Calosoma sycophanta:* its life history, behavior and successful colonization in New England. *US Dep. Agric. Bur. Entomol. Bull. 101*

25. Burmeister E-G. 1980. Funktsionsmor-

26. Butterfield JEL. 1986. Changes in lifecycle strategies of *Carabus problematicus* over a range of altitudes in Northern England. *Ecol. Entomol.* 11:17–26

27. Calow P. 1979. The cost of reproduction—a physiological approach. *Biol. Rev.* 54:23–40

28. Chambers RJ, Sunderland KD, Wyatt IJ, Vickerman GP. 1983. The effects of predator exclusion and caging on cereal aphids in winter wheat. *J. Appl. Ecol.* 20:209–24

29. Chiverton PA. 1984. Pitfall-trap catches of the carabid beetle *Pterostichus melanarius*, in relation to gut contents and prey densities, in insecticide treated and untreated spring barley. *Entomol. Exp. Appl.* 36:23–30

30. Chiverton PA. 1988. Searching behaviour and cereal aphid consumption by *Bembidion lampros* and *Pterostichus cupreus*, in relation to temperature and prey density. *Entomol. Exp. Appl.* 47:173–82

31. Churchfield JS, Hollier J, Brown VK. 1991. The effects of small mammal predators on grassland invertebrates, investigated by field exclosure experiment. *Oikos* 60:283–90

32. Cohen AC. 1995. Extraoral digestion in predaceous terrestrial Arthropoda. *Annu. Rev. Entomol.* 40:85–103

33. Colombini I, Chelazzi L, Scapini F. 1994. Solar and landscape cues as orientation mechanisms in the beach-dwelling beetle *Eurynebria complanata* (Coleoptera, Carabidae). *Mar. Biol.* 118:425–32

34. Crowson RA. 1981. *The Biology of the Coleoptera.* London: Academic. 802 pp.

35. Darlington PJ. 1943. Carabidae of mountains and islands: data on the evolution of isolated faunas, and atrophy of wings. *Ecol. Monogr.* 13:37–61

36. Davies L. 1987. Long adult life, low reproduction and competition in two sub-Antarctic carabid beetles. *Ecol. Entomol.* 12:149–62

37. Davies MJ. 1953. The contents of the crops of some British carabid beetles. *Entomol. Mon. Mag.* 89:18–23

38. De Angelis DL, Goldstein Ra, O'Neill RV. 1975. A model for trophic interaction. *Ecology* 56:881–92

39. Deleted in proof

40. Deleted in proof

41. Deleurance S, Deleurance EP. 1964. Reproduction et cycle evolutif larvaire des *Aphenops* (*A. cerberus* Dieck, *A.*

crypticola Lindner), insectes Coleopteres cavernicoles. *C. R. Acad. Sci. Paris* 258:4369–70

42. den Boer PJ. 1970. On the significance of dispersal power for populations of carabid beetles (Coleoptera, Carabidae). *Oecologia* 4:1–28

43. den Boer PJ. 1977. Dispersal power and survival. Carabids in a cultivated countryside (with a mathematical appendix by J. Reddingius). *Misc. Pap. Landbouwhogesch. Wageningen* 14:1–190

44. den Boer PJ. 1980. Exclusion or coexistence and the taxonomic or ecological relationship between species. *Neth. J. Zool.* 30(2):278–306

45. den Boer PJ. 1986. What can carabid beetles tell us about dynamics of populations? See Ref. 52, pp. 315–30

46. den Boer PJ. 1987. On the turnover of carabid populations in changing environments. *Acta Phytopathol. Entomol. Hung.* 22:71–83

47. den Boer PJ. 1989. Comment on the article "On testing temporal niche differentiation in carabid beetles" by M. Loreau. *Oecologia* 81:97–98

48. den Boer PJ, ed. 1971. Dispersal and dispersal power of carabid beetles. *Misc. Pap. Landbouwh. Wageningen* 8:1–151

49. den Boer PJ, den Boer-Daanje W. 1990. On life history tactics in carabid beetles: Are there only spring and autumn breeders? See Ref. 173, pp. 247–58

50. den Boer PJ, Grüm L, Szyszko J, eds. 1986. *Feeding Behaviour and Accessibility of Food for Carabid Beetles.* Warsaw: Warsaw Agric. Univ. Press. 167 pp.

51. den Boer PJ, Lövei GL, Stork NE, Sunderland KD, eds. 1987. Proc. 6th European carabidologist meeting. *Acta Phytopathol. Entomol. Hung.* 22:1–458

52. den Boer PJ, Luff ML, Mossakowski D, Weber F, eds. 1986. *Carabid Beetles. Their Adaptations and Dynamics.* Stuttgart/New York: Fischer Verlag. 551 pp.

53. den Boer PJ, Thiele H-U, Weber F, eds. 1979. On the evolution of behavour in carabid beetles. *Misc. Pap. Agric. Univ. Wageningen* 18:1–222

54. den Boer PJ, van Huizen THP, den Boer–Daanje W, Aukema B, den Bieman CFM. 1980. Wing polymorphism and dimorphism as stages in an evolutionary process (Coleoptera, Carabidae). *Entomol. Gen.* 6:107–34

55. Dennison DF, Hodkinson ID. 1984. Structure of the predatory beetle community in a woodland soil ecosystem. V. Summary and conclusions. *Pedobiologia* 26:171–77

55a. de Ruiter PC, van Stralen MR, van Euwijk FA, Slob W, Bedaux JJM, Ernsting G. 1989. Effects of hunger and prey traces on the search activity of the predatory beetle *Notiophilus biguttatus*. *Entomol. Exp. Appl.* 51:87–95

56. Desender K. 1987. Heritability estimates for different morphological traits related to wing development and body size in the halobiont and wing polymorphic carabid beetle *Pogonus chalceus* Marsham (Coleoptera, Carabidae). *Acta Phythopathol. Entomol. Hung.* 22:85–101

57. Desender K, Dufrene M, Loreau M, Luff ML, Maelfait J-P, eds. 1994. *Carabid Beetles: Ecology and Evolution.* Ser. Entomol. 51. Dordrecht: Kluwer Academic. 474 pp.

58. Desender K, Maelfait J-P, D'Hulster M, Vanhercke L. 1981. Ecological and faunal studies on Coleoptera in agricultural land. I. Seasonal occurrence of Carabidae in the grassy edge of a pasture. *Pedobiologia* 22:379–84

59. Desender K, van den Broeck D, Maelfait J-P. 1985. Population biology and reproduction in *Pterostichus melanarius* Ill. (Coleoptera, Carabidae) from a heavily grazed pasture ecosystem. *Med. Fac. Landbouwwet. Rijksuniv. Gent* 50:567–75

59a. de Vries HH, den Boer PJ. 1990. Survival of populations of *Agonum ericeti* Panz. (Col., Carabidae) in relation to fragmentation of habitats. *Neth. J. Zool.* 40:484–98

60. Edwards CA, Sunderland KD, George KS. 1979. Studies of polyphagous predators of cereal aphids. *J. Appl. Ecol.* 16:811–23

61. Eisner T, Hurst JJ, Meinwald T. 1963. Defensive mechanisms of arthropods. XI. The structure, function and phenolic secretion of the glands of a chordeumid millipede and a carabid beetle. *Psyche* 70:94–116

62. Erbeling L. 1987. Thermal ecology of the desert carabid beetle *Thermophilum (Anthia) sexmaculatum* F. (Coleoptera, Carabidae). *Acta Phytopathol. Entomol. Hung.* 22:119–33

63. Ernsting G, Isaaks JA. 1991. Accelerated ageing: a cost of reproduction in the carabid beetle *Notiophilus biguttatus*. *Func. Ecol.* 5:299–303

64. Erwin TL. 1979. Thoughts on the evolutionary history of ground beetles: hypotheses generated from comparative faunal analysis of lowland forest sites in temperate and tropical regions. See Ref. 68, pp. 539–92

65. Erwin TL. 1979. A review of the natural

history and evolution of ectoparasitoid relationships in carabid beetles. See Ref. 68, pp. 479–84

66. Erwin TL. 1985. The taxon pulse: a general pattern of lineage radiation and extinction among carabid beetles. In *Taxonomy, Phylogeny and Zoogeography of Beetles and Ants*, ed. GE Ball, pp. 437–72. Dordrecht: Junk. 514 pp.

67. Erwin TL, Adis J. 1982. Amazonian inundation forests: their role as short-term refuges and generators of species richness and taxon pulses. In *Biological Diversification in the Tropics*, ed. G Prance, pp. 358–71. New York: Columbia Univ. Press. 714 pp.

68. Erwin TL, Ball GE, Whitehead DL, Halpern AL, eds. 1979. *Carabid Beetles: Their Evolution, Natural History and Classification.* The Hague: Junk. 635 pp.

69. Evans MEG. 1977. Locomotion in the Coleoptera Adephaga, especially Carabidae. *J. Zool.* 181:189–226

70. Evans MEG. 1986. Carabid locomotor habits and adaptations. See Ref. 52, pp. 59–77

71. Evans MEG, Forsythe TG. 1985. Feeding mechanisms, and their variation in form, of some adult ground-beetles (Coleoptera: Caraboidea). *J. Zool.* 206: 113–43

72. Evans WG. 1988. Chemically mediated habitat recognition in shore insects (Coleoptera: Carabidae; Hemiptera: Saldidae). *J. Chem. Ecol.* 14:1441–54

73. Eyre MD. 1994. Strategic explanations of carabid species distributions in northern England. See Ref. 57, pp. 267–75

74. Faragalla AA, Adam EE. 1985. Pitfall trapping of tenebrionid and carabid beetles (Coleoptera) in different habitats of the central region of Saudi Arabia. *Z. Angew. Entomol.* 99:466–71

75. Fiori G. 1974. Contributi alla conoscenza morfologica et etologica dei Coleotteri. X. La "sutura" elitrale. *Boll. Ist. Entomol. Univ. Bologna* 31:129–52

76. Forbes SA. 1883. The food relations of the Carabidae and the Coccinellidae. *Bull. Ill. State Lab. Nat. Hist.* 1:33–64

77. Forsyth DJ. 1972. The structure of the pygidial defence glands of Carabidae (Coleoptera). *Trans. Zool. Soc. London* 32:249–309

78. Forsythe TG. 1980. Sound production in *Amara familiaris* Duft with a review of sound production in British Carabidae. *Entomol. Mon. Mag.* 115:177–79

79. Forsythe TG. 1982. Qualitative analyses of certain enzymes of the oral defence fluids of *P. madidus. Entomol. Mon. Mag.* 118:1–5

80. Forsythe TG. 1987. *Common Ground Beetles. Naturalists' Handbook 8.* Richmond: Richmond Publishing. 74 pp.

81. Freeland WJ. 1983. Parasites and the coexistence of animal host species. *Am. Nat.* 121:223–36

82. Freitag R, Lee SK. 1972. Sound producing structures in adult *Cicindela tranquebarica* (Coleoptera: Cicindelidae) including a list of tiger beetles and ground beetles with flight wing files. *Can. Entomol.* 104:851–57

83. Gergely G, Lövei GL. 1987. Phenology and reproduction of the ground beetle *Dolichus halensis* in maize fields: a preliminary report. *Acta Phytopathol. Entomol. Hung.* 22:357–61

84. Greenslade PJM. 1963. Daily rhythms of locomotory activity in some Carabidae (Coleoptera). *Entomol. Exp. Appl.* 6:171–80

85. Grime JP. 1977. Evidence for the existence of three primary strategies in plants and its relevance to ecological and evolutionary theory. *Am. Nat.* 111: 1169–94

86. Grüm L. 1984. Carabid fecundity as affected by extrinsic and intrinsic factors. *Oecologia* 65:114–21

87. Haeck J. 1971. The immigration and settlement of Carabida in the new IJsselmeerpolders. See Ref. 48, pp. 33–52

88. Halsall NB, Wratten SD. 1988. The efficiency of pitfall trapping for polyphagous predatory Carabidae. *Ecol. Entomol.* 13:293–99

89. Hammond PM. 1979. Wing-folding mechanisms of beetles, with special reference to investigations of Adephagan phylogeny. See Ref. 68, pp. 113–80

90. Hassell MP. 1978. *The Dynamics of Arthropod Prey-Predator Systems.* Princeton: Princeton Univ. Press. 237 pp.

91. Heessen HJL. 1981. Egg mortality in *P. oblongopunctatus* (Coleoptera, Carabidae). *Oecologia* 50:233–35

92. Heessen HJL, Brunsting AMH. 1981. Mortality of larvae of *P. oblongopunctatus* (Coleoptera, Carabidae) and *Philonthus decorus* (Coleoptera, Carabidae). *Neth. J. Zool.* 31:729–45

93. Heliovaara K, Vaisanen R. 1993. *Insects and Pollution.* Boca Raton: CRC Press. 393 pp.

94. Hengeveld R. 1980. Food specialization in ground beetles: an ecological or a phylogenetic process? (Coleoptera, Carabidae). *Neth. J. Zool.* 30:585–94

95. Hengeveld R. 1980. Polyphagy, oligophagy and food specialisation in ground beetles (Coleoptera, Carabidae). *Neth. J. Zool.* 30:564–84

96. Hlavac TF. 1971. Differentiation of the carabid antenna cleaner. *Psyche* 78:51–66

97. Horne PA. 1990. Parental care in *Notonomus* Chaudoir (Coleoptera: Carabidae: Pterostichinae). *Aus. Entomol. Mag.* 17:65–69

98. Horvatovich S, Szarukán I. 1986. Faunal investigation of ground beetles (Carabidae), in the arable soils of Hungary. *Acta Agron. Hung.* 35:107–23

99. Houston WWK. 1981. The life cycles and age of *Carabus glabratus* Paykull and *C. problematicus* Herbst. (Col.: Carabidae) on moorland in northern England. *Ecol. Entomol.* 6:263–71

100. Hurka K. 1986. The developmental type of Carabidae in the temperate zones as a taxonomic character. See Ref. 52, pp. 187–93

101. Jaspar-Versali MF, Goffinet G, Jeuniaux C. 1987. The digestive system of adult carabid beetles: an ultrastructural and histoenzymological study. *Acta Phytopathol. Entomol. Hung.* 22:375–82

102. Jones MG. 1979. The abundance and reproductive activity of common carabidae in a winter wheat crop. *Ecol. Entomol.* 4:31–43

103. Kamoun S. 1991. Parasematic coloration: a novel anti-predator mechanism in tiger beetles (Coleoptera: Cicindelidae). *Coleopt. Bull.* 45:15–19

104. Kane TC, Ryan T. 1983. Population ecology of carabid cave beetles. *Oecologia* 60:46–55

105. Kanehisa K, Murase M. 1977. Comparative study of the pygidial defensive systems of carabid beetles. *Appl. Entomol. Zool.* 12:225–35

106. Kegel B. 1990. Diurnal activity of carabid beetles living on arable land. See Ref. 173, pp. 65–76

107. Kennedy P. 1994. The distribution and movement of ground beetles in relation to set-aside arable land. See Ref. 57, pp. 439–44

108. Larochelle A. 1975. A list of mammals as predators of Carabidae. *Carabologia* 3:95–98

109. Larochelle A. 1975. A list of amphibians and reptiles as predators of Carabidae. *Carabologia* 3:99–103

110. Larochelle A. 1980. A list of birds of Europe and Asia as predators of carabid beetles including Cicindelini (Coleoptera: Carabidae). *Cordulia* 6:1–19

111. Larochelle A. 1990. The food of carabid beetles. *Fabreries Suppl.* 5:1–132

112. Lawrence JF, Britton EB. 1991. Coleoptera. In *The Insects of Australia*, 2:543–683. Melbourne: Melbourne Univ. Press. 1137 pp. 2nd ed.

113. Lenski RE. 1984. Food limitation and competition: a field experiment with two *Carabus* species. *J. Anim. Ecol.* 53:203–16

114. Lindroth CH. 1949. *Die Fennoskandischen Carabiden*, Part 3, *Algemeiner Teil*. Stockholm: Bröderna Lagerström Boktrychare. 911 pp.

115. Lindroth CH. 1969. The ground beetles (Carabidae excl. Cicindelidae), of Canada and Alaska, Part 1. *Opusc. Entomol. Suppl.* 25:1–48

116. Lindroth CH. 1971. Disappearance as a protective factor. A supposed case of Batesian mimicry among beetles (Coleoptera, Carabidae and Chrysomelidae). *Entomol. Scand.* 2:41–48

117. Lindroth CH. 1979. The theory of glacial refugia. See Ref. 68, pp. 385–94

118. Loreau M. 1989. On testing temporal niche differentiation in carabid beetles. *Oecologia* 81:89–96

119. Loreau M. 1990. Competition in a carabid beetle community: field experiment. *Oikos* 58:25–38

120. Lövei GL. 1991. The ground-dwelling predatory arthropod fauna in an organic and abandoned kiwifruit orchard. In *Proc. Symp. Sustainable Agriculture and Organic Food Production, Trentham, New Zealand*, ed. IA Popay, pp. 9–14. Christchurch: NZ Inst. Agric. Sci./NZ Hort. Soc.

121. Lövei GL, Sárospataki M. 1990. Carabids in eastern European agricultural fields: a review. See Ref. 173, pp. 87–93

122. Lövei GL, Sopp PI, Sunderland KD. 1990. Digestion rate in relation to alternative feeding in three species of polyphagous predators. *Ecol. Entomol.* 15:291–300

123. Luck RF, Shepard BM, Kenmore PE. 1988. Experimental methods for evaluating arthropod natural enemies. *Annu. Rev. Entomol.* 33:367–91

124. Luff ML. 1978. Diel activity patterns of some field Carabidae. *Ecol. Entomol.* 3:53–62

125. Luff ML. 1982. Population dynamics of Carabidae. *Ann. Appl. Biol.* 101:164–70

126. Luff ML. 1987. Biology of polyphagous ground beetles in agriculture. *Agric. Zool. Rev.* 2:237–78

127. Luff ML, Eyre MD, Rushton SP. 1992. Classification and prediction of grassland habitats using ground beetles (Coleoptera, Carabidae). *J. Environ. Manage.* 35:301–15

128. Lys J-A. 1994. The positive influence of strip-management on ground beetles

in a cereal field: increase, migration and overwintering. See Ref. 57, pp. 451–55

129. Makarov KV. 1994. Annual reproduction rhythms of ground beetles: a new approach to the old problem. See Ref. 57, pp. 177–82

130. Mascanzoni D, Wallin H. 1986. The harmonic radar: a new method of tracing insects in the field. *Ecol. Entomol.* 11: 387–90

131. Metzenauer P. 1981. Pattern of the intestinal enzymes of the carnivorous ground beetle *Ptereostichus nigrita* (Paykull) (Col. Carabidae). *Zool. Anz.* 207:113–19

132. Mols PJM. 1979. Motivation and walking behaviour of the carabid beetle *Pterostichus coerulescens* L. at different densities and distributions of the prey. See Ref. 53, pp. 185–98

133. Mols PJM. 1988. Simulation of hunger, feeding and egg production in the carabid beetle *Pterostichus coerulescens* L. *Agric. Univ. Wageningen. Pap.* 88–3:1–99

134. Moore BP. 1979. Chemical defense in carabids and its bearing on phylogeny. See Ref. 68, pp. 193–203

135. Moosbeckhofer R. 1983. Laboruntersuchungen ueber den Einfluss von Diazinon, Carbofuran und Chlorfenvinphos auf die Laeufaktivitaet von *Poecilus cupreus* L. (Col., Carabidae). *Z. Angew. Entomol.* 95:15–21

136. Murdoch WW. 1966. Aspects of the population dynamics of some marsh Carabidae. *J. Anim. Ecol.* 35:127–56

137. Murdoch WW, Chesson J, Chesson PL. 1985. Biological control in theory and practice. *Am. Nat.* 125:344–66

138. Nelemans MNE. 1987. Possibilities for flight in the carabid beetle *Nebria brevicollis* (F.). The importance of food during the larval growth. *Oecologia* 72: 502–9

139. Nelemans MNE, den Boer PJ, Spee A. 1989. Recruitment and summer diapause in the dynamics of a population of *Nebria brevicollis* (Coleoptera: Carabidae). *Oikos* 56:157–69

140. New TR. 1991. *Insects as Predators.* Kensington: NSW Univ. Press. 178 pp.

141. Niemela J. 1990. Spatial distribution of carabid beetles in the Southern Finnish taiga: the question of scale. See Ref. 173, pp. 143–55

142. Niemela J. 1993. Interspecific competition in ground-beetle assemblages (Carabidae)—what have we learned. *Oikos* 66:325–35

143. Niemela J, Spence JR. 1991. Distribution and abundance of an exotic ground-beetle (Carabidae)—a test of community impact. *Oikos* 62:351–59

144. Niemela J, Spence JR. 1994. Distribution of forest dwelling carabids (Coleoptera)—spatial scale and the concept of communities. *Ecography* 17:166–75

145. Nilsson AN, Pettersson RB, Lemdahl G. 1993. Macroptery in altitudinal specialists versus brachyptery in generalists—a paradox of alpine Scandinavian carabid beetles (Coleoptera, Carabidae). *J. Biogeogr.* 20:227–34

146. Paarmann W. 1986. Seasonality and its control by environmental factors in tropical ground beetles (Col. Carabidae). See Ref. 52, pp. 157–71

147. Pakarinen E. 1994. The importance of mucus as a defence against carabid beetles by the slugs *Arion fasciatus* and *Deroceras reticulatum. J. Mollusc. Stud.* 60:149–55

148. Parmenter RR, MacMahon JA. 1988. Factors influencing species composition and population sizes in a ground beetle community (Carabidae): predation by rodents. *Oikos* 52:350–56

149. Pearson DL. 1988. The biology of tiger beetles. *Annu. Rev. Entomol.* 33:123–47

150. Pearson DL, Cassola F. 1992. Worldwide species richness patterns of tiger beetles (Coleoptera: Cicindelidae): indicator taxon for biodiversity and conservation studies. *Cons. Biol.* 6:376–91

151. Penev LD, Turin H. 1994. Patterns of distribution of the genus *Carabus* L. in Europe: approaches and preliminary results. See Ref. 57, pp. 37–43

152. Pizzolotto R. 1994. Ground beetles (Coleoptera, Carabidae) as a tool for environmental management: a geographical information system based on carabids and vegetation for the Karst near Trieste (Italy). See Ref. 57, pp. 343–51

153. Poinar GO. 1975. *Entomogenous Nematodes. A Manual and Host List of Insect-Nematode Associations.* Leiden: Brill

154. Pollet M, Desender K. 1987. Feeding ecology of grassland-inhabiting carabid beetles (Carabidae, Coleoptera) in relation to the availability of some prey groups. *Acta Phytopathol. Entomol. Hung.* 22:223–46

155. Ponomarenko AG. 1977. Mesozoic Coleoptera. *Trans. Paleontol. Inst.* Moscow 161:1–204

156. Potts GR, Vickerman GP. 1974. Studies on the cereal ecosystem. *Adv. Ecol. Res.* 8:107–97

157. Price PW. 1980. *Evolutionary Biology of Parasites.* Princeton: Princeton Univ. Press

158. Refseth D. 1984. The life cycles and growth of *Carabus glabratus* and *C. violaceus* in Budelan, central Norway. *Ecol. Entomol.* 9:449–55

159. Rijnsdorp AD. 1980. Pattern of movement in and dispersal from a Dutch forest of *Carabus problematicus* Hbst (Coleoptera, Carabidae). *Oecologia* 45:274–81

160. Roff DA. 1994. The evolution of flightlessness: Is history important? *Evol. Ecol.* 8:639–57

161. Rusdea E. 1994. Population dynamics of *Laemostenus schreibersi* (Carabidae) in a cave in Carinthia (Austria). See Ref. 57, pp. 207–12

162. Simpson SJ, Simpson CL. 1990. The mechanisms of nutritional compensation by phytophagous insects. In *Insect-Plant Interactions*, ed. EA Bernays, 2:111–60. Boca Raton, FL: CRC

163. Skuhravy V. 1959. Die Nährung der Feldcarabiden. *Acta Soc. Entomol. Cech.* 56:1–18

164. Sopp PI, Sunderland KD. 1989. Some factors affecting the detection period of aphid remains in predators using ELISA. *Entomol. Exp. Appl.* 51:11–20

165. Sopp PI, Sunderland KD, Fenlon J, Wratten SD. 1992. An improved quantitative method for estimating invertebrate predation in the field using an enzyme-linked immunosorbent assay (ELISA). *J. Appl. Ecol.* 29:295–302

166. Sota T. 1984. Long adult life span and polyphagy of a carabid beetle, *Leptocarabus kumagaii* in relation to reproduction and survival. *Res. Popul. Ecol.* 26:389–400

167. Sota T. 1985. Limitation of reproduction by feeding condition in a carabid beetle, *Carabus yaconinus*. *Res. Popul. Ecol.* 27:171–84

168. Sota T. 1987. Mortality pattern and age structure in two carabid populations with different seasonal life cycles. *Res. Popul. Ecol.* 29:237–54

169. Southwood TRE. 1978. The components of diversity. In *Diversity of Insect Faunas*, ed. LA Mound, N Waloff, pp. 19–40. Oxford: Blackwell. 204 pp.

170. Southwood TRE, Comins HN. 1976. A synoptic population model. *J. Anim. Ecol.* 45:949–65

171. Spence JR, Spence DH. 1988. Of ground beetles and man: introduced species and the synanthropic fauna of western Canada. *Mem. Entomol. Soc. Can.* 144:151–68

172. Stork NE. 1987. Adaptations of arboreal carabids to life in trees. *Acta Phytopathol. Entomol. Hung.* 22:273–91

173. Stork NE, ed. 1990. *The Role of Ground Beetles in Ecological and Environmental Studies.* Andover: Intercept. 424 pp.

174. Sturani M. 1962. Osservazioni e ricerche biologiche sul genere *Carabus* Linnaeus (sensu lato) (Coleoptera: Carabidae). *Mem. Soc. Entomol. Ital.* 41:85–202

175. Sunderland KD. 1987. A review of methods of quantifying invertebrate predation occurring in the field. *Acta Phytopathol. Entomol. Hung.* 22:13–34

176. Sunderland KD, de Snoo GR, Dinter A, Hance T, Helenius J, et al. 1995. Density estimation for invertebrate predators in agroecosystems. *Acta Jutl.* In press

177. Sunderland KD, Lövei GL, Fenlon J. 1995. Diets and reproductive phenologies of the introduced ground beetles *Harpalus affinis* and *Clivina australasiae* (Coloeptera: Carabidae) in New Zealand. *Aust. J. Zool.* 43:39–50

178. Sunderland KD, Vickerman GP. 1980. Aphid feeding by some polyphagous predators in relations to aphid density in cereal fields. *J. Appl. Ecol.* 17:389–96

179. Sustek Z. 1987. Changes in body size structure of carabid communities (Coleoptera, Carabidae) along an urbanisation gradient. *Biologia* 42:145–56

180. Symondson WOC, Liddell JE, eds. 1995. *The Ecology of Agricultural Pests: Biochemical Approaches.* Systematics Assoc. Special Vol. Ser. No. 53. London: Chapman & Hall

181. Szyszko J. 1983. *State of Carabidae (Col.) Fauna in Fresh Pine Forest and Tentative Valorisation of This Environment.* Warsaw: Warsaw Agric. Univ. Press. 80 pp.

182. Thiele H-U. 1977. *Carabid Beetles in Their Environments.* Berlin/Heidelberg: Springer-Verlag. 369 pp.

183. Thomas MB, Wratten SD, Sotherton NW. 1991. Creation of 'island' habitats in farmland to manipulate populations of beneficial arthropods: predator densities and emigration. *J. Appl. Ecol.* 28:906–17

184. Turin H, den Boer PJ. 1988. Changes in the distribution of carabid beetles in the Netherlands since 1880. II. Isolation of habitats and long-term time trends in the occurrence of carabid species with different powers of dispersal (Coleoptera, Carabidae). *Biol. Conserv.* 44:179–200

185. Turin H, Peters H. 1986. Changes in the distribution of carabid beetles in the Netherlands since about 1880. I. Introduction. See Ref. 52, pp. 489–95

186. van Dijk TS. 1972. The significance of the diversity in age composition

of *Calathus melanocephalus* L. (Coleoptera, Carabidae) in space and time at Schiermonnikoog. *Oecologia* 10:111–36

187. van Dijk TS. 1983. The influence of food and temperature on the amount of reproduction in carabid beetles. See Ref. 19, pp. 105–23

188. van Dijk TS. 1987. The long-term effects on the carabid fauna of nutrient impoverishment of a previously arable field. *Acta Phytopathol. Entomol. Hung.* 22:103–18

189. van Dijk TS. 1994. On the relationship between food, reproduction and survival of two carabid beetles: *Calathus melanocephalus* and *Pterostichus versicolor*. *Ecol. Entomol.* 19:263–70

190. van Huizen THP. 1977. The significance of flight activity in the life cycle of *Amara plebeja* Gyllh. (Coleoptera, Carabidae). *Oecologia* 29:27–41

191. van Huizen THP. 1979. Individual and environmental factors determining flight in carabid beetles. See Ref. 53, pp. 199–211

192. van Huizen THP. 1990. 'Gone with the wind': flight activity of carabid beetles in relation to wind direction and to the reproductive state of females in flight. See Ref. 173, pp. 289–93

193. Vlijm L, van Dijk TS, Wijmans YS. 1968. Ecological studies on carabid beetles. III. Winter mortality in adult *Calathus melanocephalus* (Linn.). Egg production and locomotory activity of the population which has hibernated. *Oecologia* 1:304–14

194. Wallin H. 1988. The effects of spatial distribution on the development and reproduction of *Pterostichus cupreus* L., *P. melanarius* Illiger, *P. niger* Schaller and *Harpalus rufipes* De Geer (Coleoptera, Carabidae) on arable land. *J. Appl. Entomol.* 106:483–87

195. Wallin H. 1991. Movement patterns and foraging tactics of a caterpillar hunter inhabiting alfalfa fields. *Funct. Ecol.* 5:740–49

196. Wallin H, Chiverton PA, Ekbom BS, Borg A. 1992. Diet, fecundity and egg size in some polyphagous predatory carabid beetles. *Entomol. Exp. Appl.* 65:129–40

197. Ward RD. 1979. Metathoracic wing structures as phylogenetic indicators in the Adephaga (Coleoptera). See Ref. 68, pp. 181–91

198. Wasner U. 1977. Zur Oekologie und Biologie sympatrischer *Agonum* (*Europhilus*)—Arten (Carabidae, Coleoptera). I. Individual Entwicklung und Gonadenreifung, Generationsaufbau, Eiproduktion und Fruchtbarkeit. *Zool. Jahrb. Abt. Syst. Oekol. Geogr. Tiere* 106:105–23

199. Weber F, Casale A, Lamprecht G, Rusdea E. 1994. Highly sensitive reactions of microphthalmic carabid beetles to light/dark cycles. See Ref. 57, pp. 219–25

200. Weber F, Klenner M. 1987. Life history phenomena and risk of extinction in a subpopulation of *Carabus auronitens*. *Acta Phytopathol. Entomol. Hung.* 22:321–28

201. Weseloh RM. 1993. Adult feeding affects fecundity of the predator, *Calosoma sycophanta* (Col, Carabidae). *Entomophaga* 38:435–39

202. Wheater CP. 1989. Prey detection by some predatory Coleoptera (Carabidae and Staphylinidae). *J. Zool.* 218:171–85

203. Wiedenmann RN, O'Neill RJ. 1990. Effects of low rates of predation on selected life-history characteristics of *Podisus maculiventris* (Say) (Heteroptera: Pentatomidae). *Can. Entomol.* 122:271–83

204. Williamson M. 1972. *The Analysis of Insect Populations*. London: Arnold. 180 pp.

205. Wilson EO. 1990. *Success and Dominance in Ecosystems: The Case of the Social Insects*. Oldendorf: Ecology Inst. 104 pp.

206. Wratten SD. 1987. The effectiveness of native natural enemies. In *Integrated Pest Management,* ed. AJ Burn, TH Coaker, PC Jepson, pp. 89–112. London: Academic

207. Yano K, Yahiro K, Uwada M, Hirashima T. 1989. Species composition and seasonal abundance of ground beetles (Coleoptera) in a vineyard. *Bull. Fac. Agric. Yamaguchi Univ.* 37:1–14

208. Zetto Brandmayr T. 1983. Life cycle, control of propagation rhythm and fecundity of *Ophonus rotundicollis* Fairm. et. Lab. (Coleoptera, Carabidae, Harpalini) as an adaption to the main feeding plant *Daucus carota* L. (Umbelliferae). See Ref. 19, pp. 93–103

209. Zetto-Brandmayr T. 1990. Spermophagous (seed-eating) ground beetles: first comparison of the diet and ecology of the Harpaline genera *Harpalus* and *Ophonus* (Col. Carabidae). See Ref. 173, pp. 307–16

Annu. Rev. Entomol. 1996. 41:257–86

FLORAL RESOURCE UTILIZATION BY SOLITARY BEES (HYMENOPTERA: APOIDEA) AND EXPLOITATION OF THEIR STORED FOODS BY NATURAL ENEMIES[1]

William T. Wcislo
Smithsonian Tropical Research Institute, Apartado 2072, Balboa, Republic of Panamá[2]

James H. Cane
Department of Entomology and Alabama Agricultural Research Station, Auburn University, Auburn, Alabama 36849-5413

KEY WORDS: oligolecty, specialization, parasitism, phenotypic plasticity, pollen

---

ABSTRACT

Bees are phytophagous insects that exhibit recurrent ecological specializations related to factors generally different from those discussed for other phytophagous insects. Pollen specialists have undergone extensive radiations, and specialization is not a always a derived state. Floral host associations are conserved in some bee lineages. In others, various species specialize on different host plants that are phenotypically similar in presenting predictably abundant floral resources. The nesting of solitary bees in localized areas influences the intensity of interactions with enemies and competitors. Abiotic factors do not always explain the intraspecific variation in the spatial distribution of solitary bees. Foods stored by bees attract many natural enemies, which may shape diverse facets of nesting and foraging behavior. Parasitism has evolved repeatedly in some, but not all,

---

[2]Mailing address: Smithsonian Tropical Research Institute, Unit 0948, APO AA 34002-0948.

bee lineages. Available evidence suggests that cleptoparasitic lineages are most speciose in temperate zones. Female parasites frequently have a suite of characters that can be described as a masculinized feminine form. The evolution of resource specialization (including parasitism) in bees presents excellent opportunities to investigate phenotypic mechanisms responsible for evolutionary change.

PERSPECTIVES AND OVERVIEW

The >20,000 species of bees are allied to nest-building wasps and ants (Hymenoptera: Aculeata) (77). Bees arose within an assemblage of hunting wasps (sometimes known as Sphecoidea) that abandoned arthropods as the food source for their young in favor of using floral resources. Most bees are solitary (140, 160, 164, 167, 174, 202), although they may nest in aggregations. Unless indicated otherwise, when we refer to "bees," we mean solitary bees (for social species, see 140).

After a natural history overview, we focus on the following topics:

1. Some bees nest in isolation, while others nest gregariously. These spatial associations of conspecifics are likely to influence intraspecific competition and interactions with natural enemies (cf 232).
2. Many solitary bee species are resource specialists for pollen and oil, but rarely for nectar. The patterns and mechanisms of floral-host specialization by bees are governed by factors generally different from those affecting other insect specialists (see 14, 15). The interests of bees, unlike those of herbivores, are more congruent with those of the plant.
3. Numerous parasites attack adult bees, their offspring, or their stored food. Nest architecture, and foraging and nesting behavior, are hypothetically influenced by natural enemies, but this has not been tested critically. Major enemies of bees are other bees that are either facultative or obligate parasites. The great diversity of obligately parasitic bees, with their convergent morphology, allows us to study the evolutionary origins of behavioral flexibility and novel expression of condition-sensitive traits.

For brevity, we cite only more recent or comprehensive papers and recommend that interested readers use these bibliographies to find references to earlier works.

APOIDEA—WHAT ARE BEES?

Terminology

Solitary females store food in individual cells for their offspring. Most have no regular interactions with conspecifics, including offspring (for exceptions, see e.g. 139). Cleptoparasites do not collect oil or pollen but rather invade nests of solitary bees and deposit eggs in the cells or on stored provisions (18, 224); the parasite's offspring consume the stored resources. Oligolecty describes persistent and apparently heritable provisioning behavior in which a bee species, throughout its range, restricts itself to gathering pollen from a few related plant genera (51, 118, 134, 170); these bees are specialists. Polylectic bees are generalists that collect pollen from plants of more than one family. Generalists usually are constant to a particular plant species over a short time, but the preference changes in response to past experience and perceived rewards (58, 65, 81, 221). Oligoleges can also be flower constant, but they have a narrower range of acceptable hosts (51, 217). As with other classes of specialization, the dichotomy between oligolecty and polylecty is for classification purposes only.

Overview of Natural History

Some bees excavate nests in the soil, in living or rotting wood, or even in sandstone (13), whereas others construct nests from plant or earthen materials, usually in preexisting cavities, including snail shells or tunnels made by other boring insects (104, 135, 138, 160, 167, 174, 202). Representative illustrations of bees' nests appear elsewhere (104, 167, 174, 202). Nests contain one or more cells, either distributed throughout a branched tunnel system or clustered together. The mother usually coats the walls of each cell, or sometimes the pollen mass, with a hydrophobic secretion from her Dufour's gland (see section on antiparasite strategies). She provisions it with a mixture of pollen and nectar or oils collected from flowers in an area surrounding the nest (see 196), sometimes supplemented with glandular secretions (references in 59, 95). She lays an egg on, in, or, rarely, under the mass and seals the cell.

Life-history and diapause patterns are not well documented. Most temperate bees pass an unfavorable season as immatures, usually as postdefecating larvae (prepupae) (for exceptions, see 92, 139). Others diapause as adults (examples in 157). Frequently, species are typologically classified by the number of generations per season, and until recently, individual variability was neglected in these classifications. Facultative diapause (parsivoltinism) occurs (references in 157) and may be a bet-hedging strategy related to unpredictable resource availability (156) or to the abundance of natural enemies. Physiological mechanisms of apoid diapause are not well studied (see 133).

Classification of Apoidea

The superfamily Apoidea contains more than just bees. It includes some, though not all, of the carnivorous spheciforme wasps [= Sphecidae (sensu 22)], but which spheciforme lineage is the most closely related to bees is unknown (reviewed in 5, 30). Bees constitute a monophyletic group, as evidenced by several shared derived characters (reviewed in 5, 147).

Origin and Evolutionary History

Michener (142) speculated that bees arose in the xeric interior of West Gondwanaland, presumably concurrent with, or soon after, the radiation of angiosperms (211). Early apoid history is speculative, because fossils older than those from the Cenozoic are rare or nonexistent (references in 168). Fossilized remains of bee activity include putative nest burrows or cells, as well as leaves with disc-like sections removed as if cut by megachilid bees (reviewed in 24).

Phylogeny and Classification of Apiformes

Two informal groups, referred to as short-tongued (S-T) and long-tongued (L-T), are used to classify bees (145). The phylogenetic relationships among S-T bees remain unresolved, but phylogenetic hypotheses for the major lineages of L-T bees have led to major changes in classification (171). Michener and coworkers (147, 171) provide higher taxonomic classifications for the genera mentioned here.

Diversity and Distribution

Faunal studies, though limited and idiosyncratic, show that bees abound in most habitats on all continents except Antarctica (see 142, 157, 174). Generic diversity is greatest in the Neotropical region, but bees are more speciose in xeric, temperate regions (8, 142, 175). Given that tropical angiosperms are diverse, especially in the Old World (207), the relatively depauperate tropical bee fauna is surprising. From a behavioral and ecological perspective, however, tropical bees may be more diverse relative to temperate-zone bees (175).

Several hypotheses may account for the less speciose tropical fauna, but none have been rigorously tested. All the surmised factors may play some role. The apparently lower diversity may relate to the following: competitive exclusion by honey bees (*Apis* spp.) and by the abundant and speciose social stingless bees (Meliponini) (90, 174); increased fungal attack and food spoilage, including the tendency of provisions to become hygroscopic (134); or increased predation and parasitism (cf 110).

DISTRIBUTION OF POLLEN SPECIALISTS AND CLEPTOPARASITES   The relative diversity of oligolectic bees appears low in the tropics (see 89, 142, 174). Generalist species apparently predominate in most temperate biomes, except in xeric and Mediterranean-climate temperate regions (150). However, as most surveys do not examine pollen loads and as different authors use different definitions, conclusions are tentative.

Parasitic bees are also apparently rarer at lower latitudes (224; but see 90). A reported correlation (224) between the proportion of parasitic species in a given fauna and the latitude of the region surveyed may be biased because the analysis used species as independent data points. Analyses using higher taxonomic levels are currently unreliable, because at least some genera of parasitic bees are arbitrary rather than phylogenetically based. This geographic trend in parasitism is not a collecting artifact (224), and findings implicate seasonality and host synchrony in the evolution of parasitism (see 159, 224).

HOST LOCALIZATION IN TEMPORAL ENVIRONMENTS   The apparent latitudinal decrease in relative abundance of both parasitic and oligolectic bees may be related to the common problem of localizing a specific resource in a spatially and temporally complex environment (see 15). Spatial or temporal unpredictability in flowering phenology can be substantial in tropical forests (see 12, 27). Unfortunately, long-term studies of spatial and temporal abundance of solitary tropical bees are uncommon (e.g. 219a). If synchronization with host activity is important for these specializations, then they should be relatively more common in tropical dry forests, where resource availability is more seasonal, than in wet forests (references in 12, 27). The greater percentage of oligolectic bees in Costa Rican dry forest (89), relative to bees in French Guiana (142), is consistent with these predictions. No pertinent data are available for parasitic species.

HABITAT PREFERENCE AND AGGREGATIONS OF NESTS

Many solitary bees have habitat preferences for nesting and often choose sites near flowers (e.g. 76, 118, 126, 164, 202). Insufficient information is available to assess any degree of preference: Few studies sample and describe suitable nesting habitats where floral resources occur but the bees do not. Numerous bees nest in aggregations (e.g. 62, 118, 164), which range in size from several to many nests per square meter (49). One aggregation contained an estimated 423,000 nests in 1300 m² (186). Aggregations can persist for more than 20 years (149, 176, 230) or die out a few years (126). Some bee species disperse from their natal sites and then form ephemeral aggregations at unpredictable locations (e.g. 64). Congeners exhibit different degrees of philopatry and the tendency to aggregate nests (compare 35 with 2 and 186).

Proximate Mechanisms for Gregarious Behavior

Nest aggregations might arise because of (a) limited suitable nesting habitats (e.g. 222), (b) a tendency for offspring to nest in natal areas (e.g. 146), or (c) a tendency for individuals to nest near conspecifics (e.g. 64). The mechanisms responsible for gregarious behavior are poorly understood (see 38, 146, 223). Different species prefer different soil compositions, temperatures, or moisture levels (38, 146, 202). Moreover, some species favor vertical, and others horizontal, surfaces (references in 126, 138). However, in general, no obvious abiotic factors correlate with the locations of aggregations within a suitable habitat (38, 146).

For bees with persistent aggregations, an investigator can use genetic studies of population structure (e.g. 17) to determine whether hypothesis *b* or *c* applies. The mechanisms by which females learn the characteristics of their natal habitat upon emergence have not been studied, even though Fabre (71) long ago suggested they were important. Mandibular gland secretions may serve as aggregation pheromones (e.g. 59). Indeed, anecdotal observations suggest that, after initial colonization, ephemeral aggregations form via a self-organization process resulting from the increase in chemical attraction to a site as more females nest there (e.g. 64), analogous to the growth of urban populations (6a).

Evolutionary Benefits of Gregarious Behavior

The benefits associated with gregarious behavior are obscure and may differ for perennial and annual aggregations (for costs, see section on natural enemies of bees). One hypothesized benefit is that individuals form a so-called selfish-herd as cover from enemies (223). Available evidence is equivocal (reviewed in 172; also 54, 88). Whether bees gain information about foraging sites by following conspecifics from a central place, as known in some birds (e.g. 31), needs to be investigated.

RESOURCE SPECIALIZATION

Nonparasitic bees are prone to resource specialization. Factors associated with specialization in other phytophagous insects (e.g. detoxification) (reviewed in 14, 15, 84, 104a, 212) are less relevant for bees. Resource specialization by bees differs in its mechanisms, evolutionary causes, and ecological implications.

A few species specialize on a single species of pollen host (monolecty) (40, 66, 101, 152, 170, 187, 213). Because usually only one phenotypically similar plant congener is available locally (e.g. 170), these species may simply be without access to closely related hosts (also 118, 150). Where closely related pollen hosts bloom together, specialists typically use all of them (e.g. 217).

Specialist bees will occasionally provision nests with nonpreferred pollen if their host is temporarily unavailable (127, 161, 213). They then abandon alternative plants when the host blooms (43, 148). We do not know whether emergency resources are nutritionally adequate for larval development.

Phylogenetic Origins of Oligolecty

Without an independently derived phylogeny, it is impossible to discern whether oligolectic or polylectic habits are primitive or derived within a given taxon. Like specialization in other insects and cleptoparasitism among bees, oligolecty has multiple independent evolutionary origins.

TAXONOMIC DISTRIBUTION OF OLIGOLEGES   Pollen host specialization, which is unevenly distributed among higher bee taxa, is frequent in Rophitinae, Colletidae (other than Hylaeinae), Andrenidae, Melittidae, Megachilidae [including Fideliinae (236)], and nonsocial Apinae (118, 167). Most genera within these taxa contain both oligolectic and polylectic species (99). Polylecty is frequent in the Xylocopinae, Oxaeidae, Halictinae, presumably the Euglossini, and probably the Hylaeinae (118, 167). Pollen preferences of some colletids are not well known because these bees transport pollen internally; some euryglossines are oligolectic (96).

POLYLECTIC ANCESTRY   Some oligoleges, including *Ceratina sequoiae* (52) and several halictines (e.g. 21, 69), occur in otherwise polylectic taxa. These oligolectic taxa are presumably not basal within their clades, and thus their classification supports the usual supposition that specialists descend from generalists.

OLIGOLECTIC ANCESTORS ON RELATED HOSTS   Specialization on related hosts is sometimes conserved through multiple speciation events. Diverse genera exemplify evolutionarily conserved associations (Table 1), even across enormous geographic ranges (e.g. 166). Geographic ranges overlap for some species in these conservatively oligolectic genera (references in 102), an observation at odds with the tenets of competition exclusion.

OLIGOLECTY AND HOST SWITCHING   Another pattern for entirely oligolectic lineages is that, as a group, related taxa may specialize on several unrelated plant families; however, any one species may be more restricted in its resource utilization. Oligolectic bee genera associated with multiple host families include *Diadasia* (5 plant families) (123, 124, 158), *Melitta* (8+ plant families) (41, 205, 235), *Dufourea* (17 plant families) (25, 60, 66, 127, 128, 217, 235; T Griswold, personal communication), *Micralictoides* (4 plant families) (19), *Conanthalictus* (2 plant families) (188; T Griswold, personal communication), *Andrena* (*Diandrena*) (3 plant families) (213), *Hesperapis* (including *Capi-*

Table 1 Oligolectic bee taxa whose species share a common genus of pollen host[a]

Bee family	Genera	No. of species	Pollen host genera	Plant family	Reference
Colletidae	*Leioproctus (L. con-ospermj* sp. group)	3	*Conospermum*	Proteaceae	94
Andrenidae	*Andrena (Onagrandrena)*	24	*Oenothera, Cammisonia*	Onagraceae	214
Halictidae	*Lasioglossum (Sphecodogostra)*	8	*Oenothera*	Onagraceae	21, 106a, 106b
Melttidae	*Rediviva*	~8	*Diascia*	Scrophulariaceae	33, 201
	Macropis	14	*Lysimachia*	Primulaceae	166, 220, 239
Megachilidae	*Lithurge (Lithurgopsis)*	9	*Opuntia* and other cacti	Cactaceae	144, 198
Apidae	*Peponapis, Xenoglossa*	17	*Cucurbita*	Cucurbitaceae	100, 148
	Melitoma	4	*Ipomoea*	Convolvulaceae	126

[a] Several less-studied, small, and likely monophyletic genera of New World bees are likewise composed of oligoleges that share a common genus of pollen host. These include the rophitine bee genera *Sphecodosma* (*Nama*, Hydrophyllaceae) and *Xeralictus* (*Mentzelia*, Loasaceae); three subgenera of the panurgine genus *Calliopsis, Verbenapis* (*Verbena*, Verbenaceae), *Perissander* (*Euphorbia*, Euphorbiaceae), *Micronomadopsis* (*Trifolium*, Fabaceae); and several subgenera of *Perdita* (118). In addition, several Australian bee genera may be oligolectic for genera of the Myrtaceae (45).

cola) (4 plant families) (98, 128, 199, 205) and many subgenera of *Perdita* (99, 118). How did these disparate floral associations of oligolectic lineages originate?

Escape from competition is sometimes invoked as driving such radiations (references in 14, 212). Many oligoleges, however, belong to speciose pollen-collecting guilds with 75 or more bee species (e.g. 41, 100, 128, 235). Specialization has not opened empty niches but rather has channeled bees into bustling guilds. We hypothesize that the immediate ancestors to taxonomically diversified oligolectic bee lineages were themselves oligolectic. Mechanical factors such as pollen-grain size (e.g. 124) may underlie such switches, but chemical coevolution probably does not explain them (see below). Once established, new host associations may be enforced by two mechanisms: (*a*) the requisite match of phenology and habitat with the new host and (*b*) a tendency of male bees to search host flowers for receptive females (124). Both factors could promote fixation of the new host association once the switch occurred.

CONGRUENCE BETWEEN BEE AND PLANT PHYLOGENIES   Probably few host associations among oligolectic bees result from reciprocal evolution or cospeciation (sensu 28), but relevant data are lacking. Cospeciation between bees and flowers may be disfavored for two reasons. First, most oligolectic bees do not discriminate among congeneric hosts, and so they should readily colonize new sister taxa. Examples at the population level include a *Vaccinium* specialist that readily adopts another *Vaccinium* species outside its native range (44). Second, rates of host switching and evolutionary specialization by bees may outstrip speciation rates. In the >30,000 years since the progenitor of the desert shrub *Larrea tridentata* first appeared in the fossil record of North America (16), *L. tridentata* has come to dominate the warmer deserts. It now hosts 22 oligolectic bee species from 8 genera (100) that do not occur on South American *Larrea* species (142, 197) nor on any other member of the Zygophyllaceae (197a).

Adaptations for the Maintenance of Oligolecty

MORPHOLOGICAL SPECIALIZATIONS   Morphological features used for pollen collection and transport, such as branched setae, distinguish all bees from wasps. At a course level, oligolectic bees are usually not recognizable by characters other than their limited floral niches. Body size, for instance, seems unimportant.

Pollen harvest   The only known morphological innovations for pollen harvest by specialists consist of long hairs with modified tips. Such modifications, usually on the mouthparts or forelegs, have been reported for nine genera (19, 66, 94, 161, 170, 214). Sister taxa using other plants lack such hooked hairs (e.g. 161).

Pollen transport   Hair density and pilosity may conform to extreme pollen sizes. Bees that carry tiny, dry pollen (<25 µm in diameter) often have dense scopae of plumose hairs (170, 214). Stout, unbranched, sometimes fluted hairs typify scopae of bees that carry large pollen (>100 µm in diameter), or onagraceous pollen webbed together with viscin threads (94, 118, 170, 214).

Gathering nectar   Mouthparts of specialists may be differentially elongated or shortened to reach nectar concealed in tubular corollas or in shallow, open flowers (114, 195). Polylectic bees, however, also differ in proportional tongue lengths, as reflected in the informal divisions of short-tongued and long-tongued bees (see above) (for functional differences, see 85).

Gathering floral oils   The use of floral oils has engendered a rich diversity of prominent morphological specializations in as many as ten genera (references in 33, 201, 220). Bees that collect oils often possess conspicuous setal brushes and combs on their tarsi and sometimes on abdominal sterna. Some *Rediviva* species use remarkably elongate forelegs to probe the twinned, oil-secreting, elongate spurs of *Diascia* flowers, which is one of the more persuasive cases supporting morphological coevolution among bees and flowers (201). Only two genera, *Rediviva* and *Macropsis,* are oil specialists (references in 33, 201).

BEHAVIORAL SPECIALIZATIONS   Generalists, but usually not specialists, commonly visit flowers with complex corollas (231) that require prolonged trial-and-error learning for adept handling (e.g. 115). Conversely, specialists often utilize shallow flowers with easily accessible rewards (e.g. 60, 101, 154, 205, 235).

Floral sonication is a conspicuous and widespread behavioral adaptation used to harvest pollen from anthers that shed their pollen through terminal pores or slits (32, 214). Sonication can enhance pollen harvesting (JH Cane, in preparation). Many bees sonicate by shivering their flight muscles while contacting the stamens (32, 153, 214). Others drum, stroke, or milk anthers using their legs or mandibles (41; but see 153).

The propensity to sonicate anthers does not appear related to taxonomic classifications. However, there are a few patterns. For instance, tiny bees rarely sonicate anthers. Polyleges in diverse lineages do sonicate anthers using different mechanisms, but oligoleges rarely use novel means of sonication (39). We know little of the ontogeny and evolution of sonication, its mechanics (but see 109), or the advantages of different methods of floral sonication.

PHYSIOLOGICAL SPECIALIZATION   Nectars are aqueous solutions of simple sugars that vary in concentration and are readily evaluated by bees (references in 177). Pollen also varies widely in nutritional composition, including the content of proteins (169; JH Cane, unpublished data), lipids (56), starches (9), and sterols (72). Melittophilous pollen is more protein rich than anemophilous pollen (JH Cane, unpublished data), but some plants (e.g. *Helianthus* spp.) that attract many oligoleges (e.g. 100, 101) do not produce especially proteinaceous pollen (JH Cane, unpublished data).

Bee larvae might be expected to develop best on pollen preferentially collected by their mothers. Some evidence supports a preference-performance relationship; larvae matured faster, grew larger, and survived better on pollen from their species' preferred pollen hosts than on other pollens (e.g. 83). Conversely, *Megachile rotundata* larvae grew normally on nonhost pollens such as carrot and cranberry (VJ Tepedino, unpublished data; JH Cane, unpublished data). Larvae of several oligolectic species fed and grew on nonhost pollen as well (21, 178). Even anemophilous pollen constitutes a marginally satisfactory bee diet (82). Limited evidence suggests that oligoleges do not generally specialize because of larval dietary requirements or pollen nutritional quality.

Floral oils   Some bees incorporate floral oils into provision masses for larvae (references in 33, 220). Oils are calorically rich (175) and presumably less prone to spoilage than nectar. The physiological adaptations of bee larvae [including cleptoparasites (e.g. 186)] for diets of pollen or oil are unexplored.

Toxic pollen or nectar   One hypothesis for food specialization in insects, namely evolved counter-adaptations to host chemical defenses (references in 14, 15, 84, 104a, 212), does not generally apply to the floral-host associations of oligolectic bees. Plants that chemically defend their tissues and seeds (e.g. *Solanum, Larrea, Curcubita*) usually do not defend their pollen or nectar (e.g. 56, 84, 200). Hundreds of generalist or specialist bee species collect pollen and nectar from these plants (80, 100, 101, 125, 235). Furthermore, chemically defended plants often host proportionately more species of specialized herbivores than specialized species of bees (compare 192 with 101).

Some plants have toxic nectar or pollen (7, 9, 56, 84, 200). In general, these plants (e.g. *Ranunculus* spp.) are infrequently visited by bees, and few host oligoleges (125, 209, 235). Most bees simply do not visit plants with toxic nectar or pollen.

PHENOLOGICAL SPECIALIZATION   Bees are largely diurnal. Daily foraging activities, as for many insects, seem to be limited by temperature and illumination thresholds, as well as tolerance of wind and precipitation (50), but few experimental studies have tested these observations (117). Some taxa specialize in foraging at dawn, dusk, late afternoon (vespertine), or occasionally twice during the day (69, 119, 127, 128, 149, 155), and others forage at night (174, 238).

Diel patterns for provisioning often correspond with the daily initiation, cessation, or exhaustion of nectar or pollen production at preferred floral hosts (119, 149, 154, 155). A common pattern among plants is morning floral anthesis followed by the reduction of standing pollen crops, so early-arriving bees probably always do better than late comers. The first individual species of bee to arrive at a given pollen plant is called a matinal bee (119), and these bees can diminish the standing pollen crop by over a third by the time other bees arrive (JH Cane & SL Buchmann, unpublished data).

Diel foraging specializations have several anatomical and biogeographical correlates. Bees that fly at twilight or night have enlarged ocelli and somewhat paler coloration (references in 174). Matinal temperate bees are typically robust (119) with a thick pile of thoracic hair; both features are useful for thermoregulation (203). Bees of the lowland tropics (174) or warm deserts (118, 119) often exhibit matinal habits, and species of matinal genera retain their ancestral foraging habits in other biomes (102, 126).

Oligolectic species are mainly univoltine; brood production corresponds to the annual bloom of their hosts (e.g. 170, 235). However, these species will still produce only one brood even if the host blooms twice annually (e.g. 101). Although date of first arrival at flowers is commonly used for evaluating phenological synchrony, a better measure is the yearly correspondence between the onset of bloom and bee emergence (149, 158). Oligoleges that emerge early may seek nectar at other plant species yet will delay nesting or provisioning until their preferred hosts bloom (44, 61, 149, 205).

In seasonal habitats, plants often do not bloom during a drought. Bees sometimes skip emergence under these conditions, and facultatively remain in diapause (see section on natural history). Conversely, a normally univoltine desert oligolege was observed to have a second annual generation after a freak summer rain induced its host plant to bloom (98). Unfortunately, abnormal emergence schedules of oligoleges have not been compared with those of univoltine polyleges. Detection of any special relationship between oligolecty and phenological tracking of flowers awaits additional, longer-term studies of emergence phenologies of univoltine oligoleges and polyleges at common floral hosts.

HOST RECOGNITION   The conclusions regarding floral recognition and discrimination by bees have largely been drawn from the rich literature on nectar-gathering by social bees (see 58). In general, sensory systems do not differ substantially among bees, although a *Petunia* specialist has tetrachromatic photoreceptors rather than the trichromatic ones possessed by 26 other bee species, including another specialist (163). Spectral sensitivity functions of photoreceptors do not differ with respect to ecology (163), and the psychophysics of color coding may be universal in Hymenoptera (46). Specialists and generalists have similar densities of antennal chemosensilla (228).

Nevertheless, the relevance of generalizations from social bees to solitary bees is debatable, because the behavioral context is totally different: Foraging decisions of solitary bees are not influenced by the social needs of a colony. Many eusocial species stockpile food in their nests, so individual bees can temporally dissociate nectar and pollen foraging. Solitary bees, in contrast, usually do not stockpile food and must gather appropriate resources on a continual basis. The requisite discrimination capabilities of solitary bees are not well investigated. Only one study shows that solitary bees can recognize and respond to differences in the amount of pollen gained per flower (34). Diel foraging patterns (above) indicate that such capabilities may be more widespread.

Flower choice may be innate, or it might be learned from the scent of the bees' natal provision mass (119). Choice bioassays showed that inexperienced females of a local specialist (which is polylectic across its range) can discern the scent of their local floral host from others (55). Adult *M. rotundata* selected flowers of their preferred host, even when reared on a diet of carrot pollen and nectar. This observation suggests that host recognition is heritable (VJ Tepedino, unpublished data). Field-foraging oligoleges may be constant to host petal-color morphs, but such constancy for visual cues is not universal (57, 130). Additional choice experiments are needed with naive foraging oligoleges before we can understand the origins and mechanisms of host recognition.

Pathways to Oligolecty

REPRODUCTIVE COSTS AND BENEFITS   Oligolectic bees often forage at plants that have numerous other floral visitors. Observations that females complete fewer nest cells when resources are scarce, or when abundance fluctuates on daily or seasonal time scales, point to the costs of specialization (e.g. 20, 21, 34, 117, 149, 154, 158). Oligolectic bees are sometimes more proficient than competing polyleges in acquiring pollen from host flowers (43, 204). Oligoleges that use *Vaccinium* spp. (JH Cane, unpublished data) acquire as much pollen per flower as much larger polylectic bumble bees, as true for oligoleges on *Pontederia cordata* (86). Another *Vaccinium* spp. oligolege (*Habropoda laboriosa*) harvests as much pollen per flower as polylectic competitors, but it works faster (43).

Bee species evolve oligolecty via two non–mutually exclusive paths that relate to the relative abundance of floral resources. For convenience we name these the predictable-plethora and the restricted-resource pathways.

PREDICTABLE PLETHORA   Many cases of oligolecty likely arose following this pathway. Plethoric plants are characterized by the following attributes: (*a*) production of large quantities of accessible pollen and nectar; (*b*) local patches that persist because either the plant is perennial or, if annual, it reseeds locally; (*c*) annual blooms that are reliable relative to nonhosts; and (*d*) blooms that coincide with local bee activity. Such plants (e.g. *Helianthus* spp.) are termed apparent (73). Their pollen is nutritionally adequate but may be unremarkable. These plants host taxonomically diverse guilds rich in polylectic and oligolectic bees, and such guilds may actually intensify competition among bees (170). Indeed, plants that produce a predictable plethora of pollen and nectar may act as evolutionary attractors for specialist bees. As bees are central-place foragers, if they must travel to sparse or widely spaced host patches, they will complete relatively few nest cells. Bees specialized for locally reliable flowers can devote more time to provisioning, especially when their emergence or activity coincides with peak bloom.

Bees are selective nectar foragers, but we do not know whether they recognize the nutritional qualities of pollen. Some social bees feed larvae progressively, such that hungry larvae fed deficient pollen solicit more food (219).

Most bees store a mass of provisions for their progeny and never receive direct larval feedback. Oligolectic females disregard interspecific heterogeneity in pollen quality, but we do not know whether bees that use nutritionally poor pollen compensate for the low quality by providing higher quantities, analogous to the way most bees match provision mass to sex of offspring (91). If adult bees cannot compare the nutritional value of different pollen, then taxonomic fidelity would help them gather resources adequate to support their own development, and polylecty becomes the strategy that requires evolutionary explanations.

RESTRICTED RESOURCES   Some specialists seek out hosts that are minor elements of floral communities. Such bee guilds are more depauperate, with few generalists and perhaps less competition. These specialists are more likely to possess derived traits for harvesting or using host resources. Host associations are conserved (Table 1), and some may be relictual (166). This pathway mirrors the "biochemical arms race" (212) herbivory model, except that dietary restrictions are generally governed not by antagonistic host interactions, but rather by mutual benefits of host-pollinator relationships.

NATURAL ENEMIES OF BEES

Bees store floral resources for varying lengths of time. Animals that hoard food have many enemies, and bees are no exception (218). Although the amount of food stored in a given nest will be small, an aggregation can collectively contain large quantities (compare 149 with 237). While a solitary bee forages, her nest cells remain unprotected, unlike those of social bees (140).

The known enemies of bees are too numerous to list. However, an excellent survey of West German bees illustrates their diversity (235; also 99), which embodies the following parasitic, commensal, predatory, or scavenging associates: fungi, protozoa, nematodes, thomisid spiders, diverse mites, earwigs (Dermaptera), beetles (Coleoptera, 5 families), flies (Diptera, 6 families), wasps (11 families), bees (3 families), ants, birds, and mammals (for reviews of their biology, see 29, 48, 63, 68, 78, 79, 93, 97, 106, 108, 121, 132, 224). Parasites are hypothesized to regulate bee population size, but quantitative data are scarce (see 116, 122, 206, 222, 229).

Foraging Behavior, Habitat Preference, and Parasitism Rates

PROVISIONING SPECIALIZATION   Modes of provisioning behavior influence bees' exposure to natural enemies, as does the foraging activity of other phytophagous insects (see 14). Polylectic behavior may increase mortality from natural enemies encountered on flowers [e.g. some meloid and rhipiphorid beetles (121, 129)]. Conversely, specialization may allow such bees to forage in enemy-free space. Female meloids, for example, were placed in a greenhouse with nesting bees and plants from seven genera. Beetle larvae were recovered only from flowers of *Borago* spp., and the only parasitized bee nests contained *Borago* pollen (216). At this locality, pollen specialists restricted to other plants would be free of this parasite, while *Borago* visitors would be parasitized (see also 123).

NESTING PREFERENCE   The choice of nesting location partially determines which suite of enemies will attack the nest. Some meloid beetles, for example, oviposit in cells or in nest tunnels. Their larvae attach themselves to bees active at those sites, restricting the beetles to gregariously nesting bees (68, 118, 120). Gregarious behavior presumably has the disadvantage of concentrating enemies in a local area (see 118, 223). Some anecdotal observations show that, on average, nests in aggregations are more heavily parasitized than isolated ones (e.g. 61, 226). Conversely, other studies show that 100% of the cells of isolated nests of several species can be parasitized (e.g. 111). Aggregations can be ephemeral or persistent (see section on overview of natural history), but we do not know whether persistent ones are more likely to support large parasite populations.

   This frequently discussed relationship between tendency to aggregate nests and rates of parasitism is difficult to assess for several reasons. First, the number of taxa that obligately nest in aggregations relative to those that must nest in isolation is unknown. Aggregated nests are easier to locate and are probably more frequently reported. To address this problem, we need intraspecific comparisons among the many facultatively gregarious species (e.g. 61, 126, 230). Second, the spatial scale used to define an aggregation may not be relevant to the parasites (see 173). Third, many data on parasitism rates are point estimates, and the few long-term studies that have been done demonstrate how such data may not be representative. For a hole-nesting *Osmia* species, the rates of cell parasitism by a meloid beetle were usually less than 1% from 1974 to 1989, with a maximum of 3.7% in 1976. During a drought in 1990, however, parasitism rates increased to almost 33% (216). Within a season, rates can be equally variable (e.g. 189). Fourth, comparative studies often combine different host populations or taxa, which can have different numbers of specialized parasites (see 28).

   Michener (143) hypothesized that nests distributed in two dimensions (those of ground-nesters) should be more heavily parasitized than those distributed in three dimensions (those of twig-nesters). The hypothesis incorporates considerations from foraging theory and makes two assumptions about information acquisition and processing (15): (*a*) Objects distributed in two dimensions are easier to locate than those distributed in three dimensions, and (*b*) parasites search in both kinds of habitats. Assumption *b* is true for some parasites but not for others, including intraspecific parasites (summaries in 75, 135, 174, 224), and parasites presumably have sophisticated host-searching behaviors. Whether habitat specialists counterbalance each other in tallies of total rates of parasitism is unknown.

Available data do not support Michener's hypothesis: The mean rate of parasitized cells (x) for 27 ground-nesting species was 29.9% (SE = 4.6), which did not differ significantly from the rate for 25 twig- or mud-nesting species (x = 29.2%) (WT Wcislo, submitted). Within one species, ground nests had 51.8% cells parasitized, while nests in twigs had approximately 36% cells parasitized, although these nests were at different localities (162). Many data on twig-nesting bees came from studies of trap nests, which artificially create aggregations (see 111), and parasitism rates may be lower under more natural nest densities. Proper tests should be based on comparisons within closely related taxa, such as twig- vs ground-nesting megachilids, xylocopines, or colletids (45, 67, 103), and should use data only from natural enemies that search in both habitats (e.g. some miltogrammine flies).

Strategies Against Natural Enemies

ACTIVE DEFENSE   Most solitary bees do not aggressively attack natural enemies that wander too close to nest entrances (e.g. 95, 202), except for intraspecific parasites (see below). Active defenses are subtle and include opening cells for inspection and packing those containing fungi with soil (10). This behavior may decrease offspring mortality rates, as suggested for some spheciforme wasps (70).

PASSIVE DEFENSES   Passive defenses involve physical and chemical features of nest architecture. Some solitary bees, especially those that do not nest in aggregations, often situate their nest entrances in concealed locations (e.g. under rocks, within clumps of plants). An *Osmia* species, for example, nests in a snail shell it drags to a depression and covers with debris (74). Constricted nest entrances and turrets may reduce parasitism (126, 138, 202). Nests of some solitary bees sometimes contain empty cells interspersed between provisioned cells (190, 230), which would theoretically reduce rates of parasitism (210).

The cells of most bees are lined with chemicals synthesized in the abdominal (metasomal) Dufour's gland (59, 87). The chemical composition of these secretions, which differs among taxa, includes macrocyclic lactones, terpenes, acetates, and alcohols. Chemical characterization of these compounds has greatly outpaced understanding of their functional significance. Allodapine and ceratinine bees have mandibular-gland compounds with repellent properties against ants and other enemies (references in 37, 59). Certain mandibu-lar-gland compounds have germicidal properties (references in 42, 59), and they may help fumigate nests, because secretions can be released during biting motions of the kind made during nest excavations. Some megachilid, eu-glossine, and centridine bees incorporate plant resins into nest-cell linings. Resins physically defend plants and disinfect their wounds (113), and bees may gain similar benefits (136; JH Cane, unpublished data; G Gilbert, WS Armbruster & DW Roubik, unpublished data).

Parasitic Bees and Condition-Sensitive Behavior

Close neighbors are often a bee's worst enemy: Both opportunistic (facultative) parasitism (75, 224) and permanent (obligate) parasitic behavior (18, 224) are widespread. Thus, this array of parasitic specializations provides an excellent model to study the evolution of condition-sensitive behavior (cf 224, 233, 234).

PHYLETIC DISTRIBUTION   Approximately 15% of bee genera or subgenera contain obligate parasites of other bees (167, 224). A conservative estimate would be that obligate parasitism has evolved independently at least 26 times (18, 167, 224). Parasitic bees are concentrated in the Apidae (and especially in the Nomadinae), Halictidae, and Megachilidae. The number of species in a taxon seems unrelated to the number of parasitic lineages generated by that taxon. For example, speciose families like Andrenidae lack parasitic bees, and a small taxon, Ctenoplectrini, contains a parasitic species.

Obligate parasitism is also phyletically biased within families. In Halictidae, for example, parasitic lineages occur in the Halictinae, but not in the Rophitinae or Nomiinae (141). Within Halictinae, in turn, the Nomioidini contain no known parasites, but parasitic behavior has evolved at least eight times in the cosmopolitan Halictini. Some lineages (e.g. *Sphecodes*) contain numerous parasitic species (141; WT Wcislo, in preparation). In the equally large but primarily neotropical tribe Augochlorini, parasitism has purportedly evolved at least three times (147; RW Brooks, personal communication), but no be-havioral data confirm the parasitic status of these species.

HOST-PARASITE RELATIONSHIPS   Lineages of parasitic bees have long been hypothesized to be related to, and derived from, their host lineages (18, 151, 224). This heuristic relationship is named Müller's Law (165), or Emery's Rule when applied to ants and wasps (e.g. 93). Phylogenetic studies are not yet numerous enough for statistical tests of this hypothesis, but it is valid in some cases, and certainly invalid for lineages with many host shifts (e.g. many nomadines) (references in 171).

Parasites differ in the degree of host specificity, even among congeners (e.g.

182; for other host associations, see 3, 4, 141, 167, 174, 184, 185, 202, 235).
Apparent specificity may simply represent lack of information. Cladistic studies have demonstrated generic-level host shifts for various parasites (e.g. 3, 4,
182). Hosts shifts, like some parasite-host associations (e.g. 193), tend to be
habitat specific. A few parasitic bees regularly associated with twig- or ma-
son-nesting bees have shifted to ground-nesting bees, and vice versa (e.g. 11).
L-T bees have expanded their host ranges to parasitize S-T bees, yet no S-T
bee, not even the large, cosmopolitan genus *Sphecodes,* parasitizes L-T bees
(WT Wcislo, unpublished data). The historical biogeography of bees (142) is
not well-enough known to assess whether this pattern is related to the relative
ages of these groups.

Some speciose taxa appear to be immune to bee parasites (167; WT Wcislo,
unpublished data). No known hosts occur in Stenotritidae, Xeromelissinae,
Euryglossinae, Diphaglossini, possibly Hylaeini [there is an unconfirmed report (241) of parasitic Hylaeini in Hawaii, which purportedly parasitize other
hylaeines], Ceratinini, Xylocopini, Manuelini, Apini (excluding robber species—see 174, 224), Fideliinae, or Meganomiinae. Many of these taxa are
abundant in Australia (45), which may partially account for their apparent
immunity because the rich Australian bee fauna is curiously depauperate in
parasites (except Allodapini) (224). A detailed comparative study of immune
and susceptible taxa may lend insight into the evolution of parasitic bees.

HOST RECOGNITION, ASSESSMENT, AND NEST-ENTERING BEHAVIOR   The mechanisms by which parasites recognize their host(s) have been studied in only
one species, and this report showed the importance of olfactory evaluation
(36). In addition, adult *Stelis montana* are known to ascertain whether a host
cell is suitably provisioned (215), but most species have not been studied. The
mechanisms for entry into a nest also are not well studied. Some parasites
aggressively enter nests (194), while others avoid contact with a host (e.g. 1,
194, 215).

Parasitic *Nomada* spp. males have an odor bouquet similar to that of host
females; hypothetically, these males transfer this odor to conspecific females
during mating, thus enabling the parasitic female to more easily enter the host
nest (references in 59). No behavioral data are available to support these
hypotheses. No evidence supports chemical mimesis in other parasitic bees
(see 194, 208).

Many parasites place their eggs in concealed locations within cells (representative illustrations are given in 167, 174, 202). Most nomadines oviposit in
a cell wall before the cell is fully provisioned (183). This hiding is common
in parasites. *Stelis* species oviposit within the host pollen mass, and *Coelioxys*
species oviposit into the leaf mass that surrounds the cell of its megachilid
host. When parasite larvae (e.g. Nomadinae) are in the same cell as living host

eggs or larvae, the parasites destroy the host larvae. For example, some parasite larvae undergo hypermetamorphosis to a "killer" instar, which actively crawls about the host cell and kills the host egg or larva, as well as other parasite larvae (180, 181). After dispatching its competitors, the killer larva molts to a typical grub-like bee larva. In other species, the adult parasite destroys the host egg or young larva. The parasite larvae of these taxa are not morphologically differentiated from host larvae (137) and are not active crawlers.

CONDITION-SENSITIVITY AND THE EVOLUTION OF PARASITISM   Changes in context-dependent expression are increasingly recognized as a basic feature in the evolution of diversity (e.g. 225, 233, 234). The perceptual mechanisms that animals use to assess local conditions and implement alternative behaviors are poorly understood (see 15). The frequent evolution of obligate parasitism from facultative behavior in bees allows us to redress this situation.

Facultative parasitism   Facultative parasitic behavior is generally thought to represent an evolutionary precursor to obligate parasitism (e.g. 18, 75, 224, 233). Kaitala et al (105) argue the two phenomena are unrelated, presupposing that pollen-collecting structures are lost concomitant with the origin of parasitic behavior. This presupposition is inconsistent with facts. Among parasitic *Paralictus,* for example, different species have lost pollen-collecting structures to differing degrees (141; WT Wcislo, in preparation), which shows that behavioral and morphological evolution can be uncoupled.

   Factors that induce facultative parasitic behavior may include a failure to resorb developing oocytes (references in 224). A shortage of nest sites has been widely implicated in facultative parasitism in nest-making animals (references in 224), but its relative importance is unknown for bees. A shortage of nest sites, or loss of a nest because of predation or other factors, may be less important in lineages that have open, communal groups (1, 53, 112, 179, 227) because a female could join another group. Joining behavior (but not necessarily in relation to nest loss) has been confirmed by genetic studies (53, 112). The open nature of communal living possibly helps explain why lineages with widespread communal behavior (e.g. Andrenidae, Agapostemonini) have not generated many parasitic lineages.

Obligate parasitism and expression of morphological novelty   Obligately parasitic female bees frequently (but to differing degrees) show a suite of external characters putatively related to their parasitic behavior (e.g. loss or reduction of pollen-collecting structures, loss of a toothed mandible, reinforced exoskeleton, changes in proportion of antennal structures, bright or metallic coloration) (see 141, 224, 228). Nothing is known about the evolutionary origins or development of such structures.

Some external features (e.g. loss or reduction of pollen-collecting structures, loss of toothed mandible, relative lengths of the antennal scape and flagellum) seem to represent the masculinization of the female phenotype (141, 228). Sex transfers of phenotypic development patterns have been discussed for other insects (e.g. 47, 107). Strepsiptera parasitism (191) and other pathologies (240) can induce various degrees of masculinization. In *Andrena vega* stylopization decreases the volume of the corpora allata (26) and probably reduces titers of juvenile hormone (JH). Bonetti & Kerr (23) show that male larvae treated with topical JH are feminized as adults. Thus, studies of stylops and other pathogens reveal the potential of developmental systems to express preexisting traits under novel conditions.

Obligate parasites have also undergone evolutionary changes in ovarian physiology and morphology (references in 6). Parasitic bees have either a greater number of oocytes per ovariole or a greater number of ovarioles per ovary, relative to the putative ancestors. Associated with this increase in fecundity is a relative decrease in egg size. Rozen (183) hypothesized that the smaller size of parasite eggs is related to the need to hide eggs within a host cell. Smaller egg size is also expected from the decreased parental investment per egg, according to arguments associated with r- and K-selection (131).

SUMMARY

The spatial distribution of bees influences interactions with parasites and intraspecific competitors for food. The study of many resource-specializing (oligolectic) bees allows us to make illuminating comparisons with other phytophagous insects, because bees provide an indirect service (pollination) to the plant, unlike herbivores. Hypotheses proposed to explain resource specialization among other phytophagous insects are not generally applicable to bees. Instead, diverse oligolectic (and generalist) bees utilize resources from flowers that are predictably abundant in time and space. Apart from phenological matching, these oligolectic bees usually do not have striking behavioral, physiological, or anatomical modifications associated with their specialized behaviors. Some cases of bee specialization, however, are probably explained by analogy to the biochemical arms race models proposed for herbivores. Bees that are oligolectic on rare and patchily distributed plants often possess specific behavioral or anatomical keys that allow them access to pollen, nectar, or oil that a plant has locked away. Resources stored by bees present a rich target for natural enemies. Natural history observations suggest these enemies have shaped bees' foraging behavior, nest architecture, and nest-site selection. Opportunistic, facultative parasitism is common in bees, and obligate parasitism has repeatedly evolved. The convergent phenotypes of parasitic bees yield insights into the relationship between the evolution of condition-sensitive behavior and morphology and the expression of novel combinations of traits. Comparative studies of bees provide exceptional opportunities to those interested in merging natural history studies with research addressing underlying mechanisms. We know a good deal about what bees do, but we know little about how they do it, and therefore we can only speculate about why. Significant advances in understanding "why" questions are likely to come from those areas that address the "how" questions.

ACKNOWLEDGMENTS

A collaborator, George Eickwort, was killed in an automobile accident during the planning stages of this review. We dedicate this paper to his memory. WTW thanks J Bronstein, M Engel, D Roubik, and D Yanega for comments on sections of the manuscript, discussions, or publications; R Chapman for encouragement at the outset; F Wcislo for help with Russian translations; D Conlon for help with editorial matters; and the Smithsonian Tropical Research Institute for general research funds. JHC is grateful for the insights, experience, and criticisms generously shared by RL Minckley, CD Michener, PF Torchio, VJ Tepedino, T Griswold, BA Alexander, JG Rozen, Jr., T Roulston, and RW Thorp.

Literature Cited

1. Abrams J, Eickwort GC. 1981. Nest switching and guarding by the communal sweat bee *Agapostemon virescens* (Hymenoptera: Halictidae). *Insectes Soc.* 28:105–16

2. Alcock J, Jones CE, Buchmann SL. 1976. The nesting behavior of three species of *Centris* bees (Hymenoptera: Anthophoridae). *J. Kans. Entomol. Soc.* 49:469–74

3. Alexander BA. 1990. A cladistic analysis of the nomadine bees (Hymenoptera: Apoidea). *Syst. Entomol.* 15:121–52

4. Alexander BA. 1991. Nomada phylogeny reconsidered (Hymenoptera: Anthophoridae). *J. Nat. Hist.* 25:315–30

5. Alexander BA. 1992. An exploratory analysis of cladistic relationships within the superfamily Apoidea, with special reference to sphecid wasps (Hymenoptera). *J. Hymenoptera Res.* 1:26–61

6. Alexander BA, Rozen JG Jr. 1987. Ovaries, ovarioles, and oocytes in parasitic bees (Hymenoptera: Apoidea). *Pan-Pac. Entomol.* 63:155–64

6a. Allen PM, Sanglier M. 1978. Dynamic model of urban growth. *J. Soc. Biol. Struct.* 1:265–80

7. Atkins EL. 1992. Injury to honey bees by poisoning. In *The Hive and the Honey Bee,* ed. JM Graham, pp. 13–21. Hamilton, IL: Dadant & Sons. 6th ed.

8. Ayala R, Griswold TL, Bullock SH. 1993. The native bees of México. In *Biological Diversity of México, Origins and Distributions,* ed. TP Ramamoorthy, R Bye, A Lot, J Fa, pp. 179–225. Sunderland, MA: Sinauer. 812 pp.

9. Baker HG, Baker I. 1979. Starch in angiosperm pollen grains and its evolutionary significance. *Am. J. Bot.* 66:591–600

10. Batra SWT, Bohart GE. 1970. Alkali

bees: response of adults to pathogenic fungi in brood cells. *Science* 165:607–8

11. Batra SWT, Schuster JC. 1977. Nests of *Centris, Melissodes*, and *Colletes* in Guatemala (Hymenoptera: Apoidea). *Biotropica* 9:135–38

12. Bawa KS. 1990. Plant-pollinator interactions in tropical rain forests. *Annu. Rev. Ecol. Syst.* 21:399–422

13. Bennett B, Breed MD. 1985. The nesting biology, mating behavior and foraging ecology of *Perdita opuntiae* (Hymenoptera: Andrenidae). *J. Kans. Entomol. Soc.* 58:185–94

14. Bernays EA, Chapman RF. 1994. *Host-Plant Selection by Phytophagous Insects*. New York: Chapman & Hall. 312 pp.

15. Bernays EA, Wcislo WT. 1994. Sensory capabilities, information processing, and resource specialization. *Q. Rev. Biol.* 69:187–204

16. Betancourt JL, Van Devender TR, Martin PS, eds. 1990. *Packrat Middens: the Last 40,000 Years of Biotic Change*. Tucson: Univ. Ariz. Press. 467 pp.

17. Blows MW, Schwarz MP. 1991. Spatial distribution of a primitively social bee: Does genetic population structure facilitate altruism? *Evolution* 45:680–93

18. Bohart GE. 1970. The evolution of parasitism among bees. *Utah State Univ. Fac. Honor Lec.* 41:1–30

19. Bohart GE, Griswold TL. 1987. A revision of the Dufoureine genus *Micralictoides* Timberlake (Hymenoptera: Halictidae). *Pan-Pac. Entomol.* 63:178–93

20. Bohart GE, Youssef NN. 1972. Notes on the biology of *Megachile (Megachiloides) umatillensis* Mitchell (Hymenoptera: Megachilidae) and its parasites. *Trans. R. Entomol. Soc. London* 124:1–19

21. Bohart GE, Youssef NN. 1976. The biology and behavior of *Evylaeus galpinsiae* Cockerell (Hymenoptera: Halictidae). *Wasmann J. Biol.* 34:185–234

22. Bohart RM, Menke AS. 1976. *Sphecid Wasps of the World*. Berkeley: Univ. Calif. Press. 695 pp.

23. Bonetti AM, Kerr WE. 1985. Sex determination in bees. XX. Estudo da açao genica em *Melipona marginata* e *Melipona compressipes* a partir de analise morfométrica. *Rev. Brasil. Genet.* 4:629–38

24. Boucot AJ. 1990. *Evolutionary Palaeobiology of Behavior and Coevolution*. Amsterdam: Elsevier. 725 pp.

25. Bouseman JK. 1976. *Dufourea monardae* (Viereck) in Illinois and Michigan

(Hymenoptera: Apoidea). *J. Kans. Entomol. Soc.* 49(4):531–32

26. Brandenburg J. 1956. Das endokrine system des Kopfes von Andrena vega Pz. (Ins. Hymenopt.) und Wirkung der Stylopisation (Stylops, Ins. Strepsipt.). *Z. Morphol. Oekol. Tierre* 45:343–64

27. Bronstein JL. 1995. The plant/pollinator landscape. In *Mosaic Landscapes and Ecological Processes*, ed. L Fahrig, L Hansson, G Merriam. New York: Chapman & Hall. In press

28. Brooks DR, McLennan DA. 1993. *Parascript: Parasites and the Language of Evolution*. Washington, DC: Smithson. Inst. 429 pp.

29. Brothers DJ. 1972. Biology and immature stages of *Pseudomethoca frigida*, with notes on other species (Hymenoptera: Mutillidae). *Univ. Kans. Sci. Bull.* 50:1–38

30. Brothers DJ, Carpenter JM. 1993. Phylogeny of Aculeata: Chrysidoidea and Vespoidea. *J. Hymenoptera Res.* 2:227–301

31. Brown CR. 1986. Cliff swallow colonies as information centers. *Science* 234:83–5

32. Buchmann SL. 1983. Buzz pollination in angiosperms. See Ref. 104b, pp. 73–113

33. Buchmann SL. 1987. The ecology of oil flowers and their bees. *Annu. Rev. Ecol. Syst.* 18:343–69

34. Buchmann SL, Cane JH. 1989. Bees assess pollen returns while sonicating *Solanum* flowers. *Oecologia* 81:289–94

35. Callan E McC. 1977. Observations on *Centris rufosuffusa* Cockerell (Hymenoptera: Anthophoridae) and its parasites. *J. Nat. Hist.* 11:127–35

36. Cane JH. 1983. Olfactory evaluation of *Andrena* host nest suitability by kleptoparasitic *Nomada* bees. *Anim. Behav.* 31:138–44

37. Cane JH. 1986. Predator deterrence by mandibular gland secretions of bees (Hymenoptera: Apoidea). *J. Chem. Ecol.* 12:1295–309

38. Cane JH. 1992. Soils of ground-nesting bees (Hymenoptera: Apoidea): texture, moisture, cell depth and climate. *J. Kans. Entomol. Soc.* 64:406–13

39. Cane JH, Buchmann SL. 1989. Novel pollen-harvesting behavior by the bee *Protandrena mexicanorum* (Hymenoptera: Andrenidae). *J. Insect Behav.* 2:431–36

40. Cane JH, Eickwort GC, Wesley FR, Spielholz J. 1983. Foraging, grooming and mate-seeking behaviors of *Macropis nuda* (Hymenoptera, Melittidae) and use of *Lysimachia ciliata* (Primulaceae) oils

in larval provisions and cell linings. *Am. Midl. Nat.* 110:257–64

41. Cane JH, Eickwort GC, Wesley FR, Spielholz J. 1985. Pollination ecology of *Vaccinium stamineum* (Ericaceae: Vaccinioideae). *Am. J. Bot.* 72:135–42

42. Cane JH, Gerdin S, Wife G. 1983. Mandibular gland secretions of solitary bees (Hymenoptera: Apoidea): potential for nest cell disinfection. *J. Kans. Entomol. Soc.* 56:199–204

43. Cane JH, Payne JA. 1988. Foraging ecology of the bee *Habropoda laboriosa* (Hymenoptera: Anthophoridae), an oligolege of blueberries (Ericaceae: *Vaccinium*) in the southeastern United States. *Ann. Entomol. Soc. Am.* 81:419–27

44. Cane JH, Payne JA. 1993. Regional, annual and seasonal variation in pollinator guilds: intrinsic traits of bees (Hymenoptera: Apoidea) underlie their patterns of abundance at *Vaccinium ashei* (Ericaceae). *Ann. Entomol. Soc. Am.* 86:577–88

45. Cardale JC. 1993. *Zoological Catalogue of Australia,* Vol. 10, *Hymenoptera: Apoidea,* ed. WWK Houston, GV Maynard. Canberra: AGPS. 406 pp.

46. Chittka L, Beier W, Hertel H, Steinmann, Menzel R. 1992. Opponent colour coding is a universal strategy to evaluate the photoreceptor inputs in Hymenoptera. *J. Comp. Physiol. A* 170:545–63

47. Clarke C, Clarke FMM, Collin SC, Gill ACL, Turner JRG. 1985. Male-like females, mimicry, and transvestism in butterflies (Lepidoptera: Papilionidae). *Syst. Entomol.* 10:257–83

48. Clausen CP. 1940. *Entomophagous Insects.* New York: McGraw-Hill. 688 pp.

49. Cockerell TDA. 1933. The excessive abundance of certain bees. *Am. Nat.* 67:1–3

50. Corbet SA. 1990. Pollination and the weather. *Isr. J. Bot.* 39:13–30

51. Cripps C, Rust RW. 1989. Pollen preferences of seven *Osmia* species (Hymenoptera: Megachilidae). *Environ. Entomol.* 18:133–38

52. Daly HV. 1973. Bees of the genus *Ceratina* in America north of Mexico (Hymenoptera: Apoidea). *Univ. Calif. Publ. Entomol.* 74:1–114

53. Danforth BN, Neff JL, Barretto-Ko P. 1995. Nestmate relatedness in a communal bee, *Perdita texana,* (Hymenoptera: Andrenidae), based on DNA fingerprinting. *Evolution.* In press

54. Danforth BN, Visscher PK. 1993. Dynamics of a host-cleptoparasite relationship: *Holcopasites ruthae* as a parasite of *Calliopsis pugionis* (Hymenoptera:

55. Anthophoridae, Andrenidae). *Ann. Entomol. Soc. Am.* 86:833–40

55. Dobson HEM. 1987. Role of flower and pollen aromas in host-plant recognition by solitary bees. *Oecologia* 72:618–23

56. Dobson HEM. 1988. Survey of pollen and pollenkitt lipids: chemical cues to flower visitors? *Am. J. Bot.* 75:170–82

57. Dobson HEM. 1994. Floral volatiles in insect biology. In *Insect-Plant Interactions,* ed. EA Bernays, 5:47–81. Boca Raton, FL: CRC

58. Dukas R, Real LA. 1993. Cognition in bees: from stimulus reception to behavioral change. In *Insect Learning,* ed. DR Papaj, AC Lewis, pp. 343–73. New York: Chapman & Hall

59. Duffield RM, Wheeler JW, Eickwort GC. 1984. Sociochemicals of bees. In *Chemical Ecology of Insects,* ed. WJ Bell, RT Cardé, pp. 287–428. Sunderland, MA: Sinauer

60. Eckhart VM. 1992. Spatio-temporal variation in abundance and variation in foraging behavior of the pollinators of gynodioecious *Phacelia linearis* (Hydrophyllaceae). *Oikos* 64:573–86

61. Eickwort GC. 1973. Biology of the European mason bee, *Hoplitis anthocopoides* (Hymenoptera: Megachilidae), in New York state. *Search* 3:1–31

62. Eickwort GC. 1981. Presocial insects. In *Social Insects,* ed. HR Hermann, 2:199–280. New York: Academic

63. Eickwort GC. 1993. Evolution and life-history patterns of mites associated with bees. In Mites: Ecological and Evolutionary Analyses of Life History Patterns, ed. M. Houck, pp. 218–51. New York: Chapman & Hall

64. Eickwort GC, Eickwort KR, Linsley EG. 1977. Observations on nest aggregations of the bees *Diadasia olivacea* and *D. diminuta* (Hymenoptera: Anthophoridae). *J. Kans. Entomol. Soc.* 50:1–17

65. Eickwort GC, Ginsberg HS. 1980. Foraging and mating behavior in Apoidea. *Annu. Rev. Entomol.* 25:421–46

66. Eickwort GC, Kukuk PF, Wesley FR. 1986. The nesting biology of *Dufourea novaeangliae* (Hymenoptera: Halictidae) and the systematic position of the Dufoureinae based on behavior and development. *J. Kans. Entomol. Soc.* 59:103–20

67. Eickwort GC, Matthews RW, Carpenter J. 1981. Observations on the nesting behavior of *Megachile rubi* and *M. texana* with a discussion of the significance of soil nesting in the evolution of megachilid bees (Hymenoptera: Megachilidae). *J. Kans. Entomol. Soc.* 54:557–70

68. Erickson EH, Enns WR, Werner FG. 1976. Bionomics of the bee-associated Meloidae (Coleoptera): bee and plant hosts of some Nearctic beetles—a synopsis. *Ann. Entomol. Soc. Am.* 69:959–70

69. Estes JR, Thorp W. 1975. Pollination ecology of *Pyrrhopappus carolinianus* (Compositae). *Am. J. Bot.* 62(2):148–59

70. Evans HE, West Eberhard MJ. 1970. *The Wasps.* Ann Arbor: Univ. Mich. Press. 265 pp.

71. Fabre JH. (1915) 1920. *Bramble-bees and Others.* New York: Dodd, Mead. 456 pp. (Transl.)

72. Feldlaufer MF, Buchmann SL, Lusby WR, Weirich GF, Svoboda JA. 1993. Neutral sterols and ecdysteroids of the solitary cactus bee *Diadasia rinconis* Cockerell (Hymenoptera: Anthophoridae). *Arch. Insect Biochem. Physiol.* 23: 91–98

73. Feeny P. 1976. Plant apparency and chemical defense. *Adv. Phytopathol.* 10: 1–40

74. Ferton C. 1923. *La Vie des Abeilles et des Guêpes,* ed. E Rabaud, F Picard. Paris: Étienne Chiron. 376 pp.

75. Field J. 1992. Intraspecific parasitism as an alternative reproductive tactic in nest-building wasps and bees. *Biol. Rev.* 67:79–126

76. Frankie GW, Newstrom L, Vinson SB, Barthell JF. 1993. Nesting-habitat preference of selected *Centris* bee species in Costa Rican dry forest. *Biotropica* 25:322–33

77. Gauld ID, Bolton B, eds. 1988. *The Hymenoptera.* Oxford: Oxford Univ. Press. 332 pp.

78. Giblin-Davis RM, Norden BB, Batra SWT, Eickwort GC. 1990. Commensal nematodes in the glands, genitalia, and brood cells of bees (Apoidea). *J. Nematol.* 22:150–61

79. Gilliam M, Buchmann SL, Lorenz, Schmalzel RJ. 1990. Bacteria belonging to the genus *Bacillus* associated with three species of solitary bees. *Apidologie* 21:99–105

80. Graenicher S. 1909. Wisconsin flowers and their pollination. *Bull. Wisc. Nat. Hist. Soc.* 7:19–77

81. Grant V. 1994. Modes and origins of mechanical and ethological isolation in angiosperms. *Proc. Natl. Acad. Sci. USA* 91:3–10

82. Greenberg L. 1982. Year-round culturing and productivity of a sweat bee, *Lasioglossum zephyrum* (Hymenoptera: Halictidae). *J. Kans. Entomol. Soc.* 55: 13–22

83. Guirguis GN, Brindley WA. 1974. Insecticide susceptibility and response to selected pollens of larval alfalfa leafcutting bees, *Megachile pacifica* (Panzer) (Hymenoptera: Megachilidae). *Environ. Entomol.* 3:691–94

84. Harborne JB. 1993. *Introduction to Ecological Biochemistry.* New York: Academic. 318 pp. 4th ed.

85. Harder LD. 1983. Functional differences of the proboscides of short- and long-tongued bees (Hymenoptera, Apoidea). *Can. J. Zool.* 61:1580–86

86. Harder LD, Barrett SCH. 1993. Pollen removal from tristylous *Pontederia cordata:* effects of anther position and pollinator specialization. *Ecology* 74: 1059–72

87. Hefetz A. 1987. The role of Dufour's gland secretions in bees. *Physiol. Entomol.* 12:243–53

88. Hefetz A, Tengö J. 1992. Dispersed versus gregarious nesting strategies in the mason bee *Chalicodoma siculum. J. Zool.* 226:529–37

89. Heithaus ER. 1979. Flower-feeding specialization in wild bee and wasp communities in seasonal neotropical habitats. *Oecologia* 42:179–94

90. Heithaus ER. 1979. Community structure of neotropical flower visiting bees and wasps: diversity and phenology. *Ecology* 60:190–202

91. Helms KR. 1994. Sexual size dimorphism and sex ratios in bees and wasps. *Am. Nat.* 143:418–34

92. Herbst P. 1922. Zur Biologie der Gattung *Chilicola* Spin. (Apidae, Hymen.). *Entomol. Mitt. Zool. Mus. Hamburg* 11: 63–68

93. Hölldobler B, Wilson EO. 1990. *The Ants.* Cambridge, MA: Harvard Univ. Press. 732 pp.

94. Houston TF. 1989. *Leioproctus* bees associated with Western Australian smoke bushes (*Conospermum* spp.) and their adaptations for foraging and concealment (Hymenoptera: Colletidae: Paracolletini.). *Rec. West. Aust. Mus.* 14:275–92

95. Houston TF. 1991. Ecology and behavior of the bee *Amegila dawsoni* Rayment with notes on a related species (Hymenoptera: Anthophoridae). *Rec. West. Aust. Mus.* 15:535–54

96. Houston TF. 1992. Three new, monolectic species of *Euryglossa* (*Euhesma*) from Western Australia (Hymenoptera: Colletidae). *Rec. West. Aust. Mus.* 15: 719–28

97. Hull FM. 1973. *Bee Flies of the World.* Washington, DC: Smithson. Inst. 687 pp.

98. Hurd PD Jr. 1957. Notes on the autum-

nal emergence of the vernal desert bee, *Hesperapis fulvipes* Crawford. *J. Kans. Entomol. Soc.* 30:10

99. Hurd PD Jr. 1979. Superfamily Apoidea. In *Catalog of Hymenoptera in North America North of Mexico,* ed. KV Krombein, PDJ Hurd, DR Smith, BD Burks, 2:1741–2209. Washington, DC: Smithson. Inst.

100. Hurd PD Jr, LaBerge WE, Linsley EG. 1980. Principal sunflower bees of North America with emphasis on the southwestern United States (Hymenoptera: Apoidea). *Smithson. Contrib. Zool.* 310: 1–158

101. Hurd PD Jr, Linsley EG. 1975. The principal *Larrea* bees of the southwestern United States (Hymenoptera: Apoidea). *Smithson. Contrib. Zool.* 193: 1–74

102. Hurd PD Jr, Linsley EG, Whitaker TW. 1971. Squash and gourd bees (Peponapis, Xenoglossa) and the origin of the cultivated *Cucurbita. Evolution* 25: 218–34

103. Hurd PD Jr, Moure JS. 1963. A classification of the large carpenter bees (Xylocopini) (Hymenoptera: Apoidea). *Univ. Calif. Publ. Entomol.* 29:1–365

104. Iwata K. 1976. *Evolution of Instinct: Comparative Ethology of Hymenoptera.* New Delhi: Amerind. 535 pp.

104a. Jaenike J. 1990. Host specialization in phytophagous insects. *Annu. Rev. Ecol. Syst.* 21:243–73

104b. Jones CE, Little RJ, eds. 1983. *Handbook of Experimental Pollination Biology.* New York: Van Nostrand Reinhold. 558 pp.

105. Kaitala V, Smith BH, Getz WM. 1990. Nesting strategies of primitively eusocial bees: a model of nest usurpation during the solitary state of the nesting cycle. *J. Theor. Biol.* 144:445–71

106. Kathirithamby J. 1989. Review of the order Strepsiptera. *Syst. Entomol.* 14: 41–92

106a. Kerfoot WB. 1967. The lunar periodicity of *Specodagastra texana,* a nocturnal bee (Hymenoptera: Halictidae). *Anim. Behav.* 15:479–86

106b. Kerfoot WB. 1967. Nest architecture and associated behavior of the nocturnal bee *Specodagastra texana* (Hymenoptera: Halictidae). *J. Kans. Entomol. Soc.* 40:84–93

107. Kerr WE, da Cunha RA. 1990. Sex determination in bees. XXVI. Masculinism of workers in the Apidae. *Rev. Brasil. Genet.* 13:479–89

108. Kimsey LS, Bohart RM. 1990. *The Chrysidid Wasps of the World.* Oxford: Oxford Univ. Press. 652 pp.

109. King MJ. 1993. Buzz foraging mechanism of bumble bees. *J. Apic. Res.* 32: 41–49

110. Kojima J-i. 1993. A latitudinal gradient in intensity of applying ant-repellant substance to the nest petiole in paper wasps (Hymenoptera: Vespidae). *Insectes Soc.* 40:403–21

111. Krombein KV. 1967. *Trap-Nesting Wasps and Bees.* Washington, DC: Smithson. Inst. 570 pp.

112. Kukuk PF, Sage GK. 1994. Reproductivity and relatedness in a communal halictine bee *Lasioglossum (Chilalictus) helichalceum. Insectes Soc.* 41:443–56

113. Langenheim JH. 1990. Plant resins. *Am. Sci.* 78:16–24

114. Laroca S, Michener CD, Hofmeister RM. 1989. Long mouthparts among "short-tongued" bees and the fine structure of the labium in *Niltonia* (Hymenoptera, Colletidae). *J. Kans. Entomol. Soc.* 62:400–10

114a. LaSalle J, Gauld ID, eds. 1993. *Hymenoptera and Biodiversity.* Wellingford: CAB Int. 348 pp.

115. Laverty TM. 1994. Bumble bee learning and flower morphology. *Anim. Behav.* 47:531–45

116. Lin N, Michener CD. 1972. Evolution of sociality in insects. *Q. Rev. Biol.* 47:131–59

117. Lind H. 1984. Nest-provisioning cycle and daily routine of behaviour in *Dasypoda plumipes* (Hym., Apidae). *Entomol. Medd.* 36:343–72

118. Linsley EG. 1958. The ecology of solitary bees. *Hilgardia* 27:543–99

119. Linsley EG. 1978. Temporal patterns of flower visitation by solitary bees, with particular reference to the southwestern United States. *J. Kans. Entomol. Soc.* 51:531–46

120. Linsley EG, MacSwain JW. 1942. Bionomics of the meloid genus *Hornia* (Coleoptera). *Univ. Calif. Publ. Entomol.* 7:189–206

121. Linsley EG, MacSwain JW. 1952. Notes on the biology and host relationships of some species of *Nemognatha* (Coleoptera: Meloidae). *Wasmann J. Biol.* 10: 91–102

122. Linsley EG, MacSwain JW. 1952. Notes on some effects of parasitism upon a small population of *Diadasia bituberculata* (Cresson). *Pan-Pac. Entomol.* 28: 131–35

123. Linsley EG, MacSwain JW. 1957. The nesting habits, flower relationships, and parasites of some North American species of *Diadasia* (Hymenoptera: Anthophoridae). *Wasmann J. Biol.* 15: 199–235

124. Linsley EG, MacSwain JW. 1958. The significance of floral constancy among bees of the genus *Diadasia* (Hymenoptera: Anthophoridae). *Evolution* 12: 219–23

125. Linsley EG, MacSwain JW. 1959. Ethology of some *Ranunculus* insects with emphasis on competition for pollen. *Univ. Calif. Publ. Entomol.* 16:1–46

126. Linsley EG, MacSwain JW, Michener CD. 1980. Nesting biology and associates of *Melitoma* (Hymenoptera, Anthophoridae). *Univ. Calif. Publ. Entomol.* 90:1–45

127. Linsley EG, MacSwain JW, Raven PH. 1963. Comparative behavior of bees and Onagraceae. I. *Oenothera* bees of the Colorado desert. *Univ. Calif. Publ. Entomol.* 33:1–24

128. Linsley EG, MacSwain JW, Raven PH, Thorp RW. 1973. Comparative behavior of bees and Onagraceae. V. *Camissonia* and *Oenothera* bees of cismontane California and Baja California. *Univ. Calif. Publ. Entomol.* 71:1–68

129. Linsley EG, MacSwain JW, Smith RF. 1952. The life history and development of *Rhipiphorus smithi* with notes on their phylogenetic significance. (Coleoptera, Rhipophoridae). *Univ. Calif. Publ. Entomol.* 9:291–314

130. Lunau K. 1992. A new interpretation of flower guide colouration: absorption of ultraviolet light enhances colour saturation. *Plant Syst. Evol.* 183:51–65

131. MacArthur RH, Wilson EO. 1967. *The Theory of Island Biogeography.* Princeton: Princeton Univ. Press. 203 pp.

132. Maeta Y, MacFarlane RP. 1993. Japanese Conopidae (Diptera): their biology, overall distribution, and role as parasites of bumble bees (Hymenoptera, Apidae). *Jpn. J. Entomol.* 61:493–509

133. Maeta Y, Saito K, Hyodo K, Sakagami SF. 1993. Diapause and non-delayed eusociality in a univoltine and basically solitary bee, *Ceratina japonica* (Hymenoptera, Anthophoridae). I. Diapause termination by cooling and application of juvenile hormone analog. *Jpn. J. Entomol.* 61:203–11

134. Malyshev SI. 1935. The nesting habits of solitary bees. *Eos* 11:201–309

135. Malyshev SI. 1968. *Genesis of the Hymenoptera and the Phases of Their Evolution.* London: Methuen. 319 pp.

136. Messer AC. 1985. Fresh dipterocarp resins gathered by megachilid bees inhibit growth of pollen-associated fungi. *Biotropica* 17:175–76

137. Michener CD. 1953. Comparative morphology and systematic studies of bee larvae with a key to the families of hymenopterous larvae. *Univ. Kans. Sci. Bull.* 35:987–1102

138. Michener CD. 1964. Evolution of nests of bees. *Am. Zool.* 4:227–39

139. Michener CD. 1964. The bionomics of *Exoneurella,* a solitary relative of *Exoneura* (Hymenoptera: Apoidea: Ceratinini). *Pac. Insects* 6:411–26

140. Michener CD. 1974. *The Social Behavior of the Bees.* Cambridge, MA: Harvard Univ. Press. 404 pp.

141. Michener CD. 1978. The parasitic groups of Halictidae. *Univ. Kans. Sci. Bull.* 51:291–339

142. Michener CD. 1979. Biogeography of the bees. *Ann. Mo. Bot. Gard.* 66:277–347

143. Michener CD. 1985. From solitary to eusocial: Need there be a series of intervening species? In *Experimental Behavioral Ecology and Sociobiology,* ed. B Hölldobler, M Lindauer, pp. 293–305. Stuttgart: Fischer

144. Michener CD. 1988. The genus *Lithurge* in the Antilles (Hymenoptera, Megachilidae). *Folia Entomol. Méxicana* 76: 159–64

145. Michener CD, Greenberg L. 1980. Ctenoplectridae and the origin of long-tongued bees. *Zool. J. Linn. Soc.* 69: 188–203

146. Michener CD, Lange RB, Bigarella J, Salamuni R. 1958. Factors influencing the distribution of bees' nest in earth banks. *Ecology* 39:207–17

147. Michener CD, McGinley RJ, Danforth BN. 1994. *The Bee Genera of North and Central America.* Washington, DC: Smithson. Inst. 209 pp.

148. Michener CD, Rettenmeyer CW. 1956. The ethology of *Andrena erythronii* with comparative data on other species (Hymenoptera, Andrenidae). *Univ. Kans. Sci. Bull.* 37:645–84

149. Minckley RL, Wcislo WT, Yanega D, Buchmann SL. 1994. Behavior and phenology of a specialist bee (*Dieunomia*) and sunflower (*Helianthus*) pollen availability. *Ecology* 73:1406–19

150. Moldenke AR. 1979. Host-plant coevolution and the diversity of bees in relation to the flora of North America. *Phytology* 43:357–419

151. Müller H. 1872. Anwendung der Darwinschen Lehre auf Bienen. *Verh. Naturhist. Ver. Preuss. Rheinl. Westfäl.* 29:1–96

152. Neff JL, Rozen JG Jr. 1995. Foraging and nesting biology of the bee *Anthemurgus passiflorae* (Hymenoptera: Apoidea) and descriptions of its immature stages, with observations on its

floral host, *Passiflora lutea* (Passifloriaceae). *Am. Mus. Novit.* In press

153. Neff JL, Simpson BB. 1988. Vibratile pollen-harvesting by *Megachile mendica* Cresson (Hymenoptera: Megachilidae). *J. Kans. Entomol. Soc.* 61: 242–44

154. Neff JL, Simpson BB. 1990. The roles of phenology and reward structure in the pollination biology of wild sunflower (*Helianthus annuus* L., Asteraceae). *Isr. J. Bot.* 39:197–216

155. Neff JL, Simpson BB. 1991. Nest biology and mating behavior of *Megachile fortis* in central Texas (Hymenoptera: Megachilidae). *J. Kans. Entomol. Soc.* 64:324–36

156. Neff JL, Simpson BB. 1992. Partial bivoltinism in a ground-nesting bee: the biology of *Diadasia rinconis* in Texas (Hymenoptera, Anthophoridae). *J. Kans. Entomol. Soc.* 65:377–92

157. Neff JL, Simpson BB. 1993. Bees, pollination systems and plant diversity. See Ref. 114a, pp. 143–68

158. Neff JL, Simpson BB, Dorr LJ. 1982. The nesting biology of *Diadasia afflicta* Cress. (Hymenoptera: Anthophoridae). *J. Kans. Entomol. Soc.* 55:499–518

159. Nonacs P, Tobin JE. 1992. Selfish larvae: development and the evolution of parasitic behavior in the Hymenoptera. *Evolution* 46:1605–20

160. O'Toole C, Raw A. 1991. *Bees of the World.* New York: Facts on File. 192 pp.

161. Parker FD. 1978. Biology of the bee genus *Proteriades* Titus (Hymenoptera: Megachilidae). *J. Kans. Entomol. Soc.* 51:145–73

162. Parker FD. 1986. Nesting, associates, and mortality of *Osmia sanrafaelae* Parker. *J. Kans. Entomol. Soc.* 59:367–77

163. Peitsch D, Fietz A, Hertel H, de Souza J, Ventura DF, Menzel R. 1992. The spectral input systems of hymenopteran insects and their receptor-based colour vision. *J. Comp. Physiol. A* 170:23–40

164. Plateaux-Quénu C. 1972. *La Biologie des Abeilles Primitives.* Paris: Masson et Cie. 200 pp.

165. Popov VV. 1945. Parazitizm pchelinykh, ego osobennosti i evolutsiya. *Zh. Obshch. Biol.* 6:183–203; *US Dept. Agric. Transl. TT76–59121*

166. Popov VV. 1958. Special features of the correlated evolution of *Macropis, Epeoloides* (Hymenoptera, Apoidea) and *Lysimachia* (Primulaceae). *Entomol. Obozr.* 37:433–51

167. Radchenko VG, Pesenko YA. 1994. *Biologiia pchel (Hymenoptera, Apoidea).*

Saint Petersburg: Rossiiskais Acad. Nauk. Zool. Inst. 350 pp.

168. Rasnitsyn AP, Michener CD. 1991. Miocene fossil bumblebee from the Soviet far east with comments on the chronology and distribution of fossil bees (Hymenoptera: Apidae). *Ann. Entomol. Soc. Am.* 84:583–89

169. Rayner CJ, Langridge DF. 1985. Amino acids in bee-collected pollens from Australian indigenous and exotic plants. *Aust. J. Exp. Agric.* 25:722–26

170. Robertson C. 1925. Heteroptropic bees. *Ecology* 6:412–36

171. Roig-Alsina A, Michener CD. 1993. Studies of the phylogeny and classification of long-tongued bees. *Univ. Kans. Sci. Bull.* 55:123–62

172. Rosenheim JA. 1990. Density-dependent parasitism and the evolution of aggregated nesting in the solitary Hymenoptera. *Ann. Entomol. Soc. Am.* 83:277–86

173. Rosenheim JA, Meade T, Powch IG, Schoenig SE. 1989. Aggregation by foraging insect parasitoids in response to local variations in host density: determining the dimensions of a host patch. *J. Anim. Ecol.* 58:101–17

174. Roubik DW. 1989. *Ecology and Natural History of Tropical Bees.* Cambridge: Cambridge Univ. Press. 514 pp.

175. Roubik DW. 1992. Loose niches in tropical communities: Why are there so few bees and so many trees? In *Effects of Resource Distribution on Animal-Plant Interactions,* ed. MD Hunter, T Ohgushi, PW Price, pp. 327–54. New York: Academic

176. Roubik DW, Michener CD. 1985. Nesting biology of *Crawfordapis* in Panam (Hymenoptera, Colletidae). *J. Kans. Entomol. Soc.* 57:662–71

177. Roubik DW, Yanega D, Aluja M, Buchmann SL, Inouye DW. 1995. On optimal nectar foraging by some tropical bees (Hymenoptera: Apoidea). *Apidologie.* In press

178. Rozen JG Jr. 1963. Notes on the biology of *Nomadopsis*, with descriptions of four new species (Apoidea, Andrenidae). *Am. Mus. Novit.* 2142:1–17

179. Rozen JG Jr. 1989. Life history studies of the "primitive" panurgine bees (Hymenoptera: Andrenidae: Panurginae). *Am. Mus. Novit.* 2962:1–27

180. Rozen JG Jr. 1989. Morphology and systematic significance of first instars of the cleptoparasitic bee tribe Epeolini (Anthophoridae: Nomadinae). *Am. Mus. Novit.* 2957:1–19

181. Rozen JG Jr. 1991. Evolution of cleptoparasitism in anthophorid bees as re-

284   WCISLO & CANE

vealed by their mode of parasitism and
first instars (Hymenoptera: Apoidea).
Am. Mus. Novit. 3029:1–36
182. Rozen JG Jr. 1992. Systematics and host
relationships of the cuckoo bee genus
Oreopasites (Hymenoptera: Anthophoridae: Nomadinae). *Am. Mus. Novit.*
3046:1–56
183. Rozen JG Jr. 1992. Biology of the bee
Ancylandrena larreae (Andrenidae: An-
dreninae) and its cleptoparasite *Hex-
epeolus rhodogyne* (Anthophoridae:
Nomadinae) with a review of egg deposition in the Nomadinae (Hymenoptera:
Apoidea). *Am. Mus. Novit.* 3038:1–15
184. Rozen JG Jr. 1994. Biology and immature stages of some cuckoo bees belonging to Brachynomadini, with descriptions of two new species
(Hymenoptera: Apidae: Nomadinae).
Am. Mus. Novit. 3089:1–23
185. Rozen JG Jr. 1994. Revision of the
cleptoparasitic bee tribe Protepeolini,
including biologies and immature stages
(Hymenoptera: Apoidea: Apidae). *Am.
Mus. Novit.* 3099:1–38
186. Rozen JG Jr, Buchmann SL. 1990. Nesting biology and immature stages of the
bees *Centris caesalpiniae*, *C. pallida*,
and the cleptoparasite *Ericrocis lata*
(Hymenoptera: Apoidea: Anthophoridae). *Am. Mus. Novit.* 2985:1–30
187. Rozen JG Jr, Jacobson NR. 1980. Biology and immature stages of *Macropis
nuda*, including comparisons to related
bees (Apoidea, Melittidae). *Am. Mus.
Novit.* 2702:1–11
188. Rozen JG Jr, McGinley RJ. 1976. Biology of the bee genus *Conanthalictus*
(Halictidae, Dufoureinae). *Am. Mus.
Novit.* 2602:1–6
189. Rust RW, Thorp RW. 1973. The biology
of *Stelis chlorocyanea*, a parasite of
Osmia nigrifrons (Hymenoptera: Megachilidae). *J. Kans. Entomol. Soc.* 46:
548–62
190. Sakagami SF, Laroca S. 1971. Observations on the bionomics of some neotropical xylocopine bees, with
comparative and biofaunistic notes
(Hymenoptera, Anthophoridae). *J. Fac.
Sci. Hokkaido Univ. Ser. VI Zool.* 18:57–
127
191. Salt G. 1928. The effects of stylopization on Aculeate Hymenoptera. *J. Exp.
Zool.* 48:223–33
192. Schultz JC, Otte D, Enders F. 1977.
Larrea as a habitat component for desert
arthropods. In *Creosote Bush: Biology
and Chemistry of Larrea in New World
Deserts*, ed. TJ Mabry, JH Hunziker,
DRJ Difeo, pp. 176–208. Stroudsburg:
Dowden, Hutchinson & Ross

193. Shaw SR. 1988. Euphorine phylogeny:
the evolution of diversity in host-utili-
zation by parasitoid wasps (Hymenoptera: Braconidae). *Ecol. Entomol.* 13:
323–35
194. Sick M, Ayasse M, Tengö J, Engels W,
Lübke G, Francke W. 1994. Host-parasite relationships in six species of *Sphecodes* bees and their halictid hosts: nest
intrusion, intranidal behavior, and
Dufour's gland volatiles (Hymenoptera:
Halictidae). *J. Insect Behav.* 7:101–17
195. Silveira FA. 1993. The mouthparts of
Ancyla and the reduction of the
labiomaxillary complex among long-
tongued bees (Hymenoptera: Apoidea).
Entomol. Scand. 24:293–300
196. Simpson BB, Neff JL. 1983. Evolution
and diversity of floral rewards. See Ref.
104b, pp. 142–59
197. Simpson BB, Neff JL. 1987. Pollination
ecology in the arid Southwest. *Aliso*
11:417–40
197a. Simpson BB, Neff JL, Moldenke AR.
1977. Reproductive systems of *Larrea*.
In *Creosote Bush: Biology and Chemistry of Larrea in New World Deserts*,
ed. TJ Mabry, JH Hunziker, DRJ Difeo,
pp. 92–114. Stroudsburg, PA: Dowden,
Hutchinson & Ross
198. Snelling RR. 1983. The North American
species of the bee genus *Lithurge*,
Hymenoptera: Megachilidae. *Contrib.
Sci. Los Angeles Co. Mus. Nat. Hist.*
343:1–11
199. Stage GI. 1966. *Biology and systematics
of the American species of the genus
Hesperapis Cockerell*. PhD dissertation.
Univ. Calif., Berkeley. 461 pp.
200. Stanley RG, Linskens HF. 1974. *Pollen:
Biology, Biochemistry, Management.*
Heidelberg: Springer-Verlag. 307 pp.
201. Steiner KE, Whitehead VB. 1991. Oil
flowers and oil bees: further evidence
for pollinator adaptation. *Evolution* 45:
1493–501
202. Stephen WP, Bohart GE, Torchio PF.
1969. The biology and external morphology of bees. *Oreg. State Univ.
Agric. Exp. Stn. Publ.* 140 pp.
203. Stone GN. 1994. Patterns of evolution
of warm-up rates and body temperatures
in flight in solitary bees of the genus
Anthophora. *Funct. Ecol.* 8:324–35
204. Strickler K. 1979. Specialization and
foraging efficiency of solitary bees.
Ecology 60:998–1009
205. Struck M. 1994. Flowers and their insect
visitors in the arid winter rainfall region
of southern Africa: observations on permanent plots. Insect visitation behaviour. *J. Arid Environ.* 28:51–74
206. Szymas B. 1991. The parasitic entomo-

fauna as a factor limiting populations of solitary bees. *Przegl. Zool.* 35:307–13

207. Takhtajan AL. 1986. *Floristic Regions of the World.* Berkeley: Univ. Calif. Press. 522 pp.

208. Tengö J, Sick M, Ayasse M, Engels W, Svensson BG, et al. 1992. Species specificity of Dufour's gland morphology and volatile secretions in kleptoparasitic *Sphecodes* bees (Hymenoptera: Halictidae). *Biochem. Syst. Ecol.* 20:351–62

209. Tepedino VJ, Knapp AK, Eickwort GC, Ferguson DC. 1989. Death camas (*Zigadenus nuttallii*) in Kansas: pollen collectors and a florivore. *J. Kans. Entomol. Soc.* 62:411–12

210. Tepedino VJ, McDonald LL, Rothwell R. 1979. Defense against parasitization in mud-nesting Hymenoptera: Can empty cells increase net reproductive output? *Behav. Ecol. Sociobiol.* 6:99–104

211. Thomas BA, Spicer RA. 1987. *The Evolution and Paleobiology of Land Plants.* London: Dioscorides. 309 pp.

212. Thompson JN. 1994. *The Coevolutionary Process.* Chicago: Univ. Chicago Press. 376 pp.

213. Thorp RW. 1969. Systematics and ecology of bees of the subgenus *Diandrena* (Hymenoptera: Andrenidae). *Univ. Calif. Publ. Entomol.* 52:1–146

214. Thorp RW. 1979. Structural, behavioral, and physiological adaptations of bees (Apoidea) for collecting pollen. *Ann. Mo. Bot. Gard.* 66:788–812

215. Torchio PF. 1989. Biology, immature development, and adaptive behavior of *Stelis montana*, a cleptoparasite of *Osmia* (Hymenoptera: Megachilidae). *Ann. Entomol. Soc. Am.* 82:616–32

216. Torchio PF, Bosch J. 1992. Biology of *Tricrania stansburyi*, a meloid beetle cleptoparasite of the bee *Osmia lignaria propinqua* (Hymenoptera: Megachilidae). *Ann. Entomol. Soc. Am.* 85:713–21

217. Torchio PF, Rozen JG Jr, Bohart GE, Favreau MS. 1967. Biology of *Dufourea* and of its cleptoparasite, *Neopasites* (Hymenoptera: Apoidea). *J. NY Entomol. Soc.* 75:132–46

218. Vander Wall SB. 1990. *Food Hoarding in Animals.* Chicago: Univ. Chicago Press. 445 pp.

219. Velthuis HH. 1992. Pollen digestion and the evolution of sociality in bees. *Bee World* 73:77–89

219a. Vinson SB, Frankie GW, Barthell J. 1993. Threats to the diversity of solitary bees in a Neotropical dry forest in Central America. See Ref. 114a, pp. 53–81

220. Vogel S. 1974. *Ölblumen und Olsam-*

melnde Bienen. Wiesbaden: Franz Steiner. 267 pp.

221. Waser NM. 1986. Flower constancy: definition, cause, and measurement. *Am. Nat.* 127:593–603

222. Watmough RH. 1983. Mortality, sex ratio and fecundity in natural populations of large carpenter bees (*Xylocopa* spp.). *J. Anim. Ecol.* 52:111–25

223. Wcislo WT. 1984. Gregarious nesting of a digger wasp as a "selfish herd" response to a parasitic fly (Hymenoptera: Sphecidae; Diptera: Sacrophagidae) [sic]. *Behav. Ecol. Sociobiol.* 15: 157–60

224. Wcislo WT. 1987. The role of seasonality, host-synchrony, and behaviour in the evolutions and distributions of nest parasites in Hymenoptera (Insecta), with special reference to bees (Apoidea). *Biol. Rev.* 62:515–43

225. Wcislo WT. 1989. Behavioral environments and evolutionary change. *Annu. Rev. Ecol. Syst.* 20:137–69

226. Wcislo WT. 1990. Parasitic and courtship behavior of *Phalacrotophora halictorum* (Diptera: Phoridae) at a nesting site of *Lasioglossum figueresi* (Hymenoptera: Halictidae). *Rev. Biol. Trop.* 38: 205–09

227. Wcislo WT. 1993. Communal nesting in a North American pearly-banded bee, *Nomia tetrazonata*, with notes on nesting behavior in *Dieunomia heteropoda* (Hymenoptera: Halictidae: Nomiinae). *Ann. Entomol. Soc. Am.* 86: 813–21

228. Wcislo WT. 1995. Sensilla numbers and antennal morphology of parasitic and non-parasitic bees (Hymenoptera: Apoidea). *Int. J. Insect Morphol. Embryol.* 24:63–81

229. Wcislo WT, Minckley RL, Leschen RAB, Reyes S. 1994. Rates of parasitism by natural enemies of a solitary bee, *Dieunomia triangulifera* (Hymenoptera, Coleoptera, and Diptera) in relation to phenologies. *Sociobiology* 23: 265–73

230. Wcislo WT, Wille A, Orozco E. 1993. Nesting biology of tropical solitary and social sweat bees, *Lasioglossum (Dialictus) figueresi* Wcislo and *L. (D.) aeneiventre* (Friese) (Hymenoptera: Halictidae). *Insectes Soc.* 40:21–40

231. Weberling F. 1989. *Morphology of Flowers and Inflorescences.* Cambridge: Cambridge Univ. Press. 405 pp. (Transl.)

232. West-Eberhard MJ. 1979. Sexual selection, social competition, and evolution. *Proc. Am. Philos. Soc.* 123: 222–34

233. West-Eberhard MJ. 1986. Alternative adaptations, speciation, and phylogeny (a review). *Proc. Natl. Acad. Sci. USA* 83:1388–92

234. West-Eberhard MJ. 1992. Behavior and evolution. In *Molds. Molecules and Metazoans,* ed. PR Grant, HS Horn, pp. 57–79. Princeton: Princeton Univ. Press

235. Westrich P. 1989. *Die Wildbienen Baden-Württembergs,* Band I. Stuttgart: Eugen Ulmer GmbH. 431 pp.

236. Whitehead VB. 1984. Distribution, biology and flower relationships of fideliid bees of southern Africa (Hymenoptera, Apoidea, Fideliidae). *S. Afr. J. Zool.* 19:87–90

237. Winston ML. 1987. *The Biology of the Honey Bee.* Cambridge, MA: Harvard Univ. Press. 281 pp.

238. Wolda H, Roubik DW. 1986. Nocturnal bee abundance and seasonal activity in a Panamanian forest. *Ecology* 67:426–33

239. Wu Y-r, Michener CD. 1986. Observations on Chinese *Macropis* (Hymenoptera: Apoidea: Melittidae). *J. Kans. Entomol. Soc.* 59:42–48

240. Wülker W. 1964. Parasite-induced changes of internal and external sex characters in insects. *Exp. Parasitol.* 15:561–97

241. Zimmerman EC. 1970. Adaptive radiation in Hawaii with special reference to insects. *Biotropica* 2:32–38

Annu. Rev. Entomol. 1996. 41:287–308
Copyright © 1996 by Annual Reviews Inc. All rights reserved

PREDATORY BEHAVIOR OF JUMPING SPIDERS

R. R. Jackson and S. D. Pollard

Department of Zoology, University of Canterbury, Christchurch, New Zealand

KEY WORDS: salticids, salticid eyes, *Portia*, predatory versatility, aggressive mimicry

ABSTRACT

Salticids, the largest family of spiders, have unique eyes, acute vision, and elaborate vision-mediated predatory behavior, which is more pronounced than in any other spider group. Diverse predatory strategies have evolved, including araneophagy, aggressive mimicry, myrmicophagy, and prey-specific prey-catching behavior. Salticids are also distinctive for development of behavioral flexibility, including conditional predatory strategies, the use of trial-and-error to solve predatory problems, and the undertaking of detours to reach prey. Predatory behavior of araneophagic salticids has undergone local adaptation to local prey, and there is evidence of predator-prey coevolution. Trade-offs between mating and predatory strategies appear to be important in ant-mimicking and araneophagic species.

INTRODUCTION

With over 4000 described species (11), jumping spiders (Salticidae) compose the largest family of spiders. They are characterized as cursorial, diurnal predators with excellent eyesight. Although spider eyes usually lack the structural complexity required for acute vision, salticids have unique, complex eyes with resolution abilities without known parallels in animals of comparable size (98). Salticids are the end-product of an evolutionary process in which a small silk-producing animal with a simple nervous system acquires acute vision, resulting in a diverse array of complex predatory strategies.

Here, we begin by discussing how salticid eyes work and then review the predatory strategy of *Portia*—a tropical genus whose members are jacks of all spider trades. In many ways, this spider is the ultimate salticid. Using the exceptionally complex predatory behavior of *Portia* spp. as a baseline, we provide in the second half of the review a survey of the various predatory behaviors of salticid spiders.

287

0066-4170/96/0101-0287$08.00

THE SALTICID EYE

Salticids have four pairs of eyes, but it is the pair of very large anterior median eyes (known as the principal eyes) that stare back when you look at the spider's face. Located on either side of the principal eyes are three pairs of smaller secondary eyes, which are highly proficient motion detectors (19, 41, 93, 94). Immediately to either side of the principal eyes are the anterior lateral eyes, which face forward in most species and have binocular overlap in front of the spider. They probably share a role with the principal eyes in range finding and in controlling the pursuit of prey (32, 33, 42). The next pair of secondary eyes, the posterior medians, are very small and apparently degenerate in most salticids, although they are large, well-developed motion detectors in some of the primitive spartaeines and lyssomanines (97, 122, 123). The most rearward-directed secondary eyes, the posterior laterals, have the widest fields of view. With its combination of four, or sometimes six, functional secondary eyes, the salticid's vision covers virtually the entire 360° ambit around the spider (96).

The principal eyes are the most interesting because they provide for acute vision (32, 48, 91, 92), allowing the salticid to identify mates, rivals, and predators from distances of 30 body lengths or more away (71). In typical predatory stalk and leap sequences (16, 18, 30, 33, 34), the salticid turns so its principal eyes face the prey. Next, it stalks the prey until it is a few body lengths away, lowers its body, and slowly crawls forward. Now, worthy of its name, the jumping spider attaches a dragline, raises its forelegs and makes an accurate, visually mediated leap onto the prey. Vision also plays an important role in other aspects of salticid behavior: They display to their image in a mirror (33, 42, 71); discriminate between the images of prey and conspecifics shown on a television screen (10); and respond appropriately to visual cues from motionless mates, rivals, and prey (18, 42, 80).

Salticid eyes, especially the principal eyes, are constructed very differently from the more familiar vertebrate and insect eyes (97). The retinas of the principal eyes have a four-layer, tiered arrangement. Light entering through the corneal lens passes successively through layers 4, 3, and 2 before reaching layer 1, which in cross section has a distinctive boomerang shape. Layer 2 is roughly the same in shape, whereas layers 3 and 4 more closely approximate a circle (91).

Layer 1 forms only an approximate layer, because this set of photoreceptors is not entirely in one plane. Instead, the receptors are arranged in a staircase so that receptors closer to the periphery of layer 1 are closer to the corneal lens; those in the central region are farthest from the lens (1a, 6). A primary function of the tiered arrangement of the retina as a whole, and the staircase arrangement of layer 1 in particular, is apparently to compensate for chromatic aberration and an inability to focus by changing eye tube length (8).

In the central area of layer 1 (the fovea) (1a), receptors are packed especially close together (about 1 μm apart). They appear to be primarily responsible for shape recognition, because this is the only region that can process a sharply focused image (8).

The part of the salticid eye seen from outside is the corneal lens, behind which lies the eye tube. While the tubes of the secondary eyes are shallow and fixed in place, each principal eye has a long eye tube and, because of a set of six attached muscles, is capable of precise, complex rotary and side-to-side movements (92). However, the principal eyes cannot focus images by elongating and shortening the tube. Because of the tiered arrangement of the retina, the salticid principal eye receives a sharp image in the fovea of layer 1 at distances ranging from little more than a body length away to infinity (8). Because the retina of the principal eye is at the end of a long eye tube, the eye has a large focal length, which gives the spider a telephoto lens system. A second lens, just in front of the retina, increases the magnifying power of the eye and turns these eyes into miniature Galilean telescopes (129).

The distance between the receptors in the layer 1 fovea (~1 μm) appears to be optimal for resolution, given the details of the rest of the optical system. If the receptors were any closer together, then the image obtained would be degraded by quantum effects (8, 97).

The principal eyes appear to be capable of color vision (5, 12, 15, 89, 106, 131), which is most likely dichromatic.

One of the great challenges for future research will be to understand how the salticid's unique eyes enable these spiders to distinguish between different types of prey, webs, and other parts of the environment. The fovea of layer 1 contains at most only a few hundred receptors (8), and an eye with so few components cannot be operating on the same principles as the much larger eye of vertebrates (97).

The principal eye is an active eye, and this is probably the key to understanding shape perception. Yet the only detailed information we have on precisely how the salticid eye tube moves is from Land's (92) work on *Metaphidippus* spp., which described four modes of movement: spontaneous activity, saccades, tracking, and scanning. Scanning, which takes place only after the salticid is oriented so that an image is projected onto the fovea of layer 1, is the most complex movement and should be the target for future studies of shape perception.

THE COMPLEX PREDATORY STRATEGY OF *PORTIA*

Predatory Versatility

A versatile predator has a conditional strategy consisting of a repertoire of predatory tactics, each specific to different circumstances or different types of prey (13). The predatory versatility of *Portia* spp. may represent the most pronounced example from arthropods, if not from all nonhuman animals. Five species of *Portia* have been studied (72–74, 76): *P. africana* and *P. schultzi* from Kenya; *P. albimana* from Sri Lanka; *P. labiata* from Malaysia and Sri Lanka; and *P. fimbriata* from Australia, Malaysia, and Sri Lanka. In these species, each spider is a cursorial predator and a web-builder, as well as a predator that invades alien webs, where it uses aggressive mimicry to trick, then catch, the resident spider. In addition, *Portia* spp. prey on the resident spider's eggs and eat insects ensnared in the alien web. The three chapters of the *Portia* predatory strategy—hunting in the open, using its own prey-catching web, and making predatory forays into the webs of other spiders—each features intricate stories of predatory versatility.

In the discussion below, unless otherwise noted, we use the generic name *Portia* to refer to any and all *Portia* species.

Web Invasion and Aggressive Mimicry

The types of webs built by spiders are highly diverse, ranging from sparsely woven three-dimensional webs, to highly organized two-dimensional orb webs, to densely woven sheet webs (115). Some spiders enhance the stickiness of their web by secreting special substances (i.e. glue) onto the structural lines. Cribellate spiders are species that lay a very fine wool of sticky threads across the structural threads of the web. In contrast, some ecribellate spiders string droplets of fluid glue along the structural threads of the web at regular intervals. Generally, cursorial spiders and spiders that build nonsticky webs adhere to sticky webs, and spiders that build ecribellate sticky webs tend to adhere to cribellate webs, and vice versa (RR Jackson, unpublished data). *Portia* is an exception. It spins nonsticky webs, yet it can walk across and capture prey on virtually any type of web, including both cribellate and ecribellate sticky webs (65).

When *Portia* walks onto another spider's web, it enters the other spider's perceptual world, as the web is an extension and critical component of the web-building spider's sensory system (29, 130). On the web, intimate contact with the other spider's sensory system is often dangerous for *Portia*. When the resident spider detects something wrong, instead of fleeing, it may actively defend itself. Then the tables may be turned, and the intended prey becomes the predator.

After entering another spider's web, *Portia* usually does not simply stalk or chase down its victim but instead sends vibratory signals across the silk. The resident spider may respond to these signals in the same way it responds to the vibrations caused by a small insect becoming ensnared in the web. When the duped spider gets close, *Portia* lunges and catches it. A system of this sort, in which a predator (in the present example, *Portia*) deceives its victim (e.g. a web-building spider) by mimicking prey (e.g. a small insect ensnared in a web) is called aggressive mimicry (28).

For *Portia,* aggressive mimicry involves pronounced behavioral complexity because *Portia* has an exceptionally diverse repertoire of vibratory signals. The spider can make the signals by manipulating, plucking, and slapping the silk with one or any combination of its eight legs and two palps, all of which can be moved in different ways. *Portia* also signals by flicking its abdomen up and down and can combine abdominal movements with virtually any of the appendage movements (72, 73, 86). Many of these signaling behaviors appear to be evolutionary modifications of grooming behaviors (76). The web-building spider has acute abilities to detect and discriminate between vibratory signals transmitted over the silk in its web. How the prey spider interprets these web-borne vibrations varies considerably among species and also with the sex, age, previous experience, and feeding state of the spider (53, 101, 130). Yet *Portia* uses aggressive mimicry to catch just about every kind of web-building spider imaginable, as long as it is about one tenth to twice *Portia*'s size (72, 73; RR Jackson, unpublished data).

The question of how *Portia* chooses, from its large repertoire of signals, the appropriate signals for hunting a particular prey spider has driven a research program at the University of Canterbury, Christchurch, New Zealand, carried out in collaboration with Stimson Wilcox from the State University of New York in the United States. A computer-based system was developed for recording and playing back signals on webs, much as if we could listen and talk to spiders in their own language. This work, which is still in progress, has indicated that the key to the success of *Portia* at victimizing so many different types of spiders is an interplay of two basic ploys (83, 86): (*a*) the use of specific preprogrammed signals when cues from some of the more common prey species are detected; and (*b*) the flexible adjustment of signals for different prey species according to feedback from the victims.

Trial-and-Error Behavior

The first ploy, using preprogrammed tactics, is consistent with the popular image of spiders as animals that are hard-wired and governed by instinct, but the second ploy is based on the use of trial and error to derive signals, an unexpectedly flexible behavior for a spider (see 105). To illustrate how trial and error works, let us look at what happens when *Portia* goes into the web of a species of web-building spider for which it does not have a preprogrammed tactic. *Portia* first presents the resident spider with a range of different vibratory signals. When one of these signals eventually elicits an appropriate response from the victim (e.g. it behaves as though *Portia* were a small insect caught in the web), *Portia* ceases to vary its signals and concentrates on producing the signal that elicits the response (86). However, communication between predator and prey is often more subtle.

Aggressive mimicry for *Portia* is a dangerous way to make a living, especially when facing a large and powerful spider in a web, and simply to pretend to be prey and provoke a full-scale predatory attack would probably not be a successful tactic. Instead, *Portia*'s strategy appears geared toward finely controlling the victim's behavior (59, 65–69). *Portia* may make signals that draw the victim in slowly. In contrast, the signals may keep the victim calm while *Portia* moves in slowly for the kill. Calming effects appear to be achieved through monotonous repetition of a habituating signal (RR Jackson & RS Wilcox, unpublished data), as though *Portia* were putting its victim to sleep with a vibratory lullaby derived from trial and error.

When using trial and error, *Portia* associates success with a particular signal and remembers to keep using it. This is at least a simple kind of learning (see 105). Learning is not unique to *Portia*. Some typical insectivorous salticids learn to avoid ants, and some appear to improve with practice their performance of the stalk-and-leap routine against their normal insect prey (18, 26, 31, 42, 100). Salticids will also acquire aversion to models paired with electric shock (18).

However, *Portia*'s trial-and-error behavior is not only an example of learning: The wide range of signals generated and the ability to identify and remember successful ones in a variety of contexts gives these salticids problem-solving capabilities (39).

Smoke-Screen Behavior

Another example of flexibility in *Portia*'s predatory strategy is the smoke-screen tactic. In the field, investigators have noticed that whenever the wind blows, movement of the web masks nearly all other signals going across the silk. Interestingly, when the wind blows, *Portia* is most likely to walk rapidly toward the spider in the web. Laboratory experiments using fans to generate artificial wind demonstrated that *Portia* deliberately chooses to approach its victim when a breeze provides a vibratory "smoke screen" to hide behind (128a). Also, if the wind does not blow, *Portia* can make its own vibratory camouflage. While walking across the web, *Portia* masks the faint vibrations caused by its steps by adding large-scale vibrations that simulate a breeze (128a; RS Wilcox & RR Jackson, unpublished data). *Portia* is selective; it uses opportunistic and self-generated smoke screens when hunting spiders, but not, for instance, when stalking insects caught in the webs or preying upon the eggsacs of other spiders when masking would be irrelevant (RS Wilcox & RR Jackson, unpublished data).

Detouring

In the field, sometimes *P. fimbriata* stops, looks at a web, then turns and walks away, only to approach the web later from another direction. This behavior is especially distinct on webs of the Queensland spider *Argiope appensa*, which builds orb webs on tree trunks. Walking straight from the tree into the web would seem easy for *P. fimbriata*, but *A. appensa* has a dramatic defense. This spider is exceedingly sensitive to anything foreign touching its web, and hence it rarely gives *P. fimbriata* time to enter the web and start signaling. If *A. appensa* is sure the intruder is an insect prey, it attacks; otherwise, it pumps the web fibers (79) by rapidly, repeatedly flexing its legs. It thus sets the web in motion, which either drives or throws *P. fimbriata* off (67).

In its natural environment, the rain forest, *P. fimbriata* often walks up the tree trunk toward *A. appensa,* stops, looks around, then goes off in a different direction, later reappearing above the web. Vines and other vegetation, which usually grow near the tree, often extend out above the web. After looking at the web, the vine, and the neighboring vegetation, *P. fimbriata* sometimes moves away, perhaps to where the web is completely out of view, crosses the vegetation, and comes out on the vine above the web. From above the web, *P. fimbriata* drops on its own silk line alongside, but without touching, the web of the *A. appensa.* Then, when parallel with the spider in the web, *Portia* swings in to make a kill (67, 84).

In the laboratory, experimental evidence shows that *P. fimbriata* makes deliberate, planned detours (116, 117). For example, if presented with a choice of two routes on artificial vegetation, only one of which leads to a prey spider, *P. fimbriata* consistently walks past the inappropriate path to take the appropriate one, even when that path initially leads away from the prey to where the prey is temporarily out of view (117; MS Tarsitano & RR Jackson, unpublished data).

Predation by Queensland's P. fimbriata on Cursorial Salticids

The habitat of *P. fimbriata* in Queensland, Australia, is unique in that it has a superabundance of cursorial salticids (73). The predatory behavior of this *Portia* species appears to be specially adapted to this locally abundant prey.

Although strictly cursorial salticids do not spin prey-catching webs, they do spin shelters out of silk (nests) that are usually densely woven, tubular, and not much larger than the resident spider. A salticid that finds a conspecific inside a nest may court or threaten the resident spider by making vibratory signals on the silk (66, 112). *P. fimbriata* responds to nests of nonconspecific salticid spiders with vibratory signals (nest probing), to which the resident reacts by poking its front out of the nest, only to be grabbed and eaten (73).

P. fimbriata also catches salticids out in the open, away from their nests, by using a special type of trickery, known as cryptic stalking (72), which capitalizes on the unusual appearance of *Portia* spp. Markings, tufts of hairs, and long, spindly legs give *Portia* the appearance of detritus in a web (120), which presumably protects it from visually hunting predators. Normally, *Portia* locomotion consists of a slow, choppy gait that renders the genus difficult to recognize even when moving. When inactive in a web, *Portia* adopts a special posture, the cryptic rest posture, with palps retracted beside the chelicerae and legs retracted beside and under the body. This positioning blurs the outlines of these appendages into the contours of the body (72).

When cryptically stalking a salticid away from webs, *P. fimbriata* moves even more slowly than usual, often remaining undetected until too late for the victim to escape. However, as salticid secondary eyes are excellent movement detectors, sometimes the victim suddenly swivels around to see what is coming up on it. The Queensland *P. fimbriata* compensates: It freezes in its tracks and stays motionless until the salticid turns away again. When the salticid takes a look, it apparently perceives a piece of detritus. Another consistent component of cryptic stalking by *P. fimbriata* is that it retracts its palps, as in the cryptic rest posture. Experiments have confirmed that hiding the outlines of palps is important because these outlines are cues by which the salticid can recognize *P. fimbriata* as a predator (SD Pollard & RR Jackson, unpublished data).

Interactions between *P. fimbriata* and *Euryattus* (species undetermined) illustrate the evolution of a prey-specific predatory behavior for use against a single species. *Euryattus* sp. is sympatric with *P. fimbriata* in Queensland but is not known to be sympatric with other populations of *Portia*. *Euryattus* females are unusual salticids because, instead of making a tubular silk nest, they suspend a rolled-up dead leaf by heavy silk guylines from a rock ledge, tree trunk, or the vegetation in the rain forest and use this as a nest (51). *Euryattus* males go down guylines onto leaves and court by suddenly flexing their legs and making the leaf rock back and forth. *Euryattus* females then come out of their leaves to mate with or drive away conspecific males. Unlike any other *Portia* studied, *P. fimbriata* from Queensland also goes down guylines onto the leaves and makes the leaf rock by suddenly flexing legs, apparently simulating the courtship behavior of *Euryattus* males (83). *Euryattus* females that come out of their leaves when "courted" by *P. fimbriata* are attacked and eaten.

Coevolution of P. fimbriata and Euryattus sp.

The Queensland *P. fimbriata* is not always successful at deceiving and catching *Euryattus* sp. Sometimes the strategy fails because the *Euryattus* female detects an approaching *P. fimbriata* and drives it away, either before or after it reaches the leaf. To drive *P. fimbriata* away, the *Euryattus* spider comes out of the rolled-up leaf, then suddenly and violently strikes, leaps at, or charges toward the *Portia* individual. Sometimes *Euryattus* sp. leaps and bangs into *P. fimbriata* (usually head-on) and knocks it away, after which *Euryattus* sp. swings down on its dragline, then climbs back to the leaf. Once attacked, the *Portia* spider flees and the *Euryattus* survives (83).

Observations of thousands of interactions between *P. fimbriata* and many different species of salticids (72, 73; RR Jackson, unpublished data) have shown that *Euryattus* sp. is more effective than other salticids at recognizing and defending itself against a stalking *Portia*. Frequent predation by *P. fimbriata* on *Euryattus* sp. has apparently resulted in *Euryattus* sp. evolving special abilities to recognize and defend itself against *P. fimbriata*, which suggests coevolution between these two species.

Interpopulation variation in *Euryattus* behavior supports the coevolution hypothesis. *P. fimbriata* is absent from a second *Euryattus* habitat sampled about 15 km away from the location where *Euryattus* sp. and *P. fimbriata* are sympatric. In tests using laboratory-reared spiders, allopatric *Euryattus* sp. only rarely evaded or attacked stalking *P. fimbriata*, and *P. fimbriata* hunted allopatric more efficiently than sympatric *Euryattus* sp. (85).

SPARTAEINE SALTICIDS, A PRIMITIVE GROUP

Behaviorally, the Spartaeinae, the subfamily to which *Portia* belongs, is a collection of unusual salticids. This subfamily of primarily tropical African, Asian, and Australasian species is of special interest because of morphological characters (i.e. presence of female palpal claws and unreduced posterior medial eyes) that are regarded as plesiomorphic for salticids (114, 123, 128). Most strikingly, retinal ultrastructure of the principal and especially the secondary eyes of spartaeines tends to be less organized than that of typical salticids. Findings from extensive comparative and ontogenetic studies consistently indicate that the eyes of the Spartaeinae (and Lyssomaninae: see below) represent a remarkable series of plesiomorphic states leading up to the condition prevailing in advanced salticids (1a–4, 7, 8).

Among salticids, only ten species (in four genera) are known to use vibratory aggressive mimicry in conjunction with araneophagic web invasion, and all of these are spartaeines. Besides the five studied species of *Portia*, this group includes *Brettus adonis, Brettus cingulatus* (75), and *Gelotia lanka* (60) from Sri Lanka; *Cyrba algerina* (61, 75) from southern Europe; and *Cyrba ocellata* (61) from Australia, Kenya, Sri Lanka, and Thailand.

Brettus spp., *Cyrba* spp., and *G. lanka* have not been studied as thoroughly as *Portia* spp., but all these genera exhibit some aspects of aggressive mimicry in common. None of these spiders are exclusively web invaders; each also catches prey away from webs. *G. lanka*, like *Portia*, not only invades webs, but also builds them. Like *Portia*, the other spiders are armed with large repertoires of vibratory signals, and preliminary evidence (60, 61, 75) indicates that each uses a trial-and-error tactic similar to that of *Portia*.

Behavioral studies have been carried out on another three spartaeine genera. *Cocalus gibbosus*, from Queensland, invades webs and eats spiders in addition to catching prey away from webs, but does not practice aggressive mimicry (62). *Phaeacius malayensis* and *Phaeacius wanlessi*, from Singapore and Sri Lanka, respectively, are specialized ambush predators that neither build nor invade webs (75, 63). *Spartaeus spinnimanus* and *Spartaeus thailandica*, from Singapore and Thailand, respectively, are web-building, but not web-invading, salticids (78).

Although *Brettus, Cocalus, Cyrba*, and *Phaeacius* spp. do not build webs, they do build aberrant, web-like silk edifices in which to molt and oviposit (60–63, 75). These structures contrast with the tightly woven tube-like nests typically built by salticids (53). Also, the way that all spartaeine species attack their prey is atypical for salticids. All spartaeines studied do not perform the typical stalk-and-leap sequence and usually lunge at rather than leap upon their prey. *G. lanka* and *Portia, Brettus, Cocalus*, and *Cyrba* spp., the web invaders, all have cuticles that do not adhere to sticky webs; they differ from the nonspartaeine web-invading salticids in this respect. All of the web-invading spartaeines prey not only on the resident spider, but on its eggs as well, and also take insects from the other spider's web (65).

These odd salticids appear to be evolutionary experiments that branched off in the early history of the family, before the majority of salticids got locked into a path toward becoming insect hunters. However, the spartaeines are not the only unusual salticids. Another salticid subfamily, Lyssomaninae, also exhibits a predominance of plesiomorphic morphological characters. This subfamily contains seven genera and about 85 species of primarily tropical salticids (36, 122). *Chinoscopus* and *Lyssomanes* are New World genera, whereas *Asemonea, Goleba, Macopaeus, Onomastus*, and *Pandisus* are Old World genera. Details concerning predatory behavior are available for species from four of the genera: *Lyssomanes, Asemonea, Goleba*, and *Onomastus*.

Compared with the spartaeines, the lyssomanines do not appear to have diversified very much in their predatory behavior. For instance, no evidence supports web invasion. Yet like the spartaeines, these spiders are quite unusual. They do not adopt typical stalk-and-leap sequences, and they usually make contact with the prey by lunging instead of leaping. Although the lyssomanines do not build large prey-catching webs comparable to the webs of *G. lanka, Spartaeus*, and *Portia* spp., neither do they build tightly woven silk nests comparable to those of the majority of salticids. Instead, they spin flimsy sheets under leaves, which they use for rudimentary webs. When an insect contacts the silk, the lyssomanine rushes out and grabs it (40, 52, 64).

PREDATORY VERSATILITY IN NONSPARTAEINE WEB-BUILDING SALTICIDS

Simaetha paetula and *Simaetha thoracica,* from tropical Queensland, resemble *Portia* spp. in that they are versatile predators, but they have some interesting idiosyncrasies. *Simaetha* spp. build a large prey-catching web. However, spiders of this genus, unlike *Portia* spiders, often also build a typically salticid tube-like nest. Nests and webs may be built alone or the nest may be incorporated into the web (52).

Besides building their own prey-catching webs, *Simaetha* spp. often live within the colonies of a social web-building cribellate spider, *Phryganoporus* (formerly *Badumna*) *candidus* (17). *Simaetha* spp. glean insects from the edges of the alien webs and incorporate their own nest, web, or web-nest combination within the alien communal web. Although cribellate silk adheres to their cuticles, *Simaetha* spp. avoid becoming prey of the social spiders by moving carefully along vegetation and old, no-longer-sticky silk mixed in among the fresh, sticky silk of the web.

Female *Euryattus* sp., the leaf-hanging salticids preyed on by *P. fimbriata* in Queensland (see above), are not the only unusual members of this species. The juveniles are also unusual because they build prey-catching webs (51). The webs of *Euryattus* and *Simaetha* spp. are nonsticky, three-dimensional space webs that lack the funnel shape of webs of *Portia* spp. and are not as densely spun as the sheet web of *Spartaeus* spp. However, another salticid builds a dense sheet web. *Pellenes arciger,* from southern France, builds a large silk sheet in the vegetation (99). The predatory behavior of this species has not been studied, but the large web it spins is probably used as a prey-capture device.

The spinning of *Plexippus paykulli* illustrates the blurry distinction between a nest and a web. The nest of this species is a tube surrounded by a dense tangled array of silk that forms a sticky layer over the tube. Insects, such as grasshoppers, coming into contact with nests of *P. paykulli* tend to become stuck for several seconds, or even minutes. *P. paykulli* responds to ensnared insects by coming out of the nest and leaping onto the prey or by walking across the nest and over to the insect to attack it (77).

All of the web-building salticids studied are versatile predators that not only use webs for predation, but also catch prey cursorially.

PREY-SPECIFIC PREDATORY BEHAVIOR OF ANT-EATING SALTICIDS

Most salticids avoid ants, which generally bite, sting, and taste bad, but an interesting minority routinely eat these heavily defended prey. The most thoroughly studied are three euophrynes, *Corythalia canosa, Habrocestum pulex,* and *Zendora* (formerly *Pystira*) *orbiculata* (14, 24, 81), and six heliophanines, *Chrysilla lauta, Siler semiglaucous, Natta rufopicta,* and another three undescribed species of *Natta* (82).

Predatory behavior used against ants varies among the species. The six heliophanines are remarkably similar to each other but differ from each of the three euophrynes. Among the euophrynes, *Z. orbiculata* differs considerably in behavior from *C. canosa* and *H. pulex. C. canosa* and *H. pulex* usually maneuver to attack ants head on. Heliophanines also often attack head on, but they attack from directly behind as well. *Z. orbiculata* attacks ants from just about any orientation. However, this species, unlike the other ant-eating salticids, also frequently positions itself facing down on ant-infested tree trunks and ambushes ants by lunging down on them instead of actively pursuing them. *C. canosa* usually holds its cephalothorax elevated while pursuing, attacking, and starting to feed upon ants. The heliophanines, in contrast, tend to hold the first pair of legs, but not their cephalothoraxes, elevated. When the euophrynes attack, they usually hold on, but the heliophanines often stab ants then back away (81, 82).

Although the ant-eating heliophanines and euophrynes are behaviorally specialized on ants, their diet is not restricted to ants. They attack other prey in typical salticid stalk-and-leap sequences (81, 82).

SUBTLE PREDATORY VERSATILITY IN *PHIDIPPUS*

Phidippus is a genus of common, sometimes large (e.g. *Phidippus regius* can reach 22 mm in body length) salticids in North America. At least one species, *Phidippus audax,* is an important insect predator in agroecosystems (132). In studies of predatory behavior, *Phidippus* spp. have generally been described as typical insectivorous salticids, but there is more to these spiders than the usual stalk-and-leap routine.

In nature, the studied species of *Phidippus* appear to prey opportunistically on a diverse assortment of arthropods, but these spiders' diets seem to be biased especially toward caterpillars and flies (49, 124, 126). Although not so striking as the predatory versatility of *Portia* spp., the predatory behavior of *Phidippus* spp. apparently consists of two different prey-specific strategies (25, 26, 35). Each appears to be especially efficient for catching particular prey.

Phidippus spp. approach the two types of prey differently and leap on them from different distances, stalking closer to caterpillars than to flies before attacking. Typically, upon seeing a moving caterpillar, a *Phidippus* spider approaches rapidly to within 10–12 mm, then pauses and watches it. If the caterpillar continues to move, the spider circles until it is directly in front, stalks forward a few millimeters, then leaps and pins the caterpillar's head down. In interactions with flies, *Phidippus* spp. approach by walking rapidly, pause when 25–30 mm away, then leap. Attacks on flies are initiated from just about any direction, but regardless of the direction of attack, the spiders almost always capture the flies by biting their thoraxes near the wing bases (25, 26). Also in *Phidippus* spp. and some other salticids, the spider's approach and method of attack are influenced by the size and speed of the prey (16, 18, 32, 35, 37, 47). These salticids move around and attack large prey from the rear, but attack small prey from any orientation. Prey that is stationary or moving only slowly is stalked slowly, but rapidly moving prey is chased.

Observations of the predatory behavior of *Phidippus* spp. suggest that prey-specific predatory behavior may be more common than previously supposed. Perhaps when we compare salticids interspecifically, the question to ask is not whether the species exhibits predatory versatility, but instead how pronounced that versatility is. The same lesson may apply to the individual behavioral flexibility demonstrated by the detours taken by *Portia* spp. to reach prey. *Phidippus* spp. have not been seen to undertake detours as long as those of *Portia* spp. However, the length of *Portia*'s detours makes sense as a hunting strategy against web-building spiders, because these are sedentary victims. An active insect is unlikely to stay put long enough to allow for a long detour, but *Phidippus* spp. do undertake short detours to reach insect prey (45). A common European insectivorous salticid, *Evarcha blancardi,* also takes short detours to reach prey (42), which suggests that detouring may be a widespread ability in the salticids.

Whether *Phidippus* spp. undertake longer detours when the prey are web-building spiders has not been tested. This question is appropriate because *Phidippus* spp. are, in fact, web-invading araneophagic spiders (90, 119). However, rather than practice aggressive mimicry like *Portia* spp., *Phidippus* spp. leap into the web to catch the spider. Similar leaping attacks on web-building spiders may be widespread among salticids (9, 27, 51, 56–58, 113).

Givens (38) demonstrated that *P. audax* males feed less often and take smaller prey than do females. *Portia* spp. appear to exhibit a similar male-female trend (73), which may be widespread in the Salticidae.

PREDATORY STRATEGY OF *MYRMARACHNE*

Myrmarachne is a large genus of predominantly tropical ant-like species (121). Generally these species do not eat ants, but instead, like the majority of salticids, feed on a wide range of arthropod prey. However, their prey-catching methods are unusual (54, 87).

In a typical predatory sequence, a *Myrmarachne* spider runs up to the prey, taps it with the first pair of legs, then attacks it by lunging rather than leaping (54, 87). These sequences are unique among salticids that have been studied

(112) and appear especially appropriate for an ant-mimicking species because, by adopting this style of predation, the spider can usually capture prey with only a minor disruption in its ant-like walking gait. Ants often tap each other and other animals with their antennae, and *Myrmarachne* spp. may tap their prey to maintain Batesian mimicry (21) during predatory sequences: If *Myrmarachne* spp. did not tap their prey, the spiders' predators might more readily recognize them. However, is there a cost? Does tapping alert potential prey or give them an opportunity to flee before being attacked?

How serious the disadvantage of warning the prey might be to *Myrmarachne* spp. probably varies considerably with the type of prey, and this variation may have influenced the natural diet of *Myrmarachne* spp. These spiders apparently feed on a wide range of arthropod prey, including other salticids, but they seem to prefer insects that are slow to take flight, such as moths (87). Interestingly, whether tapping alarms the prey is not clear. It may even have a calming effect. Moths, and even salticids, sometimes stay more or less stationary when tapped by *Myrmarachne* spp.

Myrmarachne spp. also feed on a prey item that cannot flee—the eggs of other spiders. *G. lanka and Portia, Brettus,* and *Cyrba* spp. also eat spider eggs. *Myrmarachne* adults get at eggs by using their fangs to tear open nests of cursorial spiders, especially those of other salticids, including other *Myrmarachne* spp. Small juveniles of *Myrmarachne* spp. enter nests of other spiders and feed on eggs one at a time (54, 87). This tactic is also used by adults of *Phyaces comosus* (55), a minute, highly cryptic salticid from Sri Lanka. Other salticid species feed on insect eggs (43, 44, 88, 103, 104, 125, 127). Nectar is another stationary, but energy-rich, salticid food source (22, 108, 111; SD Pollard & RR Jackson, unpublished data).

SEXUAL DIMORPHISM AND TRADE-OFFS BETWEEN MATING AND PREDATORY STRATEGIES IN *MYRMARACHNE* AND *PORTIA*

In salticids, as in insects (118), intrasexual competition for access to potential mates has probably been the primary selection pressure responsible for the evolution of secondary sexual characteristics in males. In males of *Myrmarachne* spp., we find some of the most dramatic examples of these often bizarre and, from the perspective of survival, incongruous features, which are usually exaggerated forms of structures found on conspecific females.

In *Myrmarachne plataleoides,* one of the most sexually dimorphic species in this genus, the differences in male and female predatory behavior can be attributed to enlarged chelicerae in males (109, 110). In this species, the female's chelicerae hang down at right angles to the horizontal plane, as in most salticids. The male's chelicerae are about five times longer than the female's and project forward, parallel with the body's horizontal plane. During intrasexual conflict, males first spread apart the elongated basal segments of their chelicerae to expose their extra-long fangs, then approach, make contact, and push against each other in contests of strength, as described in detail in another sexually dimorphic *Myrmarachne* species, *M. lupata* (50). The male and female *M. plataleoides* are both convincing mimics of weaver ants (*Oecophylla smaragdina*). Juvenile *M. plataleoides* males have short chelicerae like those of females. During the final molt, the male emerges sexually mature, equipped with fully elongated chelicerae (102). Mature females and juveniles of both sexes are armed with venom-injecting glands, but the male's fangs have no openings (109, 110). The venom glands are at the base of each chelicera, close to the spider's eyes, and in males, a continuous duct from there to the tips of the fangs would have to be about five times longer than the ducts in females. Even if the male could organize an intact duct of this length in its final molt, it is unlikely that it could generate, by squeezing the venom glands, sufficient pressure to eject venom from the distant fang tips.

M. plataleoides males have apparently made evolutionary adjustments in prey capture to compensate for the inability to envenomate. Unlike females, males hold prey down while making repeated stabs with the long fangs. This method is less effective than the female's venom-based style of prey capture, and many prey manage to escape before males can stab them to death (54, 87). In addition, the modified male chelicerae reduce the effectiveness of feeding on captured prey. *M. plataleoides* females puncture prey with their fangs, then suck nutrients out from the holes. However, the male's long fangs push through both sides of the prey's body so that the tips point back into the spider's mouth. The male then sucks from the large holes in its skewered prey. Potential nutrients can leak out of the prey's body, and the viscera that remain in place increase in viscosity through evaporative fluid loss (see 107). Consequently, males take longer to feed compared with females and extract less food (SD Pollard, unpublished data).

Although *M. plataleoides* males are less efficient than females at catching active prey, they appear as efficient at oophagy (87). In fact, the male's large chelicerae might be an asset when it removes silk from egg sacs and reaches into the nest for the eggs.

The elongated chelicerae of males of *Myrmarachne* spp. have apparently not jeopardized their ability to mimic ants. *M. plataleoides* is especially interesting. Weaver ant colonies have major workers that forage and minor workers that care for the eggs and larvae inside the nest. Major workers commonly carry minors from one subnest to another by holding the smaller ants' abdomens in their mandibles. The minor worker being carried holds its legs against the side of its body (46). Remarkably, the *M. plataleoides* male, with chelicerae that simulate the minor worker, closely resembles this duo. The ability of *M. plataleoides* males to maintain the illusion of being ant-like, by mimicking ants carrying nest mates, may have facilitated evolution of large chelicerae. In fact, *M. plataleoides* may be only an extreme example of how *Myrmarachne* males generally resemble ants carrying nest mates, food, or other objects.

The behavior of ant-eating salticids (see above) supports this hypothesis. These salticids prefer to attack ants that are carrying something in their mandibles (RR Jackson & SD Pollard, unpublished data), presumably because such ants cannot readily use their mandibles for defense. Also, the ant-eating salticids more readily stalk *Myrmarachne* males than females, which suggests that the ant-hunters initially mistake *Myrmarachne* males for ants with occupied mandibles. Stalked *Myrmarachne* spp. of both sexes, however, usually escape unharmed because, by briefly displaying to the ant-hunter, they communicate that they are really salticids and risky prey to attack.

Perhaps not only secondary, but also primary sexual characters have had evolutionary effects on the foraging behavior of salticid males. In spiders, the males' palps, being gonopods (i.e. primary sexual structures, or genitalia), are considerably enlarged compared with the females' palps (29). Palps of salticid males may also have secondary characters—conspicuous hairs and markings that are absent from the females'. In Queensland *P. fimbriata*, the males' palps appear to have a foraging cost in relation to cryptic stalking, the tactic by which this population of *Portia* catches cursorial salticids (72). Cryptic stalking depends primarily on concealment, and one of its consistent components is for *P. fimbriata* to pull back its palps so that their outlines blur into the contours of the body. However, the male of the Queensland *P. fimbriata* is less effective at catching salticids, apparently because his enlarged palps reveal him as a predator to his visually competent prey. Moreover, in experiments where the males' palps were removed, their capture efficiency became indistinguishable from that of females (SD Pollard & RR Jackson, unpublished data). Evidently, in this population of *Portia,* primary sexual characters compromise a predatory tactic. We know the secondary characters (i.e. markings) are not responsible for the males' failures because juvenile males, which have enlarged palps but not the markings, are also less effective than females (adult and juvenile) at catching salticids. Also, primary sexual characters are usually not exaggerated sufficiently to have costs comparable to those of secondary sexual characters (20). However, a combination of the spider's method of sperm transfer, the relation of intersexual selection in salticids to visual displays, and the acute vision of both the cryptically stalking predator and its prey appears to have resulted in a unique adaptive trade-off in males of Queensland *P. fimbriata*.

CUES FOR PREDATORY DECISIONS

The cues typical salticids use for distinguishing between their insect prey and other objects such as mates, rivals, enemies, and irrelevant stimulation have been investigated extensively. Shape, symmetry, presence of legs and wings, size, and style of motion (short, jerky movements) are some of the more important features by which these salticids appear to recognize their prey (12, 18, 23, 32, 34, 42). Appreciating predatory versatility forces us to go beyond the question of how the salticid recognizes prey, but for salticids with complex predatory strategies, we have little information about the cues that influence the different components of the strategy. Most of what we know concerns the cues that govern *Portia*'s decisions about whether to enter a web, whether to make signals once in a web, and whether to persist at signaling (70).

In eliciting web entry, visual cues are effective, but volatile chemicals from the web are not. Seeing a spider in a web increases *Portia*'s inclination to enter the web. After web entry, cues from webs of prey spiders are sufficient to elicit signaling behavior, even in the absence of other cues coming directly from the prey. In contrast, volatile chemical cues from prey spiders are not important. Once *Portia* is on a web and signaling, seeing a moving spider and detecting vibrations on the web encourage it to persist in signaling. On the basis of visual cues alone, *Portia* can distinguish between quiescent spiders, insects, and eggsacs (70).

These studies of cues highlight how far we remain from fully understanding the functioning of the salticid visual system. Although salticid eyes are large and complex for a spider, this animal is no primate (95). The principal eye lens is only a few millimeters in diameter. The numbers of receptors in the salticid eye and neurons in the salticid brain are limited. How so small a visual system, with so few components, can perform these perceptual feats is currently a mystery.

ACKNOWLEDGMENTS

We thank Tracey Robinson and Helen Spinks for their help in the preparation of the manuscript. Parts of the research summarized was supported by grants to RRJ from the National Geographic Society (2330-81, 3226-85) and the United States–New Zealand Cooperative Program of the National Science Foundation (BNS 8617078), and a Killam Memorial Postdoctoral Fellowship held by SDP in the Department of Entomology at the University of Alberta, Edmonton, Canada.

Literature Cited

1. Barth FG, ed. 1985. *Neurobiology of Arachnids*. Berlin-Heidelberg-New York: Springer-Verlag
1a. Blest AD. 1987. Comparative aspects of the retinal mosaics of jumping spiders. In *Arthropod Brain: Its Evolution, Development, Structure, and Functions*, ed. AP Gupta, pp. 203–29. New York: Wiley & Son
2. Blest AD. 1988. Post-embryonic development of the principal retina of a jumping spider. I. The establishment of receptor tiering by conformational changes. *Philos. Trans. R. Soc. London (Biology)* 320:489–504
3. Blest AD, Carter M. 1987. Morphogenesis of a tiered principal retina and the evolution of jumping spiders. *Nature* 328:152–55
4. Blest AD, Carter M. 1988. Post-embryonic development of the principal retina of a jumping spider. II. The acquisition and reorganization of rhabdomeres and growth of the glial matrix. *Philos. Trans. R. Soc. London (Biology)* 320:505–15
5. Blest AD, Hardie RC, McIntyre P, Williams DS. 1981. The spectral sensitivities of identified receptors and the function of retinal tiering in the principal eyes of jumping spiders. *J. Comp. Physiol. A* 145:227–39
6. Blest AD, McIntyre P, Carter M. 1988. A re-examination of the principal retinae of *Phidippus johnsoni* and *Plexippus validus* (Araneae: Salticidae): implications of optical modelling. *J. Comp. Physiol. A* 162:47–56
7. Blest AD, O'Carroll DC. 1989. The evolution of the tiered principal retinae of jumping spiders (Araneae: Salticidae). In *Neurobiology of Sensory Systems*, ed. RN Singh, NJ Strausfeld, pp. 155–70. New York: Plenum
8. Blest AD, O'Carroll DC, Carter M. 1990. Comparative ultrastructure of layer I receptor mosaics in principal eyes of jumping spiders: the evolution of regular arrays of light guides. *Cell Tiss. Res.* 262:445–60
9. Bristowe WS. 1941. *The Comity of Spiders*, Vol. 2. London: Ray Soc.
10. Clark DL, Uetz GW. 1990. Video image recognition by the jumping spider, *Maevia inclemens* (Araneae: Salticidae). *Anim. Behav.* 40:884–90
11. Coddington JA, Levi HW. 1991. Systematics and evolution of spiders (Araneae). *Annu. Rev. Ecol. Syst.* 22:565–92
12. Crane J. 1949. Comparative biology of salticid spiders at Rancho Grande, Venezuela. Part IV. An analysis of display. *Zoologica* 34:159–215
13. Curio E. 1976. *The Ethology of Predation*. Berlin: Springer-Verlag. 250 pp.
14. Cutler B. 1980. Ant predation by *Habrocestum pulex* (Hentz) (Araneae: Salticidae). *Zool. Anz.* 204:97–101
15. De Voe RD. 1975. Ultraviolet and green receptors in principal eyes of jumping spiders. *J. Gen. Physiol.* 66:193–208
16. Dill LM. 1975: Predatory behavior of the zebra spider *Salticus scenicus* (Araneae, Salticidae). *Can. J. Zool.* 53:1284–89
17. Downes MF. 1993. The life history of *Badumna candida* (Araneae: Amaurobioidea). *Aust. J. Zool.* 41:441–66
18. Drees O. 1952. Untersuchungen uber die angeborenen Verhaltensweisen bei Springspinnen (Salticidae). *Z. Tierpsychol.* 9:169–207
19. Duelli P. 1978. Movement detection in the posterolateral eyes of jumping spiders (*Evarcha arcuata*, Salticidae). *J. Comp. Physiol.* 124:15–26
20. Eberhard WG. 1985. *Sexual Selection and Animal Genitalia*. Cambridge, MA: Harvard Univ. Press
21. Edmunds M. 1974. *Defence in Animals: a Survey of Anti-Predatory Defences*. New York: Longman
22. Edmunds M. 1978. On the association between *Myrmarachne* spp. (Salticidae) and ants. *Bull. Br. Arachnol. Soc.* 4:149–60
23. Edwards GB. 1980. Experimental demonstration of the importance of wings to prey evaluation by a salticid spider. *Peckhamia* 4:1–9
24. Edwards GB, Carroll JF, Whitcomb WH. 1974. *Stoidis aurata* (Araneae: Salticidae), a spider predator of ants. *Fla. Entomol.* 57:337–46
25. Edwards GB, Jackson RR. 1993. Use of prey-specific predatory behaviour by North American jumping spiders (Araneae: Salticidae) of the genus *Phidippus*. *J. Zool. London* 229:709–16
26. Edwards GB, Jackson RR. 1994. The role of experience in the development of predatory behaviour in *Phidippus regius*, a jumping spider (Araneae, Salticidae) from Florida. *NZ J. Zool.* 21:269–77
27. Enders F. 1975. The influence of hunting manner on prey size, particularly in spiders with long attack distances (Araneidae, Linyphiidae, and Salticidae). *Am. Nat.* 109:737–63
28. Endler JA. 1981. An overview of the

relationships between mimicry and crypsis. *Zool. J. Linn. Soc.* 16:25–31

29. Foelix RF. 1982. *Biology of Spiders.* Cambridge, MA: Harvard Univ. Press. 306 pp.

30. Forster LM. 1977. A qualitative analysis of hunting behaviour in jumping spiders (Araneae: Salticidae). *NZ J. Zool.* 4:51–62

31. Forster LM. 1977. Some factors affecting feeding behaviour in jumping spiders (Araneae: Salticidae). *NZ J. Zool.* 4:435–43

32. Forster LM. 1979. Visual mechanisms of hunting behaviour in *Trite planiceps*, a jumping spider (Araneae: Salticidae). *NZ J. Zool.* 6:79–93

33. Forster LM. 1982. Vision and prey-catching strategies in jumping spiders. *Am. Sci.* 70:165–75

34. Forster LM. 1985. Target discrimination in jumping spiders (Araneae: Salticidae) See Ref. 1, pp. 249–74

35. Freed AN. 1984. Foraging behaviour in the jumping spider *Phidippus audax:* bases for selectivity. *J. Zool. London* 202:49–61

36. Galiano ME. 1976. Commentaries sobre la categoria sistematica del taxon Lyssomanidae (Araneae). *Rev. Mus. Argent. Cienc. Nat. Bernardino Rivadavia Inst. Nac. Invest. Cienc. Nat. Cienc. Geol.* 5:59–70

37. Gardner BT. 1965. Observations on three species of *Phidippus* jumping spiders (Araneae: Salticidae). *Psyche* 72:133–47

38. Givens RP. 1978. Dimorphic foraging strategies of a salticid spider (*Phidippus audax*). *Ecology* 59:309–21

39. Griffin DR. 1981. *The Question of Animal Awareness.* New York: Rockefeller Univ. Press. 2nd ed.

40. Hallas, SEA, Jackson RR. 1986. A comparative study of Old and New World lyssomanines (Araneae, Salticidae): utilisation of silk and predatory behaviour of *Asemonea tenuipes* and *Lyssomanes viridis. NZ J. Zool.* 13:543–51

41. Hardie RC, Duelli P. 1978. Properties of single cells in posterior lateral eyes of jumping spiders. *Z. Naturforsch.* 33:156–58

42. Heil KH. 1936. Beiträge zur Physiologie und Psychologie der Springspinnen. *Z. Vgl. Physiol.* 23:125–49

43. Hensley SD. 1971. Management of sugarcane borer populations in Louisiana, a decade of change. *Entomophaga* 16:133–46

44. Hensley SD, Long WH, Roddy LR, McCormick WJ, Concienne EJ. 1961. Effects of insecticides on predaceous arthropod fauna of Louisiana sugarcane fields. *J. Econ. Entomol.* 54:146–49

45. Hill DE. 1979. Orientation by jumping spiders of the genus *Phidippus* (Araneae: Salticidae) during the pursuit of prey. *Behav. Ecol. Sociobiol.* 5:301–22

46. Hölldobler B, Wilson EO. 1990. *The Ants.* Cambridge, MA: Harvard Univ. Press. 732 pp.

47. Hollis JH, Branson BA. 1964. Laboratory observations on the behavior of the salticid spider *Phidippus audax* (Hentz). *Trans. Kans. Acad. Sci.* 67:131–48.

48. Homann H. 1928. Beträge zur Physiologie der Spinnenaugen. I. Untersuchungsmethoden, II. Das Sehvermöen der Salticiden. *Z. Vgl. Physiol.* 7:201–68

49. Jackson RR. 1977. Prey of the jumping spider *Phidippus johnsoni* (Araneae: Salticidae). *J. Arachnol.* 5:145–49

50. Jackson RR. 1982. The biology of ant-like jumping spiders: intraspecific interactions of *Myrmarachne lupata* (Araneae, Salticidae). *Zool. J. Linn. Soc.* 76:293–319

51. Jackson RR. 1985. The biology of *Euryattus* sp. indet., a web-building jumping spider (Araneae, Salticidae) from Queensland: utilization of silk, predatory behaviour, and intraspecific interactions. *J. Zool. London B* 1:145–73

52. Jackson RR. 1985. The biology of *Simaetha paetula* and *S. thoracica*, web-building jumping spiders (Araneae, Salticidae) from Queensland: co-habitation with social spiders, utilization of silk, predatory behaviour and intraspecific interactions. *J. Zool. London* B1:175–210

53. Jackson RR. 1986. Web building, predatory versatility, and the evolution of the Salticidae. See Ref. 114a, pp. 232–68

54. Jackson RR. 1986. The biology of ant-like jumping spiders (Araneae, Salticidae): prey and predatory behaviour of *Myrmarachne* with particular attention to *M. lupata* from Queensland. *Zool. J. Linn. Soc.* 88:179–90

55. Jackson RR. 1986. The biology of *Phyaces comosus* (Araneae: Salticidae), predatory behaviour, antipredator adaptations and silk utilization. *Bull. Br. Mus. Nat. Hist. Zool.* 50:109–16

56. Jackson RR. 1988. The biology of *Jacksonoides queenlandica*, a jumping spider (Araneae: Salticidae) from Queensland: intraspecific interactions, web-invasion, predators, and prey. *NZ J. Zool.* 15:1–37

57. Jackson RR. 1988. The biology of *Tauala lepidus*, a jumping spider (Araneae: Salticidae) from Queensland: dis-

play and predatory behaviour. *NZ J. Zool.* 15:347–64

58. Jackson RR. 1989. The biology of *Cobanus mandibularis*, a jumping spider (Araneae: Salticidae) from Costa Rica: intraspecific interactions, predatory behaviour and silk utilisation. *NZ J. Zool.* 16:383–92

59. Jackson RR. 1990. Predator-prey interactions between jumping spiders (Araneae, Salticidae) and *Pholcus phalangioides* (Araneae, Pholcidae). *J. Zool. London* 220:553–59

60. Jackson RR. 1990. Predatory and silk utilisation behaviour of *Gelotia* sp. indet. (Araneae: Salticidae: Spartaeinae), a web-invading aggressive mimic from Sri Lanka. *NZ J. Zool.* 17:475–82

61. Jackson RR. 1990. Predatory versatility and intraspecific interactions of *Cyrba algerina* and *C. ocellata*, web-invading spartaeine jumping spiders (Araneae, Salticidae). *NZ J. Zool.* 17:157–68

62. Jackson RR. 1990. Predatory and nesting behaviour of *Cocalus gibbosus*, a spartaeine jumping spider (Araneae: Salticidae) from Queensland. *NZ J. Zool.* 17:483–90

63. Jackson RR. 1990. Ambush predatory behaviour of *Phaeacius malayensis* and *Phaeacius* sp. indet., spartaeine jumping spiders (Araneae: Salticidae) from tropical Asia. *NZ J. Zool.* 17:491–98

64. Jackson RR. 1990. Comparative study of lyssomanine jumping spiders (Araneae: Salticidae): silk use and predatory behaviour of *Asemonea*, *Goleba*, *Lyssomanes*, and *Onomastus*. *NZ J. Zool.* 17:1–6

65. Jackson RR. 1992. Eight-legged tricksters: spiders that specialize at catching other spiders. *BioScience* 42:590–98

66. Jackson RR. 1992. Conditional strategies and interpopulation variation in the behaviour of jumping spiders. *NZ J. Zool.* 19:99–111

67. Jackson RR. 1992. Predator-prey interactions between web-invading jumping spiders and *Argiope appensa* (Araneae, Araneidae), a tropical orb-weaving spider. *J. Zool. London* 228:509–20

68. Jackson RR. 1992. Predator-prey interactions between web-invading jumping spiders and two species of tropical web-building pholcid spiders, *Psilochorus sphaeroides* and *Smeringopus pallidus*. *J. Zool. London* 227:531–36

69. Jackson RR. 1992. Predator-prey interactions between web-invading jumping spiders and a web-building spider, *Holocnemus pluchei* (Araneae, Araneidae). *J. Zool. London* 228:589–94

70. Jackson RR. 1995. Cues for web invasion and aggressive mimicry signalling in *Portia* (Araneae, Salticidae). *J. Zool. London* 236:131–49

71. Jackson RR, Blest AD. 1982. The distances at which a primitive jumping spider, *Portia fimbriata*, makes visual discriminations. *J. Exp. Biol.* 97:441–45

72. Jackson RR, Blest AD. 1982. The biology of *Portia fimbriata*, a web-building jumping spider (Araneae, Salticidae) from Queensland: utilization of webs and predatory versatility. *J. Zool. London* 196:255–93

73. Jackson RR, Hallas SEA. 1986. Comparative biology of *Portia africana*, *P. albimana*, *P. fimbriata*, *P. labiata*, and *P. schultzi*, araneophagic web-building jumping spiders (Araneae: Salticidae): utilisation of silk, predatory versatility, and intraspecific interactions. *NZ J. Zool.* 13:423–89

74. Jackson RR, Hallas SEA. 1986. Capture efficiencies of web-building jumping spiders (Araneae, Salticidae): Is the jack-of-all-trades the master of none? *J. Zool. London* 209:1–7

75. Jackson RR, Hallas SEA. 1986. Predatory versatility and intraspecific interactions of spartaeine jumping spiders (Araneae: Salticidae): *Brettus adonis*, *B. cingulatus*, *Cyrba algerina* and *Phaeacius* sp. indet. *NZ J. Zool.* 13:491–520

76. Jackson RR, Hallas SEA. 1990. Evolutionary origins of displays used in aggressive mimicry by *Portia*, a web-invading, araneophagic jumping spider (Araneae, Salticidae). *NZ J. Zool.* 17:7–23

77. Jackson RR, Macnab AM. 1989. Display, mating and predatory behaviour of the jumping spider *Plexippus paykulli*, (Araneae, Salticidae). *NZ J. Zool.* 16:151–68

78. Jackson RR, Pollard SD. 1990. Web-building and predatory behaviour of *Spartaeus spinimanus* and *Spartaeus thailandicus*, primitive jumping spiders (Araneae, Saltidicae) from Southeast Asia. *J. Zool. London* 220:561–67

79. Jackson RR, Rowe RJ, Wilcox RS. 1993. Anti-predator defences of *Argiope appensa* (Araneae, Araneidae), a tropical orb-weaving spider. *J. Zool. London* 229:121–32

80. Jackson RR, Tarsitano MS. 1993. Responses of jumping spiders to motionless prey. *Bull. Br. Arachnol. Soc.* 9:105–9

81. Jackson RR, van Olphen A. 1991. Prey-capture techniques and prey-preferences of *Corythalia canosa* and *Pystira orbiculata*, ant-eating jumping spiders

(Araneae, Salticidae). *J. Zool. London* 223:577–91

82. Jackson RR, van Olphen A. 1992. Prey-capture techniques and prey preferences of *Chrysilla, Natta* and *Siler,* ant-eating jumping spiders (Araneae, Salticidae) from Kenya and Sri Lanka. *J. Zool. London* 227:163–70

83. Jackson RR, Wilcox RS. 1990. Aggressive mimicry, prey-specific predatory behaviour and predator-recognition in the predator-prey interactions of *Portia fimbriata* and *Euryattus* sp., jumping spiders from Queensland. *Behav. Ecol. Sociobiol.* 26:111–19

84. Jackson RR, Wilcox RS. 1993. Observations in nature of detouring behaviour by *Portia fimbriata,* a web-invading aggressive mimic jumping spider from Queensland. *J. Zool. London* 230:135–39

85. Jackson RR, Wilcox RS. 1993. Predator-prey co-evolution of *Portia fimbriata* and *Euryattus* sp., jumping spiders from Queensland. *Mem. Qld. Mus.* 33:557–60

86. Jackson RR, Wilcox RS. 1993. Spider flexibly chooses aggressive mimicry signals for different prey by trial and error. *Behaviour* 127:21–36

87. Jackson RR, Willey MB. 1994. Comparative study of the predatory behaviour of *Myrmarachne,* ant-like jumping spiders (Araneae, Salticidae). *Zool. J. Linn. Soc.* 110:77–102

88. Jennings DT, Houseweart MW. 1978. Spider preys on spruce budworm egg mass. *Entomol. News* 89:183–86

89. Kaestner A. 1950. Reaktion der Hüpfspinnen (Salticidae) auf unbewegte farblose und farbige Gesichtsreize. *Zool. Beitr.* 1:12–50

90. Lamore DH. 1958. The jumping spider *Phidippus audax* Hentz and the spider *Conopistha trigona* Hentz, as predators of the basilica spider *Allepeira lemniscata* Walckenaer, in Maryland. *Proc. Entomol. Soc. Wash.* 60:286

91. Land MF. 1969. Structure of the retinae of the principal eyes of jumping spiders (Salticidae: Dendryphantinae) in relation to visual optics. *J. Exp. Biol.* 51:443–70

92. Land MF. 1969. Movements of the retinae of jumping spiders (Salticidae: Dendryphantinae) in response to visual stimuli. *J. Exp. Biol.* 51:471–93

93. Land MF. 1971. Orientation by jumping spiders in the absence of visual feedback. *J. Exp. Biol.* 54:119–39

94. Land MF. 1972. Stepping movements made by jumping spiders during turns

mediated by the lateral eyes. *J. Exp. Biol.* 57:15–40

95. Land MF. 1974. A comparison of the visual behavior of a predatory arthropod with that of a mammal. In *Invertebrate Neurons and Behavior,* ed. CAG Wiersma, pp. 411–18. Cambridge, MA: MIT Press

96. Land MF. 1985. Fields of view of the eyes of primitive jumping spiders. *J. Exp. Biol.* 119:381–84

97. Land MF. 1985. The morphology and optics of spider eyes. See Ref. 1, pp. 53–78

98. Land MF, Fernald RD. 1992. The evolution of eyes. *Annu. Rev. Neurosci.* 15:1–29

99. Lopez A. 1986. Construction de toiles en 'Voile de bataeu' par une araignee salticide languedocine. *Bull. Soc. Arch. Beziers* 2:65–68

100. Manly BFJ, Forster L. 1979. A stochastic model for the predatory behaviour of naive spiderlings (Araneae: Salticidae). *Biom. J.* 21:115–22

101. Masters WM, Markl HS, Moffat AM. 1986. Transmission of vibrations in a spider's web. See Ref. 114a, pp. 49–69

102. Mathew AP. 1954. Observations on the habits of the two spider mimics of the red ant, *Oecophylla smaragdina* (Fabr.). *J. Bombay Nat. Hist. Soc.* 52:249–63

103. McDaniels SG, Sterling WL. 1982. Predation of *Heliothis virescens* (F.) eggs on cotton in east Texas. *Environ. Entomol.* 11:60–66

104. McDaniels SG, Sterling WL, Dean DA. 1981. Predators of tobacco bud-worm larvae in Texas cotton. *Southwest. Entomol.* 6:102–8

105. Mitchell RW. 1986. A framework for discussing deception. In *Deception: Perspectives on Human and Nonhuman Deceit,* ed. RW Mitchell, NS Thompson, pp. 3–40. Albany, NY: State Univ. NY Press

106. Peaslee AG, Wilson G. 1989. Spectral sensitivity in jumping spiders (Araneae, Salticidae). *J. Comp. Physiol.* 164:359–63

107. Pollard SD. 1989. Constraints affecting partial prey consumption by a crab spider, *Diaea* sp. indet. (Araneae: Thomisidae). *Oecologia* 81:392–96

108. Pollard SD. 1993. Little murders. *Nat. Hist.* 102(10):58–65

109. Pollard SD. 1994. Consequences of sexual selection on feeding in male jumping spiders. *J. Zool. London* 234:203–8

110. Pollard SD. 1995. Samurai spiders. *Nat. Hist.* 104(3):44–47

111. Pollard SD, Beck MW, Dodson GN

1995. Why do male crab spiders drink nectar? *Anim. Behav.* 49:1443–8

112. Richman DB, Jackson RR. 1992. A review of the ethology of jumping spiders (Araneae, Salticidae). *Bull. Br. Arachnol. Soc.* 9:33–37

113. Robinson MH, Valerio CE. 1977. Attack on large or heavily defended prey by tropical salticid spiders. *Psyche* 84:1–10

114. Rodrigo AG, Jackson RR. 1992. Four jumping spider genera of the *Cocalodes*-group are monophyletic with genera of the Spartaeinae (Araneae: Salticidae). *NZ Nat. Sci.* 19:61–67

114a. Shear WA, ed. 1986. *Spiders: Webs, Behavior, and Evolution.* Stanford, CA: Stanford Univ. Press

115. Shear WA. 1994. Untangling the evolution of the web. *Am. Sci.* 82:256–66

116. Tarsitano MS, Jackson RR. 1992. Influence of prey movement on the performance of simple detours by jumping spiders. *Behaviour* 123:106–20

117. Tarsitano MS, Jackson RR. 1994. Jumping spiders make predatory detours requiring movement away from prey. *Behaviour.* 131:65–73

118. Thornhill R, Alcock J. 1983. *The Evolution of Insect Mating Systems.* Cambridge, MA: Harvard Univ. Press

119. Tolbert WW. 1975. Predator avoidance behaviors and web defensive structures in the orb weavers *Argiope aurantia* and *Ariope trifasciata* (Araneae, Araneidae). *Psyche* 82:29–52

120. Wanless FR. 1978. A revision of the spider genus *Portia* (Araneae: Salticidae). *Bull. Br. Mus. Nat. Hist.* 34:83–124

121. Wanless FR. 1978. A revision of the spider genera *Belippo* and *Myrmarachne* (Araneae: Salticidae) in the Ethiopian region. *Bull. Br. Mus. Nat. Hist. Zool.* 33:1–139

122. Wanless FR. 1980. A revision of the spider genera *Asemonea* and *Pandisus* (Araneae: Salticidae). *Bull. Br. Mus. Nat. Hist. Zool.* 39:213–57

123. Wanless FR. 1984. A review of the spider subfamily Spartaeinae nom.n. (Araneae: Salticidae) with descriptions of six new genera. *Bull. Br. Mus. Nat. Hist. Zool.* 46:135–205

124. Warren LO, Peck WB, Tadic M. 1967. Spiders associated with the fall webworms, *Hyphantria cunea* (Lepidoptera: Arctiidae). *J. Kans. Entomol. Soc.* 40:382–95

125. Whitcomb WH, Bell K. 1964. Predaceous insects, spiders, and mites of Arkansas cotton fields. *Bull. Ark. Agric. Exp. Stn.* 690:1–84

126. Whitcomb WH, Exline H, Hunter RC. 1963. Spiders of the Arkansas cotton field. *Ann. Entomol. Soc. Am.* 56:653–60

127. Whitcomb WH, Tadic M. 1963. Araneida as predators of the fall webworm. *J. Kans. Entomol. Soc.* 36:186–90

128. Wijesinghe DP. 1992. A new genus of jumping spider from Borneo with notes on the spartaeine palp (Araneae: Salticidae). *Raffles Bull. Zool.* 40:9–19

128a. Jackson RR. 1995. Cues for web invasion and aggressive mimicry signalling in *Portia* (Araneae, Salticidae). *J. Zool. London* 236:131–49

129. Williams DS, McIntyre P. 1980. The principal eyes of a jumping spider have a telephoto component. *Nature* 288:578–80

130. Witt PN. 1975. The web as a means of communication. *Biosci. Commun.* 1:7–23

131. Yamashita S, Tateda H. 1976. Spectral sensitivities of jumping spiders' eyes. *J. Comp. Physiol.* 105:29–41

132. Young OP, Edwards GB. 1990. Spiders in United States field crops and their potential effect on crop pests. *J. Arachnol.* 18:1–27

Annu. Rev. Entomol. 1996. 41:309–24

DISCONTINUOUS GAS EXCHANGE IN INSECTS

John R. B. Lighton
Department of Biological Sciences, University of Nevada at Las Vegas, Las Vegas,
Nevada 89154-4004

KEY WORDS: ventilation, metabolic rate, tracheae, spiracles

ABSTRACT

Many insects exchange respiratory gases cyclically and discontinuously. A typical discontinuous gas exchange cycle (DGC) starts with a closed-spiracle (C) phase, during which little external gas exchange takes place, followed by a fluttering-spiracle (F) phase, which is usually dominated by diffusive oxygen uptake. The DGC is terminated by an open-spiracle (O) phase, during which accumulated CO_2 escapes. This review critically examines the applicability of the DGC to insect gas exchange in general, discusses the primary mechanisms of gas exchange in the F and O phases, evaluates the widespread hypothesis that the DGC lowers respiratory water loss rates adaptively, and proposes new hypotheses concerning the evolutionary genesis of the DGC in insects and other tracheate arthropods.

Introduction to the Field

The behavioral and other epiphenomena that we rightly admire in insects, and to which most entries in this volume are devoted, vanish milliseconds after oxidative phosphorylation stops and the currency that neuronal circuitry and motor effectors require is no longer minted. In large measure, we ignore this fact because natural selection has fine-tuned insect gas delivery and elimination systems to—from our perspective—an eerie state of irrelevance. When faced with insects that seldom display performance limitations imposed by gas-exchange systems, we conclude quite erroneously that these systems are functionally irrelevant to the epiphenomena alluded to above. Such is not the case. The gas exchange capacities and strategies of insects occupy a strategic crossroads between physiology, ecology, behavior, and evolution. For example, leaf-cutting ants allocate 25% of total body weight to mandibular adductors. Thus empowered, leaf-cutters slice the flesh of hard leaves to create their famous semicircular leaf fragments. This behavior increases whole-body gas

309

exchange rates 30-fold above resting levels (48). Most vertebrates can barely sustain a 12-fold increase; many even less (49). Had the gas exchange capabilities of the ants been similarly proscribed by their respiratory physiology, leaf-cutting as we know it would not exist; fungus-garden tending in the absence of a dependable, abundant substrate resource would not have evolved; and the evolutionary history of the Attine ants, and perhaps of neotropical forests, would have been profoundly different. The tracheal system of insects is arguably the foundation of their terrestrial success, because it permits a huge scope for aerobic metabolism; however, this advantage is associated with an allied disadvantage.

Teleologically speaking, by eliminating respiratory pigments and opting instead for direct supply of O_2 to the tissues, insects have avoided the extensive, expensive circulatory arborization characteristic of vertebrates. As such, they are capable of aerobic metabolic performance well in excess of vertebrate limits, especially in the area of flight. With the benefits of enhanced gas flux rate capacity comes a dismaying corollary, because insects may now exacerbate their already water-challenged state (caused primarily by large surface area to volume ratios) with high rates of water vapor efflux from the tracheal system. The tracheal system of most insects is therefore caught between the Scylla of insufficient gas flux rates for peak activity levels and the Charybdis of excessive water vapor efflux rates when inactive. Efficient modulation, by the spiracles, of gas flux rates both into and out of the tracheal system is therefore an absolute requirement for long-term osmotic homeostasis in the terrestrial environments that insects have so successfully invaded. As a secondary effect of active flux rate modulation, a curious gas exchange cycle has developed in many insect groups that is characterized by substantial discontinuities in external gas exchange rates, by the effective intracyclic decoupling of O_2 uptake and CO_2 emission, and by a division—like Caesar's Gaul—into three distinct parts or phases. This gas exchange strategy is the subject of this review.

A necessarily brief and idiosyncratic history of the field would open with the discovery a half century ago (references in 46) that diapausing moth pupae emit CO_2 in discrete bursts. In the 1950s and 1960s (4, 46) sundry speculative works pondered mechanisms, the investigators often erroneously concluding that cellular respiration itself was discontinuous. Schneiderman and his collaborators first unravelled the nature of the discontinuous gas-exchange cycle, here called the DGC, in diapausing lepidopteran pupae (reviewed in 15, 19, 30, 43, 44, 51, 52).

The DGC begins (Figure 1) with the closed-spiracle (C) phase. During this period, the pupa consumes O_2 from its endotracheal stores. Cellular respiration, meanwhile, produces CO_2 that is largely buffered by bicarbonate and proteins in the hemolymph. One gas species is removed and not fully replaced. Thus,

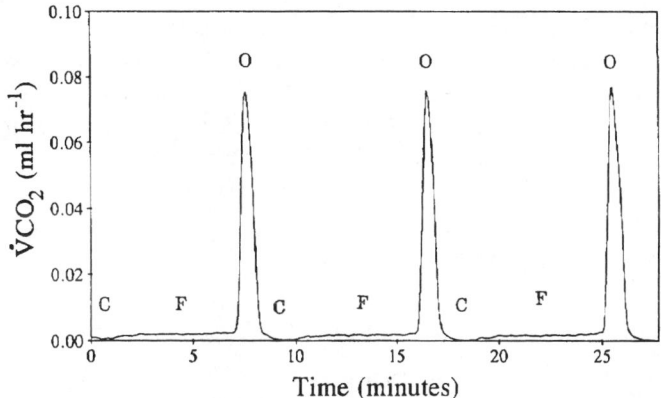

Figure 1   Three discontinuous gas-exchange cycles (DGCs) in an adult ant. These DGCs show the same division into three distinct phases as those of the lepidopteran pupae. They were measured as rate of CO_2 release, or $\dot{V}CO_2$, in an individual ant, *Camponotus vicinus,* mass 0.03 g, at 25°C. The C, F, and O phases marked on the graph have been validated in ants through direct spiracular observation (36).

pO_2 falls and endotracheal pressure tends toward negative. Finally, a central nervous system (CNS)–mediated endotracheal O_2 setpoint of about 5 kPa is reached; the spiracular closer muscles are periodically inactivated; and the spiracular valves start to flutter (F phase). Mass inflow of air occurs down the pressure gradient, supplying the pupa with O_2 sufficient to meet steady-state tissue respiration requirements. This phase is useful (10) because of the resistance it imposes to outward diffusion of H_2O molecules from the H_2O-saturated interior of the tracheae. Opinions vary concerning the importance of diffusion, as opposed to convection, in the F phase (e.g. 16), but many authorities hold F phase gas exchange to be largely (see especially 19) or entirely (53) convective. Meanwhile, CO_2 continues to accumulate, and although still buffered in the hemolymph (see 3), pCO_2 rises until it reaches an endotracheal partial pressure of about 4–6 kPa. At this point, peripherally mediated inactivation of the spiracular closer muscles occurs; the spiracles open (O phase); and the CO_2 escapes. Thus, for two of the three phases of the DGC, respiratory water loss is either zero (C phase) or very low (F phase). Allied phenomena were sporadically documented in adult insects (46, 55).

Nevertheless, diapausing pupae are specialized, and the relevance of the DGC theories to adult insects remained doubtful. Most textbooks repeated the Kroghian dogma (20) that diffusion through partly open spiracles characterized gas exchange in small, inactive insects. Spiracles were generally assumed to exist, in Miller's delightful phrase, to keep out small strolling arthropods (43).

This belief prevailed despite persuasive evidence that spiracles are under more active and stringent control to prevent water loss (8, 15, 55). Miller (44) first drew wide attention to Paul Kestler's pioneering work (17, 18), which suggested that DGCs of moth pupae and cockroaches shared several common features. Kestler followed respiratory water losses over the DGC using real-time measurements and demonstrated extremely low rates of H_2O efflux from the spiracular system during the F phase, as predicted [but not measured, except qualitatively by Kanwisher (16)] by the workers on lepidopteran pupae. This work was later exhaustively summarized (19).

Occurrence and Nature of the DGC in Adult Insects

Convincing demonstrations of DGCs in adult insects were limited in the mid-1980s to Kestler's work on cockroaches (17–19; but see 2, 23). This work also plainly demonstrated that cockroaches developed subatmospheric endotracheal pressures during the F phase. Hence, Kestler reasonably concluded that the F phase in adult insects was characterized primarily by inward convective mass flow (19). However, how the tok-tok beetle, *P. striatus* (24), could sustain significant negative endotracheal pressures was difficult to imagine because of the extreme flaccidity of the abdominal tergites beneath its fused subelytral chamber and the fact that it possessed air sacs (JRB Lighton, personal observations). However, the role of diffusion vs convection in the F phase could be tested by using sensitive flow-through O_2 respirometry in addition to the CO_2-based respirometry used by Kestler and others. A purely convective mass inflow into the tracheal system of an insect measured external to the animal is exactly equivalent to a slow withdrawal, at an equivalent flow rate, of all gas species from the air stream flowing into the respirometry chamber. Such a flow does not change relative partial pressures and thus is undetectable in a measurement paradigm that estimates rates of O_2 uptake from just such changes.

A purely convective F phase of exactly such an undetectable nature forms the basis of some recent theoretical analyses of the DGC in adult insects (53). Significantly, the first quantitative publication on real-time O_2 and CO_2 exchange during the DGC in any adult insect (24) showed that the F phase rate of O_2 uptake ($\dot{V}O_2$) was equivalent to tissue-level (mitochondrial) $\dot{V}O_2$. This was assumed to be the case within pupae (21) and cockroaches (19), but in *P. striatus*, the F phase $\dot{V}O_2$ was measured externally by flow-through respirometry. Therefore, its F phase $\dot{V}O_2$ had to be primarily diffusive, because it could be measured in the first place and because it was similar to overall $\dot{V}O_2$. If mongrel convection that contains a diffusive component is taking place, then some in-diffusion of O_2 and out-diffusion of N_2 must occur (19) and would reduce detected O_2 concentration, although not sufficiently to mimic whole-animal $\dot{V}O_2$, which in such a case comprises total O_2 inflow by both diffusion and convection. As inferred from $\dot{V}O_2$ recordings and the absence of a pupa-like uniformly hard exoskeleton, a significant role for negative endotracheal pressure, and thus for pure inward mass flow, was unlikely in *P. striatus'* F phase.

Meanwhile, CO_2 was emitted from *P. striatus* during the F phase at a rate far slower than tissue catabolism, inferred from $\dot{V}O_2$, would dictate. These data could suggest the presence of negative endotracheal pressure, save that tissue and hemolymph CO_2 capacitance exceed O_2 capacitance by more than an order of magnitude because of the far greater solubility of CO_2 and accompanying hemolymph bicarbonate ion and carbamino buffering. Therefore, partial pressure gradients driving out CO_2 diffusion from the tracheal system must be far lower than similar gradients driving in O_2, which by analogy with pupal systems are approximately 15–18 kPa vs approximately 4–6 kPa for CO_2 (21). Given a measured whole-animal tissue-level respiratory quotient (RQ) ($\dot{V}CO_2/\dot{V}O_2$) of 0.83 (24), a rough calculation—of sufficient precision given uncertain endotracheal gas concentrations—yields an F phase short-term RQ (respiratory exchange ratio, or RER) of approximately 0.2, which does not differ significantly from the measured RER of 0.15 ± 0.06 SD (24). In *P. striatus,* therefore, a largely or purely convective F phase is again ruled out by experimental evidence.

The O phase, in contrast, is undoubtedly convective, as expected from the large CO_2 storage capacity of a 3-g beetle. Repeated abdominal pulsations maximized the replenishment of low-pCO_2 air into which CO_2 could diffuse from the tracheoles; even so, the O phase lasted a remarkable (in the light of later data from much smaller insects) 39% of the total DGC.

Given that the F phase of an insect as large as *P. striatus* appeared to be diffusion-dominated, smaller insects—usually thought to rely entirely on continuous diffusion (20)—should be able to readily sustain a DGC even in the presence of a higher mass-specific metabolic rate. At a small body mass, the effective surface area of the tracheoles vs respiring tissue mass could be maximized, and an O phase dominated by diffusion rather than convection might be expected. At this time (1988), however, no DGCs had been detected in insects smaller than *Periplaneta americana.* The textbook consensus held that such insects exchanged respiratory gases by simple, continuous diffusion (20, 49).

The consensus hypothesis that small insects would not exhibit a DGC was tested with the formicine ant *C. vicinus,* body mass 0.02–0.09 g, and disproved (25). The small size of these ants precluded real-time $\dot{V}O_2$ measurement, but $\dot{V}CO_2$ clearly demonstrated regular CO_2 bursts (O phases). The C and F phases could not be distinguished in this early investigation owing to the comparatively low resolution of then-current CO_2 analyzers.

After the work on *Camponotus vicinus,* further comparative work on other insects attempted to correlate observed DGC parameters with presumed environmental stresses. Thus, studies showed that the Namib Desert dune ant, *Camponotus detritus,* cycles through its DGC far more slowly than *C. vicinus,* which could be interpreted as an adaptive strategy to reduce O phase frequency and thus respiratory water loss (26). Equivalent phenomena were demonstrated in the Saharan thief ant, *Cataglyphis bicolor* (40). In addition, Namib Desert tenebrionid beetles were shown to exhibit a DGC, in accord with the consensus paradigm linking the DGC to selective pressure for reducing respiratory water loss (27). Interestingly, this DGC displayed a purely convective F phase consisting of intermittent abdominal ventilatory movements. The O phases of the ants did not, on the basis of present evidence, utilize convection (30), but those of the (much larger) beetles did (24, 27).

Some progress was made in elucidating other areas of interest, such as the modulation and mass scaling of the DGC (see below). However, it had become clear that a simplistic a posteriori comparative paradigm was not going to yield satisfactory answers to the central questions of the origins and mechanism of the DGC. A more rigorous experimental approach was needed, but greater rigor required more precise instrumentation and a suitably stable experimental preparation that displayed a classic three-phase DGC. The first requirements—CO_2 analyzers sensitive enough to detect the F phase, together with baselining capabilities and software to optimally exploit their capabilities—were soon in place (28), followed by methods for real-time measurement of infrared absorption (38) and water-loss rates based on continuous mass measurement (29). Finally, the central problem of activity—which normally disrupts the DGC (24, 25)—was solved by using beheaded ants as the experimental preparation, because beheading them had little or no effect on the ants' DGC (36).

Initial investigations clearly demonstrated the importance of spiracular control in reducing respiratory water loss, which could increase total water-loss rates by as much as threefold during the O phase (38). Of more significance was the nature and modulation of the C and F phases, which comprise the great majority—typically >80%—of the ant's DGC. These DGC phases are characterized by low spiracular conductance, and thus low respiratory water-loss rates, and are at the crux of understanding the significance of the DGC with respect to respiratory water-loss rates.

Two important issues, largely settled in the case of diapausing pupae but still unclear in the case of adult insects, now needed to be examined: the nature of the stimulus that initiates the F phase and the nature of the dominant gas-exchange modality in the F phase. The role of convection and diffusion in the F phase could be estimated, as alluded to above, via flow-through O_2 and CO_2 respirometry, but this method has been impractical for use with individual ants. Most laboratories experience significant problems when measuring the $\dot{V}O_2$ of even small rodents; however, the technique can be utilized with individual ants if sufficient care is taken (for details, see my review of this field in the *Journal of Insect Physiology* next year). These experiments are still ongoing in my laboratory, but results from a different investigation into the nature of the F phase can be summarized here.

Manipulation of ambient pO_2 by utilizing hypoxic in addition to normoxic flow-through respirometry represents another powerful method for resolving convection and diffusion in the F phase and was the approach taken in a recent study using beheaded *C. vicinus* as the experimental animals (37). This approach hinges on the prediction that predominantly convective vs diffusive F phase gas exchange should affect the duration of the F phase (and the DGC) in opposite ways under hypoxic conditions. The duration of a convective F phase should decrease with diminishing pO_2, because elevated endotracheal pN_2 should cause premature termination of the F phase, owing to premature dissolution of the pressure gradient required for sufficient inward mass flow of air to meet tissue O_2 requirements. Snyder and others also made this prediction in recent theoretical treatments of the subject (53; see also 22). A diffusive F phase, in contrast, could modulate trans-spiracular conductance to accommodate the lower pO_2 values. In so doing, CO_2 emission rate would have to be increased during the F phase [F phase CO_2 emission is detectable even in normoxic conditions (Figure 1), as validated by simultaneous respirometry and spiracular observation (36)]. This increase in F phase $\dot{V}CO_2$ would, in turn, delay the hypercapnic trigger that initiates the O phase, thereby increasing F phase duration as pO_2 decreased. At the same time, a diminishing endotracheal store of O_2 would prematurely terminate the C phase, if the F phase is initiated by a hypoxic trigger, as in diapausing pupae.

At least in the case of *C. vicinus,* our results plainly disprove the hypothesis that the F phase is primarily convective in nature, as Snyder and others predicted (53). Decreasing ambient pO_2 values substantially increased both F phase $\dot{V}CO_2$ and F phase durations, as predicted for a largely diffusive F phase. At the same time, external hypoxia decreased C phase durations, as predicted for hypoxic F phase initiation.

Further evidence for a primarily diffusive F phase comes from estimations of F phase RER (37). These calculations are necessarily indirect, derived from estimates of tissue-level $\dot{V}O_2$, which are derived in turn from whole-DGC $\dot{V}CO_2$ and RQ. Nevertheless, they agree well with the predicted value of ~0.2 (see above). Moreover, they increase markedly with decreasing pO_2, as predicted by the increase in spiracular conductance required to compensate for lower pO_2 values in a diffusive F phase.

Are these findings relevant to all insects that exhibit a DGC? Probably not. In diapausing pupae, inter-O phase duration does decrease with decreasing

pO_2 (5, 50). At least part of this decrease could arise from mechanisms similar to those predicted for primarily convective F phases (53). Meanwhile, Kestler's work with *Periplaneta americana* argues for a primarily convective or mongrel convective-diffusive F phase (19), and trans-spiracular pressure gradients occur in ants (36). Hence, an at least partly convective F phase seems most likely. However, these intra-ant pressures could occasionally be consistently positive (36), which would make them rather difficult to interpret. These questions are being further investigated by manipulating the density and kinematic viscosity of the ambient air, which allow more detailed tests of mechanisms and control strategies (JRB Lighton, in preparation).

Two Factors Affecting the DGC: Body Mass and Metabolic Rate

In mammals, $\dot{V}O_2$ (units: volume/time) scales with mass to approximately the 0.75 power, while lung volume (and thus ventilation volume) scales with mass to the 1.0 power (49). Ventilation rate therefore scales with mass to the $(0.75 - 1.00) = -0.25$ power, yielding the observed inverse relation between body mass and ventilation rate. The situation in insects is quite different, primarily because the DGC effectively decouples O_2 uptake and CO_2 emission, making CO_2 accumulation the primary determinant of O phase initiation and thus DGC frequency. Recent studies have shown, in a representative group of insects, that CO_2 emission volumes and $\dot{V}CO_2$ share very similar mass-scaling exponents, which leads to a mass-scaling exponent for DGC frequency that is not significantly different from 0 (27). Thus, DGC frequency is independent of body mass, at least in those insects for which we have the requisite data.

The situation with respect to the modulation of the DGC to accommodate different rates of gas exchange is less clear. In the first determination of temperature-induced modulation of the DGC in an adult insect (*C. vicinus*), O phase frequency increased with temperature with a Q_{10} of 3.05, and O phase volume decreased with a Q_{10} of 0.61, yielding a combined Q_{10} of 1.86, equivalent to the Q_{10} determined by integrative, constant-volume respirometry (25). This relation parallels the temperature-mediated decline in O phase CO_2 emission volumes in the harvester ant genus *Pogonomyrmex* (47). By loose analogy with tetrapod systems, these insects increased ventilation rate and decreased tidal volume with rising body temperature and thus $\dot{V}CO_2$. However, this analogy misses two crucial points: The effective decoupling of O_2 and CO_2 exchange dynamics in insects (which makes these CO_2 exchange dynamics more closely analogous to tetrapod ventilation) and the effect of temperature on factors other than $\dot{V}CO_2$.

Temperature is, of course, an expedient means of varying overall $\dot{V}CO_2$ in ectotherms. However, it does introduce complications, notably the steeply negative relation between CO_2 solubility and temperature, which is exacerbated by the negative relation between hemolymph pH and temperature (6). These effects will necessarily conspire to alter hemolymph CO_2-buffering capacity with increasing temperature. Thus, the discovery that temperature and $\dot{V}CO_2$ did not affect O phase CO_2 emission volumes in the arid-region ant *C. bicolor* remains surprising (40). This effect minimizes DGC frequency in *C. bicolor,* which must either be able to modulate the hypercapnic setpoint of the O phase upwards or offset the normal temperature-dependent reduction in CO_2-buffering capacity by some other means, thereby yielding the observed constancy of O phase CO_2 volumes. Which of the two methods *C. bicolor* employs is unknown, as is the mechanism. Also unknown is the frequency of such phenomena in insects that exhibit the DGC and the significance of these possible adaptations with respect to respiratory water loss. Similarly underexplored is the effect on the DGC of varying metabolic demand at a single temperature (a situation more strictly analogous to tetrapod ventilation control), mostly because significant elevation of metabolic rate by activity generally disrupts the DGC (25).

A single investigation has examined single-temperature DGC modulation in worker and alate castes of the ant *Messor julianus* (31); this study showed that both O phase volume and DGC frequency were modulated with $\dot{V}CO_2$, but that frequency was modulated some 30-fold more in worker, compared with alate, castes. Again, more information on DGC modulation at single temperatures is obviously required.

The DGC in Noninsect Arthropods

Insects are not the only tracheate arthropods. Tracheation has evolved independently in several other arthropod groups, most notably acari (ticks and mites), solphugids, pseudoscorpions, and harvestpeople. We have clearly established that ticks exhibit a pronounced DGC (35); its detailed nature is under investigation. Interestingly, the giant red velvet mite *Dinothrombium magnificum* appears to be a continuous gas exchanger (32). Perhaps the most satisfactory investigation of noninsect DGCs concerns solphugids. These highly active, aberrant, archaic, tracheate arachnids are common denizens of xeric regions and exhibit a three-phase DGC almost exactly analogous to that of insects (34), except that—like mammals—their $\dot{V}O_2$ scales with mass to a lower exponent than CO_2 emission (ventilation) volume per DGC. Therefore, DGC frequency increases in smaller individuals. Even their F phase, as shown by simultaneous O_2 and CO_2 respirometry, is functionally equivalent to the insect F phase and is primarily diffusive in nature (34). Gas exchange in solphugids and insects therefore represents a fascinating example of convergent evolution, not only in the development of analogous tracheal systems, but in the details of analogous control mechanisms.

The DGC and Control of Respiratory Water Loss

The DGC may have evolved as a means of restricting respiratory water loss (19, 44). Respiratory water-loss rates, if excessive, can exert a disastrous influence on the fitness of small animals such as insects (8, 55), and development of mechanisms to restrict this loss will be strongly favored by natural selection. To argue that respiratory water loss comprises an insignificant proportion of total water-loss rates in most insects, and is therefore of limited importance, is tendentious. Using hypercapnia to force spiracular opening causes large increases in water-loss rates (55), and real-time recordings of water-loss rates during the DGC clearly demonstrate a severalfold increase in total water-loss rates during the O phase (38). Thus, the perceived insignificance of respiratory water loss is clearly false, as discussed at the start of this review. Instead, the selection pressure to minimize the respiratory water-loss rate has resulted in the reduction of such rates to seemingly insignificant levels. However, reduction to levels well below cuticular water-loss rates should be subject to a lesser selective pressure (38). Thus, insects with high cuticular permeabilities, such as grasshoppers (11), may exhibit lower relative respiratory water-loss rates than insects with lower cuticular permeabilities. These relative rate differences may explain the well-developed DGCs in xeric ants with low cuticular permeabilities (26).

The inevitable association of the DGC with minimization of respiratory water loss begs the question of the absence of the DGC in many insects, including several species with very low total water-loss rates that are highly successful in xeric habitats, such as *Eleodes* and *Tenebrio* species (JRB Lighton, unpublished data).

Clearly, lack of a DGC does not connote absence of spiracular control, nor does lack of a DGC connote unusually high respiratory water-loss rates (consider *Tenebrio* and *Eleodes* species). Continuous gas exchange is simply an alternative stratagem, and it may actually confer advantages, as indicated by its present existence in cases where the DGC is definitely known to be the primitive condition, as we shall shortly see.

In seeking to explain the patchy distribution of the DGC among insects, both logic and a rudimentary comparative exercise (1, 13) lead to similar conclusions. First the logic: The DGC is characterized by extreme fluctuations of trans-spiracular O_2 and CO_2 partial pressure gradients. Considering gas exchange only, such gradient fluctuations are needlessly dramatic if the insect lives in normoxic and normocapnic conditions. However, they are useful, if not essential, when the insect is exposed to hypoxic and/or hypercapnic conditions (31). Regarding the comparative approach, a decade's worth of investigation (JRB Lighton, both published and unpublished data) into insect gas exchange has shown that the DGC was weak or absent among insects seldom if ever exposed to hypoxic or hypercapnic conditions, such as many surface-dwelling insect species (*Eleodes* spp. and all true bugs investigated so far). However, it was pronounced in species that may be exposed to such conditions (burrowing carabid beetles, ants, and subsurface-dwelling psammophilous Old World tenebrionids).

Thus, the evolutionary genesis of the DGC may be chthonic (derived from gas-exchange contingencies in underground conditions characterized by hypercapnia and hypoxia); it may be hygric (derived from selective pressures to minimize respiratory water-loss rates); or it may be some combination of the two.

We are now faced with the challenge of distinguishing between the chthonic/hygric DGC genesis hypotheses. One approach is to turn the DGC on and off and examine the respiratory water-loss consequences. Unfortunately, this approach is difficult, except in one system that might not utilize a true three-phase DGC but that definitely merits further exploration (39). The time-honored comparative method (1) remains a viable alternative.

However, a vexing problem in any comparative study is controlling for phylogenetic effects (13). To tackle the problem of the likely genesis of the DGC, we (31) used not only an interspecific, congeneric approach but an intraspecific and intraindividual approach (in the sense that ant colonies can be considered single reproductive entities). The subject species were *Messor pergandei,* famed for its ability to thrive in hyperarid habitats in the southwestern United States (54), and *M. julianus,* a congener from a more mesic area in northern Mexico.

For the intraspecific comparisons, we utilized caste differentiation within colonies. In many ant species, ontogenic programs of caste differentiation result in workers and alates that are phenotypically very distinct and that clearly reflect distinct colony-level selective pressures (12, 38), some of which are likely to affect the ants' respiratory physiology. Fertilized female alate (queen) ants undergo a stage of subterranean isolation referred to as the claustral period (12), during which they lay eggs, raise their first brood, and do not feed. A reasonable (if unproven) hypothesis is that their sealed burrows become significantly hypoxic and hypercapnic. In contrast, the worker caste probably does not experience similar conditions, because the nests of most social insects are fairly well ventilated (41). As a consequence of nest ventilation, and because many workers forage from the nest in desiccating environments and have a much larger surface area:volume ratio than the larger-bodied queens, workers may experience significant water stress (7 and references therein).

Between species, we compared the DGCs of mesic with those of xeric ants, and within species, we compared the DGCs of castes subject to hypercapnic and hypoxic conditions, as well as possible water stress, with the DGCs of those more likely to be subject to water stress alone. We hypothesized that if the DGC was primarily a mechanism for conserving respiratory water (hygric genesis), it would be preferentially expressed in those species and castes subject to more desiccating conditions. If, on the contrary, the DGC was primarily a mechanism for successfully exchanging respiratory gases in an environment with shallow trans-spiracular diffusion gradients (chthonic genesis), it should be preferentially expressed in the alate caste (irrespective of environment). Moreover, if the DGC was not only in some way unnecessary for efficient respiratory water conservation but weakly disadvantageous in xeric normoxic environments, we hypothesized that it may be reduced or absent in the worker caste (recall the examples of *Eleodes* and *Tenebrio* species, which suggest a tendency for the DGC to be lost if it is not required).

Our findings were in accord with the chthonic genesis—that the DGC acts primarily to facilitate gas exchange in hypoxic and hypercapnic environments. Alates of both *Messor* species showed a pronounced DGC; however, the mesic *M. julianus* workers barely exhibited a functional DGC, and the xeric *M. pergandei* workers showed no discontinuity or cyclicity in gas exchange (31). Moreover, this characteristic of *M. pergandei* gas exchange was not unique to the location of the colonies we used (Phoenix, Arizona). The workers of colonies from Zzyzx in the East Mojave also exhibited continuous gas exchange (31; JRB Lighton & FD Duncan, in preparation).

Thus, the insight, seldom before disputed, that the DGC restricts respiratory water loss, is simplistic at best. At present the chthonic genesis looks more and more attractive. Again, further corroboration comes inter alia from the absence of the DGC in several highly xeric surface-dwelling tenebrionid beetle species, as well as from the presence of a DGC almost exactly analogous (but not homologous) to the insect DGC in fossorial tracheate arachnids such as solphugids (34).

Ticks may at first appear to be an exception that proves this rule, because they display a pronounced DGC even though their environment is normoxic. However, if continuous and thus intermittent-convective gas exchange (see below) is the most viable alternative to the DGC, it must of necessity exact an energetic penalty when compared with a purely passive gas-exchange system. Ticks, which are all-or-nothing feeders and are able to survive on somatic lipid reserves for up to several years between blood meals (14), have extraordinarily low metabolic rates (33) and may not be able to afford this slight extra expense. Significantly, ticks solve their water-relations problems by means of active water-vapor uptake (45), which would conceivably relax selective pressure on respiratory water-loss rates.

Also relevant to this question is some recent work by Hadley & Quinlan on grasshoppers (11). They showed that desiccated grasshoppers, against all reasonable hygric-genesis and adaptationist expectations, were much less apt to engage in DGCs than hydrated grasshoppers in far less ultimate need of reducing respiratory water loss. Although these authors did not state it, the reason seems obvious: The DGC places demands on the CO_2-buffering capacity of any insect displaying it and must perforce be marginalized if that insect is dehydrated, which reduces hemolymph reserves and thus whole-body CO_2-buffering capacity. Additional evidence supporting this supposition is found in another tracheate arthropod, the Rocky Mountain wood tick, *Dermacentor andersoni,* which loses its DGC capability when dehydrated (9).

In essence, the DGC can certainly restrict respiratory water loss, but respiratory water loss was not necessarily the proximate selective force that led to its evolution. A paradox emerges, one of many in this field: From the standpoint of reducing respiratory water loss, the DGC is primarily feasible only when it is not urgently required.

The principal advantage of abandoning the DGC may therefore lie in greater constancy of the insect's internal environment, particularly with respect to proton balance at low body-hydration levels. Yet exactly how this alternate gas-exchange strategy functions remains obscure; the nature of nondiscontinuous gas exchange in species from xeric habitats plainly requires further research. This system is likely, given a priori considerations, to possess a prominent convective component, because maximizing the diffusion gradient for O_2 and CO_2 will minimize the spiracular conductance required to effect the required respiratory gas fluxes and thus reduce water-vapor effusion from the water-saturated endotracheal space.

If, by analogy with pupal DGC partial pressures, endotracheal O_2 partial pressures are in the region of 3–6 kPa, the partial-pressure gradient across the spiracles is approximately 15–18 kPa for O_2. However, in the steady state a similar diffusion gradient is required for CO_2 in the absence of outward convection, and such an endotracheal pCO_2 would effect a catastrophic breakdown of pH regulation. Thus, unless moderate trans-spiracular partial pressure differences of a few kilopascals are maintained continuously (with a corresponding water-loss penalty), periodic convective ventilation is required. Such a response is in fact found in *Eleodes* spp. (JRB Lighton, personal observations). The tradeoff between steady-state diffusive vs intermittent-convective ventilation is further complicated by the presence of buffering gas reservoirs, such as the subelytral space in many New World xeric tenebrionids. In addition, the psammophilous Old World owners of just such reservoirs exhibit very pronounced DGCs (24, 27).

It is obvious that much work remains to be done in the field of discontinuous gas exchange in insects and other arthropods. In light of the paucity of data, more descriptive work is urgently needed. Such work, however, should ideally be placed within a conceptual framework constructed from respiratory physiology, evolutionary theory, ecology, phylogenetic analysis, metabolic biochemistry, energetics, and behavioral biology. It is only when viewed from the many perspectives provided by this framework that the field of insect gas exchange can achieve coherence. We are still a long way from this goal.

ACKNOWLEDGMENTS

I thank the US National Science Foundation (grants BSR 9006265 and IBN 9306537), the US National Institutes of Health (grant RO1 A36345-01), and a David and Lucile Packard Fellowship for generous financial support. I am also grateful for the stimulating and constructive interactions I have had with many colleagues both in and out of my field, especially George Bartholomew, Gideon Louw, David Berrigan, Jon Harrison, Frances Duncan, Dana Garrigan, Karel Slama, Paul Kestler, Raul Suarez, Tim Casey, Martin Feder, Peter Hochachka, Mike Dickinson, and Barbara Joos. Jon Harrison and Barbara Joos also made valuable comments on this manuscript.

Literature Cited

1. Bartholomew GA. 1987. Interspecific comparison as a tool for ecological physiologists. See Ref. 8a, pp. 11–34
2. Bartholomew GA, Lighton JRB, Louw, GN. 1985. Energetics of locomotion and patterns of respiration in tenebrionid beetles from the Namib desert. *J. Comp. Physiol.* 155:155–62
3. Bridges CR, Scheid P. 1982. Buffering and carbon dioxide dissociation of body fluids in the pupa of the silkworm moth *Hyalophora cecropia*. *Resp. Physiol.* 48:183–98
4. Buck J. 1962. Some physical aspects of insect respiration. *Annu. Rev. Entomol.* 7:27–56
5. Burkett BN, Schneiderman HA. 1974. Roles of oxygen and carbon dioxide in the control of spiracular function in cecropia pupae. *Biol. Bull. Woods Hole* 147:274–93
6. Cameron JN. *The Respiratory Physiology of Animals.* Oxford: Oxford Univ. Press
7. Duncan FDD, Lighton JRB. 1994. Water relations in nocturnal and diurnal foragers of the desert honeypot ant *Myr-*

mecocystus: implications for colony-level selection. *J. Exp. Zool.* 270:350–59
8. Edney EB. 1977. *Water Balance in Land Arthropods.* Berlin: Springer Verlag
8a. Feder ME, Bennett AF, Burggren WW, Huey RB, eds. 1987. *New Directions in Ecological Physiology.* Cambridge: Cambridge Univ. Press
9. Fielden LJ, Lighton JRB. 1995. Effects of water stress and humidity on ventilation in the tick *Dermacentor andersoni* (Acari: Ixodidae). *Physiol. Zool.* In press
10. Gould SJ, Lewontin RC. 1979. The spandrels of San Marco and the Panglossian paradigm: a critique of the adaptationist programme. *Proc. R. Soc. London* 205:581–98
11. Hadley NF, Quinlan MC. 1993. Discontinuous CO_2 release in the eastern lubber grasshopper *Romalea guttata* and its effect on respiratory transpiration. *J. Exp. Biol.* 177:169–80
11a. Herreid CF II, ed. 1981. *Locomotion and Energetics in Arthropods.* New York: Plenum
12. Hölldobler B, Wilson EO. 1990. *The*

Ants. Cambridge, MA: Harvard Univ. Press

13. Huey RB. 1987. Phylogeny, history, and the comparative method. See Ref. 8a, pp. 76–97

14. Jaworski DC, Sauer JR, Williams JP, McNew RW, Hair JA. 1984. Age-related effects on water, lipid, hemoglobin and critical equilibrium humidity in unfed adult lone star ticks (Acari, Ixodidae). *J. Med. Entomol.* 21:100–4

15. Kaars C. 1981. Insects—spiracle control. See Ref. 11a, pp. 367–90

16. Kanwisher JW. 1966. Tracheal gas dynamics in pupae of the *Cercopia* silkworm. *Biol. Bull. Woods Hole* 130:96–105

17. Kestler A. 1978. Atembewegungen und Gasaustauch bei der Ruheatmung adulter terrestrischer Insekten. *Verh. Dtsch. Zool. Ges.* 1978:269

18. Kestler A. 1980. Saugventilation verhindert bei Insekten die Wasserabgabe aus dem Tracheensystem. *Verh. Dtsch. Zool. Ges.* 1980:306

19. Kestler A. 1985. Respiration and respiratory water loss. In *Environmental Physiology and Biochemistry of Insects,* ed. KH Hoffmann, pp. 137–83. Berlin: Springer Verlag

20. Krogh A. 1920. Studien über Tracheenrespiration. 2. Über Gasdiffusion in den Tracheen. *Pflugers Arch.* 179:95–112

21. Levy RI, Schneiderman HA. 1966. Discontinuous respiration in insects. II. The direct measurement and significance of changes in tracheal gas composition during the respiratory cycle of silkworm pupae. *J. Insect Physiol.* 12:83–104

22. Loudon C. 1989. Tracheal hypertrophy in mealworms: design and plasticity in oxygen supply systems. *J. Exp. Biol.* 147:217–35

23. Lighton JRB. 1985. Minimum cost of transport and ventilatory patterns in three African beetles. *Physiol. Zool.* 58:390–99

24. Lighton JRB. 1988. Simultaneous measurement of oxygen uptake and carbon dioxide emission during discontinuous ventilation in the tok-tok beetle, *Psammodes striatus. J. Insect Physiol.* 34:361–67

25. Lighton JRB. 1988. Discontinuous carbon dioxide emission in a small insect, the formicine ant *Camponotus vicinus. J. Exp. Biol.* 134:363–76

26. Lighton JRB. 1990. Slow discontinuous ventilation in the Namib dune-sea ant, *Camponotus detritus* (Hymenoptera, Formicidae). *J. Exp. Biol.* 151:71–82

27. Lighton JRB. 1991. Ventilation in Namib Desert tenebrionid beetles: mass scaling, and evidence of a novel quantized flutter phase. *J. Exp. Biol.* 159:249–68

28. Lighton JRB. 1991. Measurements on insects. In *Concise Encyclopedia on Biological and Biomedical Measurement Systems,* ed. PA Payne, 201–208. Oxford: Pergamon

29. Lighton JRB. 1992. Simultaneous measurement of CO_2 emission and mass loss in two species of ants. *J. Exp. Biol.* 173:289–93

30. Lighton JRB. 1994. Discontinuous ventilation in terrestrial insects. *Physiol. Zool.* 67:142–62

31. Lighton JRB, Berrigan D. 1995. Questioning paradigms: Caste-specific ventilation in harvester ants, *Messor pergandei* and *M. julianus* (Hymenoptera: Formicidae). *J. Exp. Biol.* In press

32. Lighton JRB, Duncan FD. 1995. Metabolic effects of temperature, mass and activity in an unsegmented arthropod, the velvet mite *Dinothrombium pandorae. J. Insect Physiol.* In press

33. Lighton JRB, Fielden LJ. 1995. Mass scaling of standard metabolism in ticks: a valid case of low metabolic rates in sit-and-wait strategists. *Physiol. Zool.* In press

34. Lighton JRB, Fielden LJ. 1995. Gas exchange in wind spiders: evolution of convergent respiratory strategies in solphugids and insects. *J. Insect Physiol.* In press

35. Lighton JRB, Fielden L, Rechav Y. 1993. Characterization of discontinuous ventilation in a non-insect, the tick *Amblyomma marmoreum* (Acari: Ixodidae) *J. Exp. Biol.* 180:229–45

36. Lighton JRB, Fukushi T, Wehner R. 1993. Ventilation in *Cataglyphis bicolor:* regulation of CO_2 release from the thoracic and abdominal spiracles. *J. Insect Physiol.* 39:687–99

37. Lighton JRB, Garrigan D. 1995. Ant breathing: testing regulation and mechanism hypotheses with hypoxia. *J. Exp. Biol.* In press

38. Lighton JRB, Garrigan D, Duncan FD, Johnson RA. 1993. Respiratory water loss during discontinuous ventilation in female alates of the harvester ant, *Pogonomyrmex rugosus. J. Exp. Biol.* 179:233–44

39. Lighton JRB, Lovegrove BG. 1990. A temperature-induced switch from diffusive to convective ventilation in the honeybee. *J. Exp. Biol.* 154:509–16

40. Lighton JRB, Wehner R. 1993. Venti-

lation and respiratory metabolism in the thermophilic desert ant, *Cataglyphis bicolor* (Hymenoptera, Formicidae). *J. Comp. Physiol.* 163:11–17

41. Lüscher M. 1961. Air conditioned termite nests. *Sci. Am.* 238:138–45

42. Deleted in proof

43. Miller PL. 1964. Respiration: aerial gas transport. In *The Physiology of Insecta*, ed M Rockstein III, pp. 558–617. New York: Academic

44. Miller PL. 1981. Ventilation in active and in inactive insects. See Ref. 11a, pp 367–90

45. Needham GR, Teel PD. 1986. Water balance in ticks between blood meals. In *Morphology, Physiology and Behavioral Biology of Ticks*, ed. JR Sauer, JA Hair, pp. 100–51. Chichester: Ellis Horwood

46. Punt A, Paarser WJ, Kuchlein J. 1957. Oxygen uptake in insects with cyclic carbon dioxide release. *Biol. Bull. Woods Hole* 112:108–17

47. Quinlan MC, Lighton JRB. 1995.Respiratory physiology and water relations of three species of *Pogonomyrmex* harvester ants (Hymenoptera: Formicidae). *Physiol. Zool.* Submitted

48. Roces F, Lighton JRB. 1995. Larger bites of leaf-cutting ants. *Nature* 373: 392–93

49. Schmidt-Nielsen K. 1980. *Animal Physiology: Adaptation and Environment.* Cambridge: Cambridge Univ. Press. 2nd ed.

50. Schneiderman HA, Williams CM. 1955. An experimental analysis of the discontinuous respiration of the *Cecropia* silkworm. *Biol. Bull. Woods Hole* 109: 123–43

51. Slama K. 1988. A new look at insect respiration. *Biol. Bull. Woods Hole* 175: 289–300

52. Slama K. 1994. Regulation of respiratory acidemia by the autonomic nervous system (coelopulse) in insects and ticks. *Physiol. Zool.* 67:163–74

53. Snyder GK, Sheafor B, Scholnick D, Farrelly C. 1995. Gas exchange in the insect tracheal system. *J. Theor. Biol.* In press

54. Tevis L. 1958. Interrelations between the harvester ant *Veromessor pergandei* (Mayr) and some desert ephemerals. *Ecology* 39:695–704

55. Wigglesworth VB. 1965. *The Principles of Insect Physiology.* London: Methuen

Annu. Rev. Entomol. 1996. 41:325–52

GEOGRAPHIC STRUCTURE OF INSECT POPULATIONS: Gene Flow, Phylogeography, and Their Uses

George K. Roderick

Center for Conservation Research and Training, 3050 Maile Way, Gilmore 409, University of Hawaii, Honolulu, Hawaii 96822

KEY WORDS: population genetics, diversity, variation, demography, conservation, metapopulation

ABSTRACT

Geographic structure of populations is a fundamental component of ecology and evolution that combines both demographic and genetic processes, such as gene flow and migration, genetic drift, selection, and·population extinction. Recent advances in both molecular biology and theory have revolutionized the field and have not only expanded the availability of data but also facilitated accessibility and interpretation of current data. These new techniques allow analysis of genetic similarity among populations to be coupled with phylogeography and the distribution of genotypes within and among populations relative to the history of those genotypes. The numerous case studies described herein illustrate the growing impact of geographic structure on insect science, as well as the importance of insect model systems for understanding general concepts in ecology and evolution.

INTRODUCTION

These are exciting times for population biologists. New molecular genetic techniques have allowed fresh insights in such fields as conservation biology, agriculture, fisheries, medicine, and evolutionary biology and have revitalized the study of all taxonomic groups (56, 69). In this review, I focus on the new implications of population genetics for the study of the geographic structure of insect populations.

Analysis of geographic structure combines population genetics with demography (149). Although the synthesis of these two disciplines is by no means

325

0066-4170/96/0101-0325$08.00

complete (64), the value of the approach is already apparent; genetic methods are providing novel understandings of demographic and evolutionary processes, and in turn, demographic and evolutionary processes are illuminating the nature of genetic variation (171, 172). I consider the geographic structure of insect populations, first in terms of the special interest it holds for entomologists and second as a model for examining related questions in other organismal groups and fields of research. The discussion emphasizes the different approaches taken to study and test for geographic structure, in particular the distinction between methods that measure variation among populations and those that examine the diversity and history of genotypes (alleles, haplotypes). Aspects of insect science that appear especially promising for the study of geographic structure include incorporation of genetics into multi- or metapopulation demographic models, evolutionary ecology of species interactions, and evolution at species boundaries. In addition, I provide a brief update on the modern genetic "toolbox"; further information on this subject can be found in other recent publications (6, 14, 40, 54, 66, 69, 70, 73, 94, 100, 118, 119, 126, 142, 143, 148–150, 176).

WHAT IS GEOGRAPHIC STRUCTURE?

Geographic structure is the distribution and abundance of genotypes within and among populations. This definition encompasses two distinct but related components: demographic structure and genetic structure (149). Demographic structure concerns processes that influence the number and distribution of phenotypic classes of individuals. Phenotypic classes can be age groups, sexes, life-history variants, etc, and the processes include birth, death, and immigration and emigration. Direct observations and manipulations can be used to study demography and do not require knowledge of genetics.

Genetic structure can be described as the distribution of genetic variation and the result of migration, selection, mutation, genetic drift, and related factors. Because demography figures in these genetic processes, genetic and demographic structures are necessarily tightly coupled. For this reason, genetic structure can be used to infer both demographic processes and demographic history. Genetic data sometimes constitute the best—or only—data for a particular aspect of demography, such as migration among multiple populations (147). In the literature, geographic structure and population structure are used interchangeably; I have used geographic structure to avoid confusion with age structure of populations.

THE CURRENT GENETIC TOOLBOX

We can examine genetic variation using an array of molecular genetic tools. However, each technique is useful for a specific set of problems, and to be of value, the limitations of each must be understood. The following survey covers studies of nuclear and mitochondrial genetic material separately. Although some of the techniques and analyses can be applied to both the nuclear and mitochondrial genomes, this division emphasizes the differences in inheritance, effective population size, and potential for recombination.

Nuclear Genetic Variation

Nuclear markers offer the promise of nearly unlimited loci for the analysis of geographic structure and systematics. I outline techniques that are used extensively, or have great potential for use, in studies of insects at the population level.

ALLOZYMES   For over three decades, allozyme electrophoresis has proven itself a useful technique in the analysis of geographic structure of insects and other organisms and is still the most accessible of all genetic methods for such studies (90). Details of electrophoretic methods are not outlined here (for specific information, see 126). However, important limitations of allozyme electrophoresis are worth noting: (*a*) With some exceptions (169), insects must be alive or deep-frozen before use. (*b*) The allozymes of some insects [especially many parasitic Hymenoptera (45)] show little variability. (*c*) Bands that comigrate are assumed to be homologous. (*d*) Bands may be present that do not follow Mendelian inheritance. Mating studies can test this assumption. (*e*) At any given locus, allozyme electrophoresis reveals only a subset of the actual genetic variation (90). (*f*) The patterns of ancestry and descent among different alleles are impossible to define (40). For these last two reasons, the analysis of allozymes cannot reveal historical relationships among alleles themselves. However, when an answer regarding allele frequencies is sufficient, allozyme electrophoresis is well suited for addressing questions of geographic structure (90). The technique is inexpensive, fast, and can give insight into multiple loci.

DNA AMPLIFICATION AND PCR   Most techniques for examining nuclear variation rely on the polymerase chain reaction (PCR) to amplify sufficient quantities of DNA. PCR makes use of the enzyme DNA polymerase to construct complementary DNA sequences beginning at a site where another piece of DNA (a primer) has annealed to the template molecule (for further details, see 48, 118). Simply stated, a PCR machine carries out successive temperature cycles (typically 35–55 cycles), each of which first denatures DNA (typically 94°C), allows primers to anneal to the template (typically 45–60°C), and finally permits DNA extension through the action of DNA polymerase (typically 72°C). If the DNA sequences between sites of primer annealing are short enough to be extended in the time specified, they are duplicated with each cycle. Five techniques for analyzing variation in nuclear DNA are widely used (or show potential) in the examination of geographic structure: microsatellites, randomly amplified polymorphic DNA (RAPDs), restriction fragment–length polymorphisms (RFLPs), introns, and internal transcribed spacer (ITS) regions.

Microsatellites   Microsatellites are repeated short sequences of DNA that occur throughout the genomes of many organisms, including insects (66, 125, 174). Because repeat units are readily added to or lost from microsatellite DNA, the sequence length of these regions rapidly evolves. Microsatellites are amplified (via PCR) using DNA primers that flank a particular repeat region, and the resulting size variants can be visualized through electrophoresis. Because the flanking regions are often known, homology can be verified via sequencing. Microsatellites offer a valuable pool of genetic variation that should be particularly useful when other methods show insufficient variability. The drawback to the method is that to locate the repeated sequences, a genomic library must be constructed or obtained commercially (118). This step can be omitted if primers known to work in one species can be used to amplify variable microsatellite loci in related species.

The analysis of microsatellite data to infer geographic structure differs slightly from that for allozyme loci because, unlike allozymes, microsatellites do not conform to an infinite allele model with low mutation (174). Slatkin (150) recently developed a measure of population subdivision that does account for the evolution of microsatellites and is otherwise analogous to measures used for allozymes (see below). Like allozyme data, microsatellite data cannot be used to establish relationships of alleles within loci [unless one assumes a model for their evolution (174)].

RAPDs   In the RAPD procedure, a single nucleotide primer (8–10 base pairs long) is used to amplify random sections of nuclear DNA (11, 12, 66). The products are visualized on a gel that separates bands corresponding to the size of a particular piece of amplified DNA. As for allozymes and microsatellites, the analysis of RAPD bands relies on allele frequencies within populations. A tree based on frequency similarity can be constructed to reflect the bands shared among populations.

The interpretation of RAPD data is limited by poor repeatability, a lack of codominance, and finally the possibility of nonheritable or nonhomologous elements (11, 118)—a problem the technique shares with the allozyme and RFLP (see below) methods.

In RAPDs, the DNA amplifications constitute the final data. As a result, experimental variation can be high, especially when low annealing temperatures are used. Often, many bands appear in any given RAPD amplification, and only the most visible bands will be consistent enough for use as markers.

The method uses randomly constructed primers, and as a result, some matches to the template DNA will not be very good. Although this problem can be partly solved by consistent procedures (same PCR machine, same laboratory, same batches of chemicals, etc), RAPDs may not be suitable as a general technique for which repeatability is crucial.

The allozyme and microsatellite techniques, which show codominance, allow one to distinguish heterozygotes. In contrast, RAPD banding patterns exhibit dominance; both homozygotes and heterozygotes will produce a band. Thus, the frequency of heterozygotes is difficult to determine from RAPD banding patterns. However, using traditional genetics theory, one can estimate allele frequencies from the frequency of homozygotes (WC Black, unpublished computer program).

Bands may appear that are not Mendelian and thus not suitable for examining genotypes within and among populations. This result may arise either when the bands do not reflect true genetic variation or when they are not homologous. Bands from two individuals that appear to be of similar size but are different sequences of DNA will be mistakenly identified as homologous. As in the allozyme technique, mating studies can be used to examine the inheritance of bands. Homology of RAPD bands can be assessed by sequencing or hybridization studies. However, RAPD techniques are generally used because they provide information quickly and easily; time-consuming checks of heritability and homology are rarely performed. Despite these limitations, RAPDs have potential in mapping genetic loci, examining familial relatedness, and identifying variable sites for further analysis (66).

RFLPs   Regions of nuclear DNA, isolated through PCR or other means, can be digested with restriction enzymes that cut the DNA at specific four- or six-base sequences, generating RFLPs. Similar to the RAPD method, RFLP techniques can exploit an enormous amount of genetic variation. In some cases, known restriction sites can be targeted. For example, in a comprehensive study of *Drosophila melanogaster,* four-base cutters were used to diagnose variation within particular genes (9). The analysis revealed a previously unknown geographic structure, showing that restriction sites were not shared between US and African fly populations. The limitations of RFLPs have been noted above.

Nuclear introns (and EPIC PCR)   Introns are noncoding sequences within nuclear genes that have recently shown potential for population studies (119, 144). Because introns are not subject to the same selective constraints as exons, which are the coding regions, bases within introns likely evolve at least as fast as nuclear DNA silent sites (bases at which a change does not affect the amino acid sequence). Exon-primed intron-crossing (EPIC) primers can be designed to anneal to highly conserved nuclear genes and therefore provide homologous and heritable loci for examining geographic structure in a broad range of organisms (118). For example, introns have been used successfully for the study of many insect groups, including Mediterranean fruit flies, whiteflies, mealybugs, and parasitic Hymenoptera (GK Roderick, FX Villablanca & SR Palumbi, unpublished data), spiders (GK Roderick, SR Palumbi & RG Gillespie, unpublished data), marine invertebrates such as sea urchins and shrimp (SR Palumbi, unpublished data), and marine mammals (119, 144). Prior to sequencing, amplified introns are cloned to avoid simultaneous sequencing of two alleles from a possible heterozygote. When the sequence variation within the organism of question is understood, restriction enzymes can be chosen that cut the introns at diagnostic positions, thus eliminating the need to clone and sequence all the individuals required for an extensive population survey (119).

Preliminary work indicates that introns will be useful for reconstructing phylogenies of alleles within insect species as well as among closely related species (118, 130). However, intron sequences are probably too variable to be used to reconstruct relationships among more distantly related insect species or genera (14). How the rate of intron evolution compares with that of silent-site substitutions in nuclear genes or in parts of the mitochondrial genome is not yet known (118). The limitations of the method are that (a) all taxa do not have introns at the same location, (b) some introns may not be sufficiently variable within species, and (c) cloning and sequencing is required, at least initially.

ITS regions   Internal transcribed spacer (ITS) regions within ribosomal RNA genes can vary between conspecific individuals in the same population and therefore might be used in the analysis of geographic structure (see 176). These ribosomal genes are organized in clusters of tandemly repeated units, and concerted evolution is believed to result in homogenization of individual repeats. ITS regions can be amplified by universal primers, much like introns in the method described above. The primary limitation of ITS regions in the study of populations is that much variation may exist within individuals, thereby making identification of homologous DNA difficult or impossible. For this reason, the utility of ITS regions for population studies must be reassessed for each species.

Mitochondrial Genetic Variation

Mitochondrial DNA (mtDNA) has been used extensively in population studies of many insect species because it evolves rapidly and, unlike nuclear DNA, it lacks recombination (5, 6, 142). Many regions are conserved enough so that primers can be used in different insect taxa. Variation in mtDNA can be examined by means of sequences (142) or RFLPs (57) in a manner similar to that described for nuclear loci. As for nuclear DNA, actual sequences of mtDNA contain more information than do restriction fragments and therefore should be more useful for inferring the histories of haplotypes or populations. In some species, mtDNA has revealed significant variation among populations (62, 107, 139, 156), whereas in others it has shown less variability (107, 143). For population-level studies, variation at unconstrained sites appears to be more useful than that within ribosomal genes (142). With the development of methods to amplify greater lengths of DNA and eventually the whole insect mitochondrion, mtDNA will continue to be an essential tool for population studies of insects.

The mitochondrial genome has several important characteristics that affect both analysis and interpretation of data. First, this genome is inherited as a single entity, and because of lack of recombination, its genes represent a single genetic locus (5). Therefore, although different mtDNA genes may reveal different levels of variation, they cannot be interpreted as independent loci. Second, the effective population size of mtDNA is smaller than that of nuclear DNA because of its haploid nature and maternal inheritance. Within a population, the frequency of a single mtDNA haplotype should fluctuate more rapidly than the frequencies of nuclear DNA alleles, and thus mtDNA should be more sensitive to founder events and small population sizes than nuclear DNA (5, 10, 39). As a consequence, the probable loss or gain of a mtDNA haplotype will be great for small populations (114), and only a little migration among populations would be sufficient to fix one mtDNA lineage in all populations (166). The smaller effective population size may make mtDNA a superior molecule for population-specific markers and for systematic studies at higher taxonomic levels when a taxon-specific marker is desired (39, 142).

A third feature of mtDNA stems from its usual matrilineal inheritance in insects (5). When mtDNA is used to study gene flow, the consequences of dispersal by males will be missed (119). Therefore, because male and female insects frequently differ in migratory and colonization ability (38, 129), mtDNA results may lead to an erroneous interpretation of the history of colonists relative to their source population (130). In contrast, information from nuclear DNA would permit the documentation of colonization by both males and females, and a comparison of sex-specific data provides a method to examine behavioral differences between the sexes (10, 153).

GEOGRAPHIC STRUCTURE AND GENE FLOW

Approaches for the study of geographic structure have been lumped into two categories: direct and indirect methods. In the study of migration, direct methods are those that use actual observations of movements of individuals whereas indirect methods use genetic data to infer movement (146). Direct methods to study gene flow have one advantage in that they are based on actual observations. However, they have several disadvantages:

1. Casual observations of dispersal or dispersal capability can be misleading because the ultimate fate of dispersers is difficult to measure and because observations are often limited to much shorter distances than individuals can actually move (146).
2. Following migration, the reproductive success of migrants is assumed to equal that of residents. In fact, immigrant reproductive success may be lower than that of residents for many reasons, including difficulties in finding mates or appropriate habitats. Immigrant success could also be higher. In a laboratory study of *Tribolium castaneum,* immigrants had a mating advantage over residents (84).
3. Direct measurements from mark-and-recapture studies are necessarily limited in both space and time—marked individuals must be recaptured in sufficient numbers and individuals that leave the area of study, or disperse by unexpected means, will not be recorded. Because gene flow can result from various independent and stochastically triggered dispersal events, direct measurements made over only a few generations for only one population may miss important episodes of gene flow (147).
4. Marking and monitoring of individuals for more than a few migration events is often impractical, even for insects that can be mass-reared (but see 27, 42).

The second approach to determine geographic structure is to infer it indirectly from genetic data obtained from the techniques outlined above (146, 147). Genetic data can be analyzed in two distinct ways (Figure 1). Tradition-

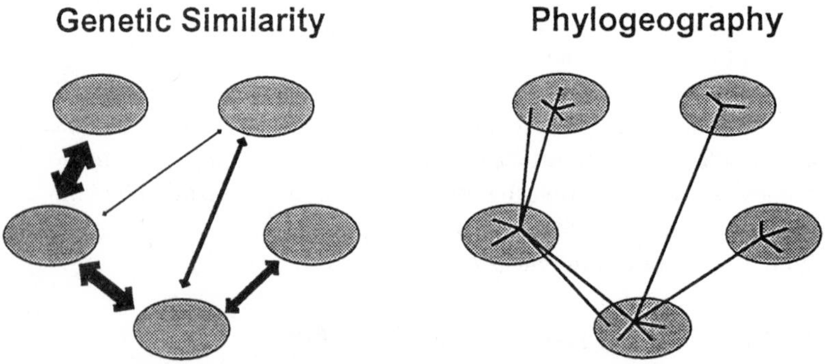

Genetic Similarity   Phylogeography

Figure 1 Two approaches to examine geographic structure based on genetic similarity (*left*) and phylogeography (*right*). Ovals represent populations. Width of arrows (*left*) denotes relative genetic similarity between populations, and lines (*right*) show genealogical relationships between individuals in different populations. See text for appropriate analysis for each.

ally, populations are treated as genetic units, and measures of genetic similarity and population subdivision are calculated among populations. Alternatively, relationships among genotypes (phylogeny of alleles or haplotypes) from one or more population(s) can be examined relative to their geographical location. This second method is termed *phylogeography* (7) and requires DNA sequence variation (by means of sequencing or RFLPs) to establish relationships between alleles or haplotypes at the same locus. Although the distinction between the two approaches may seem trivial, the methods used and possible conclusions that can be drawn are very different. Other methods, including those based on spatial correlations, have been proposed to examine geographic structure (47, 155) but are not considered further here (see 151).

A priori, one might expect estimates of migration and dispersal from direct and indirect results to yield similar estimates of the extent of gene flow. However, a comparison of the two sets of data can be as important as the actual values and may provide information on the importance of the frequency of dispersal events (149). For example, a direct observation of infrequent dispersal would lead to an estimate of low dispersal distances, even when more extensive gene flow may be occurring. This phenomenon is illustrated in the now classic studies of *Drosophila pseudoobscura* (27, 42) and the checkerspot butterfly, *Euphydryas editha* (46).

Genetic Similarity

Wright (187) recognized the value of using genetic similarity to infer geographic structure, leading to F_{ST} and related indices as measures of population subdivision (147, 149). This approach requires that the population as a unit be defined before the analysis, which may be problematic for some questions and some insects (40). Also, sufficient individuals (genotypes) must be examined to acquire an accurate picture of the nature of genetic diversity within and between populations (126). Numerous other measures of genetic similarity and distance are more difficult to interpret in the context of gene flow (reviewed in 146, 190); hence, only F_{ST} and related measures are described below.

F_{ST} represents the standardized variance in allele frequencies among local populations; it ranges between 0, indicating no genetic differentiation, and 1, meaning complete differentiation (189). F_{ST} has several other interpretations (146, 149). For example, it can be expressed as the absence of heterozygotes in subpopulations relative to what is expected in the total population: $F_{ST} = 1 - H_S/H_T$, where H_S is the average heterozygosity within subpopulations and H_T is the estimated total heterozygosity (146). Though F_{ST} was first derived for two alleles at one locus, similar measures, such as Nei's G_{ST} (10, 22, 113), describe values for more alleles at many loci. In a slightly different approach, Weir & Cockerham (183, 184) proposed a between-population component of variance, θ. Variations on both the F_{ST} (190) and θ (183, 184) approaches can be used to test for nonrandom mating or for hierarchical structure.

Both F_{ST} and θ have been used widely for the analysis of allozyme data. In simulations, θ appears to perform better than F_{ST} (or G_{ST}) especially for moderate and large levels of gene flow (see 149). For the analysis of microsatellite data, Slatkin (150) developed the measure R_{ST}, which is analogous to θ and represents the fraction of total variance of allele size that is between populations. F_{ST}-like parameters can also be obtained for RFLPs or DNA sequence data (95, 165): For an infinite allele model, Hudson et al (76) propose that $F_{ST} = 1 - H_w/H_b$, where H_w is the mean number of differences between different sequences sampled from the same subpopulation and H_b is the mean number of differences between sequences sampled from two different subpopulations. The theory used to estimate gene flow from measures of genetic variation based on RAPD data is not as well developed as that for allozyme or RFLP data, largely because of the problem of dominance (discussed above).

Whether significant geographic structure exists among populations can be assessed by testing whether F_{ST} is significantly greater than zero. Confidence limits can be obtained through bootstrapping or other resampling techniques (76, 183). An alternative test for significant geographic structure is to compare allele frequencies among populations (see 133, 146).

GENE FLOW FROM F_{ST} Indirect methods to estimate gene flow based on genetic similarity among populations do not suffer from many of the limitations of direct observations (see above). However, indirect methods do rely on several assumptions that must be, but rarely are, evaluated critically. Four major assumptions have been summarized and discussed (32, 149): (*a*) Gene flow is random with respect to the genotypes studied; (*b*) the rate of gene flow for the alleles studied exceeds the rate of allele-frequency change resulting from selection; (*c*) the rate of gene flow exceeds a minimum level to offset the effects of genetic drift; and (*d*) the populations are at genetic equilibrium, such that loss of alleles by drift is balanced by the gain of alleles from migration. The first two assumptions are likely to be met for allozymes or markers that are either neutral or under weak selection (85). Also, similar estimates across independent loci would be unlikely to result from similar patterns of selection at all loci (147; but see 83).

The last two assumptions are more difficult. For assumption *c*, Wright (187) showed that any gene flow among populations will prevent complete fixation and that genetic drift will lead to substantial genetic differentiation only if gene flow does not exceed a certain minimum level. This condition can be expressed approximately as $m \gg 1/4N_e$ or $N_e m \gg 1/4$, where m is the average rate of immigration and N_e is the local effective population size (146, 147). The term

N_em is especially useful and represents the number of migrants that leave and enter a population each generation. $N_em \gg 1/4$ can be evaluated by means of field observations (see 117). The last assumption states that the populations are in genetic equilibrium. At equilibrium, the balance between drift and migration will hold over a range of population sizes (147). For example, in large populations, the loss of alleles by drift will occur relatively slowly (allele loss is inversely proportional to population size), and any one migrant will have little effect on allele frequencies. In small populations, the effect of genetic drift is larger, and the immigration or emigration of any individual will probably have a greater effect.

A major problem is to determine whether populations in agricultural systems are close enough to genetic equilibrium to validate this assumption (32). In populations that have not reached genetic equilibrium, genetic similarity could be a result of a recent shared ancestry or of high gene flow. For example, in a study of Colorado potato beetles, *Leptinotarsa decemlineata,* populations on potatoes in Maryland and North Carolina were significantly more similar to each other than were populations on wild hosts (*Solanum rostratum* and *Solanum elaeagnifolium*) in Arizona and Mexico. This observation most likely reflects a recent shared ancestry in the agricultural setting and not greater gene flow (PA Follett & GK Roderick, unpublished data). These alternatives might be tested by estimating the departure from genetic equilibrium from the magnitude of the difference between levels of gene flow determined via direct vs indirect measures (149).

For neutral nuclear alleles, F_{ST} can be used to estimate gene flow with the equation $F_{ST} \approx 1/(4N_em + 1)$, where N_e and m are defined as above (189); a modified relationship holds for haploid data (see 10, 76). This result assumes (*a*) genetic equilibrium; (*b*) an "island model," in which every local population is equally likely to exchange migrants with any other; and (*c*) neutrality, as noted earlier. The island-model assumption appears to be the least restrictive. With only slight modification, this equation is also valid for two-dimensional stepping-stone models (30, 31, 145), and Slatkin (149, 154) suggests a method to distinguish between the island and stepping-stone models. Two-dimensional stepping-stone structures probably represent agricultural settings better (18, 19). Neutrality can be assessed by comparing results obtained from different loci (146). For sequence data of single loci, tests of neutrality are also possible (75, 164).

An estimate of N_em, the number of migrants per generation, can be obtained by rewriting the equation above in terms of F_{ST} or θ (for other methods, see 149). Thus, for any given value of F_{ST} or θ, various combinations of the effective population size, N_e, and the migration rate, $m,$ will yield the same level of migration, N_em. By estimating N_em, we can compare the magnitude of gene flow without an exact estimate of the average effective population size, N_e. The migration rate, m, can also be inferred if N_e can be estimated from census data (147).

Phylogeography

Phylogeography is the study of relationships among genotypes relative to the populations from which they were collected (7). This approach can be used both to assess geographic structure and to estimate gene flow (4, 6, 7, 148). Phylogeography is attractive for several reasons. First, the populations need not be defined prior to analysis. In fact, demes might be determined a posteriori based on the resulting relationships among genotypes. Second, the history or genealogy of alleles may be inferred from sequence variation. Such an analysis might be used to trace the origins of colonization events or to establish the directionality of gene flow. Third, phylogeography can be used to test for monophyly of populations or groups of populations (3, 50; but see 51), which might be useful in establishing whether an invasion was caused by more than one colonization event (58, 130).

The phylogeographic approach requires an allele or haplotype tree. When a reasonable outgroup is available, the allele tree or genealogy can be reconstructed using standard phylogenetic techniques (4, 54). As with most phylogenetic techniques, choice of a suitable outgroup is critical. Unfortunately, sister groups are often difficult to identify for intraspecific relationships (40). To circumvent this problem, Crandall & Templeton (28, 29) have proposed using coalescent theory to generate intraspecific trees or networks. This theory looks backward in time and asks at what point two different genes coalesce to a common ancestor. The coalescence method may also allow such trees to be rooted (21).

GENE FLOW FROM PHYLOGEOGRAPHY   One can estimate gene flow from an intraspecific allele or haplotype tree by scoring the minimum number of migration events necessary to explain the observed geographic distribution of alleles. Using simulations, Slatkin & Maddison (148, 153) have shown that the number of migration events is related to $N_e m$ for a given number of individuals sampled per population. As with the other methods outlined above, confidence intervals can be readily obtained (153). In comparing this cladistic method for estimating $N_e m$ with that using F_{ST} (described above), Hudson et al (68) found that the approach based on F_{ST} was more accurate when recombination is possible (for example, with nuclear loci), but that the cladistic method was better when recombination does not occur (as with mtDNA); when migration rates were very low ($N_e m = 0.1$), the method based on F_{ST} was better regardless of recombination.

INSECTS AS MODEL SYSTEMS

Given that insect species likely number in the millions (49), we should not be surprised that many have evolved unique or unusual features. Some of these characteristics make insects ideal models for determining the broader roles of geographic structure in evolution and ecology. Below, I outline a limited set of such model systems that have been especially important or that appear most promising. Clearly, this treatment cannot include all the estimates of F_{ST} and N_em that have been calculated, or every intraspecific phylogeny (for such data see 6, 32, 74, 94, 105, 146).

Habitat Discontinuities in Space and Time

Genetic variation among populations and patterns of gene flow may reflect underlying discontinuities in available habitat. For insects and other organisms, these discontinuities can be in either space and/or time (1) and provide a null-model with which to test additional hypotheses. For example, Tamura et al (167) studied two subspecies of Drosophila sulfurigaster: albostrigata, which has widespread populations on the Asian mainland, and bilimbata, which inhabits offshore islands. They found that mtDNA haplotype and inter-population nucleotide diversity were greater in the island-inhabiting subspecies, which is consistent with spatial subdivision influencing mtDNA polymorphism. In addition to real islands, several other habitat features have been examined with respect to their effects on the geographic structure of insects. These features include mountains and outcrops (57, 86), caves (80), stream drainages (77, 124), and host plants (102, 104).

Temporal variation in habitat availability may also affect the geographic structure of populations (103, 180). For example, in a study of the forked fungus beetle Bolitotherus cornutus, Whitlock (186) examined extinction and recolonization processes and found that the combination of processes increased the genetic variation among populations relative to that predicted by long-term genetic equilibrium.

Genetics and Demography

The importance of the link between genetics and demography has been recognized for some time—migration between subdivided populations is a central feature of Wright's (188) shifting balance theory (179). Yet recent demographic models, while incorporating a spatial component (65, 168), do not yet integrate fully population genetics (64). Genetics and demography have been combined in the application of sequence data to test hypotheses about the recent demographic history of populations. In one approach, exponentially and gradually increasing population sizes are assumed to produce different patterns of pairwise genetic distances between individuals, which might be used to distinguish between different historical hypotheses (152). In another approach, genealogy and coalescence (see above) can be used to infer whether populations have been increasing or decreasing in recent evolutionary time (112) and to estimate relative effective population size (MK Kuhner & J Felsenstein, unpublished data). Because of their small size and rapid generation times, many insect systems are ideal for testing such theory (see 179).

Dispersal Polymorphisms

Many insects display dispersal polymorphisms that are often evident as wing dimorphisms (131). Such systems can be used to evaluate the effect of dispersal and migration on geographic structure as well as to determine whether the magnitude of dispersal can indeed be inferred from genetic data (129, 181). For example, in a comparison of two species of waterstriders, Zera (191) found that populations of a short-winged (brachypterous) species, *Aquarius remigis,* were more structured than those of a long-winged (macropterous) species, *Limnoporus canaliculatus.* Preziosi & Fairbairn (124) discovered that populations of *A. remigis* were significantly differentiated between streams but not within streams, where individuals can disperse on the water surface. Yet dispersal ability itself does not always explain geographic structure. For example, Liebherr (91) compared the structure of populations of five carabid beetle species and found that a fully winged species exhibited as much genetic variation as a species with vestigial wings, whereas habitat subdivision and elevation were better predictors of genetic variation. For some species, the interaction of habitat parameters and dispersal ability may determine geographic structure. For example, habitat duration may influence genetic subdivision because of its effect on the prevalence of wing dimorphism: Reduced wings are more common in long-lasting habitats (38, 132). This hypothesis might be best tested with populations of the same wing-dimorphic species that occupy habitats differing in temporal stability.

Invasion Biology

Invasions of insects are often well documented because of their noticeable economic and social impact. For understanding the biology of invading species and for designing control strategies, a genetic approach to geographic structure is necessary (20). Recently, population genetic approaches have been used to study invasions of the Mediterranean fruit fly, *Ceratitis capitata* (108); the Africanized honey bee, *Apis mellifera* (137, 140); the mosquito *Aedes albopictus* (78, 79); and a possible new sweetpotato whitefly, *Bemisia tabaci* (15, 121). Important questions regarding invasions concern the origins of colonists, whether the invasions involve more than one colonization event, and the spread of the current infestation (130). Insect invasions also allow investigators to test the theories and methods used for the study of geographic structure, including whether a particular colonization event can be detected (58), the validity of molecular genetic methods in analyses of the demographic history of populations (98, 112, 152), and the reliability of specific trapping and detection methods (81).

Resistance to Insecticides and Transgenic Plants

A major environmental and economic concern is the continued development of resistance by insects to chemical insecticides; undoubtedly resistance will also develop to transgenic plants that produce insecticides, such as those expressing proteins of the bacterium *Bacillus thuringiensis* (163). Gene flow and the geographic structure of populations are central to models that predict the rate of resistance development (19, 60, 99, 162). Empirical studies have also provided evidence that migration contributes to the development of insecticide resistance. In a pioneering study, Daly (34) used genetic evidence to suggest that the cotton bollworm *Helicoverpa armigera* in Australia was not sedentary, as previously thought, but likely disperses long distances under suitable conditions; this pattern of dispersal and a corresponding large effective population size were consistent with the spread of resistance to DDT. For some species, e.g. *Heliothis virescens* (87), estimates of migration are very high (N_em = 135), whereas estimates for others, e.g. the diamondback moth, *Plutella xylostella* (18), are more moderate (N_em = 6–9). In the pesticide-resistant mite *Typhlodromus pyri*, migration between apple orchards and nearby blackberries was low (N_em = 2.1) (44) but was still high enough to prevent populations from diverging genetically through drift alone. Although these figures give an approximate estimate of migration, they may not be sufficiently precise to be useful in the management of resistance.

Despite the practical limitations of gene flow estimates in resistance management, insecticide resistance offers the opportunity to explore processes of selection and adaptation in different environments. Patterns of resistance variation within and among insect populations are a function of primary parameters—insect movement and regimens of insecticide application. For many species, variability is greater between than within populations, which may be attributable to the different local selective pressures (spraying patterns, amounts, etc). For example, Daly (33) suggested that resistance frequencies in *H. armigera* in Australia declined each spring as a result of immigration of susceptible individuals into insecticide-treated populations. In nine populations of *L. decemlineata* in Maryland, variability in insecticide resistance differed on a small scale, even among neighboring farms (41), whereas variation in allozymes showed little geographic structure (GK Roderick & PA Follett, unpublished data). Caprio (18) found a similar result for diamondback moths in Hawaii, suggesting that local selection for insecticide resistance is sufficiently strong to overwhelm the effects of dispersal. However, this result is not universal. In 18 pear psyllid (*Psylla* spp.) populations on the US West Coast, Unruh (173) found that allozyme variation was greater at local levels (within river drainages) than between regions.

In some species, the mechanisms of resistance are known, which enables more careful studies of the balance between selection and gene flow. For example, M Taylor (submitted) examined molecular variation in a target locus (sodium channel) for resistance in *H. virescens* and found significantly more geographic structure than that predicted by allozymes. This result is consistent with selection favoring one or more mutations in or near this locus.

Complex Life Cycles and Parthenogenesis

Insects offer numerous models with which to study the evolution of complex life cycles as well as conditions associated with sexual reproduction. For example, Martin & Simon (100, 143) have used variation in mtDNA to infer the history of 13- and 17-year broods of periodical cicadas in the genus *Magicicada*. Their data suggest that some 17-year cicadas became 13-year cicadas. This change isolated the 13-year brood from the parent brood but allowed new genetic exchange with a previously existing 13-year brood.

Parthenogenesis occurs in many insect species and can influence geographic structure, perhaps leading to reproductive isolation. In a study of three aphid species, Loxdale (92, 93) attributed the greater structuring of populations of *Sitobion fragariae* to the facts that this species is asexual and that both primary and secondary hosts are readily available. In another example, Via (175) found that clones of pea aphids attained higher fitness, as indicated by differential survival and high reproductive rate, when reared on the host crop of their progenitors; thus, pea aphid clones exhibit a significant level of local adaptation. In some situations, parthenogenesis may cause unusual patterns of structure: in three species of *Trichogramma* wasps, high levels of gene flow occurred from the sexual to the parthenogenetic fraction of the population (159). At the extreme, when species are completely parthenogenetic, geographic structure will be complete and the analysis might be more akin to that of the evolution of species rather than populations (115).

Evolution of Social Behavior

The insects illustrate a range of social interactions leading from primitive social behavior to ensociality. Several workers have exploited this feature to determine the role of geographic structure in the evolution of social behavior (see 35, 134, 158). For example, Ross (135) examined social evolution in a study of the sphecid wasp, *Microstigmus comes*. In this analysis, geographic structure at the subpopulation and microgeographic scales that was low compared with the high genetic relatedness among female nest mates evinced social evolution via a subsocial route. Other studies have focused on behavioral shifts associated with the number of queens in a colony by comparing gene flow between polygynous and monogynous populations (136, 161).

EVOLUTIONARY ECOLOGY OF SPECIES INTERACTIONS

Usually the study of geographic structure is limited to populations of one species, but such studies also offer great potential for testing hypotheses about the evolution of species interactions. In some interactions, such as insect-plant coevolution, population differentiation has been studied extensively; in others the role of geographic structure is only beginning to be explored.

Insect–Host Plant Interactions

The role of geographic structure of insect populations in the interactions between herbivorous insects and their host plants has received much attention and is not reviewed extensively here (for genetic effects of pollinators, see 123). Perhaps the best known example of geographic structure influencing the evolution of host-plant use involves the case for sympatric speciation in tephritid fly species in the genus *Rhagoletis*. Bush and colleagues (17, 106) have proposed that shifts by *Rhagoletis pomonella* to new hosts allow for reproductive isolation in sympatry. Recent work supports the idea that host fidelity in *Rhagoletis* can operate as a mechanism for premating isolation (see 53) but that populations also diverge with respect to local environmental conditions, including seasonality (52). Although it is only one example, this work illustrates the value of genetic information (mostly from allozymes) for understanding aspects of interactions among species. Now that sequence data can be obtained for population samples, a promising avenue of research is the use of phylogeography to reconstruct the history of insect genotypes associated with different host species (see 16). The same approach could be used to trace the host origins of insects that feed on hybrid plants.

Predation and Parasitism

Spatial structure has long been recognized as an important feature affecting the stability of predator-prey interactions (see 82, 168), yet consideration of a population-genetic component in these interactions is relatively new. For example, in classical biological-control theory, genetics figures in the choice of natural enemies and their subsequent success (71, 72, 111, 128). However, field tests of theory have been rare, and until the advent of PCR, they were often difficult. One problem has been the lack of allozyme variability in many parasitic Hymenoptera; RAPDs and other methods have exposed more variability (45).

Some predators, such as nabid bugs (61) and coccinellid beetles (88, 157), do exhibit sufficient allozymic variability for population studies. Of the studies that have been conducted on predators, gene flow appears to be tightly linked to behavior. For example, Coll et al (24) found levels of gene flow among populations of the coccinellid *Coleomegilla maculata* to be an order of magnitude higher than those of its most abundant prey item, *L. decemlineata,* which suggests that the predator can easily track populations of its prey. Riechert (127), in a study of a desert spider, *Agelenopsis aperta,* suggested that unidirectional gene flow from arid to riparian habitats restricts adaptations, including changes in prey-catching behavior in the riparian habitat.

Several recent studies have examined the evolution of insect-predator interactions. Brower (13) studied butterfly wing patterns and mtDNA sequences for 14 races in South America to investigate possible Müllerian mimicry in the butterfly *Heliconius erato.* Though sequence variation was low—roughly corresponding to a maximum of 200,000 years—evidence indicated simultaneous radiations of races in two geographically separated regions, which suggested independent evolutionary origins of the mimicry systems (see also 96). In another study of predator-prey interactions, Brown et al (16) used mtDNA to investigate the history of host-plant use in the goldenrod ball gallmaker, *Eurosta solidaginis.* They found two clades (each a single haplotype) and confirmed that populations on *Solidago gigantea* stemmed from eastern *Solidago altissima* populations. This result supported prior behavioral, ecological, and allozyme work indicating that escape from natural enemies facilitated the shift to the derived host (182).

The reciprocal geographic structures of insects and their pathogens (as broadly defined) have not been well studied, although models for other associations have been developed (2). Two studies illustrate the role of geographic structure in the evolution of host-parasite interactions. Herre (67) studied 11 species of Panamanian fig wasps and their parasitic nematodes. Species characterized by geographic structures that allowed increased opportunities for parasite transmission harbored more virulent species of nematodes. Shykoff & Schmid-Hempel (141) provide evidence from the bumblebee *Bombus terrestris,* which indicates that genetic relatedness is associated with enhanced transmission of intestinal trypanosomes.

Interspecific Competition

Studies of the role of geographic structure in interactions between species of the same trophic level are surprisingly few, especially given the literature and interest in intraspecific competition (36). In two sympatric delphacid planthoppers, *Prokelisia dolus* and *Prokelisia marginata,* mixed-species nymphal crowding differentially affected the production of long-winged individuals. High densities of *P. dolus* cause a migratory response in *P. marginata* (37). The diversity of spatial and temporal patterns of interspecific competition (36, 160) would appear to offer fertile ground for numerous studies of geographic structure.

Insect Symbionts and Other Interactions

The coevolution of insects and their symbionts has been examined from a phylogenetic perspective for such groups as Homoptera and their bacterial endosymbionts (109) and attine ants and their associated fungi (23). Symbiotic interactions at the population level can also be studied using phylogenies. For example, bacterial endosymbionts from closely related species of mealybugs and whiteflies can be distinguished on the basis of variations in their DNA (HS Costa & GK Roderick, unpublished data). Endosymbionts are likely to receive increasing attention in the future. For example, *Wolbachia* bacteria, other microbes, and yeasts affect numerous fitness attributes, such as sex determination (159, 185), fertility (178), and host-plant adaptation (43).

Much research has examined the geographic structure of insects, especially mosquitoes, as potential disease vectors (e.g. 89, 138). For example, the mosquito *A. albopictus* was recently introduced into the United States. A study of allozymes over three years (79) showed that the geographic structure of this species in the United States was similar to that in its native habitats, reflecting a large founding population and/or a rapid rate of growth following colonization (see also 78).

EVOLUTION AT SPECIES BOUNDARIES

Studies of hybridization and speciation in the Insecta are numerous, and geographic structure has figured in much of this work (see 63). For example, in the study of hybrid zones, geographic structure forms the basis for work on cline width relative to migration (see 8). Information from multiple loci can be used to trace introgression events and examine linkage disequilibria. A compilation of data from hybrid zones of *Heliconius* butterflies (96) exemplifies the link between geographic structure and hybridization. Using data from hybrid zones and distribution patterns of races and species, Mallet (96) disputes the importance of Pleistocene refugia as a source of initial divergence (see 13) and argues that warning-color evolution must have been partly responsible for race formation. This work implies that conservation efforts (see also below) should not be based on presumed refugia but rather on high species diversity or on species that deserve particular attention.

Recent molecular tools have facilitated new ways to investigate the process of speciation (26). For example, Hey and colleagues (68) have used genealogical models and data from multiple loci to examine speciation in a *Drosophila melanogaster* complex in the context of natural selection. Although these few studies cannot represent all research done in this area, they do show that geographic structure still plays an active role in the study of the origins of insect species, as it has since Darwin's time.

CONSERVATION BIOLOGY

Insects are rarely big or showy and therefore are not often considered worthy of conservation. However, these ideas are changing. Insects and kin are recognized as major elements of ecosystems worldwide (49), and conservation interests include not only endangered and threatened species (and habitats), but also native natural enemies that may be responsible for current levels of pest control or population regulation. As with most of the world's biota, the major threats to native insects appear to be the effects of urbanization and cultivation caused by an ever-increasing human population, as well as the impact of alien predators (59, 72).

Insects can be used for examining theories pertinent to conservation biology. For example, the relationship between the loss of genetic diversity and small effective population size, N_e, is an important consideration in biological conservation (55, 97, 116, 117): If the rate of population extinction is much larger than the rate of migration, then the effective population size, N_e, can be less than expected (101). Models and empirical work (103, 186) have shown that genetic diversity also depends on the patterns of extinction and recolonization, which can either increase or decrease the genetic differentiation of local populations. Both outcomes can lead to loss of genetic diversity from an entire array of populations. Directionality and source of migrants are also important (25, 120, 122): The effects on overall genetic diversity differ according to whether populations act as sources or sinks, or whether colonists come from propagule pools (limited sources) or migrant pools (many populations) (103). As noted earlier, an overall interpretation of the extent of movement can be obtained from average levels of gene flow based on measures of population subdivision such as F_{ST} or θ. Phylogeography provides a powerful tool with which to determine the directions and sources of colonists.

The definition of population units for management purposes is a critical issue for conservation biologists. Because management is concerned with identification of taxonomic units in their current state, Moritz (110) proposed a distinction between evolutionary significant units (ESUs) and management units (MUs): ESUs are groups of individuals that are monophyletic in mtDNA yet may differ from one another in allele frequencies at nuclear loci. MUs differ significantly at both mtDNA and nuclear loci, regardless of phylogenetic distinctiveness of alleles. Vogler et al (177) have used phylogeographic techniques to determine taxonomic units for conservation in an endangered (as defined by the US Endangered Species Act) tiger beetle, *Cicindela dorsalis*. MtDNA and allozyme variation placed populations into evolutionarily distinct units and showed that differences between subspecies were equivalent to those between related *Cicindela* species. Because of the close link between geographic structure and conservation biology, workers in this field have much to contribute to the betterment of the world's condition. Too seldom do active scientists take a stand, even when we are the sole experts (see 56, 72).

SUMMARY

Many of the recent advances in the study of geographic structure of insect populations resulted from advances in technology, especially the ability to collect sequence data from both mitochondrial and nuclear DNA. With recent theoretical developments, interpreting data has become easier and more relevant. These developments have also led to new ways of thinking about populations, such as new measures of population subdivision and the study of phylogeography. To date, much work has been descriptive and has had the goal of reconciling genetic information with that from field manipulations and natural-history studies. However, geographic structure is also being used to test current and new hypotheses in many areas of evolution and ecology. The incorporation of geographic structure into demographic models holds special promise. Undoubtedly, the knowledge of geographic structure will continue to advance our understanding of the evolution of species interactions as well as the origins of species themselves.

ACKNOWLEDGMENTS

I thank R Gillespie, P Follett, E Metz, G Oxford, S Palumbi, and F Villablanca for discussion, comments, and active collaboration. This work was supported by grants from the USDA (93-37302-9129, 94-37312-0769, 58-1275-1-127, 58-1275-4-046), California Department of Food and Agriculture (93-0477), International Rice Research Institute & Rockefeller Agricultural Biotechnology (910002,8), and the Pacific Biomedical Research Center and University Research Council of the University of Hawaii.

Literature Cited

1. Andrewartha HG, Birch LC. 1954. *The Distribution and Abundance of Animals*. Chicago: Univ. Chicago Press
2. Antonovics J. 1994. The interplay of numerical and gene-frequency dynamics in host-pathogen systems. See Ref. 125a, pp. 129–45
3. Archie JW. 1989. A randomization test for phylogenetic information in systematic data. *Syst. Zool.* 38:239–52
4. Avise JC. 1989. Gene trees and organismal histories: a phylogenetic approach to population biology. *Evolution* 43: 1192–208
5. Avise JC. 1991. Ten unorthodox perspectives on evolution prompted by comparative population genetic findings on mitochondrial DNA. *Annu. Rev. Genet.* 25:45–69
6. Avise JC. 1994. *Molecular Markers, Natural History, and Evolution*. London: Chapman & Hall
7. Avise JC, Arnold J, Ball RM, Bermingham E, Lamb T, et al. 1987. Intraspecific phylogeography: the mitochondrial DNA bridge between population genetics and systematics. *Annu. Rev. Ecol. Syst.* 18:489–522
8. Barton NH, Gale KS. 1993. Hybrids and hybrid zones: historical perspective. See Ref. 63a, pp. 13–45
9. Begun DJ, Aquadro CF. 1993. African and North American populations of *Drosophila melanogaster* are very different at the DNA level. *Nature* 365: 548–50
10. Birky CW, Fuerst P, Maruyama T. 1989. Organelle gene diversity under migration, mutation, and drift: equilibrium expectations, approach to equilibrium, effects of heteroplasmic cells, and comparison of nuclear genes. *Genetics* 121: 613–27
11. Black WC. 1993. PCR with arbitrary primers: approach with care. *Insect Mol. Biol.* 2:1–6
12. Black WC, DuTeau NM, Puterka GJ, Nechols JR, Pettorini JM. 1992. Use of the random amplified polymorphic DNA polymerase chain reaction (RAPD-PCR) to detect DNA polymorphisms in aphids (Homoptera: Aphididae). *Bull. Entomol. Res.* 82:151–59
13. Brower AVZ. 1994. Rapid morphological radiation and convergence among races of the butterfly *Heliconius erato* inferred from patterns of mitochondrial DNA evolution. *Proc. Natl. Acad. Sci. USA* 91:6491–95
14. Brower AVZ, DeSalle R. 1994. Practi-
    cal and theoretical considerations for choice of a DNA sequence region in insect molecular systematics, with a short review of published studies using nuclear gene regions. *Ann. Entomol. Soc. Am.* 87:702–16
15. Brown JK, Frohlich DR, Rosell RC. 1995. The sweetpotato or silverleaf whiteflies: biotypes of *Bemisia tabaci* or a species complex? *Annu. Rev. Entomol.* 40:511–34
16. Brown JM, Abrahamson WG, Way PA. 1995. Mitochondrial DNA phylogeography of host races of the goldenrod ball gallmaker, *Eurosta solidaginis* (Diptera: Tephritidae). *Evolution.* In press
17. Bush G. 1994. Sympatric speciation in animals: new wine in old bottles. *Trends Ecol. Evol.* 9:285–88
18. Caprio MA, Tabashnik BE. 1992. Allozymes used to estimate gene flow among populations of diamondback moth (Lepidoptera: Plutellidae) in Hawaii. *Environ. Entomol.* 21:808–16
19. Caprio MA, Tabashnik BE. 1992. Gene flow accelerates local adaptation among finite populations: simulating the evolution of insecticide resistance. *J. Econ. Entomol.* 85:611–20
20. Carey JR. 1991. Establishment of the Mediterranean fruit fly in California. *Science* 253:1369–73
21. Castelloe J, Templeton AR. 1994. Root probabilities for intraspecific gene trees under neutral coalescent theory. *Mol. Phylogenet. Evol.* 3:102–13
22. Chakraborty R, Leimar O. 1987. Genetic variation within a subdivided population. In *Population Genetics and Fisheries Management*, ed. N Ryman, F Utter, pp. 80–120. Seattle: Univ. Wash. Press
23. Chapela IH, Rehner SA, Schultz TR, Mueller UG. 1994. Evolutionary history of the symbiosis between fungus-growing ants and their fungi. *Science* 266: 1691–94
24. Coll M, Garcia de Mendoza L, Roderick GK. 1994. Population structure of a predatory beetle: the importance of gene flow for intertrophic level interactions. *Heredity* 72:228–36
25. Costa JT, Ross KG. 1994. Hierarchical genetic structure and gene flow in macrogeographic populations of the Eastern tent caterpillar. *Evolution* 48: 1158
26. Coyne JA. 1992. Genetics and speciation. *Nature* 355:511–15
27. Coyne JA, Boussy IA, Prout T, Bryant

SH, Jones JS, et al. 1982. Long distance migration of *Drosophila. Am. Nat.* 119: 589–95

28. Crandall KA. 1994. Intraspecific cladogram estimation: accuracy at higher levels of divergence. *Syst. Biol.* 43:222–35

29. Crandall KA, Templeton AR. 1993. Empirical tests of some predictions from coalescent theory with applications to intraspecific phylogeny reconstruction. *Genetics* 134:959–69

30. Crow JF. 1986. *Basic Concepts in Population, Quantitative, and Evolutionary Genetics.* New York: Freeman

31. Crow JF, Aoki K. 1984. Group selection for a polygenic behavioral trait. *Proc. Natl. Acad. Sci. USA* 81:6073–77

32. Daly JC. 1989. The use of electrophoretic data in a study of gene flow in the pest species *Heliothis armigera* (Hübner) and *H. punctigera* Wallengren (Lepidoptera: Noctuidae). See Ref. 94, pp. 115–41

33. Daly JC. 1993. Ecology and genetics of insecticide resistance in *Helicoverpa armigera:* interactions between selection and gene flow. *Genetica* 90:217–26

34. Daly JC, Gregg P. 1985. Genetic variation in *Heliothis* in Australia: species identification and gene flow in the two pest species *H. armigera* (Hübner) and *H. punctigera* Wallengren (Lepidoptera: Noctuidae). *Bull. Entomol. Res.* 75:169–84

35. Davis SK, Strassmann JE, Hughes C, Pletscher LS. 1990. Population structure and kinship in *Polistes* (Hymenoptera, Vespidae): an analysis using ribosomal DNA and protein electrophoresis. *Evolution* 44:1242–53

36. Denno RF, McClure MS, Ott JR. 1995. Interspecific interactions in phytophagous insects: competition reexamined and resurrected. *Annu. Rev. Entomol.* 40:297–331

37. Denno RF, Roderick GK. 1992. Density-related dispersal in planthoppers: effects of interspecific crowding. *Ecology* 73:1323–34

38. Denno RF, Roderick GK, Olmstead K, Döbel HG. 1991. Density-related migration in planthoppers: the role of habitat persistence. *Am. Nat.* 138:1513–41

39. DeSalle R, Giddings LV. 1986. Discordance of nuclear and mitochondrial DNA phylogenies in Hawaiian *Drosophila. Proc. Natl. Acad. Sci. USA* 83: 6902–6

40. DeSalle R, Volger A. 1994. Phylogenetic analysis on the edge: the application of cladistic techniques at the population level. See Ref. 59a, pp. 154–74

41. Dively GP, Everich R. 1993. *Ins. Res. Newsl.* 4:14–15

42. Dobzhansky T, Wright S. 1943. Genetics of natural populations. X. Dispersion rates in *Drosophila pseudoobscura. Genetics* 32:303–24

43. Douglas AE. 1992. Microbial brokers of insect-plant interactions. In *Int. Symp. Insect-Plant Relationships, 8th,* ed. SBJ Menken, JH Visser, P Harrewijn, pp. 329–36. Dordrecht: Kluwer Academic

44. Dunley JE, Croft BA. 1994. Gene flow measured by allozymic analysis in pesticide resistant *Typhlodromus pyri* occurring within and near apple orchards. *Exp. Appl. Acarol.* 18:201

45. Edwards OR, Hoy MA. 1993. Polymorphism in two parasitoids detected using random amplified polymorphic DNA polymerase chain reaction. *Biol. Control* 3:243

46. Ehrlich PR, White RR, Singer MC, McKechnie SW, Gilbert LE. 1975. Checkerspot butterflies: a historical perspective. *Science* 188:221–28

47. Epperson BK. 1993. Spatial and space-time correlations in systems of subpopulations with genetic drift and migration. *Genetics* 133:711–27

48. Erlich HA, Arnheim N. 1992. Genetic analysis using the polymerase chain reaction. *Annu. Rev. Genet.* 26:479–506

49. Erwin TL. 1986. The tropical forest canopy: the heart of biotic diversity. In *Biodiversity,* ed. EO Wilson, pp. 123–29. Washington, DC: Natl. Acad. Sci.

50. Faith DP. 1991. Cladistic permutation tests for monophyly and nonmonophyly. *Syst. Zool.* 40:366–75

51. Farris JS, Källersjö M, Kluge AG, Bult C. 1994. Permutations. *Cladistics* 10: 65–76

52. Feder JL, Bush GL. 1989. Gene frequency clines for host races of *Rhagoletis pomonella* in the midwestern United States. *Heredity* 63:245–66

53. Feder JL, Opp SB, Wlazlo B, Reynolds K, Go W, et al. 1994. Host fidelity is an effective premating barrier between sympatric races of the apple maggot fly. *Proc. Natl. Acad. Sci. USA* 91:7990–94

54. Felsenstein J. 1988. Phylogenies from molecular sequences: inference and reliability. *Annu. Rev. Genet.* 22:521–65

55. Forney KA, Gilpin ME. 1989. Spatial structure and population extinction: a study with *Drosophila* flies. *Conserv. Biol.* 3:45–51

56. Futuyma DJ. 1995. The uses of evolutionary biology. *Science* 267:41–42

57. Gerber A. 1994. *The Semiotics of Sub-*

division: An Empirical Study of the Population Structure of the Lichen Grasshopper Trimerotropis saxatilis *(Acrididae).* PhD thesis. Wash. Univ., St. Louis, MO

58. Gillespie RG, Croom H, Palumbi SR. 1994. Multiple origins of a spider radiation in Hawaii. *Proc. Natl. Acad. Sci. USA* 91:2290–94

59. Gillespie RG, Reimer N. 1993. Effects of alien predators on native Hawaiian invertebrates. *Pac. Sci.* 47:21–33

59a. Golding B, ed. 1994. *Non-Neutral Evolution: Theories and Molecular Data.* New York: Chapman & Hall

60. Gould F, Follett PA, Nault BA, Kennedy GG. 1994. Resistance management strategies for transgenic potato plants. In *Advances in Potato Pest Biology and Management,* ed. GW Zehnder, ML Powelson, RK Jansson, KV Raman, pp. 255–77. St. Paul, MN: Am Phytopathol Soc

61. Grasela JJ, Steiner WM. 1993. Population genetic structure among populations of three predaceous nabid species: *Nabis alternatus* Parshley, *Nabis roseipennis* Reuter and *Nabis americoferous* Carayon (Hemiptera: Nabidae). *Biochem. Syst. Ecol.* 21:813

62. Hale LR, Singh RS. 1991. A comprehensive study of genic variation in natural populations of *Drosophila melanogaster.* IV. Mitochondrial DNA variation and role of history vs. selection in the genetic structure of geographic populations. *Genetics* 129:103–17

63. Harrison RG. 1993. Hybrids and hybrid zones: historical perspective. See Ref. 63a, pp. 3–12

63a. Harrison RG, ed. 1993. *Hybrid Zones and the Evolutionary Process.* Oxford: Oxford Univ. Press

64. Hastings A, Harrison S. 1994. Metapopulation dynamics and genetics. *Annu. Rev. Ecol. Syst.* 25:167–88

65. Hastings A, Higgins K. 1994. Persistence of transients in spatially structured ecological models. *Science* 263:1133–36

66. Haymer DS. 1994. Random amplified polymorphic DNAs and microsatellites: What are they, and can they tell us anything we don't already know? *Ann. Entomol. Soc. Am.* 87:717–22

67. Herre EA. 1993. Population structure and the evolution of virulence in nematode parasites of fig wasps. *Science* 259:1442–45

68. Hey J, Kliman RM. 1994. Genealogical portraits of speciation in the *Drosophila melanogaster* species complex. See Ref. 59a, pp. 208–16

69. Hillis D, Moritz C, eds. 1995. *Molecular Systematics.* Sunderland, MA: Sinauer. 2nd ed.

70. Hoelzel AR. 1992. *Molecular Genetic Analysis of Populations.* Oxford: IRL/Oxford Univ. Press

71. Hopper KR, Roush RT, Powell W. 1993. Management of genetics of biological control introductions. *Annu. Rev. Entomol.* 38:27–51

72. Howarth FG. 1991. Environmental impacts of classical biological control. *Annu. Rev. Entomol.* 36:485–511

73. Hoy MA. 1994. *Insect Molecular Genetics.* San Diego: Academic

74. Hsiao TH. 1989. Estimation of genetic variability amongst the Coleoptera. See Ref. 94, pp. 143–80

75. Hudson RR, Kreitman M, Aguade M. 1987. A test of neutral molecular evolution based on nucleotide data. *Genetics* 116:138–51

76. Hudson RR, Slatkin M, Maddison WP. 1992. Estimation of levels of gene flow from DNA sequence data. *Genetics* 132:583

77. Jackson JK, Resh VH. 1992. Variation in genetic structure among populations of the caddisfly *Helicopsyche borealis* from three streams in northern California, USA. *Freshw. Biol.* 27:29–42

78. Kambhampati S. 1991. Mitochondrial DNA variation within and among populations of the mosquito *Aedes albopictus. Genome* 34:288–92

79. Kambhampati S, Black WC, Rai KS, Sprenger D. 1990. Temporal variation in genetic structure of a colonizing species: *Aedes albopictus* in the USA. *Heredity* 64:281–87

80. Kane TC, Barr TC, Badaracca WJ. 1992. Cave beetle genetics: geology and gene flow. *Heredity* 63:277–86

81. Kaneshiro KY. 1993. Introduction, colonization, and establishment of exotic insect populations: fruit flies in Hawaii and California. *Am. Entomol.* 39:23–29

82. Kareiva PM. 1987. Habitat fragmentation and the stability of predator-prey interactions. *Nature* 326:388–90

83. Karl SA, Avise JC. 1992. Balancing selection at allozyme loci in oysters: implications from nuclear RFLP's. *Science* 256:100–2

84. Kaufman B, Wool D. 1992. Gene flow by immigrants into isolated recipient populations: a laboratory model using flour beetles. *Genetica* 85:163–71

85. Kimura M. 1983. *The Neutral Theory of Evolution.* Cambridge: Cambridge Univ. Press

86. King P. 1987. Macro- and microgeographic structure of a spatially subdi-

vided beetle species in nature. *Evolution* 41:401–16

87. Korman AK, Mallet J, Goodenough JL. 1993. Population structure in *Heliothis virescens* (Lepidoptera: Noctuidae): an estimate of gene flow. *Ann. Entomol. Soc. Am.* 86:182

88. Krafsur ES, Obrycki JJ, Flanders RV. 1992. Gene flow in populations of the seven-spotted lady beetle *Coccinella septempunctata. J. Hered.* 83:440

89. Lanzaro GC, Narang SK, Seawright JA. 1990. Speciation in an anopheline (Diptera: Culicidae) mosquito: enzyme polymorphism and the genetic structure of populations. *Ann. Entomol. Soc. Am.* 83:578

90. Lewontin RC. 1991. Electrophoresis in the development of evolutionary genetics: milestone or millstone. *Genetics* 128:657–62

91. Liebherr JK. 1988. Gene flow in ground beetles (Coleoptera: Carabidae) of differing habitat preference and flight-wing development. *Evolution* 42:129–37

92. Loxdale HD. 1990. Estimating levels of gene flow between natural populations of cereal aphids (Homoptera: Aphididae). *Bull. Entomol. Res.* 80:331–38

93. Loxdale HD, Brookes CP. 1990. Temporal genetic stability within and restricted migration (gene flow) between local populations of the blackberry-grain aphid *Sitobion fragariae* in south-east England. *J. Anim. Ecol.* 59:497–514

94. Loxdale HD, den Hollander J, eds. 1989. *Electrophoretic Studies on Agricultural Pests.* Oxford: Clarendon

95. Lynch M, Crease TJ. 1990. The analysis of population survey data on DNA sequence variation. *Mol. Biol. Evol.* 7: 377–94

96. Mallet J. 1993. Speciation, raciation, and color pattern evolution in *Heliconius* butterflies: evidence from hybrid zones. See Ref. 63a, pp. 226–60

97. Mallet J. 1995. The genetics of diversity at and below the species level. In *Biodiversity: Biology of Numbers and Difference,* ed. KJ Gaston. Oxford: Blackwell. In press

98. Mallet J, Korman A, Heckel D, King P. 1993. Biochemical genetics of *Heliothis* and *Helicoverpa* (Lepidoptera: Noctuidae) and evidence for a founder event in *Heliocoverpa zea. Ann. Entomol. Soc. Am.* 86:189–97

99. Mallet J, Porter P. 1992. Preventing insect adaptation to insect-resistant crops: Are seed mixtures or refugia the best strategy? *Proc. R. Soc. London Ser. B* 250:165–69

100. Martin A, Simon C. 1990. Differing levels of among-population divergence in the mitochondrial DNA of periodical cicadas related to historical biogeography. *Evolution* 44:1066–80

101. Maruyama T, Kimura M. 1980. Genetic variability and effective population size when local extinction and recolonization of subpopulations are frequent. *Proc. Natl. Acad. Sci. USA* 77:6710–14

102. McCauley DE. 1991. The effect of host plant size variation on the population structure of a specialist herbivore insect, *Tetraopes tetraophthalmus. Evolution* 45:1675–84

103. McCauley DE. 1991. Genetic consequences of local population extinction and recolonization. *Trends Ecol. Evol.* 6:5–8

104. McCauley DE. 1993. Gene flow distances in natural populations of *Tetraopes tetraophthalmus. Evolution* 37:1239–46

105. McCauley DE, Eanes WF. 1987. Hierarchical population structure analysis of the milkweed beetle, *Tetraopes tetraophthalmus* (Forster). *Heredity* 58:193–201

106. McPheron BA. 1990. Genetic structure of apple maggot fly (Diptera: Tephritidae) populations. *Ann. Entomol. Soc. Am.* 83:568

107. McPheron BA, Gasparich GE, Han H, Steck GJ. 1994. Mitochondrial DNA restriction map for the Mediterranean fruit fly, *Ceratitis capitata. Biochem. Gen.* 32:25–33

108. McPheron BA, Sheppard WS, Steck GJ. 1995. Genetic research and the origin, establishment, and spread of the Mediterranean fruit fly. In *The Medfly in California: Defining Critical Research,* ed. JG Morse, RL Metcalf, JR Carey, RV Dowell, pp. 93–107. Riverside, CA: Univ. Calif., Cent. Exotic Pest Res.

109. Moran N, Baumann P. 1994. Phylogenetics of cytoplasmically inherited microorganisms of arthropods. *Trends Ecol. Evol.* 9:15–20

110. Moritz C. 1994. Defining 'Evolutionarily Significant Units' for conservation. *Trends Ecol. Evol.* 100:373–75

111. Narang SK, Tabachnick WJ, Faust WJ. 1993. Complexities of population genetic structure and implications for biological control programs. In *Applications of Genetics to Arthropods of Biological Control Significance,* pp. 19–52. Boca Raton, FL: CRC

112. Nee S, Holmes EC, Rambaut A, Harvey PH. 1995. Inferring population history from molecular phylogenies. *Philos. Trans. R. Soc. London.* In press

113. Nei M. 1973. Analysis of gene diversity

in subdivided populations. *Proc. Natl. Acad. Sci. USA* 20:3321–23

114. Neigel JE, Avise JC. 1985. Phylogenetic relationships of mitochondrial DNA under various demographic models of speciation. In *Evolutionary Processes and Theory*, ed. S Karlin, E Nevo, pp. 515–34. New York: Academic

115. Normark B. 1994. *Phylogeny and Evolution of Parthenogenesis in the* Aramigns tessellatus *Complex* (*Coleoptera: Curculionidae*). PhD thesis. Cornell Univ., Ithaca, NY

116. Nunney L, Campbell KA. 1993. Assessing minimum viable population size: demography meets population genetics. *Trends Ecol. Evol.* 8:234–39

117. Nunney L, Elam DR. 1994. Estimating effective population size from ecological data. *Conserv. Biol.* 8:175–84

118. Palumbi SR. 1995. Nucleic acids. III. The polymerase chain reaction. See Ref. 69

119. Palumbi SR, Baker CS. 1994. Contrasting population structure from nuclear intron sequences and mtDNA of humpback whales. *Mol. Biol. Evol.* 11: 426–35

120. Pashley DP, Johnson SJ, Sparks AN. 1985. Genetic population structure of migratory moths: the fall armyworm (Lepidoptera: Noctuidae). *Ann. Entomol. Soc. Am.* 78:756–62

121. Perring TM, Cooper AD, Rodriquez RJ, Farrar CA, Bellows TS Jr. 1993. Identification of a whitefly species by genomic and behavioral studies. *Science* 259:74–77

122. Peterson MA. 1995. Phenological isolation, gene flow and developmental differences among low- and high-elevation populations of *Euphilotes enoptes* (Lepidoptera: Lycaenidae). *Evolution.* 49:446–55

123. Philipp M, Madsen HES, Siegismund HR. 1992. Gene flow and population structure in *Armeria maritima. Heredity* 69:32–42

124. Preziosi RF, Fairbairn DJ. 1992. Genetic population structure and levels of gene flow in the stream dwelling waterstrider, *Aquarius* (= *Gerris*) *remigis* (Hemiptera: Gerridae). *Evolution* 46: 430–44

125. Queller DC, Strassmann JE, Hughes CR. 1993. Microsatellites and kinship. *Trends Ecol. Evol.* 8:285–88

125a. Real LA, ed. 1994. *Ecological Genetics.* Princeton: Princeton Univ. Press

126. Richardson BJ, Baverstock PR, Adams M. 1986. *Allozyme Electrophoresis.* New York: Academic

127. Riechert SE. 1993. Investigation of potential gene flow limitation of behavioral adaptation in an arid land spider. *Behav. Ecol. Soc.* 32:355

128. Roderick GK. 1992. Post-colonization evolution of natural enemies. In *Thomas Say Publications in Entomology: Proceedings*, ed. WC Kauffman, JR Nechols, pp. 71–86, Lanham, MD: Entomol. Soc. Am.

129. Roderick GK, Caldwell RF. 1992. An entomological perspective on animal dispersal. In *Animal Dispersal*, ed. NC Stenseth, WZ Lidicker, pp. 274–90. London: Chapman & Hall

130. Roderick GK, Villablanca FX. 1995. Genetic and statistical analysis of colonization. In *Economic Fruit Flies: A World Assessment of Their Biology and Management. Proc. Int. Symp. on Fruit Flies of Economic Importance, 4th, Tampa*, ed. BA McPheron, GJ Steck. Delray Beach, FL: St Lucie Press

131. Roff DA. 1990. The evolution of flightlessness in insects. *Ecol. Monogr.* 60: 389–421

132. Roff DA. 1994. Habitat persistence and the evolution of wing dimorphism in insects. *Am. Nat.* 144:772–98

133. Roff DA, Bentzen P. 1989. The statistical analysis of mitochondrial DNA polymorphisms: χ^2 and the problem of small samples. *Mol. Biol. Evol.* 6:539–45

134. Ross KG. 1993. The breeding system of the fire ant *Solenopsis invicta*: effects of colony genetic structure. *Am. Nat.* 141:554–76

135. Ross KG, Mathews RW. 1989. Population genetic structure and social evolution in the sphecid wasp *Microstigmus comes. Am. Nat.* 134:574–98

136. Ross KG, Shoemaker DD. 1993. An unusual pattern of gene flow between the two social forms of the fire ant *Solenopsis invicta. Evolution* 47:1595–605

137. Schiff NM, Sheppard WS. 1993. Mitochondrial DNA evidence for the 19th century introduction of African honey bees into the United States. *Experientia* 49:530–32

138. Seawright JA, Suarez MF. 1991. Genetic structure of natural populations of *Anopheles albimanus* in Columbia. *J. Am. Mos. Control Assoc.* 7:437

139. Sheppard WS, Steck GJ, McPheron BA. 1992. Geographic populations of the medfly may be distinguished by mitochondrial DNA variation. *Experientia* 48:1010

140. Sheppard WS, Rinderer TE, Mazzoli JA, Stelzer JA, Shimanuki H. 1991. Gene flow between African- and Euro-

pean-derived honey bee populations in Argentina. *Nature* 349:782–84

141. Shykoff JA, Schmid-Hempel P. 1991. Genetic relatedness and eusociality: parasite-mediated selection on the genetic composition of groups. *Behav. Ecol. Soc.* 28:371–76

142. Simon C, Fratti F, Beckenbach A, Crespi B, Liu H, et al. 1994. Evolution, weighting, and phylogentic utility of mitochondrial gene sequences and a compilation of conserved polymerase chain reaction primers. *Ann. Entomol. Soc. Am.* 87:651–701

143. Simon C, McIntosh C, Denniega J. 1993. Standard restriction fragment length analysis of the mitochondrial genome is not sensitive enough for phylogentic analysis or identification of 17-year cicada broods: the potential for a new technique. *Ann. Entomol. Soc. Am.* 86:142–52

144. Slade RW, Moritz C, Heideman A. 1994. Multiple nuclear-gene phylogenies: application to pinnepeds and comparison with a mitochondrial DNA gene phylogeny. *Mol. Biol. Evol.* 11: 341–56

145. Slatkin M. 1977. Gene flow and genetic drift in a species subject to frequent local extinctions. *Theor. Popul. Biol.* 12:253–62

146. Slatkin M. 1985. Gene flow in natural populations. *Annu. Rev. Ecol. Syst.* 16: 393–430

147. Slatkin M. 1987. Gene flow and the geographic structure of natural populations. *Science* 236:787–92

148. Slatkin M. 1994. Cladistic analysis of DNA sequence data from subdivided populations. See Ref. 125a, pp. 18–34

149. Slatkin M. 1994. Gene flow and population structure. See Ref. 125a, pp. 3–17

150. Slatkin M. 1995. A measure of population subdivision based on microsatellite allele frequencies. *Genetics* 139:457–62

151. Slatkin M, Arter HE. 1991. Spatial autocorrelation methods in population genetics. *Am. Nat.* 138:499–517

152. Slatkin M, Hudson D. 1991. Pairwise comparisons of mitochondrial DNA sequences in stable and exponentially growing populations. *Genetics* 129:555–62

153. Slatkin M, Maddison WP. 1989. A cladistic measure of gene flow inferred from the phylogenies of alleles. *Genetics* 123:603–13

154. Slatkin M, Maddison WP. 1990. Detecting isolation by distance using phylogenies of genes. *Genetics* 126:249–60

155. Sokal RR, Jacquez GM, Wooten MC. 1989. Spatial autocorrelation analysis of

migration and selection. *Genetics* 121: 845–55

156. Sperling FAH, Hickey DA. 1994. Mitochondrial DNA sequence variation in the spruce budworm species complex (*Choristoneura:* Lepidoptera). *Mol. Biol. Evol.* 11:656–65

157. Steiner WWM, Grasela JJ. 1993. Population genetics and gene variation in the predator *Coleomegilla maculata* (De Greer) (Coleoptera: Coccinellidae). *Ann. Entomol. Soc. Am.* 86:309–21

158. Stille M, Stille B. 1992. Intra- and inter-nest variation in mitochondrial DNA in the polygynous ant *Leptothorax acervorum* (Hymenoptera: Formicidae). *Insectes Soc.* 39:335–40

159. Stouthamer R, Kazmer DJ. 1994. Cytogenetics of microbe-associated parthenogenesis and its consequences for gene flow in *Trichogramma* wasps. *Heredity* 73:317–27

160. Strong DR, Lawton JH, Southwood R. 1984. *Insects on Plants.* Oxford: Blackwell

161. Sundstrom L. 1993. Genetic population structure and sociogenetic organisation in *Formica truncorum* (Hymenoptera: Formicidae). *Behav. Ecol. Soc.* 33:345–54

162. Tabashnik BE. 1994. Delaying insect adaptation to transgenic plants: seed mixtures and refugia reconsidered. *Proc. R. Soc. London Ser. B* 225:7–12

163. Tabashnik BE. 1994. Evolution of resistance to *Bacillus thuringiensis. Annu. Rev. Entomol.* 39:47–80

164. Tajima F. 1989. Statistical method for testing the neutral mutation hypothesis by DNA polymorphism. *Genetics* 123: 585–95

165. Takahata N, Palumbi SR. 1985. Extranuclear differentiation and gene flow in the finite island model. *Genetics* 109: 441–57

166. Takahata N, Slatkin M. 1984. Mitochondrial gene flow. *Proc. Natl. Acad. Sci. USA* 81:1764–67

167. Tamura K, Aotsuka T, Kitagawa O. 1991. Mitochondrial DNA polymorphisms in the two subspecies of *Drosophila sulfurigaster:* relationships between geographic structure of population and nucleotide diversity. *Mol. Biol. Evol.* 8:104–14

168. Taylor AD. 1991. Studying metapopulation effects in predator-prey systems. *Biol. J. Linn. Soc.* 42:305–23

169. Taylor DJ, Finston TL, Hebert PDN. 1994. The 15% solution for preservation. *Trends Evol. Ecol.* 9:230

170. Deleted in proof

171. Travis J. 1994. Slouching toward ma-

turity: the progress of ecological genetics. *Ecology* 75:1851–52

172. Travis J, Mueller LD. 1989. Blending ecology and genetics: progress toward a unified population biology. In *Perspectives in Ecological Theory*, ed. J Roughgarden, RM May, SA Levin, pp. 101–24. Princeton: Princeton Univ. Press

173. Unruh TR. 1990. Genetic structure among 18 West Coast pear *Psylla* populations: implications for the evolution of resistance. *Bull. Entomol. Soc. Am.* 36:37

174. Valdes AM, Slatkin M, Freimer NB. 1993. Allele frequencies at microsatellite loci: the stepwise mutation model revisited. *Genetics* 133:737–49

175. Via S. 1991. The genetic structure of host plant adaptation in a spatial patchwork: demographic variability among reciprocally transplanted pea aphid clones. *Evolution* 45:827–52

176. Vogler AP, DeSalle R. 1994. Evolution and phylogenetic information content of the ITS-1 region in the tiger beetle *Cicindela dorsalis*. *Mol. Biol. Evol.* 11:393–405

177. Vogler AP, DeSalle R, Asmann T, Knisley CB, Schultz TD. 1993. Molecular population genetics of the endangered tiger beetle *Cicindela dorsalis* (Coleoptera: Cicindelidae). *Ann. Entomol. Soc. Am.* 86:142–52

178. Wade MJ, Chang NW. 1995. Increased male fertility in *Tribolium confusum* beetles after infection with the intracellular parasite *Wolbachia*. *Nature* 373:72–74

179. Wade MJ, Goodnight CJ. 1991. Wright's shifting balance theory: an experimental study. *Science* 253:1015–18

180. Wade MJ, McCauley DE. 1988. Extinction and colonization: their effects on the genetic differentiation of local populations. *Evolution* 42:995–1005

181. Wagner DL, Liebherr JK. 1992. Flightlessness in insects. *Trends Ecol. Evol.* 7:216–20

182. Warring GL, Abrahamson WG, Howard DJ. 1990. Genetic differentiation among host-associated populations of the gallmaker *Eurosta solidaginis* (Diptera: Tephritidae). *Evolution* 44:1648–55

183. Weir BS. 1990. *Genetic Data Analysis.* Sunderland, MA: Sinauer Associates

184. Weir BS, Cockerham CC. 1984. Estimating F-statistics for the analysis of population structure. *Evolution* 38:1358–70

185. Werren J. 1994. Genetic invasion of the insect body snatchers. *Nat. Hist.* 103:36–38

186. Whitlock MC. 1992. Nonequilibrium population structure in forked fungus beetles: extinction, colonization, and the genetic variance among populations. *Am. Nat.* 139:952–70

187. Wright S. 1931. Evolution in Mendelian populations. *Genetics* 16:97–159

188. Wright S. 1938. Size of population and breeding structure in relation to evolution. *Science* 87:430–31

189. Wright S. 1951. The genetical structure of populations. *Ann. Eugen.* 15:323–54

190. Wright S. 1978. *Variability Within and Among Natural Populations.* Chicago: Univ. Chicago Press

191. Zera A. 1981. Genetic structure of two species of waterstriders (Gerridae: Hemiptera) with differing degrees of winglessness. *Evolution* 35:218–25

Annu. Rev. Entomol. 1996. 41:353–74
Copyright © 1996 by Annual Reviews Inc. All rights reserved

SEMIOCHEMICAL PARSIMONY IN THE ARTHROPODA

Murray S. Blum

Department of Entomology, University of Georgia, Athens, Georgia 30602

KEY WORDS: pheromones, allomones, multifunctional sex pheromones, alarm pheromonal parsimony, antimicrobial defense allomones

ABSTRACT

A wide variety of arthropods have adapted their own semiochemicals to subserve multiple functions in diverse contexts. Semiochemicals, the pheromones and allomones, have been detected in arthropod species in six orders, and it has been clearly established that these compounds are used with great parsimony. The versatility of these invertebrates in using these natural products for an incredible diversity of functions emphasizes the significance of semiochemicals in the evolutionary biology of the Arthropoda.

Multifunctional pheromones have proved to be especially characteristic of the queens of eusocial species. Compounds such as the queen substance of the honey bee, *Apis mellifera*, possess unrelated primer and releaser functions for the workers and act as a sex attractant for drones. Females of other hymenopterous species exploit the secretions of sting-associated glands as sex pheromones, whereas a variety of nonhymenopterous species have adapted components in diverse defensive secretions to function as sex pheromones.

The alarm pheromones of many arthropods are also used as defensive allomones, activity inhibitors, cryptic alarm pheromones, aggregative attractants, robbing agents, digging agents, trail pheromones, and antimicrobial agents. Defensive allomones also possess some of these parsimonious roles; in addition, however, some of these compounds possess highly distinctive roles, such as functioning as lethal attractants for prey or, in the aquatic milieu, cuticular wetting agents. Clearly, the availability of a variety of pheromones and allomones has enabled arthropods to evolve an elegant semiochemical parsimony with which to exploit the biological milieu.

0066-4170/96/0101-0353$08.00

PERSPECTIVES AND OVERVIEW

Endowed with a diversity of biosynthetic pathways, arthropodous species produce a vast array of natural products (10) that serve numerous important functions. Among these pheromones and allomones, the semiochemicals (65), have been detected in numerous orders in the Crustacea, Scorpionida, Chilopoda, Diplopoda, Arachnida, and Insecta, and these natural products are most likely produced by species in other classes and orders. Semiochemicals are clearly characteristic of arthropods, and unrelated species sometimes produce the same, often highly distinctive, compounds in their exocrine glands (9). Although this structural identicalness does not imply that a compound is produced via the same biosynthetic pathway in unrelated organisms, it does indicate that its production is highly adaptive. This article therefore analyzes the adaptive significance of exocrine products generated by invertebrates.

The identification of arthropod semiochemicals often occurs in the absence of any studies on the adaptive value of these compounds, and investigations that do examine the roles of these products almost invariably focus on single functions that are readily amenable to laboratory studies. Generally, the possibility that a natural product may serve more than one function is not considered, notwithstanding the fact that such parsimony was recognized 25 years ago (7). Because the number of candidate compounds available to regulate multiple systems may be biosynthetically finite, strong adaptation pressures encourage the use of single natural products for many purposes. Furthermore, evolution probably rarely favors characters with single functions, and as the examples in the next sections illustrate, arthropods have exploited their natural products with considerable parsimonious versatility.

TERMINOLOGICAL EXACTITUDE

Semiochemicals have been divided into two broad classes depending on whether they are utilized in intra- or interspecific contexts. Compounds used as communicative agents between individuals of the same species are classified as pheromones (24). These semiochemicals are considered to be adaptations of the emitter that have resulted from favorable behavioral or developmental responses produced in a receiver of the same species (65). Allomones, in contrast, are utilized as interspecific stimuli, but like pheromones, their secretion is considered to be adaptively favorable to the emitter (24, 65). For example, defensive secretions of arthropods may be fortified with allomonal compounds whose functions are eminently favorable to the producer organism (8). On the other hand, transspecific chemical messengers are adaptively favorable to the perceiver, rather than the emitter, and have been labeled kairomones (24, 65, 112).

A few studies have shown that arthropod allomones possess a wide range of biological activities; consequently, this review attempts to present each one in a separate section. This treatment should clearly demonstrate the evolutionary elegance of semiochemical parsimony. In all cases, the discussion is limited to semiochemicals that presumably have been synthesized de novo by arthropod species.

MULTIFUNCTIONAL PHEROMONES

The term eusocial is used to describe the higher social bees and wasps, and all ants and termites, to emphasize that they possess certain characteristics regarded as hallmarks of insect societies. These include (a) cooperation in caring for the young, (b) reproductive division of labor within groups that include both sterile individuals and reproductives, and (c) overlap of two or more generations that contribute to colony labor. The structures of these societies have become more complex as additional integrative mechanisms have evolved to regulate the various social interactions that occur in large, cohesive populations. These hymenopterans and isopterans are endowed with numerous exocrine glands (86) that transmit a variety of messages both in and out of the colonial milieu. On the other hand, the total information capacity of the exocrine system could be sufficiently finite as to be behaviorally limiting. However, this limitation could be largely overcome by pheromones that serve more than one function. Indeed, pheromonal parsimony appears to have been a driving force in expanding the horizons of eusociality.

Multifunctional Queen Pheromones

Pheromonal parsimony seems to reach its apogee in the honey bee, *Apis mellifera*, in terms of the response of workers to the mandibular-gland products of the queen. One of these glandular constituents, the queen substance or (*E*)-9-oxo-2-decenoic acid (9-ODA) (4, 29), is accompanied by other compounds that may increase its activity (84). Parsimony notwithstanding, 9-ODA is a very unusual pheromone in possessing well-developed primer activity in the hive and releaser activity in and outside the hive. The ability of queen substance to regulate diverse worker and drone behaviors illustrates how various societal activities can be integrated with a limited pheromonal repertoire.

9-ODA possesses two primer activities that are of considerable reproductive significance for both virgin queens and workers. Construction of queen cells by workers is inhibited by 9-ODA when this acid is accompanied by other mandibular-gland constituents (84). Another mandibular-gland constituent, (*E*)-9-hydroxy-2-decenoic acid (9-HDA), synergizes the pheromonal activity of 9-

ODA, although neither compound is active alone (27). Immature queens in sealed cells also inhibit queen-cell construction by workers (21). These mandibular-gland constituents cannot act as primers for worker activity if the workers cannot obtain the pheromone through contact with the queen.

The queen's ability to inhibit ovarian development in workers provides evidence for a second primer function for the pheromonal mélange generated in her mandibular glands (85). Ovarian inhibition occurs only if the workers have physical contact with the queen (26). Because pure 9-ODA or 9-HDA plus 9-ODA does not possess the same inhibitory activity as a queen or her mandibular-gland secretion (27, 104), other glandular constituents must be required for maximum activity. The compound 9-ODA, produced by reproductives in the genus *Apis,* also possesses at least three releaser activities, two of which occur outside the hive. In addition to its role as a reproductive primer in the colonial milieu, 9-ODA is a key element in the attraction of workers to the queen and the formation of retinues (99). Maximum retinue development occurs if 9-ODA, constituting more than two thirds of the mixture, is accompanied by four additional mandibular-gland constituents.

Swarming of honey bees is closely regulated by a medley of pheromones produced by both the workers and the queen. Worker-generated pheromones from the Nasonov-gland secretion are crucial for regulating the movement and formation of swarms (79). Scouts mark cavity entrances and utilize this secretion to recruit other scouts for both long- and short-range orientation to the cavity (95). Swarm movement to the cavity is then enhanced by the secretion of the 9-ODA–rich mandibular-gland exudate of the queen, which now functions as a short-range attractant for the workers. By itself, 9-ODA is inactive.

A third releaser function involves the other caste, the drones. 9-ODA is a powerful sex attractant that readily attracts drones when presented on an elevated lure (47). Notwithstanding the required synergists for 9-ODA activities in other pheromonal contexts, this compound, by itself, appears to be highly attractive to drones in the aerial milieu.

SEXUAL ACTIVITIES OF DEFENSIVE SECRETIONS

Some arthropods release sexual pheromones from glands that almost certainly evolved for nonsexual functions. This adaptation reflects the evolution of a noncommunicative gland to also subserve the function of a social organ. In essence, this example constitutes extreme semiochemical parsimony in which an entire secretion, not originally identified with signaling, has evolved to serve a critical sexual function. Moreover, in at least some cases, the sex attractants clearly constitute known defensive allomones. Thus, certain compounds in arthropod defensive arsenals have evolved to function as sex pheromones while retaining their original allomonal roles (e.g. as repellents). The widespread occurrence of such exocrine parsimony in the Insecta demonstrates that the conversion of noncommunicative glands into pheromonal organs has arisen independently in diverse insect lines.

Romaleid grasshoppers produce a defensive exudate in glands surrounding the metathoracic tracheal trunks, and this secretion is stored in the tracheal lumina (109). It is produced from the third instar on and is one of the defensive mainstays of these large acridids. Moreover, efficacy as a deterrent may be augmented by sequestered natural plant products in addition to compounds synthesized de novo by the insect (17). However, besides having a defensive role, the secretion of the adult female of *Taeniopoda eques* is fortified with a potent primer sex pheromone that can be detected 16–18 days after adult eclosion (109). In an expression of sexual synchrony, adult males become responsive to the sex pheromone about 16 days after their eclosion.

The synthesis of sex pheromones in defensive glands has been detected in several species of Heteroptera. In the mullein bug, *Campylomma verbasci,* a species in the family Miridae, butyl butyrate and (*E*)-crotyl butyrate have been identified as the female-produced sex pheromones (100). These compounds are typical defensive constituents generated in the metathoracic scent glands of heteropterans, and their chemistry and functions clearly illustrate the extent of semiochemical parsimony in *C. verbasci.* In contrast, males of the alydid *Riptortus clavatus* have evolved a sexual secondary function vis-à-vis the metathoracic defensive gland. Attraction is achieved with a mixture of three glandular constituents, (*E*)-2-hexenyl (*E*)-2-hexenoate, (*E*)-2-hexenyl (*Z*)-3-hexenoate, and myristyl isobutyrate, three compounds that probably are quite active as defensive allomones (66).

Compounds in the defensive gland of a beetle, the staphylinid *Aleochara curtula,* have clearly been adapted to serve as releasers of sexual behavior while retaining their defensive activities. The tergal-gland secretion of both sexes is fortified with a mixture of hydrocarbons, aliphatic aldehydes, and 1,4-benzoquinones (88). In combination with an aphrodisiac female sex pheromone from the epicuticular lipids, low concentrations of the tergal-gland secretion or of a mixture of three glandular constituents—(*Z*)-4-tridecene, dodecanal, and (*Z*)-5-tetradecenal—readily release male copulatory behavior (87). At higher concentrations, the defensive roles of some of the tergal-gland constituents (e.g. undecane, toluquinone) are pronounced, and the male copulatory response is inhibited.

Female hymenopterans have exploited the secretions of the sting-associated glands as releasers of sexual activity in males. These glands, the poison (venom) and Dufour's glands, are characteristic of female ants, wasps, and bees, and in the first two groups, sexual pheromonal parsimony has been detected. In these Hymenoptera, the sting thus emerges as an important organ of sexual communication.

Among queen ants, both the poison and Dufour's gland have secondarily evolved to function as releasers of male sexual behavior. Virgin queens of *Formica lugubris* attract males with undecane, a Dufour's-gland constituent that makes up ~90% of the volatiles (108). Minor concomitants of the C_{11} hydrocarbon are tridecane and (Z)-4-tridecene, but neither of these hydrocarbons is very active compared with undecane. Undecane, a compound utilized as an alarm pheromone by several ant species (68), has evolved to function as an alarm pheromone by *F. lugubris* as well. In addition, undecane is used as a defensive allomone by numerous arthropods (9), but semiochemical parsimony in virgin *F. lugubris* is particularly evident.

Queens of various ant species in the large subfamily Myrmicinae also attract males with poison-gland or Dufour's-gland secretions (56). As the poison gland has clearly evolved as a defensive organ, the secondary utilization of this sting-associated structure as a pheromone dispenser emphasizes the semiochemical parsimony of this system. Venom dispensed by virgin females in the genus *Xenomyrmex* is highly attractive to males, releasing both excitement and copulatory behavior in the latter (55). Similarly, virgin queens of *Harpagozenus sublaevis* are highly evident in attracting males with their poison gland secretion (25).

THE PARSIMONIOUS VERSATILITY OF ALARM PHEROMONES

Alarm pheromones, which are generally produced in much greater quantities than other classes of pheromones, seem ideally suited for use in multiple contexts (11). Furthermore, because these compounds are secreted during traumatic stimulation, often functioning as deterrents, they possess many of the same properties as defensive allomones, from which they are believed to be derived. Evolution of a preprogrammed behavioral response by a species to its own defensive allomone can be visualized as an elaboration of the already-expressed responses to environmental disturbances (11). Arthropods have clearly exploited these compounds as regulatory agents par excellence.

Alarm Pheromones as Defensive Allomones

The fact that some alarm pheromones are structurally identical to well-known defensive compounds (10, 106) raises the possibility that the latter are used both pheromonally and allomonally in confrontational contexts. Research on arthropods from various orders has documented this functional duality as a major example of semiochemical parsimony that has independently evolved in mites and insects and attests to the adaptive significance of this phenomenon.

The parsimonious utilization of selected exocrine products was suggested by the frequent identification, in nonsocial species, of natural products that were structurally congruent with well-known alarm pheromones. For example, 2-heptanone, a powerful releaser of alarm for dolichoderine ants (20), is a very effective defensive allomone for cockroaches in the genus *Platyzosteria* (107) and beetles in the genus *Dyschirius* (77). Another characteristic ant alarm pheromone, 4-methyl-3-heptanone (73), is the major compound in the defensive arsenals of several species of opilionids (15, 74). Similarly, cockroaches in the genus *Neostylopyga,* when disturbed, secrete 6-methyl-5-hepten-2-one (9), another typical alarm pheromone of dolichoderine ants (102).

The demonstration that many species of formicine ants (*Formica* spp.) utilize formic acid, the defensive hallmark of this subfamily, as a powerful releaser of alarm behavior (71) further demonstrates that some arthropods use natural products with major functional dualities. The evolution of pheromonal-allomonal compounds is also particularly well developed in several families of Heteroptera, which use typical defensive allomones such as (*E*)-2-hexenal and (*E*)-2-octenal as key releasers of alarm behavior (10). This semiochemical parsimony is well expressed in immature heteropterans that form compact aggregations that are very vulnerable to predation. Secretion of these α,β-unsaturated aldehydes from the larval dorsoabdominal glands causes the aggregations to disperse while unleashing highly effective defensive allomones at adversaries (60, 67, 72). The availability of compounds functioning both communicatively and defensively was a sine qua non in the evolution of sociality in selected heteropterous subfamilies.

Eusociality, as exemplified by numerous hymenopterous species and all isopterous species, is characterized by widespread semiochemical parsimony in which diverse defensive allomones also act as alarm pheromones (10). In ants and termites, characteristic monoterpenes, such as 6-methyl-5-hepten-2-one (102), citral, citronellal (48), and α-pinene and limonene (106), serve as both chemical releasers of alarm behavior and effective chemical agents of deterrency. Similarly, in ants and bees, 2-heptanone functions as an allomone-pheromone (20, 97), as does 4-methyl-3-heptanone and 3-octanone in many formicids (36, 73).

Inhibitors of Activity

Eusocial species may use alarm pheromones in special contexts to suppress normal colonial activities. Foraging workers of *A. mellifera* may mark nectar-depleted flowers with 2-heptanone, an alarm pheromone derived from the mandibular glands (97). The presence of this methyl ketone on flowers inhibits visitations by other bee workers (81). 2-Heptanone and another honey bee pheromone, isopentyl acetate, inhibit the secretion of the products of the Nasonov gland, a social organ that is especially important during colony fission or swarming (78). The pheromone secretion attracts workers and queens in reproductive swarms, and the inhibition of its secretion may represent a mechanism for rejecting foreign queens.

Alarm pheromones can also inhibit normal foraging activities of ant workers. Workers of the fungus-growing ant *Atta texana* cut leaf fragments and then carry them from the host plants to the nest on well-marked trails. If exposed to their cephalic alarm pheromone, 4-methyl-3-heptanone (73), the leaf-laden workers drop their forage and return to the nest, not resuming their activities until the alarm-releasing stimulus is no longer present (80). This behavior could be highly adaptive if a predator attacked the ants on their crowded foraging trails.

Cryptic Alarm Pheromones

Solitary arthropods, moving over the ground surface, may frequently confront aggressive ants that can rapidly recruit additional attacking workers. However, some nonant arthropods have evolved a novel form of chemical warfare to blunt the en masse attacks of the abundant ants. These solitary invertebrates unleash copious amounts of allomonal deterrents at their formicid tormentors, and these compounds are often structurally congruous with the alarm pheromones of the ants. In essence, for the nonant producers, these defensive allomones also constitute cryptic pheromones that can temporarily disperse the ants by causing an exaggerated alarm response (11). Thus, ants encountering cryptic pheromones, secreted in large quantities, may exhibit frenetic and exaggerated behaviors (19, 110), which often permit the solitary arthropods to escape. The disparate species of arthropods that produce cryptic pheromones—the opilionids, mutillids, cockroaches, and beetles— are all capable of great cursorial speed, enabling them to escape easily from the ants.

4-Methyl-3-heptanone, an alarm pheromone produced by formicid species in five subfamilies (9), has independently evolved as a cryptic alarm pheromone used by opilionids (15, 74) and mutillids (94) as an antipredatory device against ants. Other ketones, probably representing powerful olfactants, have also evolved as cryptic pheromones. 2-Heptanone, a typical ketonic releaser of alarm behavior (20), is contained in the defensive exudates of cockroaches (*Platyzosteria* spp.) (107) and beetles (*Dyschirius* spp.) (77). This further attests to the frequent structural congruency of the defensive allomones of solitary arthropods and ant alarm pheromones. The production of a widespread formicid alarm pheromone, 6-methyl-5-hepten-2-one, by another cockroach species, *Neostylopyga rhombifolia* (9, 12), further emphasizes the frequency with which nonsocial arthropods have evolved compounds that are identical to important behavioral releasers generated by ants.

Aggregative Attractants

Compact aggregations of some species of insects may result from exposure to alarm pheromones. In particular, the individuals of certain eusocial species may aggregate around an assailant that has been marked with a highly attractive alarm pheromone. The chemical releaser of alarm behavior thus functions as a beacon for aggressive workers that exhibit group behavior in response to the pheromonal stimulus.

The ability of alarm pheromones to simultaneously function as recruitment stimuli is probably widespread in ants (48) and has been detected in some bees and wasps as well (73). Alarm pheromones function as key recognition markers (5) for labeled adversaries and prey (2). However, in the absence of traumatic stimuli, low concentrations of alarm releasers may promote the formation of aggregations.

n-Undecane, a hydrocarbon utilized as a defensive allomone and alarm pheromone by many species of ants, orients excited workers of *Camponotus pennsylvanicus* to sources marked by their major alarm pheromone, formic acid (3). In contrast, very low concentrations of n-undecane can result in long-term aggregations of workers. Similarly, habitual feeding aggregations of the dolichoderine ant *Iridomyrmex pruinosus* are labeled with their alarm pheromone, 2-heptanone (13).

Some of these aggregative attractants actually constitute trifunctional agents, and this parsimonious phenomenon may be far more common than previously recognized. Compounds such as 2-heptanone and n-undecane have been identified as both defensive allomones and alarm pheromones in various eusocial species, and their additional roles as aggregative attractants further demonstrate their functional versatility as behavior-regulating natural products.

Alarm Pheromones as Robbing Agents

Some ants and bees have evolved a robbing modus operandi that is based primarily on the utilization of natural products to subdue individuals of the host species. Although these robbing lifestyles have evolved independently in ants and bees, in all cases where the basis for this phenomenon has been elucidated, alarm pheromones are used as effective disarming agents.

Workers of the stingless bee *Lestrimellita limao* utilize their mandibular-gland secretions to successfully raid the nests of several species of other stingless bees. *L. limao*'s main alarm pheromone, citral, dominates the initial invaders' glandular secretions and attracts other *L. limao* workers to vulnerable nests (14). Resistance by the host workers quickly ceases, primarily as a result of the disorienting effect of citral. In essence, this monoterpene functions as an attractant during raids, while constituting the chemical disarming agent for raid-susceptible species.

Ants have also adapted alarm pheromones to function as agents of chemical warfare during attacks on the nests of other species. Workers of two species of *Formica* utilize compounds synthesized in their Dufour's glands as the means of subduing the workers of raid-susceptible species (92). Prey workers are chemically disarmed with alkyl acetates sprayed from the hypertrophied glands of the raiders, and organized resistance of the disoriented host workers disappears.

Releasers of Digging Behavior

Several studies have shown that eusocial insects with terrestrial nests initiate digging behavior after exposure to their alarm pheromones. This activity, which may represent displacement behavior, can be highly adaptive in certain contexts.

Workers of *Pogonomyrmex badius*, which nest in sandy soils, initiate typical digging behavior after exposure to their alarm pheromone (110). Because sandy structures are highly susceptible to cave-ins, the purpose of this digging behavior might be to rescue buried ant workers.

High concentrations of 2-heptanone, the pygidial gland alarm pheromone of the dolichoderine ant *Conomyrma pyramicus,* release intense digging behavior in workers (18). This activity appears considerably adaptive because it can result in the burial of pheromone-treated objects that are highly stimulatory and consequently very disruptive of colonial activity.

Alarm Pheromones as Trail Pheromones

Stingless bees lay chemical trails with mandibular-gland secretions, which are also utilized as alarm pheromones. Given that some of these exocrine compounds are well-known defensive allomones, this cephalic exudate is almost certainly trifunctional.

Workers of *Trigona subterranea* generate chemical trails with a secretion that is dominated by citral (14). When the secretion is deposited in the vicinity of the nest, sustained alarm behavior results. Intense alarm behavior also occurs if workers of *T. subterranea* are exposed to pure citral. Similar results were obtained when workers of another *Trigona* species were challenged with several compounds that constitute part of their trail-pheromone blend (69).

Antimicrobial Alarm Pheromones

Many species of arthropods live in environments that should be eminently conducive to the proliferation of pathogenic microorganisms. Ground-dwelling ants, for example, live in subterranean nests characterized by the high relative humidities and ambient temperatures favored by bacteria and fungi. Arthropods have evolved mechanisms for limiting the growth of these pathogens.

Monoterpenes, alarm pheromones of mites in several generalized families, are fungitoxic against several fungal species. Citral, the alarm releaser of the mite *Carpoglyphus lactis* (64), suppresses the growth of fungi in the culture media in which the mites develop (83). In the absence of the mites, the medium is rapidly overgrown with an *Aspergillus* species.

An examination of the antifungal properties of three insect alarm pheromones demonstrated that these natural products possessed considerable activity against various fungi. Citral, 2-heptanone, and 4-methyl-3-heptanone were evaluated for their efficacy against fungi belonging to 10 genera that included plant, human, and insect pathogens (35). Citral, which was considerably more fungitoxic than the other two alarm pheromones, was as active as some commercial fungicides against dermatophytes but was less active against plant and insect pathogens.

(*E*)-2-hexenal and (*E*)-decenal, the alarm pheromones of the southern green stink bug, *Nezara viridula,* inhibit germination of spores of the entomopathogenic fungus *Metarrhizium anisopliae* (22). This fungus is a very virulent insect pathogen, and the fungistatic activities of the alarm pheromones should be highly adaptive.

THE DIVERSE ROLES OF DEFENSIVE ALLOMONES

Studies on the defensive allomones of arthropods have demonstrated that these compounds have a surprising variety of often unexpected roles and constitute elegant agents of semiochemical parsimony. Among arthropods, many of the so-called defensive compounds have, from a structural standpoint, proven to be the most interesting and novel natural products. As this section illustrates, the evolution of these compounds and their biosynthetic pathways has provided arthropods with a dazzling selection of natural products that can be manipulated parsimoniously.

Antimicrobial Defensive Allomones

Many of the highly reactive defensive allomones of arthropods are excellent antimicrobial agents as well as deterrent compounds. They are often produced in relatively large quantities as weapons against both pro- and eukaryotic adversaries.

The polydesmid millipedes *Oxidus gracilis, Euryurus maculatus,* and *Pseudopolydesmus erasus* produce phenol and guaiacol in their cyanogenic reaction chambers (43). These compounds are, in themselves, well-known deterrents, and phenol is a common bacteriostatic agent. The two phenolics may prevent the growth of soil bacteria that could penetrate the millipede's canal and gain access to the reaction chamber. Furthermore, volatiles such as phenol and guaiacol could provide a slight effusion pressure sufficient enough to restrict the entry of free-floating fungal spores.

The nests of army ants, which are found in the warm humid environments that favor fungal growth, are relatively free of fungi as long as ants are present. The relative absence of fungi has been attributed to skatole (3-methylindole), a characteristic mandibular-gland product of these ants (23). This compound inhibits the growth of *Aspergillus parasiticus,* an entomopathogenic fungus, and of the bacterium *Escherichia coli.* As skatole also repels insectivorous snakes, the semiochemical parsimony of this compound is considerable.

The defensive system of the springtail *Tetrodontophora bielanensis* reflects a very elaborate semiochemical parsimony. When molested, these very large springtails discharge pseudocellular fluids that are fortified with at least three novel pyridopyrazines (39). These alkaloids, which effectively deter soil arthropods such as carabid beetles, are potent antimicrobials as well. In addition, these compounds possess a very perceptible warning odor that enables predators to recognize *T. bielanensis* prior to contact as a potentially toxic prey.

Saturniid larvae utilize exocrine secretions and hemolymph against microorganisms, invertebrates, and vertebrate predators. The scolus secretions and hemolymph of *Eudia pavonia, Saturnia pyri, Eupackardia calleta,* and *Attacus atlas* were evaluated against bacteria and fungi, predatory ants, and two species of birds, Chinese quail (*Coturnix chinensis*) and jackdaws (*Corvus monedula*) (38). The secretions are very effective avian deterrents and also readily repel ants. Volatile compounds in the secretion (e.g. benzonitrile) are believed to envelop the larvae in a cloud of antimicrobial agents while providing an allomonal defense against eukaryotes as well. The presence of another deterrent allomone, phenylacetaldehyde, in the hemolymph provides these larvae with a second internal defense.

Chrysomelid larvae in the subfamily Galerucinae biosynthesize highly reactive 1,8-dihydroxy-9,10-anthroquinones (52, 57), which are outstanding defensive allomones. Larvae of *Xanthogaleruca luteola* and *Galeruca tanaceti* were highly repellent to myrmicine ants, and these immatures were generally avoided by two species of tits (*Parus* spp.) (53).

Defensive Allomones Adapted for Aquatic Functions

Life in an aquatic environment poses problems for arthropods, problems that are not normally encountered on land. Here too, the parsimonious utilization of natural products has enabled aquatic invertebrates to meet the special demands of their environment.

When molested, some staphylinid beetles in the genus *Stenus* utilize the defensive compounds generated in their paired pygidial glands as escape agents. If molested when moving across the water surface, the beetles discharge the compounds, thereby lowering the surface tension of the water, and the blast rapidly propels the beetle from the site of the disturbance (93). This phenomenon, given the German name Entspannungsschwimmen, reflects the utilization of spreading agents that are highly surface active. These compounds are also used as antimicrobial agents and defensive allomones.

The diminutive beetles in the Haliplidae and Hydroporinae also utilize defensive allomones synthesized in the pygidial glands to solve a water-related problem. If these beetles spend a sustained period of time away from the water, their cuticle becomes hydrophobic, and they cannot readily penetrate the water surface film (40). However, exudation and spreading the pygidial-gland secretion over the insect's cuticle renders it wettable.

Defensive Allomones as Aggregation Pheromones

In at least a few cases, potent chemical deterrents have been adapted to function as pheromonal aggregating agents. As with some other bifunctional natural products, interspecific activity is maximal with high concentrations, whereas effective intraspecific activity is associated with low concentrations of the exocrine compound(s).

Adults of the tenebrionid beetle *Blaps sulcata* form diurnal aggregations under thick rocks and remain relatively inactive until dusk. In a sense, these aggregations constitute a collective chemical defense, because each individual is armed with a powerful chemical deterrent. The paired abdominal sternal glands of *B. sulcata* generate 1,4-benzoquinone, 2-methyl-1,4-benzoquinone, and 2-ethyl-1,4-benzoquinone, as well as their corresponding hydroquinones (59). The highly reactive 1,4-quinones are highly effective defensive allomones, and an aggregation of quininoid producers should constitute a formidable deterrent to predators. Moreover, in *B. sulcata,* these defensive allomones have also been adapted to function as aggregation pheromones (62). Low concentrations of the 1,4-quinones, when applied to the undersides of rocks, attract adults into compact aggregations. This dual function of the glandular compounds is thus especially adaptive.

Defensive Allomones as Lethal Attractants

Although prey acquisition with defensive allomones does not appear to be common in the Arthropoda, its occurrence certainly represents a novel example of semiochemical parsimony. For example, adults of the staphylinid beetle *Leistotrophus versicolor* feed on flies associated with dung and carrion. When these dipterans are not available, the beetles mark the substrate with abdominal secretions and, while quivering the abdomen, evert the defensive glands and secrete volatile products that appear to constitute selective fly attractants (46). These defensive allomones of *L. versicolor* attract only flies in the families

Phoridae and Drosophilidae, but these abundant Neotropical dipterans provide a rich resource for the staphylinids. The reason for this specificity remains unknown.

A Defensive-Offensive Allomone

Because many defensive allomones possess considerable toxicological activities, these compounds would seem eminently suited to function as offensive allomones as well. Although such compounds have not been subjected to great analytical scrutiny, this type of parsimony may not be uncommon.

Many silphid carrion beetles discharge a highly malodorous anal exudate originating from a secretory diverticulum that opens into a blind sac close to the rectum. Although the anal secretions of relatively few species of these beetles have been analyzed, a potpourri of compounds with considerable biological activity has already been identified (75). One species, *Ablattaria laevigata,* attacks gastropods as preferred prey (50) and utilizes its defensive chemical arsenal to kill them. The contents of the blind sac, emptied over the body of a snail, rapidly immobilize it because of the high sensitivity of the snail to the beetle's glandular toxins. The evolution of a natural product that is simultaneously a defensive and an offensive allomone provides its producer with a wide array of adaptive options.

Defensive Allomones as Spacing Agents

Competition for resources can be of great significance, especially because it can affect individual fitness and ultimately reproductive success. Thus, different species of arthropods have evolved mechanisms for ensuring that competing herbivores are excluded from each other's feeding areas. This exclusion can be implemented with secretions that also act as defensive allomones to deter interspecific predators. Such exocrine products also function as critical spacing agents by repelling ovipositing individuals. Significantly, this semiochemical parsimony reflects the use of defensive allomones for both intra- and interspecific repellency. Although this adaptation has so far been detected only in coleopterans, this phenomenon probably occurs in other orders as well.

Larvae of the chrysomelid *Plagiodera versicolora* repelled conspecific larvae with allomones secreted from eversible defensive glands (90). The deterrent effects of the chrysomelid defensive compounds, monoterpene aldehydes and lactones, are quite pronounced, larval repellency persisting for 3–18 h (51). In addition, the larval secretion of *P. versicolora* repelled ovipositing females and thus prevented the development of intraspecific competition for the foliar resource. This defensive exudate also possesses interspecific repellency against larvae of the lepidopteran *Nymphalis antiopa,* both repelling and reducing feeding of these immatures.

The repellent secretions of *Gastrophysa viridula, Phaedon cochleariae,* and *Plagiodera versicolora* are synthesized de novo by the larvae of these three species. In contrast, salicylaldehyde, the deterrent generated by *Phratora vitellinae,* is derived from salicin in the insect's host plant. The secretions of *G. viridula, P. cochleariae,* and *P. vitellinae* act as oviposition deterrents against conspecific females. In all four species of chrysomelids, adult feeding was inhibited by the defensive secretions of their conspecific larvae. When larvae of *P. versicolora* and *P. vitellinae* occur on the same host plant (*Salix* sp.), their secretions possess strong interspecific repellent activities for the adults of both species (51). Similarly, the defensive products of the larvae of another chrysomelid, *Phyllodecta vulgatissima,* repel conspecific adults and significantly deter oviposition (54).

PHEROMONES AS INTERSPECIFIC INHIBITORS

Many scolytid bark beetles aggregate in host trees, in which mating, feeding, and reproduction can occur. Exploitation of suitable host trees requires the development of beetle numbers sufficient to overcome the tree's defenses and establish a population that can use the resource optimally. Obviously, maximal exploitation of a host tree by one scolytid species cannot be fully effective if a second species is also attacking the tree. As bark beetles utilize aggregation pheromones to attract individuals to selected trees, interference with the pheromonal signals could inhibit aggregation.

Bark beetles frequently synthesize aggregative pheromones from monoterpenes present in the host's oleoresin. Unique terpenoid attractants have been identified in *Ips* and *Dendroctonus* species, and these pheromones are responsible for some of the observed intra- and interspecific mutual interruptions (111). The aggregative pheromones produced by *Ips paraconfusus*—ipsenol, ipsdienol, and *cis*-verbenol—inhibited the attraction of *Ips latidens* to their common host, ponderosa pine (113). Mutual interruption of species in different genera, attacking the same tree, emphasizes that aggregative inhibition is not strictly intrageneric. Verbenone, one of the volatiles generated by *Dendroctonus brevicomis,* disrupts attack of *I. paraconfusus* under field conditions (28).

The widespread occurrence of mutual interference among bark beetle species documents the evolutionary importance of olfactory "jamming" as a means of preventing other species from sharing available host-plant resources. Olfactory manipulation of competing species occurs via several mechanisms, which emphasizes the significance of mutual interference. Interspecific interference may reflect the biosynthesis by one species of a volatile compound(s) that inhibits the aggregation of another (113). In addition, a bark beetle species may perceive an unnatural enantiomer of its own pheromone and consequently avoid the emission source (6). Mutual interference may also result when individuals of a species emit a sufficient concentration of a pheromonal blend to jam the olfactory chemoreceptors of another species competing for the same food resource.

SEMIOCHEMICAL PARSIMONY OF ALLOMONES WITH GREAT BIOLOGICAL ACTIVITIES

Some arthropods synthesize novel natural products that perturb both sensory and metabolic systems, with highly unusual behavioral and physiological consequences. Although these exocrine constituents are still often regarded as, for example, monofunctional agents of deterrency or unpalatability, recent examinations of the modes of action of single compounds (89, 114) have clearly shown that multifunctionality is widespread. Determination of the different roles for which these compounds have evolved constitutes one of the real challenges in arthropod biology.

Multifunctional Venom Alkaloids

Unlike the proteinaceous venoms of ants, bees, and wasps, venoms of fire ants in the genus *Solenopsis* are ~95% alkaloids. These compounds, which are generally present as mixtures of 2,6-dialkylpiperidines (70), possess a remarkable diversity of activities.

Injection of the alkaloids by stinging results in dermal necrosis and the formation of pruritic and sterile pustules in humans (30). Fire ant stings also cause considerable algogenicity, which results from the liberation of histamine from mast cells by the alkaloids (91). The dialkylpiperidines also possess powerful lytic activity; for example, they instantly hemolyze rabbit erythrocytes (1). Moreover, dialkylpiperidines have pharmacological activities, including inhibition of ATPases (63), reduction of mitochondrial respiration and uncoupling of oxidative phosphorylation (33), and blocking of neuromuscular junctions (114). The ability of these compounds to attack such a diversity of targets underscores their semiochemical parsimony. The great multifunctionality of these alkaloids is further emphasized by their diverse roles in various ecological contexts.

The piperidines possess powerful antifungal, antibacterial, phytotoxic, and · insecticidal properties (18). Numerous bacterial (61) and fungal (34) species are inhibited by the poison-gland compounds. The toxicities of the alkaloids to termites is comparable to those of commercial insecticides (45). Fire ant dialkylpiperidines exhibit well-developed repellency against other ant species

(16), which should allow fire ants to successfully compete with other ants for critical resources. Fire ant workers disperse venom through the air (i.e. by gaster flagging) as a means of repelling heterospecifics, whereas workers tending brood in the nest apply venom to larvae, presumably as an antibiotic (82). Fire ant queens (of *Solenopsis invicta*) deposit poison-sac contents on oviposited eggs, and the concentration of alkaloid is high enough to inhibit the growth of entomopathogenic fungi (103).

The Semiochemical Parsimony of a Remarkable Anhydride

Cantharidin, the anhydride of cantharidic acid, is only known to be synthesized by coleopterous species in the families Meloidae and Oedemeridae. Meloid beetles externalize this terpenoid anhydride by reflex bleeding, or autohemorrhage, and this compound is a powerful vesicant when it contacts human skin. Cantharidin was mentioned by Hippocrates as a cure for dropsy (101) and has been used medicinally as a vesicant and mistakenly as an aphrodisiac for at least two millennia. It was also regarded as a compound of considerable medicinal value, considered indispensable in the treatment of bladder and kidney infections and stones, strangury, dropsy, and certain venereal diseases (49). Unfortunately, it is the reputation of cantharidin as a putative aphrodisiac that has persisted, as exemplified by its frequent utilization by the Marquis de Sade (44), and it has recently caused the deaths of two women (58). Unexpected effects of presumably low concentrations of this anhydride on the human male have also been documented (76, 105).

Meloids are sometimes referred to as blister beetles because of the lesions that develop on human skin after contact with the cantharidin-fortified blood of these coleopterans (49). This vesicatory property clearly makes it a powerful vertebrate deterrent (58). However, the reader may note that topical treatment with cantharidin was able to effect remission of epidermal cancer in pigs (42) and humans (41) and act as a selective herbicide (37). Cantharidin is also an effective feeding deterrent for insects (31, 32).

Cantharidin also possesses potent antifungal activities, reportedly inhibiting the growth of *Trichophyton* and *Microsporum* species (89). The antifungal activity of this terpenoid anhydride may be particularly relevant to the development of blister beetle eggs, which are incubated in warm and humid environments, which favor the growth of invasive fungi. Cantharidin's main function appears to be to protect developing meloid embryos from entomopathogenic fungi (RB Selander, personal communication). This could not be so were it not for the transfer of cantharidin by the male to the female as part of the seminal ejaculate, because the female does not synthesize this compound (98).

ACKNOWLEDGMENTS

I extend my gratitude and thanks to Konrad Dettner, Miriam Rothschild, Jeffrey R Aldrich, Justin O Schmidt, John A Pickett, and Monika Hilker, whose insights on the significance of multifunctional natural products were particularly illuminating.

Literature Cited

1. Adrouny GA, Derbes VJ, Jung RC. 1959. Isolation of a hemolytic component of fire ant venom. *Science* 130: 449

2. Ayre GL. 1968. Comparative studies on the behavior of three species of ants (Hymenoptera: Formicidae). I. Prey finding, capture, and transport. *Can. Entomol.* 100:165–72

3. Ayre GL, Blum MS. 1971. Attraction and alarm of ants (*Camponotus* spp.—Hymenoptera: Formicidae) by pheromones. *Physiol. Zool.* 44:77–83

4. Barbier J, Lederer E. 1960. Structure chimique de la substance royale de la reine d'abeille (*Apis mellifica* L.). *C. R. Acad. Sci. Paris* 251:1131–35

5. Bergström G, Löfqvist J. 1971. *Camponotus ligniperda* Latr.—A model for the composite volatile secretions of Dufour's gland in formicine ants. In *Chemical Releasers in Insects*, ed. A Tahori, pp. 195–223. New York: Gordon & Breach

6. Birch MC, Light DM, Wood DL, Browne LE, Silverstein RM, Young JC. 1980. Pheromonal attraction and allomonal interruption of *Ips pini* in California by the two enantiomers of ips- dienol. *J. Chem. Ecol.* 6:703–17

7. Blum MS. 1970. The chemical basis of insect sociality. In *Chemical Controlling Insect Behavior*, ed. M Beroza, pp. 61–94. New York: Academic

8. Blum MS. 1974. Deciphering the communicative Rosetta stone. *Bull. Entomol. Soc. Am.* 20:30–35

9. Blum MS. 1980. Arthropods and ecomones: better fitness through ecological chemistry. In *Animals and Environmental Fitness*, ed. R. Gilles, pp. 207–22. Oxford: Pergamon

10. Blum MS. 1981. *Chemical Defenses of*

Arthropods. New York: Academic. 562 pp.

11. Blum MS. 1985. Alarm pheromones. In *Comprehensive Insect Physiology, Biochemistry, and Pharmacology*, ed. GA Kerkut, LJ Gilbert, 9:193–224. Oxford: Pergamon

12. Blum MS. 1991. Semiochemists and allelochemists: a wondrous diversity of relationships with insects. In *Entomology Serving Society: Emerging Technologies and Challenges*, ed. SB Vinson, RL Metcalf, pp. 11–24. Lanham, MD: Entomol. Soc. Am.

13. Blum MS. 1972. Social insect pheromones: their chemistry and function. *Am. Zool.* 12:553–76

14. Blum MS, Crewe RM, Kerr WE, Keith LH, Garrison AW, Walker MM. 1970. Citral in stingless bees: isolation and functions in trail laying and robbing. *J. Insect Physiol.* 16:1637–48

15. Blum MS, Edgar AL. 1971. 4-Methyl-3-heptanone: identification and role in opilionid exocrine secretions. *Insect Biochem.* 1:181–88

16. Blum MS, Everett DM, Jones TH, Fales HM. 1991. Arthropod natural products as insect repellents. In *Naturally Occurring Pest Bioregulators, ACS Symp. Ser. 449*, ed. PA Hedin, pp. 14–26. Washington DC: Am. Chem. Soc.

17. Blum MS, Severson RF, Arrendale RF, Whitman DW, Escoubas P, et al. 1990. A generalist herbivore in a specialist mode. Metabolic, sequestrative, and defensive consequences. *J. Chem. Ecol.* 16:223–44

18. Blum MS, Walker JR, Callahan PS, Novak AF. 1958. Chemical, insecticidal and antibiotic properties of fire ant venom. *Science* 128:306–7

19. Blum MS, Warter SL, Monroes RS, Chidester JC. 1963. Chemical releasers

of social behavior. I. Methyl-*n*-amyl ketone in *Iridomyrmex pruinosus* (Roger). *J. Insect Physiol.* 9:881–85

20. Blum MS, Warter SL, Traynham JG. 1966. Chemical releasers of social behavior. VI. The relation of structure to activity of ketones as releasers of alarm for *Iridomyrmex pruinosus* (Roger). *J. Insect Physiol.* 12:419–27

21. Boch R. 1979. Queen substance pheromone produced by immature honeybees. *J. Apic. Res.* 18:12–15

22. Borges M, Leal SCM, Tigano-Milani MS, Valadares MCC. 1993. Efeito do feromôno de alarme do percevejo verde, *Nezara viridula* (L.) (Hemiptera: Pentatomidae), sobre o fungo entomopatogênico *Metarhizium anisopliae* (Metsch.) *Sorok. Anais Soc. Entomol. Brasil* 22:505–12

23. Brown CA, Watkins JF II, Eldridge DW. 1976. Repression of bacteria and fungi by the army ant secretion: skatole. *J. Kans. Entomol. Soc.* 52:119–22

24. Brown WL, Eisner T, Whitaker RH. 1970. Allomones and kairomones: transspecific chemical messengers. *Bioscience* 20:21–2

25. Buschinger A. 1972. Giftdrüsensekret als sexual Pheromon bei der Ameise *Harpagoxenus sublaevis*. *Naturwissenschaften* 59:313–14

26. Butler CG, Callow RK. 1968. Pheromones of the honeybee (*Apis mellifera* L.): the "inhibitory scent" of the queen. *Proc. R. Entomol. Soc. London Ser. A* 43:62–65

27. Butler CG, Callow RK, Johston NC. 1961. The isolation and synthesis of queen substance, 9-oxodec-*trans*-2-enoic acid, a honeybee pheromone. *Proc. R. Entomol. Soc.* 155:417–32

28. Byers JA, Wood DL. 1980. Interspecific inhibition of the response of the bark beetles *Dendroctonus brevicomis* Le-Conteand *Ips paraconfusus* Lanier, to their pheromones in the field. *J. Chem. Ecol.* 6:149–64

29. Callow RK, Johnston NC. 1960. The chemical constitution and synthesis of queen substances of honeybees (*Apis mellifera* L.). *Bee World* 41:152–53

30. Caro M-R, Derbes VJ, Jung R. 1957. Skin responses to the sting of the imported fire ant (*Solenopsis saevissima*). *Am. Med. Assoc. Arch. Dermatol.* 75: 475–88

31. Carrel JE, Doom JP, McCormick JP. 1986. Identification of cantharidin in false blister beetles (Coleoptera: Oedemeridae) from Florida. *J. Chem. Ecol.* 12:741–47

32. Carrel JE, Eisner T. 1974. Cantharidin:

33. Cheng EY, Cutkomp LK, Koch RB. 1977. Effect of an imported fire ant component on respiration and oxidative phosphorylation of mitochondria. *Biochem. Pharmacol.* 26:1179–80

34. Cole LK. 1974. *Antifungal, insecticidal, and potential therapeutic properties of ant venom alkaloids and ant alarm pheromones.* PhD thesis, Univ. Georgia. 155 pp.

35. Cole LK, Blum MS, Roncadori RW. 1975. Antifungal properties of the insect alarm pheromones, citral, 2-heptanone, and 4-methyl-3-heptanone. *Mycologia* 67:701–8

36. Crewe RM, Blum MS. 1970. Alarm pheromones in the genus *Myrmica* (Hymenoptera: Formicidae). *Z. Vgl. Physiol.* 70:363–73

37. Cutler HG. 1975. Cantharidin: novel effects on plants. *Plant Cell Physiol.* 16:181–84

38. Deml R, Dettner K. 1993. Biogenic amines and phenolics characterize the defensive secretion of saturniid caterpillars (Lepidoptera: Saturniidae): a comparative study. *J. Comp. Physiol. B* 163: 123–32

39. Dettner K, Scheuerlein A, Fabian P, Schulz T, Francke W. 1995. Chemical defence of the giant springtail *Tetrodontophora bielanensis* (Waga) (Insecta: Collembolla). *J. Chem. Ecol.* In press

40. Dettner K, Schwinger G. 1980. Defensive substances from pygidial glands of water beetles. *Biochem. Syst. Ecol.* 8: 89–95

41. Dubois R. 1927. Note préliminaire sur l'action de la cantharidine sur diverses productions pathologiques d'origine épithéliale. *C. R. Soc. Biol. Ser. II* 97:48

42. Dubois R, Ball MV. 1933. Traitement du cancer épithélial de la peau par la cantharidine. *Bull. Acad. Med.* 110:791–93

43. Duffey SS, Blum MS. 1977. Phenol and guaiacol: biosynthesis, detoxication, and function in a polydesmid millipede, *Oxidus gracilis. Insect Biochem.* 7:57–65

44. Dulauré JA. 1794. *Collection de liste des ci-devant Ducs, Marquis, Contes, Barons, etc. (Paris, Second Year of Liberty).* Pamphlet in the Collection of Le Musée National, Paris, France.

45. Escoubas P. 1988. *Alcalöides de fourmis: identification, toxicité et mode d'action.* PhD thesis. Univ. Pierre et Marie Curie, Paris. 145 pp.

46. Forsyth A, Alcock J. 1990. Ambushing and prey-luring as alternative foraging

tactics of the fly-catching rove beetle *Leistotrophus versicolor* (Coleoptera: Staphylinidae). *J. Insect Behav.* 3:703–18

47. Gary NE. 1962. Chemical mating attractants in the queen honey bee. *Science* 136:773–74

48. Ghent RL. 1961. *Adaptive refinements in the chemical defensive mechanisms of certain Formicinae.* PhD thesis. Cornell Univ., Ithaca NY

49. Groeneveld J. 1698. *De Tuto Cantharidum in Medicina Usu Interno.* London: Typis J. H. prostant venales apud Johannem Taylor. 135 pp.

50. Heymons R, von Lengerken H. 1932. Studien über die Lebenserscheinungen der Silphini (Coleopt.). VIII. *Ablattaria laevigata* F. *Z. Morphol. Ökol. Tiere* 24:259–87

51. Hilker M. 1989. Intra- and interspecific effects of larval secretions in some chrysomelids (Coleoptera). *Entomol. Exp. Appl.* 53:237–45

52. Hilker M, Eschbach U, Dettner K. 1992. Occurrence of anthraquinones in eggs and larvae of several Galerucinae (Coleoptera: Chrysomelidae). *Naturwissenschaften* 79:271–74

53. Hilker M, Köpf A. 1995. Evaluation of palatability of chrysomelid larvae containing anthraquinones to birds. *Oecologia.* In press

54. Hilker M, Weitzel C. 1991. Oviposition deterrence by chemical signals of conspecific larvae in *Diprion pini* (Hymenoptera: Diprionidae) and *Phyllodecta vulgatissima* (Coleoptera: Chrysomelidae). *Entomol. Gen.* 15:293–301

55. Hölldobler B. 1971. Sex pheromone in the ant *Xenomyrmex floridanus. J. Insect Physiol.* 17:1497–99

56. Hölldobler B, Wüst M. 1973. Ein Sexualpheromon bei der Pharaoameise (*Monomorium pharaonis* L.). *Z. Tierpsychol.* 32:1–9

57. Howard DF, Phillips DW, Jones TH, Blum MS. 1982. Anthraquinones and anthrones: occurrence and defensive function in a chrysomelid beetle. *Naturwissenschaften* 69:91–92

58. Howell M, Ford P. 1985. *The Beetle of Aphrodite.* New York: Random House. 302 pp.

59. Ikan R, Cohen E, Shulov A. 1970. Benzo- and hydroquinones in the defensive secretions of *Blaps sulcata* and *B. wiedemanni. J. Insect Physiol.* 16:2201–6

60. Ishiwatari T. 1974. Studies on the scent of stink bugs (Hemiptera: Pentatomidae) I. Alarm pheromone activity. *Appl. Entomol. Zool.* 9:153–58

61. Jouvanez DP, Blum MS, MacConnell JG. 1972. Antibacterial activity of venom alkaloids from the imported fire ant, *Solenopsis invicta* Buren. *Antimicrob. Agents Chemother.* 2:291–93

62. Kaufmann T. 1966. Observations on some factors which influence aggregation in *Blaps sulcata* (Coleoptera: Tenebrionidae) in Israel. *Ann. Entomol. Soc. Am.* 59:660–64

63. Koch RB, Dessaiah D, Ahmed K. 1977. Effect of piperidines and fire ant venom on ATPase activities from brain homogenate fractions and characterization of Na^+-K^+ ATPase inhibition. *Biochem. Pharmacol.* 26:983–85

64. Kuwahara Y, Matsumoto K, Wada Y. 1980. Pheromone study on acarid mites IV. Citral: composition and function as an alarm pheromone and its secretory gland in four species of acarid mites. *Jpn. J. Sanit. Zool.* 31:73–80

65. Law JH, Regnier FE. 1971. Pheromones. *Annu. Rev. Biochem.* 40:533–48

66. Leal WS, Nakamori Y, Kadosawa T. 1993. Ecological significance of *Riptortus clavatus* aggregation pheromone to adults and nymphs. *Proc. Annu. Meeting Soc. Appl. Zool., 37th,* p. 9. Kyoto: Soc. Appl. Zool.

67. Levinson HZ, Bar Ilan AR. 1971. Assembling and alerting scents produced by the bedbug *Cimex lectularius* L. *Experientia* 27:102–3

68. Löfqvist J. 1976. Formic acid and saturated hydrocarbons as alarm pheromones for the ant *Formica rufa. J. Insect Physiol.* 22:1331–46

69. Luby JM, Regnier FE, Clarke ET, Weaver EC, Weaver N. 1973. Volatile cephalic substances of the stingless bees, *Trigona mexicana* and *Trigona pectoralis. J. Insect Physiol.* 19:1111–27

70. MacConnell JG, Blum MS, Fales HM. 1971. The chemistry of fire ant venom. *Tetrahedron* 26:1129–39

71. Maschwitz U. 1964. Gefahrenalarmstoffe und Gefahrenalarmierung bei sozialen Hymenoptera. *Z. Vgl. Physiol.* 47:596–655

72. Maschwitz U, Gutmann C. 1979. Spur- und Alarmstoffe bei der gefleckten Brutwanze *Elasmucha greisea* (Heteroptera: Acanthosomidae). *Insectes Soc.* 26:101–11

73. McGurk DJ, Frost J, Eisenbraun EJ, Vick K, Drew WA, Young J. 1966. Volatile compounds in ants: identification of 4-methyl-3-heptanone from *Pogonomyrmex* ants. *J. Insect Physiol.* 12:1435–41

74. Meinwald J, Kluge JF, Carrel JE, Eisner T. 1971. Acyclic ketones in the defen-

sive secretion of a "daddy longlegs" (*Leiobunum vittatum*). *Proc. Natl. Acad. Sci. USA* 68:1467–68

75. Meinwald J, Roach B, Hicks K, Alsop D, Eisner T. 1985. Defensive steroids from a carrion beetle (*Silpha americana*). *Experientia* 41:516–19

76. Meynier J. 1893. Empoisonnement par la chair de grenouilles infestées par des insectes du genre *Mylabris* de la famille des méloides. *Arch. Med. Pharm. Mil.* 22:53–56

77. Moore BP, Brown WV. 1979. Chemical composition of the defensive secretion in *Dyschirius bonelli* (Coleoptera: Carabidae: Scaritinae) and its taxonomic significance. *J. Aust. Entomol. Soc.* 18:123–25

78. Morse R. 1972. Honey bee alarm pheromone: another function. *Ann. Entomol. Soc. Am.* 65:1430

79. Morse RA, Boch R. 1971. Pheromone concert in swarming honeybees. *Ann. Entomol. Soc. Am.* 64:1414–17

80. Moser JC. 1968. Alarm pheromones of the ant *Atta texana. J. Insect Physiol.* 14:529–35

81. Nuñez JA. 1967. Sammelbienen markieren versiegte Futteruellen durch Duft. *Naturwissenschaften* 54:322–23

82. Obin MS, Vander Meer RK. 1985. Gaster flagging by fire ants (*Solenopsis* spp.): functional significance of venom dispersal behavior. *J. Chem. Ecol.* 11:1757–68

83. Okamoto M, Matsumoto K, Wada Y, Nakano H. 1978. Studies on antifungal effect of mite alarm pheromone citral. I. Evaluation of antifungal effect of citral. *Jpn. J. Sanit. Zool.* 29:255–60

84. Pain J, Barbier M, Bogdanovsky D, Lederer E. 1962. Chemistry and biological activity of the secretions of queen and worker honeybees (*Apis mellifica* L.). *Comp. Biochem. Physiol.* 6:233–41

85. Pain J, Roger B. 1970. Variation annuelle de l'acide hydroxy-10 décène-2 oïque dans les têtes d'abeilles. *Apidologie* 1:29–54

86. Pavan M, Ronchetti G. 1955. Studi sulla morfologia esterna e anatomia interna dell'operaia *Iridomyrmex humilis* Mayr e ricerche chimiche e biologiche sulla iridomirmecina. *Atti Soc. Ital. Sci. Nat. Mus. Civ. Stor. Nat. Milano* 94:379–477

87. Peschke K. 1983. Defensive and pheromonal secretion of the tergal gland of *Aleochara curtula*. II. Release and inhibition of male copulatory behavior. *J. Chem. Ecol.* 9:13–31

88. Peschke K, Metzler M. 1982. Defensive and pheromonal secretion of the tergal gland of *Aleochara curtula*. I. The chemical composition. *J. Chem. Ecol.* 8:773–83

89. Pinetti P, Biggio P. 1968. Attivita' antimicotica in vitro di alcune sostanze ad azione alopecizzante (aceto di tallio e cantaridina). *Boll. Soc. Ital. Biol. Sper.* 44:677–79

90. Raupp JM, Milan FR, Barbosa P, Leonhardt BA. 1986. Methylcyclopentanoid monoterpenes mediate interactions among insect herbivores. *Science* 232:1408–10

91. Read GW, Lind NK, Oda CS. 1974. Histamine release by fire ant (*Solenopsis*) venom. *Toxicon* 16:361–67

92. Regnier FE, Wilson EO. 1971. Chemical communication and "propaganda" in slave-maker ants. *Science* 172:267–69

93. Schildknecht H, Berger D, Krauss D, Connert J, Gehlhaus J, Eisenbreis H. 1976. Defense chemistry of *Stenus comma* (Coleoptera: Staphylinidae). *J. Chem. Ecol.* 2:1–11

94. Schmidt JO, Blum MS. 1977. Adaptations and responses of *Dasymutilla occidentalis* (Hymenoptera: Mutillidae) to predators. *Entomol. Exp. Appl.* 21:99–111

95. Schmidt JO, Slessor KN, Winston WL. 1993. Roles of Nasonov and queen pheromones in attraction of honeybee swarms. *Naturwissenschaften* 80:573–75

96. Deleted in proof

97. Shearer DA, Boch R. 1965. 2-Heptanone in the mandibular gland secretion of the honeybee. *Nature* 206:530

98. Sierra JR, Woggon W-D, Schmid H. 1975. Transfer of cantharidin during copulation from the adult male to the female *Lytta vesicatoria* (Spanish flies). *Experientia* 32:142–44

99. Slessor KN, Kaminski L-A, King GGS, Borden JH, Winston ML. 1988. Semiochemical basis of the retinue response to queen honey bees. *Nature* 332:354–56

100. Smith RF, Pierce HD, Borden JH. 1991. Sex pheromone of the mullein bug, *Campylomma verbasci* (Meyer) (Heteroptera: Miridae). *J. Chem. Ecol.* 17:1437–47

101. Sollman T. 1949. In *A Manual of Pharmacology*, p. 137. Philadelphia: WB Saunders

102. Trave R, Pavan M. 1956. Veleni degli insetti: principi estratti dalla formica *Tapinoma nigerrimum* Nyl. *Chim. Industria* 38:1015–19

103. Vander Meer RK, Morel L. 1994. Ant queens deposit pheromones and antimi-

crobial agents on eggs. *Naturwissenschaften* 81:682–84

104. Velthuis HHW, van Es J. 1964. Some functional aspects of the mandibular glands of the queen honeybee. *J. Apic. Res.* 3:11–16

105. Vézien M. 1861. Note sur la cystide cantharidienne par l'ingestion de grenouilles qui sont nourries de coléoptères vésicants. *Rec. Mém. Med. Chirurgie Pharm. Mil.* 4:457–60

106. Vrkoč J, Ubik K, Dolejs L. 1973. On the chemical composition of the frontal gland in termites of the genus *Nasutitermes* (*N. costalis* and *N. rippertii*). *Acta Entomol. Bohemoslov.* 70:74–80

107. Wallbank BF, Waterhouse DF. 1970. The defensive secretions of *Polyzosteria* and related cockroaches. *J. Insect Physiol.* 16:2081–96

108. Walter F, Fletcher DJC, Chautems D, Cherix D, Keller L. 1993. Identification of the sex pheromone of an ant, *Formica lugubris* (Hymenoptera, Formicidae). *Naturwissenschaften* 80:30–34

109. Whitman DW. 1982. Grasshopper sexual pheromone: a component of the defensive secretion in *Taeniopoda eques. Physiol. Entomol.* 7:111–15

110. Wilson EO. 1958. A chemical releaser of alarm and digging behavior in the ant *Pogonomyrmex badius* (Latreille). *Psyche* 65:41–51

111. Wood DL. 1982. The role of pheromones, kairomones, and allomones in the host selection and colonization behavior of bark beetles. *Annu. Rev. Entomol.* 27:411–46

112. Wood DL, Browne LE, Bedard WD, Tilden PE, Silverstein RM, Rodin JO. 1968. Response of *Ips confusus* to synthetic sex pheromones in nature. *Science* 159:1373–74

113. Wood DL, Stark RW, Silverstein RM, Rodin JO. 1967. Unique synergistic effects produced by the principal sex attractant compounds of *Ips confusus* (LeConte). *Nature* 215:206

114. Yeh JZ, Narahashi T, Almon RR. 1975. Characterization of neuromuscular blocking action of piperidine derivatives. *J. Pharmacol. Exp. Ther.* 194: 373–83

Annu. Rev. Entomol. 1996. 41:375-406
Copyright © 1996 by Annual Reviews Inc.

BIOLOGICAL CONTROL WITH *TRICHOGRAMMA*: Advances, Successes, and Potential of Their Use

Sandy M. Smith

Faculty of Forestry, University of Toronto, 33 Willcocks St., Toronto, Ontario, Canada M5S 3B3

KEY WORDS: parasitoid quality and selection, mass production, distribution, release, efficacy

ABSTRACT

Major contributions to the release of *Trichogramma* for biological control of lepidopterous pests have been made in the past 20 years. Most trials have used only five species of *Trichogramma* against two pests; *Ostrinia* in corn is considered the most universally feasible. All *Trichogramma* programs must address the following four aspects to be successful commercially. Selection of the appropriate population is based on inter- and intraspecific variation, as well as on current definitions of parasitoid quality. Mass rearing is comprised of both host and parasitoid components, although major emphasis is now on developing artificial systems. Effective distribution of *Trichogramma* requires supportive extension and advanced technology. Strategies for use in the field vary according to the approach desired (inundative or inoculative), the timing, frequency and rate of release, and the multiple factors that affect release, such as the weather, crop, host, predation, pesticides, and dispersal. The past difficulty in assessing the efficacy of *Trichogramma* should be improved with new guidelines for standardizing terminology and measurements.

HISTORICAL PERSPECTIVE

The release of *Trichogramma* for biological control of lepidopterous pests has been considered for more than 100 years, although the mass rearing of these hymenopterous parasitoids was not proposed in North America until the 1920s. Flanders' work (49) inspired activity during the 1930s, but this activity quickly dissipated in the West with the rise of chemical insecticides. It was left primarily to scientists in the former USSR (from 1937) and China (from 1949)

375

to develop *Trichogramma* as biological-control agents (12, 100). In the 1960s, the Europeans and Americans revitalized research on *Trichogramma*; in the 1970s, they began mass-rearing and release (14, 62, 65, 70, 118, 127, 193).

It has been almost 20 years since an exhaustive review of inundative releases has been conducted (180). Although the genus *Trichogramma* is not the only group to be used with this approach, much of our understanding of inundative release comes from studies with these minute egg parasitoids. At the same time as Stinner's review (180), a compendium of international research was published (158) that marked the beginning of an exchange of information among scientists in North America, Europe, the former USSR, and China. This exchange resulted in an explosion of research that continues today. International symposia with published proceedings have been conducted every four years since 1982 (74, 199, 206), and informal sessions have been held at the last three International Congresses of Entomology. In addition, there have been eight international symposia on quality control (16). Wajnberg & Hassan (205) summarized current knowledge in a recent publication.

This review critically analyzes and ties together the studies from the mid-1970s. Because *Trichogramma* can be considered the *Drosophila* of the parasitoid world, the genus has generated a large volume of information. Emphasis here is on how we have used this basic information to implement successful field programs.

SCOPE OF USE

Most *Trichogramma* releases have been conducted in the past 20 years; trials before 1975 were aimed at control of lepidopterous pests in sugarcane and corn. Between 1975 and 1985, pests of cotton, sugarbeet, vineyards, cabbage, plum, apple, forests, tomato, and rice were also targeted. Since 1985, numerous crops have been investigated, and the list is growing (65). Most research—and the most consistent success—has been in corn. Inconsistent results have been found in crops such as cotton and rice, for which, despite intensive research, the use of insecticides to control multiple pest problems reduces the action of *Trichogramma* (65, 88, 101).

Today, inundative releases for control of lepidopterous pests are being investigated in more than 50 countries and are reported to be used commercially on more than 32 million ha each year (65). This hectarage is inflated, as some areas in the former USSR receive repeated applications (48). *Trichogramma* are considered commercially efficient in the former USSR and China and compete successfully with insecticides for commercial control of corn borer in Europe, although some of the costs are subsidized (127). Eastern bloc countries and some Asian and South American countries have a longer history, whereas Australia and countries in North and Central America have only recently begun to investigate the potential of *Trichogramma* (38, 54, 170). In North America, their niche is in organically grown crops and in areas where pesticide resistance has developed.

Despite the plethora of crops and countries where inundative releases have been made, surprisingly few pest and *Trichogramma* species have been studied. Most trials have been initiated against the key pests, *Ostrinia* spp. and *Helicoverpa zea*, with infrequent releases against various pyralids, tortricids, noctuids, oleuthrids, and pierids, in decreasing order (Table 1). Even in the former USSR, *Trichogramma* are used to control only about seven pest species (59). Most trials have used five species of *Trichogramma*, in decreasing order: *evanescens, dendrolimi, pretiosum, brassicae* (=*maidis*), and *nubilale*. An additional ten species have been used infrequently (Table 1). In general, the first three species have dominated the studies because of their plasticity in habitat and host selection (101).

Most papers address specific research questions, although several describe commercial successes (Table 1). To reach successful commercial application, all *Trichogramma* programs must address four issues: selection of the appropriate population to release, a system for mass rearing, distribution of the parasitoid, and a strategy for field release.

SELECTION OF PARASITOIDS TO RELEASE

Interspecific and Intraspecific Variation

The selection of the most appropriate parasitoid for release starts with the best species. This process is difficult, because there is considerable interspecific variation in the more than 145 species known, and the taxonomy of the genus is poorly understood (147). Members of the genus are polyphagous egg parasitoids on ten orders of insects, including Lepidoptera, Coleoptera, Diptera, Heteroptera, Hymenoptera, and Neuroptera. As more species are discovered, however, increasing specialization is recognized (147). Recent molecular studies may help clarify the taxonomy of this genus (149, 163, 194).

The local species is generally selected for release on the ecological basis that it is better adapted to the proposed climate, habitat, and host conditions (66, 197). For example, at least six species of *Trichogramma* have been used around the world to control *Ostrinia* spp. In their native regions, the most common are *T. nubilale* and *T. pretiosum* in the United States, *T. ostriniae* and *T. dendrolimi* in China, and *T. evanescens* and *T. brassicae* (=*maidis*) in Europe (Table 1). Use of the local species is the basis of inundative theory and is only contraindicated when there is no native species or when preintroductory screening suggests otherwise (66). Different species of *Trichogramma* compete with each other (140, 182, 197). In diverse habitats, this competition

Table 1 Field strategies for inundative release of *Trichogramma* species against lepidopterous pests on various crops

Crop	Target genera	*Trichogramma* species	Releases Number	Intervals (days)	Timing predictor	Total rate (thousands)/ha
Corn	Ostrinia	brassicae	2 – 4	7	Degree-day, pheromone	196 – 200 wasps
		brassicae	1 – 3	7 – 10	Light trap	300 wasps
		brassicae	1 – 2	7 – 14	Light trap	150 – 300 wasps
		brassicae	4 – 9	—	Pheromone, light trap	450 – 2,800 wasps
		brassicae	3	7 – 10	Light trap	150 – 300 wasps
		brassicae	2 – 3	8 – 9	Pheromone	200 wasps
		nubilale	22	3 – 4	Plant development	56.5 females
		nubilale	1 – 3	7 – 8	Degree-day, light trap	681 – 4,400 wasps
		nubilale	1 – 3	7 – 8	Light trap, degree-day	4,400 wasps
		nubilale	3	7 – 8	Egg-laying	3,195 females
		nubilale	11	1 – 2	—	48.4 wasps
		ostriniae	4 – 5	1 – 5	Egg-laying	91 – 420 wasps
		ostriniae	1 – 2	—	Egg-laying	75 – 120 wasps
		ostriniae	1 – 2	—	Egg-laying	225 – 600 wasps
		evanescens	2 – 5	8 – 21	Egg-laying	600 – 2,000 eggs
		evanescens	3 – 4	7 – 10	Egg-laying	90 – 180 wasps
		evanescens	1 – 3	10 – 15	Light trap	76 – 1,200 wasps
		evanescens	1 – 4	—	Egg-laying	100 – 2,400 wasps
	Cnephasia	evanescens	1	—	Egg-laying	108 – 900 wasps
	Ostrinia	pretiosum	11	1 – 2	—	550 wasps[c]
	Helicoverpa	pretiosum	11 – 18	2 – 3	Light trap	13,750 – 22,500 wasps
Cotton	Helicoverpa	confusum	—	—	—	300 – 600 wasps
		pretiosum	3	7	Plant development	62.5 wasps
		pretiosum	7 – 18	2 – 4, 7	Degree-day, pheromone	46 – 956 wasps
		pretiosum	14	1 – 12	Date	760 wasps
	Helicoverpa	pretiosum	—	7	—	100 – 200 wasps
Sugarcane	Chilo	evanescens	1 – 6	10 – 30	Egg-laying	375 – 4,500 wasps
		Multiple species	—	—	—	—
	Tetramoera, Chilotraea	Multiple species	8	2 – 7	Egg-laying	100 wasps
	Diatraea	chilonis	6	10–15	Pheromone	900 wasps
Spruce	Choristoneura	minutum	1–2	6–10	Pheromone, degree-days	600 – 30,900 females
Pine	Dendrolimus	dendrolimi	4 – 5	5 – 7	Egg-laying	1,050 wasps
Forests	Dendrolimus	15 species	2 – 3	5 – 7	Egg-laying	200 – 750 wasps
Oak	Lampronadata	dendrolimi	6	5	Egg-laying	30,000 wasps
Apple	Cydia	dendrolimi	4–6	10–15	Pheromone	1,000 – 4,700 wasps
	Adoxophyes	embryophagum	4 – 6	10–15	Pheromone	1,000 – 4,700 wasps
Fruit	Adoxophyes, Grapholitha	dendrolimi	3 – 5	5	Egg-laying	1,800 – 2,300 wasps
Citrus	Cryptophlebia	toidea cryptophlebiae	29 – 33	7	Plant development	2,300 – 3,800 females[f]
Pomegranate	Deudorix	chilonis pomegra	1–4	30	—	7,020 wasps[g]
Tomato	Helicoverpa	pretiosum	6 – 10	3 – 4	Crop planting	430 – 795 wasps
	Trichoplusia	pretiosum	6–10	3 – 4	Crop planting	430 – 795 wasps
	Manduca	pretiosum	6 – 10	3 – 4	Crop planting	430 – 795 wasps
	Helicoverpa	ivelae	13	7	Plant development	1,300 wasps
Pepper	Ostrinia	nubilale	20	1	Plant development	225 – 450 females
	Ostrinia	brassicae	6 – 12	3 – 19	—	9,000 – 48,000 wasps
Grape	Eupoecilia	cacociae	1	3	Pheromone	2,200 wasps
	Lobesia	dendrolimi	1	3	Pheromone	2,200 wasps
		embryophagum	1	3	Pheromone	2,200 wasps
Peanut	Cadra	pretiosum	9 – 14	2 – 7	Date	5.8 – 35 wasps[h]
	Plodia	pretiosum	9 – 14	2 – 7	Date	5.8 – 35 wasps[h]
Sugarbeet	Autographa, Agrostisbuesi, Mamestra	dendrolimi, brassiecae, evanescens	2 – 3	30 – 50	Pheromone	120 black egg
Cabbage	Mamestra, Pieris, Plutella	evanescens	3	10 – 12	Pheromone	120 black eggs
Brussel sprouts	Mamestra	brassicae	7	14	Egg-laying	1,400 – 3,150 wasps
Cocoa	Conopomorpha	'toidea bactrae fumata	12 – 90	1 – 7	—	13,000 – 31,000 wasps[d]
Rice	Cnaphalocrocis	japonicum	4 – 9	7 – 10	Adult emergence	400 – 900 wasps

[a] Increase in egg parasitism calculated as percentage in treatment areas minus percentage in the control areas, percentage reduction calculated as values in the control areas minus values in the treated areas divided by values in the control areas.

[b] No control plots cited.

[c] Reduction in larval numbers and damage compared in parasitoid- vs insecticide-treated plots.

[d] Releases made year-round.

[e] Number per plot, plot size not specified.

[f] Number per tree per hectare.

[g] Number per tree.

[h] Number per m^3.

[i] Potential for using *Trichogramma*: 0 = no potential; 1 = possible potential with better understanding of system; 2 = biologically feasible; 3 = economically feasible; 4 = being used commercially.

% egg parasitism increase[a]	Reduction (%)[a]		Rearing host	Notes	Potential[i]	Reference
	Target larvae	Damage				
57 – 86	—	45 – 87	Ephestia	Adults fed honey	2	214
75 – 93[b]	—	70 – 82	Ephestia		3	14
60 – 70	—	47 – 72	Ephestia	Staggered emergence	3	21
15	—	—	Ephestia		0	111
—	—	38 – 94	Sitotroga		4	155
62[b]	—	—	Ephestia		2	38
40	—	—	Ostrinia		2	81
0 – 63	50 – 100	0 – 31	Ostrinia	Staggered emergence	2	153
57	97	93	Ostrinia	Sentinel eggs	2	153
57	92 – 97	31	Ostrinia	Sentinel eggs	2	109
20	—	—	Ostrinia	Staggered emergence; sentinel eggs	2	152
39 – 87	44 – 75	42 – 57	Dendrolimus, silkworm spp.		3	154
≤90[b]	99	—	—		4	207
60 – 70[b]	—	90[b]	—		4	207
45 – 74	12 – 68	19 – 71	Sitotroga		1	185
—	65 – 93	—	Sitotroga	Staggered emergence	3	62
—	39 – 100	—	Sitotroga		2	69
—	8 – 90	18 – 97	Ephestia	Staggered emergence	2	124
11 – 71	—	—	Sitotroga	Staggered emergence	2	200
0	—	—	Sitotroga		0	109
—	12 – 74	—	Sitotroga		1	122
28 – 68[b]	11 – 81[c]	—	Corcyra		4	219
2 – 18	—	—	Sitotroga	Aerial release	0	78
0 – 76	7 – 27	0 – 31	Sitotroga	Aerial release	0	88,90
40	—	—	—		2	116
49	—	—	—		0	187
10 – 55	—	28 – 66	Ephestia, Sitotroga		2	44
32 – 67[c]	—	0 – 21[c]	—		3	220
73	94	—	Corcyra		4	2
87 – 91	—	67[c]	Philosamia, Antheraea		3	106
0.7 – 83	0 – 82	0 – 56	Sitotroga	Sentinel eggs; aerial release	2	173
80 – 96[b]	98[b]	—	Antheraea		3	72
45 – 90[b]	98 – 99[b]	—	Corcyra		4	181
29[c]	71[c]	—	Antheraea		4	46
—	—	42 – 63	Sitotroga	Staggered emergence	1	67
—	—	12 – 88	Sitotroga	Staggered emergence	1	67
64 – 82	60[c]	45 – 58[c]	Antheraea		4	46
15 – 45	54 – 60	49 – 61	Cryptophlebia	Sentinel eggs	3	128
49 – 86[b]	—	—	Corcyra	Sentinel eggs	2	156
31 – 74	—	0 – 83	Trichoplusia		1	134
3 – 47	—	0 – 38	Trichoplusia		1	134
18 – 68	—	0 – 87	Trichoplusia		1	134
—	—	55 – 98	Sitotroga		2	114
47 – 74	—	—	Ostrinia	Sentinel eggs	2	34
44	—	79	Ephestia	Sentinel eggs	2	112
—	—	4 – 61	Sitotroga	Staggered emergence	0	36
—	—	4 – 29	Sitotroga	Staggered emergence	0	36
—	—	10 – 35	Sitotroga	Staggered emergence	0	36
—	13 – 42	4 – 23	Cadra		2	30
—	6 – 21	5 – 19	Cadra		1	30
77 – 92[b]	—	80 – 92[b]	Ephestia		4	38
17 – 32[b]	—	—	Ephestia		4	38
7 – 47	0 – 48	17 – 21	Sitotroga		0	142
33 – 44	—	25 – 36	Corcyra		2	104
7 – 31	—	0 – 59	Corcyra	Adults fed honey	2	13

could lead to the elimination of local species or strains when non-native species are released (71a). A survey of local species should be conducted before parasitoids are released, because natural levels of parasitism, although sometimes negligible, can be as high as 40–100% (116, 204a). The recent move by national agencies in some countries to restrict the importation of organisms for biological control makes it important that effective native species be identified, especially during the first screening.

Once the species has been selected, the population (= strain) to release must be determined. *Trichogramma* display both interspecific and intraspecific variations in biology and behavior that are strongly influenced by environmental factors. These variations complicate the selection process further, and numerous studies have focused on phenotypic differences among strains (interpopulations). These populations have been compared in terms of development, fecundity, egg absorption, sex ratio, longevity, host age selection, oviposition, host preference, and activity (137, 145, 174, 189), as well as in terms of their response to environmental conditions (135, 138, 144, 176). A few studies have also addressed the occurrence of thelytokous (100% female offspring) or deuterotokous (almost all female offspring) populations and their fitness or fecundity relative to the more common arrenotokous (sex ratio of offspring, 50–75% female) strains (179, 209). The ultimate choice of strain will depend on how it ranks in terms of those biological attributes considered advantageous for the environment into which it will be released and the type of release to be conducted (e.g. inundative or inoculative).

The final aspect of selection is that of founding populations, i.e. where and how many collections (both individuals and populations) are needed to initiate a vigorous colony. This field is one of the least studied, because it is based on the population genetics, and almost nothing is known about the genetics of *Trichogramma*. A few studies have examined the genetic variability of traits, including fecundity (150), reactive distance (32), walking behavior (105, 204a), and sex allocation (203, 205a), but the genetic base for biological differences among or within populations (intrapopulations) has been rarely examined (204). Traditional wisdom, based on balanced gene systems in diploid organisms, suggests that a minimum of 500–1000 individuals should be used to found a population with high levels of heterozygosity (192); this approach has been taken in the former USSR (59). *Trichogramma* are haplo-diploid organisms (most females arise from fertilized diploid eggs, and most males from unfertilized haploid eggs), however, and are characterized by high rates of sib-mating [an estimated 55–64% in the field (84)] and naturally low heterozygosity (163). This feature suggests that the degree of heterozygosity normally required to maintain a vigorous colony of a *Trichogramma* species might be less than expected and that healthy colonies may be founded with fewer than 500 individuals. Considerable work is needed to develop an understanding of if and how *Trichogramma* maintain sufficient levels of variation under conditions of small population size.

Parasitoid Quality

Variation within populations allows the selection of high-quality parasitoids. Populations for inundative release are often selected on the basis that those with high fecundity, emergence, sex ratio (percentage of female offspring), longevity, host preference for the target species, host-searching activity, and tolerance to local weather conditions will be best. A population with these characteristics is defined as a parasitoid of "high quality," because these traits are assumed to be ecologically important for these parasitoids when released inundatively. For those used in inoculative releases, such characteristics as development rate, oogenesis, and competitive ability are also important. Unfortunately, one strain is seldom superior in all attributes, and it is often unclear whether high-quality strains in the laboratory are synonymous with effective parasitoids in the field. In addition, some of those attributes that make the parasitoid effective in the field may not be advantageous in mass rearing (e.g. those that prefer the target species may be more difficult or impossible to rear on a factitious host), and trade-offs must be made in terms of the desired traits (192).

Numerous authors have examined individual components of quality, including fecundity (8, 39, 161, 204a, 217), development rate (8, 35, 61, 160, 174, 204a), oogenesis (182), emergence (39, 175, 204a), sex ratio (10, 145, 202), longevity (9, 10, 35, 39, 174, 175, 204a), host acceptance and preference (8, 64, 65, 137, 141, 189), host searching and activity (85, 183), and the effect of the environment (144, 165). Attempts to link these qualities have been rare, and the current recommendation is to measure a few individual variables (93). A few studies have attempted to combine these traits into a single predictive value (18, 36a, 47, 161). Only locomotion (walking), however, has been shown to be a good predictor of field efficacy in *Trichogramma* to date (18, 151, 197, 204a).

One of the more controversial measures of parasitoid quality has been size. Individuals of a *Trichogramma* species will vary in size according to the host egg in which they are reared (8, 26, 39). Body size in *Trichogramma*, as measured by hind tibial or wing length, has been generally shown to be positively correlated with biological characteristics, such as fecundity, longevity, rate of search, and flight (8, 17). To some extent, this relationship appears to depend on the physiologic state of the parasitoid. The correlation is stronger for adults that are fed honey than for those that are not (9), for those that are presented with large host eggs than for those presented with small ones (26), and for those that are reared on small host eggs than for those reared on large ones (83). In the last case, although increasing parasitoid size from larger host eggs resulted in higher levels of fecundity, the variation among individuals was so high that the authors considered size an unreliable measure of parasitoid quality. Pavlik (146) has recently found that the length of the hind tibia is a poor predictor of quality. Size therefore seems to be an uncertain predictor of quality, although further studies are warranted because it is so easy to measure.

There is a strong need to establish measurable parameters that assess parasitoid quality accurately and to standardize them internationally for the commercial use of *Trichogramma* (66). The current recommendation of the International Organization for Biological Control working group on Quality Control of Mass Reared Arthropods (IOBC-QC) is to conduct an intensive preliminary screening of candidate strains and species for all possible biological and behavioral traits (17, 93). This screening is considered important, because these populations will form the basis of the rearing and release program and should be selected carefully. The subsequent assessments of quality that must be made throughout the rearing and distribution processes should be less intensive and more focused on attributes that can be measured quickly (17). (See section on Parasitoid Rearing.)

PRODUCTION SYSTEM FOR MASS REARING

Once a parasitoid colony has been selected, the next step is to rear large numbers for release. This process has been accomplished in various ways during the past 70 years and has been the focus of considerable attention (14, 42, 49, 59, 62, 95, 100, 117, 119, 129, 201, 204a). Mass production of *Trichogramma* is a growing field, and many facilities have been established or expanded in the past five years (127).

Two types of rearing systems have evolved: those with short-term high daily output and those with long-term low daily output (17, 127). A range in production from 4 to 1000 million parasitoids/day has been found, depending on the mode of output; short-term output usually has the higher values (95, 100, 119, 129). Consistent levels of output of 100 million female parasitoids/week, although rarely specified, are not uncommon for the larger facilities (127).

Major commercial facilities are currently found in Europe (France, The Netherlands, Switzerland, Germany), the United States, Canada, and Mexico, as well as in large-scale, government-supported facilities in China, the former USSR, and Brazil (127). Numerous other smaller facilities can be found throughout the world (South and Central America, Australia, southern and eastern Europe, South Africa, India, and southern Asia) with various forms of government, private, and cooperative support (11, 45, 54, 127, 169, 170). Most of the larger facilities produce parasitoids on a year-long basis, whereas the smaller facilities produce parasitoids for local periodic releases (127). Many of these facilities sell wholesale to international suppliers; Hunter (73) recently published a list that shows 78 suppliers of *Trichogramma* in North America. Production companies are understandably hesitant to discuss their internal processes and standards (192). Leppla & Fisher (98) thought that it was important for industrial producers of beneficial insects to standardize their rearing of *Trichogramma*. They designated three essential areas that are being adopted currently by the industry: product, process, and production control. Aspects of process and production will probably remain somewhat guarded, as these deal with relationships between inputs and outputs within the facility. Examples of these aspects include weights or volumes of eggs produced for given amounts of grain used, for production box/unit, or for adult hosts; daily or total capacities of individual host or parasitoid production units; and costs per unit (66, 95, 117, 197). Product control is related to parasitoid quality and is discussed in the section on Parasitoid Rearing. In addition to these production lines, facilities must also consider internal economic aspects, such as initial capital costs, reusable components, automation, and space, as well as the health and safety issues of cleanliness and potential for allergies (G. Eden, Eden Consulting, Guelph, Ontario, Canada).

There are generally only two biological components in a mass rearing facility: the rearing host and the parasitoid. Most larger facilities have these compartmentalized into smaller units for host production and parasitization [most in the West are modeled after the boxes of Morrison & King (119)]; these units are then replicated in the facility according to the desired level of output. A common factor in almost all facilities is that they allocate at least two thirds their space and energy for the production of the rearing host(s) and the remainder for the parasitoid (17, 127).

Rearing Host

Two major biological aspects of host rearing are the species to use (including artificial host eggs) and whether the eggs can be stored to extend the production period of the facility. To date, the consequential choice in host rearing has been limited to species that produce either small or large eggs.

Flanders (49) originally proposed a small host egg, the Angoumois grain moth, *Sitotroga cerealella*, and producers in slightly more than half the countries today use this species (Table 1). Several countries, including France and Canada, have switched to the Mediterranean flour moth, *Ephestia kuehniella*, because of better production from the rearing medium, ease in mechanization, and improved sanitation conditions. The third small egg species, the rice meal moth, *Corcyra cephalonica*, is used in various Asian countries because of its local availability. No trials have compared parasitoids reared from *Corcyra* with other small egg species. Repeated studies have shown, however, that

Trichogramma emerging from either *Ephestia* or *Sitotroga* are equivalent in the field. Somewhat better performance has been noted in the laboratory by parasitoids from *Ephestia,* possibly because of their slightly larger size (15, 22, 39, 68, 124).

Approximately four large host egg species of Lepidoptera have been used to rear *Trichogramma;* three of these species are from China. The silkworm hosts are considered commercially viable because their eggs are readily available as a by-product of silk production (100), whereas *Ostrinia nubilale* has been used primarily for research (34, 81, 109, 152, 153). When reared on these large host eggs, *Trichogramma* wasps may be of much better quality (e.g. larger size and higher percentage of females) than when reared on small host eggs. Thus, facilities can continuously produce either large numbers of small (low-quality) parasitoids or low numbers of large (high-quality) parasitoids. The feasibility of using species that lay large host eggs in commercial production has not yet been compared with that of species that lay small host eggs, despite the potential for increased parasitoid quality.

A relatively recent development in *Trichogramma* rearing is in vitro production on artificial host media (57). This area has been researched in China since 1975 (100, 102). Two approaches have been taken. In the first approach, the natural insect egg hemolymph is partially replaced with egg yolk and milk solids (197). In the second approach, a completely artificial diet is created from biochemical analysis of the insect and its egg (123, 213). Eighteen species of *Trichogramma* have now been reared from egg to adult in various forms of artificial media (57). The closest system to commercial production is that developed for *T. dendrolimi* in China on the basis of insect hemolymph (57, 204a). This diet has been packaged in plastic host egg-cards (produced at 1200 egg-cards per hour), and the resultant parasitoids have been used on more than 1300 ha with parasitism equal to parasitoids from natural host eggs (41, 204a). The development of completely artificial hosts is an important goal and, when realized, will lead to major reductions in the size of facilities, the cost of the product, and changes in the strategy for implementation in the field.

An essential part of producing the rearing host is some means of storing the eggs to ensure a continuous supply; sterilization and cool temperatures are the most common features. Sterilization increases the storage and flexibility of unparasitized small host eggs (e.g. by preventing emergence of cannabalistic larvae) and is achieved by either cold storage or freezing for short periods of time or by irradiation using ultraviolet or gamma sources (14, 29, 41a, 55, 197). Bigler (14) reported a maximum storage of four weeks for irradiated *Ephestia* eggs held at 2°C and 90% RH, and Vieira & Tavares (196) recently suggested that eggs still can produce high-quality parasitoids after storage at 0.7°C and 60% RH in the dark for up to 3.5 months. A more promising approach to long-term storage is liquid nitrogen for all types of host eggs.

High-quality parasitoids have been produced both in China, where eggs of both silkworm and *Corcyra* have been stored from 8 to 30 months in liquid nitrogen and have produced high-quality parasitoids, and in the United States, where eggs of *Sitotroga* have been stored for 21 days (199).

Parasitoid Rearing

To ensure high product quality and to avoid contamination, facilities usually rear only a few parasitoid strains or species at any given time (59). Improved techniques for identifying different populations rapidly by using DNA markers are being developed, and their integration into facilities may help screen for such rearing problems (163, 194).

The ratio of the number of parasitoids to host eggs in the parasitization units is also important. High ratios may lead to superparasitism, high numbers of male progeny, and poor product quality, whereas low ratios may result in poor parasitization and inefficient use of host eggs (40, 164, 205a). The acceptance and allocation of offspring in host eggs by *Trichogramma* are influenced by the density of the host (87, 92, 139, 157), and parasitoid fecundity or clutch size is adjusted according to host availability relative to abundance (10, 50, 179), host egg size (167), and spacing between eggs (166). Ratios of females to small host eggs of 1:10 are often used to maintain parasitism of 70–80% and sex ratios of 50–80% females in rearing facilities.

Once uniform parasitism of the host eggs has been achieved through manipulation of lighting and temperature (119), their emergence must be programmed. Programming can be as simple as allowing the parasitoids to complete development at a specified temperature and photoperiod (114) or it may involve more complex manipulation of environmental conditions to achieve synchronization, long-term storage and delayed emergence. In general, storage at low temperatures (6–12°C) during the pupal stage is considered best for *Trichogramma* (76, 218), although such storage has never extended much longer than two weeks without losses in parasitoid quality (196). Those species that are more cold hardy [e.g. *T. brassicae* (=*maidis*), *ostriniae, evanescens,* and *dendrolimi*] and/or undergo diapause (initiated by temperature and photoperiod effects on the maternal generation and developing larvae, as well as possible host egg effects) can tolerate longer storage (24, 94, 218). The specific conditions that promote parasitoid storage and diapause are being pursued actively to allow rearing facilities to economize and maintain the genetic quality of their stock better (195, 198). Other factors that may affect the spread of emergence include superparasitism and intrinsic competition (40).

The maintenance of parasitoid quality is critical to the reputation of a production facility, and the quality may be compromised after rearing *Trichogramma* for many generations under uniform conditions and on an atypical host. Two important changes can occur: loss of tolerance to natural physical extremes and loss of preference for the target host. The first change has been rarely studied, although rearing the parasitoid under fluctuating temperatures has been recommended to maintain tolerance. Unfortunately, this recommendation is difficult to implement in a commercial rearing facility (17, 201).

The loss of preference for the target host is a controversial area, because this effect has been demonstrated for some *Trichogramma* species (59, 125, 188) but not for others (25, 80, 145, 215). Approaches taken to counter this effect include the setting of maximum limits for the number of generations that can be reared in the facility, periodically switching the parasitoids to different hosts, or both. The first approach is used in France, where 100 female *T. brassicae* are collected annually and maintained in isofemale lines for three generations and then mixed together for a maximum of 20–25 generations (52). In Switzerland, *T. brassicae* is reared for a maximum of six generations on *Ephestia;* if kept longer, it is switched to the target host (14). This switching to the target or any factitious host is also recommended in Germany, Australia, and the former USSR (59, 64, 169, 188), although parasitization problems can occur in the initial generation (25).

Laing & Bigler (93) and Gusev & Lebedev (59) recommend the use of standardized biological and behavioral tests to monitor quality in rearing facilities. Recent work with molecular markers may help identify genetic shifts in populations (163, 168, 174). Whether quality changes occur in continuous rearing must be determined to support recommendations for rejuvenating commercial cultures routinely.

DISTRIBUTION OF *TRICHOGRAMMA*

Although stringent controls may ensure that parasitized host eggs leave a production system in good condition, they do not guarantee that the eggs will be released this way (192). With the exception of the work by Bouse & Morrison (27), no attention has been given to the distribution (formulation, packaging, storage, and transport) of *Trichogramma* to the field. This problem exists because the rearing process and the field trials often are separated not only by time and distance but also by the persons involved. Conversely, many trials are small scale and local and are rarely repeated experimentally. Unfortunately, lack of attention to proper delivery of the product can negate all other efforts.

Several field studies cite transport and subsequent parasitoid problems as the cause of poor success (173, 185). Most shipping problems occur during hot conditions or when the rearing facilities are a considerable distance from the release sites (>200 km). Both these situations jeopardize survival or the programming of parasitoid emergence. Bouse & Morrison (27) developed an insulated, thermocouple-regulated box for shipping parasitized eggs to the field. Research is needed, however, to compare the different modes of transport (various ground routes vs air) and to determine the advantage that insulation, ventilation, or air circulation in the containers might provide. We also need to apply what we know about temperature and storing *Trichogramma* to problems during shipping.

Problems seen in the field are not surprising. Material often is packaged and shipped without detailed instructions regarding what it is and how it should be handled. Standard measures of ensuring product quality to the user must be implemented. Laing & Bigler (93) have proposed the use of labels for each shipment to guarantee biological data as well as the number of viable female parasitoids/unit. Individual countries are now setting legal requirements for such labels in the registration of these new pest control products.

Laing & Bigler (93) also have recommended that users take subsamples of parasitoid material to test its quality before release. This step is feasible when those who receive the material in the field (e.g. researchers) understand the technical and biological details of such tests. In a commercial operation (e.g. growers), however, this level of expertise may not always be available. Once in the field, there is often no extension support for the grower, and it may not be possible to keep parasitized eggs under acceptable conditions for long-term storage. Hassan (in 169) emphasized the extreme importance of cooperative extension in the successful implementation of a *Trichogramma* program. In Europe, extension agents monitor corn fields and distribute parasitoids to farmers only after the first moths are collected. Standardized procedures need to be established, and *Trichogramma* releases need to be integrated into all aspects of the farming system, with the support of consultants, users, and the public; these steps are lacking currently.

Smith (172) has presented the various methods of releasing *Trichogramma*. Almost all studies have used ground releases, usually from point sources in research trials or from commercial application where labor costs are low. Point sources have been containerized to reduce the effect of predation and inclement weather (19, 67, 153). Large-scale broadcast release from the ground has been developed only in the former USSR and the United States. These two countries, with Canada and China, are also the only ones to have released parasitized eggs in broadcast aerial applications from aircraft or helicopters, either alone or with extenders of grass seed, water, and sawdust (27, 59, 173). A new ultraviolet motorplane, which can treat at rates of 30 ha/h (195), has recently been described for use in France. In addition, a new liquid application system is currently being developed in the United States (S. Penn, Beneficial Insectary, Oak Run, California). Different modes of release technology that direct the material to the targeted location better, such as small, remote-controlled model airplanes, mechanized ground applicators, and global positioning systems, merit continued investigation. This has been a neglected area, and each crop/ pest system will have different requirements.

STRATEGIES IN THE FIELD

Approaches for Release

After large numbers of parasitoids have been reared and distributed to the field, they are ready for release. *Trichogramma* have been used in all three biological control approaches: introduction, inoculation, and inundation. Inundative releases achieve an immediate, nonsustaining reduction in the host population. In inoculative releases, however, it is the progeny of parasitoids released at the beginning of the season that have a later effect on the host population. Inundative releases have predominated in the West (14, 62, 88, 89, 90, 173), whereas countries in Asia and parts of the former USSR have put emphasis on inoculation and occasional introductions (201, 208). Warmer climates favor multivoltine pests and inoculative releases, because the parasitoids can multiply during a long growing season. Inundative releases, which are timed specifically to the ovipositional period of the pest, are more appropriate in northern climates with uni- or bivoltine host species. Several countries use both strategies through the repeated annual applications of *Trichogramma* (44, 101, 201). In China, a slight modification of the inoculative approach has been used; *Trichogramma* are released in vegetable gardens adjacent to the target crop in ratios ranging from 1:5 to 1:14 (release garden:target crop) (208). Introductions of new species have occurred in India (156), North America, and Russia (12).

The different approaches to the use of *Trichogramma* have resulted in two different perspectives: the inundative approach, which tends to view the parasitoid as a fast-acting replacement for chemical insecticides, and the inoculative/ introduction approach, which sees the parasitoid as one aspect of integrated pest management. Considerably less experimental research has gone into inoculative releases than into inundative releases because of the ecological complexity involved and lack of funding. This situation is unfortunate, because the few integrated studies with *Trichogramma* and microbials, such as *Bacillus thuringiensis* (Bt), *Nosema,* and *Beauvaria,* have been positive (54, 121, 173, 201, 212). Several authors suggest that releases of more than one species of *Trichogramma* will improve efficacy (38, 67, 169), and some countries have combined *Trichogramma* releases with those of other parasitoids (e.g. *Habrobracon*) to provide acceptable levels of control for pests in cotton, tomato, and pine (201, 212). In Colombia, releases that integrate *Trichogramma* with both Bt and *Apanteles* or *Telenomus* have reduced the use of insecticide by 50% on tomato and cassava (121).

Trichogramma releases may have an immediate effect not only on the target pest, but possibly on other insect populations as well (197). Direct effects may be seen on nontarget Lepidoptera in the crop and surrounding area, and indirect effects may be seen on the natural enemy complexes associated with them (71a). Only one published work has addressed this area. Lopez & Morrison (108) showed that predators of *Helicoverpa* in cotton where *Trichogramma* wasps were released were unaffected compared with insecticide plots. Little research has been conducted in this area, because agricultural systems tend to be self-contained and are already disturbed monocultures of low natural diversity. Boreal forests are much less disturbed; in Canada, work in the area is currently under way to assess the effect of *T. minutum* releases on nontarget Lepidoptera, as well as native larval and pupal parasitoids of the target pest (RS Bourchier & SM Smith, unpublished data). The effects on nontargets will become increasingly important, not only because of the worldwide interest in biodiversity but because producers are now being asked to address this aspect before *Trichogramma* can be registered.

Timing, Frequency, and Rates of Release

The timing, frequency, and rates of release all depend on the approach taken. With inoculative releases, relatively few parasitoids are required very early in the season, possibly independent of the ovipositional period of the pest. In contrast, inundative releases require large numbers to be synchronized closely with the start of oviposition of a uni- or bivoltine pest (172). The earlier oviposition can be predicted, the better for the rearing facility and the field program. If large numbers of parasitoids are needed in a short time, then some facilities may require several weeks' or months' notice.

Different methods, including calendar date, plant development, pheromone or light traps, egg-laying, and developmental degree-days, have been used to synchronize inundative releases with the start of host oviposition (Table 1). Plant development is the least accurate method, unless it is linked to pest phenology (34, 81). Although currently too variable to be used alone, the degree-days method (211) allows the greatest forecasting (approximately one month); this method is potentially the most valuable (122, 153, 173, 214). Light traps (14, 21, 111, 122, 155) and, where available, pheromone traps (38, 62, 89, 90, 111, 173) appear to be the best predictors, because they collect adult moths before oviposition starts (especially pheromone traps). Studies that compare trap catch, oviposition, and efficacy have shown consistently that the best results are achieved when the *Trichogramma* are released a few days before, rather than at the start of, oviposition (62, 63, 81, 173).

Synchronization with the host also means that programming of parasitoid emergence must be considered. Although most facilities ship parasitoid material ready to emerge, and the majority of releases have used material emerging within hours of release, this is not always the case. Some key strategies mix different stages of *Trichogramma* development, thereby staggering emergence, particularly if only one or a few releases can be made (Table 1; 14, 21, 62, 67, 71, 81, 152, 153, 173). This approach ensures that there are always some females actively searching throughout host oviposition (177). In practice, this approach is limited if the released eggs are exposed to predation or extreme temperatures.

The first release is usually aimed at just before host oviposition, not only to achieve high levels of parasitism but also to enable released material to multiply in the natural host eggs. Such an approach ensures that a continuous supply of superior-quality parasitoids is produced from the target pest in the field (14, 96, 176). This field multiplication reduces the need for many releases and is the basis for some of the more successful programs (14, 62, 67, 173). The release of unparasitized factitious host eggs concurrently with *Trichogramma* has also been proposed to provide for self-multiplication when the target pest is low (91).

The goal of most release programs is to maintain a level of more than 80% parasitism on freshly laid host eggs (91). When single releases and field multiplication do not achieve this goal, then multiple releases at various frequencies are used. Few studies have actually compared different frequencies of similar final release rates (155), although most use multiple releases (Table 1). The interval between releases is variable but averages 5–7 days (Table 1; 59). This interval often is based on unpublished studies of parasitoid longevity (67, 81, 173) rather than on direct measures of survival in the field.

Similarly, the actual rates of release vary considerably, even for the same pest, crop, and country (Table 1). For example, the total rates of release for *T. brassicae* (=*maidis*) alone, which is reared from small host eggs against *Ostrinia* in Europe, range from 150,000 to 2.8 million wasps/ha. Rates in the several millions of wasps/ha are generally cited in arboreal situations, such as forestry (72, 173), and in fruit or nut orchards (46, 67, 128, 156), whereas those in agricultural crops, such as corn, cotton, and tomato, range from 500 to more than 1 million wasps/ha, with averages of 200,000–600,000 wasps/ha (Table 1; 59). This range is probably related to the range in dimensional volume of the crop. China often reports lower rates than other countries, possibly because of the frequent use of large host eggs (207).

Much of the confusion in application rates results from the inconsistency with which they are reported. The rearing host often is not identified, and numbers may be cited per release rather than the total number per hectare. In addition, some individuals refer to numbers of host eggs, whereas others refer to numbers of parasitized (black) eggs, wasps/parasitoids, or females per hectare. To compare these values, assumptions must be made regarding parasitoid size and quality, parasitism rates (usually 60–70%), emergence rates (usually

80–90%), and sex ratio (usually 50–70%). This lack of uniformity makes it difficult to compare studies and provide specific information on which rate should be used.

The reason why different rates of application are selected is not usually given, although an expected ratio of female wasps:hosts on the order of 1:2 or 1:10 for egg masses and 1:20 for eggs and the volume of the crop are often starting points (4, 12, 81, 152). More emphasis has been placed on establishing the correct spacing of points for given release rates or their vertical location within the crop (1, 81, 171, 214). Higher rates generally result in better parasitism. This is not always the case, however, as many factors influence the outcome of the release.

Factors That Affect Release

Weather, the crop, host, predation, use of pesticides, and parasitoid quality all influence the release and disappearance rate (4) of *Trichogramma*. Weather is probably the most pervasive, in that it is a complex of meterologic variables that affect the development, emergence, survival, activity, and fecundity of *Trichogramma*. The most influential components are temperature and humidity; in the extreme, both these components have been linked to poor field results (89, 90, 111, 154, Hassan in 169, 173, 176, 214). Parasitoid development is directly related to temperature; thus, extremes in the field disrupt not only survival and performance but also the programming of emergence. From laboratory work, most species apparently perform best (in terms of activity and fecundity) at 20–29°C and 40–60% RH, with lower thresholds of 9°C and 25% RH and higher thresholds of 36°C and 70% RH (35, 51, 92, 107, 137, 144, 204a). Rare field studies suggest that *Trichogramma* avoid dew (85), extreme temperature (85, 92, 204a), areas of bright light intensity (72, 92, 176, 204a), heavy rain (92), and winds greater than 1.1 km/h (72, 85). Thus, if inclement weather cannot be avoided, the rate and frequency of release must be adjusted upward, with specific regard to changes in the pattern and extent of emergence.

As the bottom component in the tritrophic interaction and the principal factor in habitat location, the crop is another important variable for *Trichogramma*, as it has both physical and chemical effects. Different levels of parasitism can be found on the same host in different crops (3, 86, 113, 133), and *Trichogramma* are much more habitat-specific than host-specific (85, 132, 183). Because *Trichogramma* are thought to search for host eggs randomly, parasitism is directly proportional to the size of the plant (1), its surface area (5, 21, 34, 81, 112), and the complexity (number of planes and angles) of the plant (5, 85) and its leaf surfaces (82, 183, 186). From the chemical perspective, the plant provides volatile cues (synomones) that, although not specific or long range (132), arrest and stimulate searching and parasitism in *Trichogramma* (3, 130, 133). These factors all influence the rate of release necessary and the resulting expected level of parasitism. Andow & Prokrym (4) suggest that the rate of release should be standardized by always expressing it in terms of the surface area of the plant (e.g. numbers/m^2).

The abundance and location of the host also influence *Trichogramma* releases. Parasitism tends to be higher in areas that have more hosts (11, 86, 171), although some releases have not shown this response (36, 216). *Trichogramma* generally demonstrate either independent or Type 2 functional responses to host density (87, 157, 176), with better parasitism seen in hosts that lay eggs in clusters rather than singly (193). Most species use kairomones to locate hosts from varying distances and sizes and shapes of the eggs on closer [e.g. 1.8 mm (136)] examination. In addition to plant cues, sex pheromones (99, 131), chemicals on the wing scales of the host moths (58, 178), and chemicals on the surface of the host eggs (46) all delay flight initiation, suppress positive phototaxis, intensify searching, increase retention time, and decrease speed of movement (131). Releasing kairomones with *Trichogramma* has given both positive (in cotton; 58) and negative (in spruce forests; 77) results. Conditioning the parasitoid during mass rearing to the kairomone of the target host, however, also has been positive, although this method has been tried only to a limited extent (58, 204a).

If emergence of *Trichogramma* is delayed in the field, then high losses may occur through predation. Depending on the diversity and location of the crop system, major predators include *Geocoris, Nabis, Orius, Hippodamia, Coleomegilla, Chrysopa*, ants, spiders, and small vertebrates. Predation of released material can be significant, with losses up to 50% in corn (21, 214) and 91–98% in cotton (79). Studies suggest that anthocorid predators are more likely to accept unparasitized host eggs than those that contain pupae of *Trichogramma*, although younger stages of *Trichogramma* are equally susceptible (31, 159). It is difficult to predict which predators will be the most significant in a system, as this aspect depends extensively on the climate, structure of the plant community, and cultural practices. Attempts to reduce predation have been made by using specialized wax capsules for release points (197). The effect of predation on the efficacy of *Trichogramma* releases merits increased focus, as it significantly influences cost and success of releases.

Pesticides also have been shown to reduce the effect of *Trichogramma* significantly (89, 90, 103, 115, 169). Many studies have compared the relative toxicity of pesticides, including insecticides, fungicides, and herbicides, to *Trichogramma* in screening trials (28, 53, 100, 120). In general, parasitoids are affected more by insecticides than by the other two groups, with the greatest mortality of adult *Trichogramma* seen 5–10 days after the use of selective pesticides, 15 days after moderately toxic pesticides, and 20–30 days after toxic pesticides (33, 59, 75, 216). *Bacillus thuringiensis* does not appear to affect parasitism when fed to adult wasps but can reduce parasitism when applied to the surface of host eggs (162, 184). Immature stages within the parasitized host eggs are generally unaffected, especially by selective pesticides, whereas adults are extremely sensitive (92, 102, 184, 201). Voegelé (197) has suggested that the sensitivity of *Trichogramma* to toxic pesticides may extend up to 1 km from the application site. With a few exceptions (204a), the use of *Trichogramma* in the same crop system as toxic pesticides must be carefully planned, as the two are generally incompatible.

Parasitoid quality is the final component that affects releases. Quality (longevity, fecundity, and searching capacity) can be increased two to ten times by providing a food source to adult wasps (9, 97, 176, 201, 215). In the field, this food source may be obtained from host feeding, nectar (6, 210), and plant fluids of damaged leaves (85). Although we know little about plants as nurseries or refugia for *Trichogramma*, several countries use plantings of nectariferous plants successfully, either in the fields or in adjacent areas (101, 201). If these plants are not widely available, one way of improving parasitoid quality is to supply a sugar source (e.g. honey or molasses) with the parasitoids on release (214). This approach has also been proposed in rearing facilities, although neither approach has been undertaken commercially.

The selection for increased parasitoid quality, such as fecundity and tolerance to environmental extremes and pesticide residues, is important to the success of releases; yet, little research has been done in this area. Lopez & Morrison (107) found that continuous rearing under variable temperatures and light regimes did not produce more heat-resistant parasitoids; however, selection studies to improve tolerance to heat and activity in uniparental populations of *T. pretiosum* (7), fecundity in *T. brassicae* (148), and parasitism at 15°C in *T. minutum* (Tocheva & Smith, unpublished data) have all been positive. Work is needed to identify which traits have sufficient genetic variation to be selected for; how quickly selection can occur; how long it can last without selection pressure; and whether selection of one trait is linked to another, which would affect overall parasitoid quality.

The dispersal or movement of *Trichogramma* is important to releases from two aspects: uniformity of parasitism within a crop and reduction caused by dispersal outside the crop. These tiny wasps actually have two modes of dispersing: either on their own or through phoresy on adult moths. The latter has been largely undocumented; thus, little is known of its effect on releases. The former has been intensively studied, although only from the perspective of parasitism so that little is known about movement related to food, mates, or refugia (51, 85). Greenhouse studies suggest that there may be an early migratory phase for *T. evanescens;* on emergence, it responds to light by flying upward instead of searching for hosts (178).

The ability of *Trichogramma* to disperse on its own appears to be high within single plants but low between plants (81, 85, 171, 204a, 216), which suggests that the parasitoids avoid open areas where they lose their ability for directed flight. Most studies show that parasitism decreases with distance from the release point (18, 154, 204a, 216) with distances of 4–50 m. Upwind dispersal is usually impossible (85); thus, most studies find significantly greater parasitism downwind (20, 72, 171, 216). Although significant losses of 60–75% by parasitoids dispersing out of fields have been cited (5, 18), this occurrence generally is not considered a major factor. Vertical movement within the plant usually is related to the location of host eggs and the release point (1, 11, 126, 171, 204a). These studies suggest that few release points are necessary in crops with continuous foliage (9 points/ha); in those systems with individual plants, however, more information is needed to assess the relationship between distance of opening and disruption in movement within and outside the crop.

Efficacy of Releases

Various simulation approaches have been used to improve the efficacy of *Trichogramma* releases during the past 30 years. Most of these approaches have dealt with the timing or the number of parasitoids needed to achieve host reduction, in terms of host density (56, 91), parasitoid searching area (81), disappearance rate (5), and the population dynamics of the pest (177). Two other models have been developed to predict damaging host populations (211) and searching efficiency and parasitism relative to the field (37). The models that deal with application rates and timing suggest the following: 1. more than 80% parasitism is necessary to reduce pest populations (56, 91); 2. the rate of release will increase proportionally with leaf surface area and disappearance rate (5); and 3. the rate can be cut in half if the emergence of parasitoids is staggered (177). Field studies have verified most of these predictions. Simulation models warrant considerably more development, as all have increased our understanding and pointed to areas that need research emphasis.

One of the difficulties in assessing the efficacy of *Trichogramma* releases, as with application rates, is the variability with which they are reported (Table 1). Some studies cite only parasitism, others cite larval populations, others cite infestation level, and others cite weight or volume of produce. On top of this, some reports deal only with changes in these levels between control and treated areas, increases in parasitism and product, or reductions in pest, infestation, or damage. A particular problem in some studies is the reference to plots treated with insecticides as "control" plots. In other studies, true, untreated controls are never used (38, 46, 181, 207, 220). An additional complication occurs when the target host lays eggs in clusters (e.g. *Ostrinia, Choristoneura,* and

Dendrolimus), with reports of parasitism without reference to the cluster. Bin & Vinson (23) present a strong case for unifying the terminology in reporting such parasitism.

Most studies assess efficacy by measuring egg parasitism, usually of eggs laid by the target host in the field (Table 1). Unless these collections are made at the end of oviposition after sufficient time for all the parasitized eggs to be identified (e.g. turn black), however, this approach may underestimate parasitism (190). One way to solve this problem is to place sentinel eggs (e.g. factitious or target host eggs on cards) of known age in the field for a specific length of time (109, 152, 153, 173, 176). This approach provides a measure of daily parasitism, although recent work suggests that it also may underestimate parasitism, especially if the sentinel eggs are differentially attractive compared with the natural host eggs (37, 109). This problem points to the necessity of adjusting for this difference before final assessment.

Another fairly consistent measure of efficacy is reduction in pest damage (Table 1), as this has direct value for the grower. The least common assessment is the number of larvae/unit, possibly because of the labor required to sample (14, 21). Bigler et al (20) recently reported that comparing larval attack on treated and untreated corn underestimates real efficiency. Another less common means of evaluating the effectiveness of *Trichogramma* releases is M, an index of population trends from life tables (101, 143).

A relationship exists among egg parasitism, larval populations, and damage; however, this relationship has been rarely examined. Van Hamburg & Hassell (191) suggested that egg parasitoids were unlikely to reduce larval populations if mortality after the egg stage was density dependent. On the basis of analyzing the transformed results from 26 studies that report changes in both parasitism (%) and larval reduction (%) (Table 1), there does appear to be a loose but positive relationship between these two values ($R^2 = 0.54$; $p < 0.0001$). Although these data were collected from very different situations, they suggest that increases in egg parasitism by *Trichogramma* generally result in proportional reductions in larvae. Future studies must examine the relationships among release rates, parasitism, larval reduction, and damage reduction specifically to document the efficacy of this approach.

Trichogramma can be an effective form of pest control when compared with more traditional approaches in many parts of the world. Parasitoid releases in China, Switzerland, Canada, and the former USSR have all shown consistent levels of 60–80% parasitism, with reductions in damage of 77–92% on such crops as sugarcane, wheat, corn, and cole (101). In many cases, this level of control has been cost competitive, as it has either completely eliminated or reduced the use of insecticides and increased crop values three to eight times (101). The cost to buy or produce *Trichogramma* varies, but current retail figures in the West are approximately $0.30 US/1000 females or $200 US/million females (G. Eden, Eden Consulting). *Trichogramma* releases often cost more (e.g. up to 60%) than insecticides for the same level of control or are less effective for the same cost (21, 46, 72, 90, 128, 201). Several combined reasons can explain why *Trichogramma* is considered cost-effective, including lower threshold levels (e.g. *Helicoverpa* on cotton in the USSR vs the United States), a reduction in levels of residue or costs for insecticides, and promotion of increased natural enemy complexes that provide integrated control. As with most classical biocontrol work, the cost:benefit ratio of using *Trichogramma* can be very high. In the former USSR, the ratio has been estimated at 1:8; in three counties in China, it has been estimated at 1:25 (101).

FUTURE PROSPECTS AND NEEDS

The use of *Trichogramma* has made significant strides during the past 20 years, and this bodes well for the next decades. As with Bt and chemical insecticides, significant commercial achievements have been made with *Trichogramma* wherever major research emphasis has been placed (e.g. corn borer control). These achievements suggest that we have every chance of succeeding in those other host/parasitoid systems that remain unexplored. The taxonomy of this genera is now being worked out. This work is essential, for it is the basis on which other studies are built. Although considerable information is available on phenotypic variation, more work needs to be addressed to its genotypic base to determine whether the selection of a superstrain is possible. Similarly, although we have produced a large amount of data on parasitoid biology and behavior, we now need to condense these data into some coherent, standardized concept of parasitoid quality with appropriate means of prediction. Rearing facilities are now commercialized on a large scale, but there remains a need to examine different rearing hosts and artificial diets in automated systems to make major advancements.

One of the most important areas in the future will be the development of extension support to deliver the product to the user and allow them to get into the field in a form that can have an effect. Information regarding where, when, and how to release in different grower situations should be included with the product. This package, which will provide a service rather than a product alone, could come from either the producers, government extension, or private consulting. Best results with *Trichogramma* in the past have been based on trained personnel who provide this combination of biology and economic decision-making.

Stinner (180) concluded his review with an emphasis on integrating *Trichogramma* with other control options. The situation is no different today, except that we have more information on how to achieve this integration. All too often in the past, the genus *Trichogramma* has been approached as a replacement for chemical insecticides. Inoculative releases, in combination with selective insecticides (chemical or biological), other parasitoids, and nectariferous plants, such as refugia, need to be developed. Although release of *Trichogramma* is currently one of the most benign approaches to pest control, more attention must be paid to the population dynamics of the pest, the other natural mortality factors at work, and the native complex of natural enemies, in particular native *Trichogramma* species. This approach will ensure a better understanding of the effect of release on biodiversity and provide a minimally disruptive approach to pest management.

Finally, perhaps the greatest need is that of setting guidelines and standardizing terminology and measurements; this includes issues of taxonomy, quality control, and assessments of efficacy. Quality standards, such as seven-day fecundity and locomotion tests that are currently being set by the IOBC-QC, should be implemented with a view to incorporating new tests (e.g. DNA fingerprinting) as they arise (93). Researchers should ensure a minimal amount of information reported for all field studies, including the mode of release (ground, aerial, point with spacing); crop size (height or surface area); parasitoid programming (emergence pattern) and activity (longevity estimates); timing (first release relative to host), frequency (intervals), and rate of release (in number of females/ha); and assessment. In terms of assessment, untreated control plots, quantified absolute estimates of both final egg or egg mass parasitism (depending on target host) from natural hosts, and damage to the crop are recommended. All these types of data will help regulatory bodies produce sensible guidelines for registration, thus promoting rather than hindering the development of biological control agents such as *Trichogramma*.

ACKNOWLEDGMENTS

Appreciation is extended to A Clarke, G Eden, and S Penn for their contributions of unpublished or hard-to-locate information. Thanks also to F Bigler, R Bourchier, J Corrigan, J Laing, R Morrison, P Newton, and E Wajnberg for reviewing the manuscript; to my many post-docs and graduate students; and to J Carrow and especially D Wallace for their insights over the years.

Literature Cited

1. Ables JR, McCommas DW, Jones SL, Morrison RK. 1980. Effect of cotton plant size, host egg location, and location of parasite release on parasitism by *Trichogramma pretiosum*. *Southwest. Entomol.* 5:261–64

2. Alba MC. 1990. Utilization of *Trichogramma* for biological control of sug-

arcane borers in the Philippines. See Ref. 206, pp. 161–65

3. Altieri MA, Annamalai S, Katiyar KP, Flath RA. 1982. Effects of plant extracts on the rates of parasitization of *Anagasta kuehniella* (Lep.: Pyralidae) eggs by *Trichogramma pretiosum* (Hym.: Trichogrammatidae) under greenhouse conditions. *Entomophaga* 27:431–38

4. Andow DA, Prokrym DR. 1990. Plant structural complexity and host-finding by a parasitoid. *Oecologia* 82:162–65

5. Andow DA, Prokrym DR. 1991. Release density, efficiency and disappearance of *Trichogramma nubilale* for control of European corn borer. *Entomophaga* 36:105–13

6. Ashley TR, Gonzalez D. 1974. Effect of various food substances on longevity and fecundity of *Trichogramma*. *Environ. Entomol.* 3:169–71

7. Ashley TR, Gonzalez D, Leigh TF. 1974. Selection and hybridization of *Trichogramma*. *Environ. Entomol.* 3:43–48

8. Bai B, Cobanoglu S, Smith SM. 1995. Assessment of *Trichogramma* species for biological control of forest lepidopteran defoliators. *Entomol. Exp. Appl.* 75:135–43

9. Bai B, Luck RF, Forster L, Stephens B, Janssen JAM. 1992. The effect of host size on quality attributes of the egg parasitoid, *Trichogramma pretiosum*. *Entomol. Exp. Appl.* 64:37–48

10. Bai B, Smith SM. 1993. Effect of host availability on reproduction and survival of the parasitoid wasp *Trichogramma minutum*. *Ecol. Entomol.* 18:279–86

11. Basso C, Morey C. 1990. Biological control of the sugarcane borer *Diatraea saccharalis* Fab. (Lep. Pyralidae) with *Trichogramma* spp. (Hym. Trichogrammatidae) in Uruguay. See Ref. 206, pp. 165–71

12. Beglyarov GA, Smetnik AI. 1977. Seasonal colonization of entomophages in the U.S.S.R. See Ref. 158, pp. 283–329

13. Bentur JS, Kalode MB, Rajendran B, Patel VS. 1994. Field evaluation of the egg parasitoid, *Trichogramma japonicum* Ash. (Hym., Trichogrammatidae) against the rice leaf folder, *Cnaphalocrocis medinalis* (Guen.) (Lep., Pyralidae) in India. *J. Appl. Entomol.* 117:257–61

14. Bigler F. 1986. Mass production of *Trichogramma maidis* Pint. et Voeg. and its field application against *Ostrinia nubilalis* Hbn. in Switzerland. *J. Appl. Entomol.* 101:23–29

15. Bigler F. 1988. Quality of *Trichogramma maidis* Pint. et Voeg. reared in eggs of *Ephestia kuehniella* Zell. and *Sitotroga cerealella* Oliv. See Ref. 199, pp. 409–11

16. Bigler F, ed. 1991. *Proc. Workshop of the IOBC Global Working Group, 5th. Quality Control of Mass Reared Arthropods.* Wageningen, The Netherlands:Int. Org. for Biological Control. 205 pp.

17. Bigler F. 1994. Quality control in *Trichogramma* production. See Ref. 205, pp. 93–112

18. Bigler F, Bieri M, Fritschy A, Seidel K. 1988. Variation in locomotion between laboratory strains of *Trichogramma maidis* and its impact on parasitism of eggs of *Ostrinia nubilalis* in the field. *Entomol. Exp. Appl.* 49:283–90

19. Bigler F, Bosshart S, Waldburger M. 1989. Bisherige und neue Entwicklungen bei der biologischen Bekampfung des Maiszunslers *Ostrinia nubilalis* Hbn., mit *Trichogramma maidis* Pint. et Voeg. in der Schweiz. *Landwirtsch. Jahrb. Schweiz* 2:37–43

20. Bigler F, Bosshart S, Waldburger M, Ingold M. 1990. Einfluss der Dispersion von *Trichogramma evanescens* Westw. auf die Parasitierung der Eier des Maiszunslers, *Ostrinia nubilalis* Hbn. *Bull. Soc. Entomol. Suisse* 63:381–88

21. Bigler F, Brunetti R. 1986. Biological control of *Ostrinia nubilalis* Hbn. by *Trichogramma maidis* Pint. et Voeg. on corn for seed production in southern Switzerland. *J. Appl. Entomol.* 102:303–8.

22. Bigler F, Meyer A, Bosshart S. 1987. Quality assessment in *Trichogramma maidis* Pintureau et Voegelé reared from eggs of the factitious hosts *Ephestia kuehniella* Zell. and *Sitotroga cerealella* (Olivier). *J. Appl. Entomol.* 104:340–53

23. Bin F, Vinson SB. 1990. Efficacy assessment in egg parasitoids (Hym.): Proposal for a unified terminology. See Ref. 206, pp. 175–80

24. Boivin G. 1994. Overwintering strategies of egg parasitoids. See Ref. 205, pp. 219–45

25. Bourchier RS, Smith SM, Corrigan JE, Laing JE. 1994. Effect of host switching on performance of mass-reared *Trichogramma minutum*. *Biocontrol Sci. Technol.* 4:353–62

26. Bourchier RS, Smith SM, Song SJ. 1993. Host acceptance and parasitoid size as predictors of parasitoid quality for mass-reared *Trichogramma minutum*. *Biol. Cont.* 3:135–39

27. Bouse LF, Morrison RK. 1985. Transport, storage, and release of *Tricho-*

gramma pretiosum. Southwest. Entomol. 8:36–48 (Suppl.)

28. Brar KS, Varma GC, Shenhmar M. 1991. Effect of insecticides on *Trichogramma chilonis* Ishii (Hym.: Trichogrammatidae) an egg parasitoid of sugarcane borers and cotton bollworms. *Entomon* 16:43–48

29. Brower JH. 1982. Parasitization of irradiated eggs and eggs from irradiated adults of the Indianmeal moth (Lep.: Pyralidae) by *Trichogramma pretiosum* (Hym.: Trichogrammatidae). *J. Econ. Entomol.* 75:939–44

30. Brower JH. 1988. Population suppression of the almond moth and the Indianmeal moth (Lep., Pyralidae) by release of *Trichogramma pretiosum* (Hym., Trichogrammatidae) into simulated peanut storage. *J. Econ. Entomol.* 81:944–48

31. Brower JH, Press JW. 1988. Interactions between the egg parasite *Trichogramma pretiosum* (Hym.: Trichogrammatidae) and a predator, *Xylocoris flavipes* (Hem.: Anthocoridae) of the almond moth *Cadra cautella* (Lep.: Pyralidae). *J. Entomol. Sci.* 23:342–49

32. Bruins EBAW, Wajnberg E, Pak GA. 1994. Genetic variability in the reactive distance in *Trichogramma brassicae* after automatic tracking of the walking path. *Entomol. Exp. Appl.* 72:297–303

33. Bull DL, House VS. 1983. Effects of different pesticides on parasitism of host eggs by *Trichogramma pretiosum* Riley. *Southwest. Entomol.* 8:46–53

34. Burbutis PP, Koepke CH. 1981. European corn borer control in peppers by *Trichogramma nubilale. J. Econ. Entomol.* 74:246–47

35. Calvin DD, Knapp MC, Welch SM, Poston FL, Elzinga RJ. 1984. Impact of environmental factors on *Trichogramma pretiosum* reared on southwestern corn borer eggs. *Environ. Entomol.* 13:774–80

36. Castaneda-Samayoa O, Holst H, Ohnesorge B. 1993. Evaluation of some *Trichogramma* species with respect to biological control of *Eupoecilia ambiguella* Hb. and *Lobesia botrana* Schiff. (Lep., Tortricidae). *Z. Pflanzenkr. Pflanzenschutz* 100:599–610

36a. Cerutti F, Bigler F. 1995. Quality assessment of *Trichogramma brassicae* in the laboratory. *Entomol. Exp. Appl.* 75: 19–26

37. Chernyshev VB, Semevsky FN, Grinberg SM. 1988. Mathematical model of migration distribution and search for host eggs by *Trichogramma. Zoojiothyeckii Kyphaji* 67:57–58

38. Ciochia V. 1990. Some aspects of the utilization of *Trichogramma* sp. in Romania. See Ref. 206, pp. 181–82

39. Corrigan JE, Laing JE. 1994. Effects of the rearing host species and the host species attacked on perfomance by *Trichogramma minutum* Riley (Hym: Trichogrammatidae). *Environ. Entomol.* 23:755–60

40. Corrigan JE, Laing JE, Zubricky JS. 1995. Effects of parasitoid:host ratio and time of day of parasitism on development and emergence of *Trichogramma minutum* (Hym.: Trichogrammatidae), parasitizing eggs of *Ephestia kuehniella* (Lep.: Pyralidae). *Ann. Entomol. Soc. Am.* In press

41. Dai KJ, Ma ZJ, Zhang LW, Cao AH, Zhan QX, et al. 1990. Research on technology of industrial production of the artificial host egg of *Trichogramma.* See Ref, 206. pp. 137–41

41a. Daumal J, Boinel H. 1994. Qualité trophique des pontes d'*Ephestia kuehniella* Zell. traites aux basses temperatures pour l'elevage des Trichogrammes. *Mitt. Schweiz. Entomol. Ges.* 67:373–83

42. Daumal J, Voegelé J, Brun P. 1975. Les Trichogrammes. 2. Unité de production massive et quotidienne d'un hôte de substitution *Ephestia kuehniella* Zell. (Lep. Pyralidae). *Ann. Zool. Ecol. Anim.* 7:45–59

43. Deleted in proof

44. El-Heneidy AH, Abbas MST, Embaby MM, Ewiese MA. 1990. Utilization of *Trichogramma evanescens* West. to control the lesser sugar-cane borer, *Chilo agamemnon* Bles. in sugar cane fields in Egypt. 5. An approach towards large-scale release. See Ref. 206, pp. 187–90

45. Felkl G, Tulauan EJ, Lorenzana OT, Rendon VY, Fajardo AA. 1990. Establishment of *Trichogramma evanescens* Westw. in two corn growing areas in the Philippines. See Ref. 206, pp. 191–97

46. Feng J-G. 1988. Studies on the biological control of insect pests in fruit trees and oak trees with *T. dendrolimi* Matsumura. See Ref. 199, pp. 461–69

47. Ferreira L, Pintureau B, Voegelé J. 1979. Un nouveau type d'olfactométre. Application á la mesure de la capacité de recherche et á la localisation des substances attractives de l'hôte chez les Trichogrammes (Hym. Trichogrammatidae). *Ann. Zool. Ecol. Anim.* 11:271–79

48. Filippov NA. 1990. The present status and future outlook of biological control in the USSR. *Acta Entomol. Fenn.* 53: 11–18

49. Flanders SE. 1930. Mass production of egg parasites of the genus *Trichogramma*. *Hilgardia* 4:465–501
50. Fleury F, Boulétreau M. 1993. Effects of temporary host deprivation on the reproductive potential of *Trichogramma brassicae*. *Entomol. Exp. Appl.* 68:203–10
51. Forsse E, Smith SM, Bourchier R. 1992. Flight initiation in the egg parasitoid *Trichogramma minutum*: Effect of temperature, mates, food, and host eggs. *Entomol. Exp. Appl.* 62:147–54
52. Frandon J, Kabiri F, Pizzol J, Daumal J. 1991. Mass rearing of *Trichogramma brassicae* used against the European corn borer *Ostrinia nubilalis*. See Ref 16, pp.146–51
53. Franz JM, Bogenschutz H, Hassan SA, Huang P, Naton E, et al. 1980. Results of a joint pesticide test programme by the working group: Pesticides and beneficial arthropods. *Entomophaga* 25:231–36
54. Garcia-Roa F. 1990. Effectiveness of *Trichogramma* spp. in biological control programs in the Cauca Valley, Colombia. See Ref. 206, pp. 197–99
55. Goldstein LF, Burbutis PP, Ward DG. 1983. Rearing *Trichogramma nubilale* (Hym.: Trichogrammatidae) on ultraviolet-irradiated eggs of the European corn borer (Lep.: Pyralidae). *J. Econ. Entomol.* 76:969–71
56. Goodenough JL, Witz JA. 1985. Modeling augmentative releases of *Trichogramma pretiosum*. *Southwest. Entomol.* 8:169–89 (Suppl.)
57. Grenier S. 1994. Rearing of *Trichogramma* and other egg parasitoids on artificial diets. See Ref. 205, pp. 73–81
58. Gross HR, Lewis WJ, Nordlund DA. 1981. *Trichogramma pretiosum*: Effect of prerelease parasitization experience on retention in release areas and efficiency. *Environ. Entomol.* 10:554–56
59. Gusev GV, Lebedev GI. 1988. Present state of *Trichogramma* application and research. See Ref. 199, pp. 477–83
60. Deleted in proof
61. Harrison WW, King EG, Ouzts JD. 1985. Development of *Trichogramma exiguum* and *T. pretiosum* at five temperature regimes. *Environ. Entomol.* 14:118–21
62. Hassan SA. 1982. Mass production of *Trichogramma*: 3. Results of some research projects related to the practical use in the Federal Republic of Germany. See Ref. 74, pp. 213–19
63. Hassan SA. 1984. Massenproduktion und Anwendung von *Trichogramma*: 4. Feststellung der gunstigsten Freilas-sungstermine fur die Bekämpfung des Maiszunslers *Ostrinia nubilalis* Hübner. *Gesunde Pflanz.* 36:40–45
64. Hassan SA. 1990. A simple method to select effective *Trichogramma* strains for use in biological control. See Ref. 206, pp. 201–5
65. Hassan SA. 1993. The mass rearing and utilization of *Trichogramma* to control lepidopterous pests: Achievements and outlook. *Pestic. Sci.* 37:387–91
66. Hassan SA. 1994. Strategies to select *Trichogramma* species for use in biological control. See Ref. 205, pp. 55–73
67. Hassan SA, Kohler E, Rost WM. 1988. Mass production and utilization of *Trichogramma*: 10. Control of the codling moth *Cydia pomonella* and the summer fruit tortrix moth *Adoxophyes orana* (Lep.: Tortricidae). *Entomophaga* 33:413–20
68. Hassan SA, Langenbruch GA, Neuffer G. 1978. Der Einfluss des Wirtes in der Massenzucht auf die Qualitat des Eiparasiten *Trichogramma evanescens* bei der bekampfung des maiszunslers, *Ostrinia nubilalis*. *Entomophaga* 23:321–29
69. Hassan SA, Stein E, Dannemann K, Reichel W. 1986. Massenproduktion und anwendung von *Trichogramma*: 8. Optimierung des Einsatzes zur Bekämpfung des Maiszunslers *Ostrinia nubilalis* Hbn. *J. Appl. Entomol.* 101:508–15
70. Hawlitzky N. 1986. Etude de la biologie de la pyrale du maïs, *Ostrinia nubilalis* Hbn. en région parisienne durant quatre années et recherche d'éléments prévisionnels du début de ponte. *Acta Oecol. Appl.* 1:47–68
71. Hawlitsky N, Voegelé J. 1991. Démarche utilisée pour élaborer une stratégie de lutte biologique par lâchers d'entomophages contre un ravageur du maïs. Problemes apparus lors de la pratique et solutions apportées. *Bull. Soc. Zool. Fr.* 116:319–29
71a. Howarth FG. 1991. Environmental impacts of classical biological control. *Annu. Rev. Entomol.* 36:485–509
72. Hsiao K-J. 1980/1981. The use of biological agents for the control of the pine defoliator, *Dendrolimus punctatus* (Lep. Lasiocampidae) in China. *Prot. Ecol.* 2:297–303
73. Hunter CD. 1994. Suppliers of beneficial organisms in North America. *Calif. Environ. Prot. Agency, Dept. Pest. Reg.* PM 94–03. 30 pp.
74. Institut National de la Recherche Agronomique. 1982. Les trichogrammes. *Int. Symp. on les* Trichogrammes, *1st, An-*

tibes, France. No. 9. Paris: Les Colloques de l'INRA. 307 pp.
75. Jacobs RJ, Kouskolekas CA, Gross HR. 1984. Responses of *Trichogramma pretiosum* (Hym.: Trichogrammatidae) to residues of permethrin and endosulfan. *Environ. Entomol.* 13:355–58
76. Jalali SK, Singh SP. 1992. Differential response of four *Trichogramma* species to low temperatures for short term storage. *Entomophaga* 37:159–65
77. Jennings DT, Jones RL. 1986. Field tests of kairomones to increase parasitism of spruce budworm (Lep.: Tortricidae) eggs by *Trichogramma* spp. (Hym.: Trichogrammatidae). *Gr. Lakes Entomol.* 19:185–89
78. Johnson SJ. 1985. Low-level augmentation of *Trichogramma pretiosum* and naturally occurring *Trichogramma* spp. parasitism of *Heliothis* spp. in cotton in Louisianna. *Environ. Entomol.* 14:28–31
79. Jones SL, Morrison RK, Ables JR, Bull DL. 1977. A new and improved technique for the field release of *Trichogramma pretiosum*. *Southwest. Entomol.* 2:210–15
80. Kaiser L, Pham-Délégue MH, Masson C. 1989. Behavioural study of plasticity in host preferences of *Trichogramma maidis* (Hym.: Trichogrammatidae). *Physiol. Entomol.* 14:53–60
81. Kanour WW, Burbutis PP. 1984. *Trichogramma nubilale* (Hym. Trichogrammatidae) field releases in corn and a hypothetical model for control of European corn borer (Lep.: Pyralidae). *J. Econ. Entomol.* 77:103–7
82. Kauffman WC, Kennedy GG. 1989. Relationship between trichome density in tomato and parasitism of *Heliothis* spp. (Lep.: Noctuidae) eggs by *Trichogramma* spp. (Hym.: Trichogrammatidae). *Environ. Entomol.* 18:698–704
83. Kazmer DJ, Luck RF. 1990. The genetic-mating structure of natural and agricultural populations of *Trichogramma*. See Ref. 206, pp. 161–65
84. Kazmer DJ, Luck RF. 1995. Field tests of the size-fitness hypothesis in the egg parasitoid *Trichogramma pretiosum*. *Ecology* 76:412–25
85. Keller MA, Lewis WJ, Stinner RE. 1985. Biological and practical significance of movement by *Trichogramma* species: A review. *Southwest. Entomol.* 8:138–55 (Suppl.)
86. Kemp WP, Simmons GA. 1978. The influence of stand factors on parasitism of spruce budworm eggs by *Trichogramma minutum*. *Environ. Entomol.* 7:685–88

87. Kfir R. 1983. Functional response to host density by the egg parasite *Trichogramma pretiosum*. *Entomophaga* 28:345–53
88. King EG, Bouse LF, Bull DL, Coleman RJ, Dickerson WA, et al. 1986. Management of *Heliothis* spp. in cotton by augmentative releases of *Trichogramma pretiosum* Ril. *J. Appl. Entomol.* 101:2–10
89. King EG, Bull DL, Bouse LR, Phillips JR. 1985. Introduction: Biological control of *Heliothis* spp. in cotton by augmentative releases of *Trichogramma*. *Southwest. Entomol.* 8:1–10 (Suppl.)
90. King EG, Coleman RJ, Phillips JR, Dickerson WA. 1985. *Heliothis* spp. and selected natural enemy populations in cotton: A comparison of three insect control programs in Arkansas (1981–82) and North Carolina (1983). *Southwest. Entomol.* 8:71–98 (Suppl.)
91. Knipling EF, McGuire JU. 1968. Population models to appraise the limitations and potentialities of *Trichogramma* in managing host insect populations. *US Dep. Agric. Technol. Bull.* 1387:1–44
92. Kot J. 1979. Analysis of factors affecting the phytophage reduction by *Trichogramma* Westw. species. *Pol. Ecol. Stud.* 5:5–59
93. Laing JE, Bigler F. 1991. Quality control of mass-produced *Trichogramma* species. See Ref. 16, pp. 111–19
94. Laing JE, Corrigan JE. 1995. Diapause induction and post-diapause emergence in *Trichogramma minutum* Riley (Hym.: Trichogrammatidae): The role of host species, temperature, and photoperiod. *Can. Entomol.* 127:103–10
95. Laing JE, Eden GM. 1990. Mass-production of *Trichogramma minutum* Riley on factitious host eggs. See Ref. 175, pp. 10–25
96. Lawrence RK, Houseweart MW. 1985. Developmental rates of *Trichogramma minutum* (Hym.: Trichogrammatidae) and implications for timing augmentative releases for suppression of egg populations of *Choristoneura fumiferana* (Lep.: Tortricidae). *Can. Entomol.* 117:556–63
97. Leatemia JA, Laing JE, Corrigan JE. 1995. Effects of adult nutrition on the longevity, fecundity and offspring sex ratio of *Trichogramma minutum* Riley (Hym.: Trichogrammatidae). *Can. Entomol.* 127:245–54
98. Leppla NC, Fisher WR. 1989. Total quality control in insect mass production for insect pest management. *J. Appl. Entomol.* 108:452–61
99. Lewis WJ, Nordlund DA, Guldner RC,

Teal PEA, Tumlinson JH. 1982. Kairomones and their use for management of entomophagous insects. 13. Kairomonal activity for *Trichogramma* spp. of abdominal tips, excretion, and a synthetic sex pheromone blend of *Heliothis zea* (Boddie) moths. *J. Chem. Ecol.* 8:1323–31

100. Li L-Y. 1982. Integrated rice insect pest control in the Guangdong Province of China. *Entomophaga* 27:81–88

101. Li L-Y. 1994. Worldwide use of *Trichogramma* for biological control on different crops: A survey. See Ref. 205, pp. 37–51

102. Li L-Y, Liu W-H, Chen C-S, Han S-T, Shin J-C. 1988. In vitro rearing of *Trichogramma* sp. and *Anastatus* sp. in artificial "eggs" and the methods of mass production. See Ref. 199, pp. 339–53

103. Li SY, Sirois GM, Luczynski A, Henderson DE. 1993. Indigenous *Trichogramma* (Hym.: Trichogrammatidae) parasitizing eggs of *Rhopobota naevana* (Lep.: Tortricidae) on cranberries in British Columbia. *Entomophaga* 38:313–15

104. Lim GT. 1990. Augmentative release of *Trichogrammatoidea bactrae fumata* for the control of cocoa podborer in Sabah, Malaysia. See Ref. 206, pp. 205–8

105. Limburg H, Pak GA. 1991. Genetic variation in the walking behaviour of the egg parasite *Trichogramma*. See Ref. 16, pp. 47–55.

106. Liu ZC, Sun YR, Wang ZY, Liu JF. 1985. Role of biological control in integrated control of sugarcane insect pests. *Nat. Enemies Insects* 7:216–22

107. Lopez JD, Morrison RK. 1980. Effects of high temperatures on *Trichogramma pretiosum* programmed for field release. *J. Econ. Entomol.* 73:667–70

108. Lopez JD, Morrison RK. 1985. Parasitization of *Heliothis* spp. eggs after augmentative releases of *Trichogramma pretiosum* Riley. *Southwest. Entomol.* 8:110–37 (Suppl.)

109. Losey JE, Calvin DD. 1990. Parasitization efficiency of five species of *Trichogramma* (Hym.: Trichogrammatidae) on European corn borer eggs. See Ref. 206, pp. 209–13

110. Deleted in proof

111. Maini S, Burchi C, Gattavecchia C, Celli G, Voegelé J. 1988. *Trichogramma maidis* Pint. & Voeg. in Northern Italy. Augmentative releases against *Ostrinia nubilalis* HB. See Ref. 199, pp. 519–24

112. Maini S, Burgio G. 1990. Biological control of the European corn borer in protected pepper by *Trichogramma maidis* Pint. & Voeg. and *Bacillus thuringiensis* Berl. See Ref. 206, pp. 213–16

113. Martin PB, Lingren PD, Greene GL, Grissel EE. 1981. The parasitoid complex of three noctuids (Lep.) in a northern Florida cropping system: Seasonal occurrence, parasitization, alternate hosts, and influence of host-habitat. *Entomophaga* 26:401–19

114. McLaren IW, Rye WJ. 1983. The rearing, storage, and release of *Trichogramma ivelae* Pang & Chen (Hym.: Trichogrammatidae) for control of *Heliothis punctiger* Wall. (Lep.: Noctuidae) on tomatoes. *J. Aust. Entomol. Soc.* 22:119–24

115. Meierrose C, Araujo J. 1986. Natural egg parasitism on *Helicoverpa (Heliothis) armigera* Hbn. (Lep. Noctuidae) on tomato in south Portugal. *J. Appl. Entomol.* 101:11–18

116. Michael PJ, Woods WM. 1980. An entomological review of cotton growing in the Ord River area of western Australia. *Dep. Agric. West. Aust. Technol. Bull.* No. 48. 17 pp.

117. Morrison RK. 1985. Effective mass production of eggs of the angoumois grain moth, *Sitotroga cerealella* (Olivier). *Southwest. Entomol.* 8:28–35 (Suppl.)

118. Morrison RK, Jones SL, Lopez JD. 1978. A unified system for the production and preparation of *Trichogramma pretiosum* for field release. *Southwest. Entomol.* 3:62–8

119. Morrison RK, King EG. 1977. Mass production of natural enemies. See Ref. 159, pp. 173–217

120. Navarajan P. 1988. Toxicity of different pesticides to parasitoids of the genus *Trichogramma*. See Ref. 199, pp. 423–32

121. Navarro AM. 1988. Biological control of *Scrobipalpa* (Meyrick) by *Trichogramma* sp. in the tomato. See Ref. 199, pp. 453–58

122. Neil KA, Specht HB. 1990. Field releases of *Trichogramma pretiosum* Riley (Hym.: Trichogrammatidae) for suppression of corn earworm, *Heliothis zea* (Boddie) (Lep.: Noctuidae), egg populations on sweet corn in Nova Scotia. *Can. Entomol.* 122:1259–66

123. Nettles WC, Morrison RK, Xie Z-N, Ball D, Shenkir CA, Vinson SB. 1985. Effect of artificial diet media, glucose, protein hydrolyzates, and other factors on oviposition in wax eggs by *Trichogramma pretiosum. Entomol. Exp. Appl.* 38:121–29

124. Neuffer G. 1982. The use of *Trichogramma evanescens* Westw. in sweet-corn fields. A contribution to the bio-

logical control of the European corn borer *Ostrinia nubilalis* Hbn. in south west Germany. See Ref. 74, pp 231–38

125. Neuffer U. 1988. Vergleich von Parasitierungsleistung und Verhalten zweier Oekotypen von *Trichogramma evanescens* Westw. *J. Appl. Entomol.* 106:507–17

126. Newton PJ. 1990. Inundative releases of *Trichogrammatoidea cryptophlebiae* Nag. against *Cryptophlebia leucotreta* Mey. on citrus. See Ref. 206, pp. 217–21

127. Newton PJ. 1993. Increasing the use of trichogrammatids in insect pest management: A case study from the forests of Canada. *Pestic. Sci.* 37:381–86

128. Newton PJ, Odendaal WJ. 1990. Commercial inundative releases of *Trichogrammoidea cryptophlebiae* (Hym.: Trichogrammatidae) against *Cryptophlebia leucotreta* (Lep.: Tortricideae) in citrus. *Entomophaga* 35:545–56

129. Nikonov PV, Lebedev GI, Startchevsky IP. 1990. *Trichogramma* production in the USSR. See Ref. 206, pp. 151–53

130. Noldus LPJJ. 1989. Semiochemicals, foraging behaviour and quality of entomophagous insects for biological control. *J. Appl. Entomol.* 108:425–51

131. Noldus LP, Lewis WJ, Tumlinson JH, van Lenteren JC. 1988. Olfactometer and wind tunnel experiments on the role of sex pheromones of noctuid moths in the foraging behaviour of *Trichogramma* spp. See Ref. 199, pp. 223–38

132. Nordlund DA. 1994. Habitat location by *Trichogramma*. See Ref. 205, pp. 155–64

133. Nordlund DA, Chalfant RB, Lewis WJ. 1984. Arthropod populations, yields and damage in monocultures and polycultures of corn, beans, and tomatoes. *Agric. Ecosyst. Environ.* 12:127–33

134. Oatman ER, Platner GR. 1978. Effect of mass releases of *Trichogramma pretiosum* against lepidopterous pests on processing tomatoes in southern California, with notes on host egg population trends. *J. Econ. Entomol.* 71:869–900

135. Pak GA. 1986. Behavioural variations among strains of *Trichogramma* spp.: A review of the literature on host-age selection. *J. Appl. Entomol.* 101:55–64

136. Pak GA, Berkhout H, Klapwijk J. 1990. Do *Trichogramma* look for hosts? See Ref. 206, pp. 77–81

137. Pak GA, Buis HCEM, Heck ICC, Hermans MLG. 1986. Behavioural variations among strains of *Trichogramma* spp.: Host-age selection. *Entomol. Exp. Appl.* 40:247–58

138. Pak GA, Kaskens JWM, de Jong EJ. 1990. Behavioural variation among

strains of *Trichogramma* spp.: Host-species selection. *Entomol. Exp. Appl.* 56:91–102

139. Pak GA, Oatman ER. 1982. Biology of *Trichogramma brevicapillum*. *Entomol. Exp. Appl.* 32:61–67

140. Pak GA, Oatman ER. 1982. Comparative life table, behavior and competition studies of *Trichogramma brevicapillum* and *T. pretiosum*. *Entomol. Exp. Appl.* 32:68–79

141. Pak GA, van Dalen A, Kaashoek N, Dijkman H. 1990. Host egg chorion structure influencing host suitability for the egg parasitoid *Trichogramma* West. *J. Insect Physiol.* 36:869–75

142. Pak GA, van Heiningen TG, van Alebeek FAN, Hassan SA, van Lenteren JC. 1989. Experimental inundative release of different strains of the egg parasite *Trichogramma* in brussel sprouts. *Neth. J. Plant Pathol.* 95:129–42

143. Pang XF. 1988. Evaluation of the effectiveness of *Trichogramma* and other natural enemies. See Ref. 199, pp. 443–52

144. Pavlik J. 1992. The effect of temperature on parasitization activity in *Trichogramma* spp. (Hym., Trichogrammatidae). *Zool. Jahrb. Physiol.* 96:417–25

145. Pavlik J. 1993. Variablity in the host acceptance of European corn borer, *Ostrinia nubilalis* Hbn. (Lep., Pyralidae) in strains of the egg parasitoid *Trichogramma* spp. (Hym., Trichogrammatidae). *J. Appl. Entomol.* 115:77–84

146. Pavlik J. 1993. The size of the female and quality assessment of mass-reared *Trichogramma* spp. *Entomol. Exp. Appl.* 66:171–77

147. Pinto JD, Stouthammer R. 1994. Systematics of the trichogrammatidae with emphasis on *Trichogramma*. See Ref. 205, pp. 1–37.

148. Pintureau B. 1991. Indices d'isolement reproductif entre espèces proches de Trichogrammes (Hym.: Trichogrammatidae). *Ann. Soc. Entomol. France* 27:379–92.

149. Pintureau B. 1993. Enzymatic analysis of the genus *Trichogramma* (Hym.: Trichogrammatidae) in Europe. *Entomophaga* 38:411–31

150. Pintureau B, Babault M, Voegelé J. 1981. Etude de quelques facteurs de variation de la fécondité chez *Trichogramma maidis* Pint. et Voeg. (Hym. Trichogrammatidae). *Agronomie* 1:315–22

151. Pompanon F, Fouillet P, Allemand R, Boulétreau M. 1993. Organisation temporelle de l'activité locomotrice chez

les Trichogrammes (Hym. Trichogrammatidae): variabilité et relation avec l'efficacité du parasitisme. *Bull. Soc. Zool. Fr.* 118:141–48

152. Prokrym DR, Andow DA. 1990. Field evaluation of *Trichogramma nubilale* against *Ostrinia nubilalis* in sweet corn. See Ref. 206, pp. 231–35

153. Prokrym DR, Andow DA, Ciborowski JA, Sreenivasam DD. 1992. Suppression of *Ostrinia nubilalis* by *Trichogramma nubilale* in sweet corn. *Entomol. Exp. Appl.* 64:73–85

154. Qian Y-Q, Cao R-L, Li G-Z. 1984. Biology of *Trichogramma ostriniae* and evaluation of its effectiveness in controlling corn borer on spring corn. *Acta Entomol. Sinica* 27:287–93

155. Ravensberg WJ, Berger HK. 1988. Biological control of the European corn borer (*Ostrinia nubilalis* Hbn., Pyralidae) with *Trichogramma maidis* Pint. & Voeg. in Austria in 1980–1985. See Ref. 199, pp. 557–64

156. Rawat US, Pawar AD. 1991. Field recovery of *Trichogramma chilonis* Ishii (Hym.: Trichogrammatidae) from *Deudorix epijarbas* Moore (Lep.: Lycaenidae) in Himachal Pradesh, India. *Entomon* 16:49–52

157. Reznik SY, Umarova TY. 1991. Host population density influence on host acceptance in *Trichogramma*. *Entomol. Exp. Appl.* 58:49–54

158. Ridgway RL, Vinson SB, eds. 1977. *Biological Control by Augmentation of Natural Enemies*. New York: Plenum. 480 pp.

159. Rouechdi KA, Voegelé J. 1981. Prédation des trichogrammes par les chrysopides. *Agronomie* 1:187–90

160. Russo J, Voegelé J. 1982. Influence de la température sur quatre espèces de trichogrammes (Hym. Trichogrammatidae) parasites de la pyrale du maïs, *Ostrinia nubilalis* Hübn. (Lep. Pyralidae). I. Développement préimaginal. *Agronomie* 2:509–16

161. Russo J, Voegelé J. 1982. Influence de la température sur quatre espèces de trichogrammes (Hym. Trichogrammatidae) parasites de la pyrale du maïs, *Ostrinia nubilalis* Hübn. (Lep. Pyralidae). II. Reproduction et survie. *Agronomie* 2:517–24.

162. Salama HS, Zaki FN. 1985. Biological effects of *Bacillus thuringiensis* on the egg parasitoid *Trichogramma evanescens*. *Insect Sci. Appl.* 6:145–48

163. Sappal, N, Jeng RS, Hubbes M, Liu F-H. 1995. Restriction fragment length polymorphisms in PCR-amplified ribosomal DNA's of three *Trichogramma* species (Hym.: Trichogrammatidae). *Genome* 38:419–25

164. Schmidt JM. 1994. Host recognition and acceptance by *Trichogramma*. See Ref. 205, pp. 165–201

165. Schmidt JM, Pak GA. 1991. The effect of temperature on progeny allocation and short interval timing in a parasitoid wasp. *Physiol. Entomol.* 16:345–53

166. Schmidt JM, Smith JJB. 1985. The mechanism by which the parasitoid wasp *Trichogramma minutum* responds to host clusters. *Entomol. Exp. Appl.* 39:287–94

167. Schmidt JM, Smith JJB. 1987. Measurement of host curvature by the parasitoid wasp *Trichogramma minutum*, and its effect on host examination and progeny allocation. *J. Exp. Biol.* 129:151–64

168. Sekirov IA, Muntyan EM, Yazlovetsky IG. 1991. The use of enzyme tests to control the quality of entomophages under mass rearing with special reference to *Chrysopa carnea* Steph. See Ref. 16, pp. 152–60

169. Seymour J, Foster J, Brough E. 1994. Workshop report: Use of *Trichogramma* as a biocontrol agent in Australia. *Coop. Res. Ctr. Trop. Pest Mgmt.* Brisbane, Australia 54 pp.

170. Sithananthan S, Solayappan AR, Sankaran T. 1982. The role of trichogrammatids in biological control of sugarcane borers in India. See Ref. 74, pp. 239–51

171. Smith SM. 1988. Pattern of attack on spruce budworm egg masses by *Trichogramma minutum* (Hym.: Trichogrammatidae) released in forest stands. *Environ. Entomol.* 17:1009–15

172. Smith SM. 1994. Methods and timing of releases of *Trichogramma* to control lepidopterous pests. See Ref. 205, pp. 113–44

173. Smith SM, Carrow JR, Laing JE, eds. 1990. Inundative Release of the Egg Parasitoid, *Trichogramma minutum* (Hym.: Trichogrammatidae) against Forest Insect Pests such as the Spruce Budworm, *Choristoneura fumiferana* (Lep.: Tortricidae): The Ontario Project 1982–1986. *Mem. Entomol. Soc. Can.* 153:1–87

174. Smith SM, Hubbes M. 1986. Isoenzyme patterns and biology of *Trichogramma minutum* as influenced by rearing temperature and host. *Entomol. Exp. Appl.* 42:249–58

175. Smith SM, Hubbes M. 1986. Strains of the egg parasitoid *Trichogramma minutum* Riley: I. Biochemical and biological characterization. *J. Appl. Entomol.* 101:223–39

176. Smith SM, Hubbes M, Carrow JR. 1986. Factors affecting inundative releases of *Trichogramma minutum* against the spruce budworm. *J. Appl. Entomol.* 101: 29–39

177. Smith SM, You M. 1990. A life system simulation model for improving inundative releases of the egg parasite *Trichogramma minutum* against the spruce budworm. *Ecol. Modelling* 51: 123–42

178. Smits PH. 1982. The influence of kairomones of *Mamestra brassicae* L. on the searching behaviour of *Trichogramma evanescens* Westw. See Ref. 74, pp. 139–51

179. Stouthamer R, Luck RF. 1993. Influence of microbe-associated parthenogenesis on the fecundity of *Trichogramma deion* and *T. pretiosum. Entomol. Exp. Appl.* 67:183–92

180. Stinner RE. 1977. Efficacy of inundative releases. *Annu. Rev. Entomol.* 22:515–31

181. Sun X, Yu E. 1988. Use of *Trichogramma dendrolimi* in forest pest control in China. See Ref. 199, pp. 591–96

182. Tavares J, Voegelé J. 1990. Interspecific competition between three species of the Genus *Trichogramma* (Hym., Trichogrammatidae). See Ref. 206, pp.45–49

183. Thorpe KW. 1985. Effects of height and habitat type on egg parasitism by *Trichogramma minutum* and *T. pretiosum* (Hym.: Trichogrammatidae). *Agric. Ecosyst. Environ.* 12:117–26

184. Tipping PW, Burbutis PP. 1983. Some effects of pesticide residues on *Trichogramma nubilale* (Hym.: Trichogrammatidae). *J. Econ. Entomol.* 76: 892–96

185. Tran LC, Hassan SA. 1986. Preliminary results on the utilization of *Trichogramma evanescens* Westw. to control the Asian corn borer *Ostrinia furnacalis* Guenee in the Philippines. *J. Appl. Entomol.* 101:18–23

186. Treacy MF, Benedict JH, Segers JC, Morrison RK, Lopez JD. 1986. Role of cotton trichome density in bollworm (Lep.: Noctuidae) egg parasitism. *Environ. Entomol.* 15:365–68

187. Twine PH, Lloyd RJ. 1982. Observations on the effect of regular releases of *Trichogramma* spp. in controlling *Heliothis* spp. and other insects in cotton. *Queensl. J. Agric. Anim. Sci.* 39: 159–67

188. van Bergeijk KE, Bigler F, Kaashoek NK. 1989. Changes in host-acceptance and host-suitability as an effect of rearing *Trichogramma maidis* on a factitious host. *Entomol. Exp. Appl.* 52:229–38

189. van Dijken MJ, Kole M, van Lenteren JC, Brand AM. 1986. Host-preference studies with *Trichogramma evanescens* Westwood (Hym., Trichogrammatidae) for *Mamestra brassicae, Pieris brassicae* and *Pieris rapae. J. Appl. Entomol.* 101:64–85

190. van Driesche RG. 1983. Meaning of "percent parasitism" in studies of insect parasitoids. *Environ. Entomol.* 12:1611–22

191. van Hamburg H, Hassell MP. 1984. Density dependence and the augmentative release of egg parasitoids against graminaceous stalk borers. *Ecol. Entomol.* 9:101–8

192. van Lenteren JC. 1991. Quality control of natural enemies: Hope or illusion? See Ref. 16, pp. 1–15

193. van Lenteren JC, Glas RCG, Smits PH. 1982. Evaluation of control capabilities of *Trichogramma* and results of laboratory and field research on *Trichogramma* in the Netherlands. See Ref. 74, pp. 257–69

194. Vanlerberghe-Masutti F. 1995. Molecular identification and phylogeny of parasitic wasp species (Hym.: Trichogrammatidae) by mitochondrial DNA RFLP and RAPD markers. *Insect Mol. Biol.* 3:229–37

195. van Schelt J, Ravensberg WJ. 1990. Some aspects on the storage and application of *Trichogramma maidis* in corn. See Ref. 206, pp. 239–42

196. Vieira V, Tavares J. 1995. Rearing of *Trichogramma cordubensis* Var. & Cab. (Hym.: Trichogrammatidae) on Mediterranean flour moth cold-stored eggs. See Ref. 204a

197. Voegelé J. 1988. Reflections upon the last ten years of research concerning *Trichogramma* (Hym., Trichogrammatidae). See Ref. 199, pp. 17–33

198. Voegelé J, Pizzol J, Raynaud B, Hawlitzky N. 1986. La diapause chez les Trichogrammes et ses avantages pour la production de masse de la lutte biologique. *Meded. Fac. Landbouwwet. Rijksuniv. Gent.* 51:1033–39

199. Voegelé J, Waage J, van Lenteren JC, eds. 1988. *Trichogramma* and Other Egg Parasites. *Int. Symp. on* Trichogramma, 2nd, Guangzhou, PR China. Paris: Les Colloques de l'INRA No. 43. 644 pp.

200. von Glas M, Hassan SA. 1985. Massenproduktion und Anwendung von *Trichogramma*. 5. Bekämpfung von zwei Wicklerarten an Getreide, *Cnephasia longana* (Haw.) und *C. pumicana*

(Z.) (Lep., Tortricidae). *J. Appl. Entomol.* 99:393–99

201. Voronin KE. 1982. Biocenotic aspects of *Trichogramma* utilization in integrated plant protection control. See Ref. 74, pp. 269–75

202. Waage J. 1988. Understanding progeny and sex allocation in egg parasitoids. See Ref. 199, pp. 283–95

203. Wajnberg E. 1993. Genetic variation in sex allocation in a parasitic wasp. Variation in sex pattern within sequences of oviposition. *Entomol. Exp. Appl.* 69: 221–29

204. Wajnberg E. 1994. Intra-population genetic variation in *Trichogramma*. See Ref. 205, pp. 245–73.

204a. Wajnberg E, ed. 1995. *Trichogramma and Other Egg Parasitoids. Int. Symp., 4th, Cairo, Egypt.* Paris: Les Colloques de l'INRA No. 73. 226 pp.

205. Wajnberg E, Hassan SA, eds. 1994. *Biological Control with Egg Parasitoids.* Oxon, UK: CAB International. 286 pp.

205a. Wajnberg E, Pizzol J, Babault M. 1989. Genetic variation in progeny allocation in *Trichogramma maidis. Entomol. Exp. Appl.* 53:177–87

206. Wajnberg E, Vinson SB, eds. 1990. Trichogramma *and Other Egg Parasitoids. Proc. Int. Symp., 3rd, San Antonio, Texas.* No. 56. Paris: INRA. 246 pp.

207. Wang C-L. 1988. Biological control of *Ostrinia furnacalis* with *Trichogramma* sp. in China. See Ref. 199, pp. 609–13

208. Wang F, Zhang S, Hou S. 1988. Inoculative release of *Trichogramma dendrolimi* in vegetable gardens to regulate populations of cotton pests. See Ref. 199, pp. 613–21

209. Wang Z, Smith SM. 1994. Biological traits of thelytokous *Trichogramma minutum* Riley complex from *Zeiraphera canadensis* and their ecological implications. *Entomol. Exp. Appl.* In press

210. Wellinga S, Wysoki M. 1989. Preliminary investigation of food source preferences of the parasitoid *Trichogramma platneri* Nag. (Hym.: Trichogrammati-

dae). *Anz. Schaedlingskd. Pflanz. Umweltschutz* 62:133–35

211. Witz JA, Hartstack AW, King EG, Dickerson WA, Phillips JR. 1985. Monitoring and prediction of *Heliothis* spp. *Southwest. Entomol.* 8:56–71 (Suppl.)

212. Wu J-W, Fan H-L, Yang M-D, Lian Y-Y, Zu H-Y. 1988. Study of the operational techniques in using *Trichogramma dendrolimi* for the prevention and control of *Dendrolimus punctatus.* See Ref. 199, pp. 621–29

213. Xie Z-N, Nettles WC, Morrison RK, Irie K, Vinson SB. 1986. Effect of ovipositional stimulants and diets on the growth and development of *Trichogramma pretiosum* in vitro. *Entomol. Exp. Appl.* 42:119–24

214. Yu DS, Byers JR. 1994. Inundative release of *Trichogramma brassicae* Bezdenko (Hym.: Trichogrammatidae) for control of European corn borer in sweet corn. *Can. Entomol.* 126:291–301

215. Yu DSK, Hagley EAC, Laing JE. 1984. Biology of *Trichogramma minutum* Riley collected from apples in southern Ontario. *Environ. Entomol.* 13:1324–29

216. Yu DSK, Laing JE, Hagley EAC. 1984. Dispersal of *Trichogramma* spp. (Hym.: Trichogrammatidae) in an apple orchard after inundative releases. *Environ. Entomol.* 13:371–74

217. Zaslavskiy VA, Kvi MF. 1982. An experimental study of some factors affecting fecundity in *Trichogramma* Westw. (Hym., Trichogrammatidae). *Entomol. Rev.* 61:10–24

218. Zaslawskiy VA, Umarova TY. 1990. Environmental and endogenous control of diapause in *Trichogramma* species. *Entomophaga* 35:23–29

219. Zhou S-Z. 1988. Advance in extention of *Trichogramma* utilization in Guangdong Province of China. See Ref. 199, pp. 633–40

220. Zhou LT. 1988. Study on parasitizing efficiency of *Trichogramma confusum* Vig. in controlling *Heliothis armigera* Hub. and its modelling. See Ref. 199, pp. 641–44

Annu. Rev. Entomol. 1996. 41:407–31

THE ROLE OF NOURISHMENT IN OOGENESIS

Diana Wheeler

Department of Entomology, University of Arizona, Tucson, Arizona 85721

KEY WORDS: autogeny, parasitoids, flight muscle, migration, social insects, mating

ABSTRACT

Oogenesis in insects is typically a nutrient-limited process, triggered only if sufficient nourishment is available. This nourishment can be acquired during the larval or adult stage, depending on the insect. Timing of food intake will have major effects on mechanisms of hormonal control. When nourishment for eggs is taken primarily by adults, insufficient nutrition inhibits egg development through mechanisms such as inhibition of corpora allata, as seen in Orthoptera and Blattaria. In adult Diptera, lack of protein inhibits release of brain factors that produce reproductive competency or ovarian stimulation. Lepidoptera have been characterized as lacking substantial regulation of oogenesis because egg development is well under way at emergence. Many species for which ecological data are available do not mobilize reserves carried over from the larval stage until they feed as adults. The endocrine mechanisms underlying these systems are poorly understood. In many insects, mating and activity can affect nutritional state and therefore oogenesis. Mating can stimulate oogenesis through mobilization of reserves or through nutritional contributions by males to females. Activity, especially flight, and oogenesis can compete for energy. The flight apparatus, especially the muscle, can also compete with oogenesis for protein. Social insects exhibit extreme specializations in oogenesis; females range in fertility from completely sterile to hyperfecund. Food flow within colonies is a major factor regulating fecundity. Finally, maternal nourishment is not needed for oogenesis in parasitoids and pseudoplacental viviparous insects, which produce eggs with little or no yolk. Virtually nothing is known about the endocrine regulation of oogenesis in these insects.

PERSPECTIVES AND OVERVIEW

Most insect eggs contain large supplies of nutrients to support embryogenesis. Thus, gametogenesis in females fundamentally differs from that in most insect males, whose direct energetic investment per gamete is quite small. Because

0066-4170/96/0101-0407$08.00

oogenesis is commonly a nutrient-limited processes, any factor that affects the acquisition of nutrients will influence egg production. Food quantity and quality will play important roles, as will the pattern of food availability over the life stages. Together, these factors will either stimulate or inhibit oogenesis. Wigglesworth (210) summarized the raison d'être of hormonal regulation (p. 77):

> ...egg production, like moulting is a cyclical process. It is desirable that it should be initiated only at those times when (i) nutrition, (ii) the phase of the reproductive cycle..., and (iii) the season of the year, are all appropriate. The insect must therefore have some mechanism for restraining ovary development. A hormonal control system actuated by suitable environmental stimuli and feedback mechanisms is a natural solution.

Clearly, nutrition is the main environmental factor limiting reproduction. Season, although represented by light, photoperiod, and humidity, is related to nutrition in that insects time their activity to coincide with availability of food. The phase of the reproductive cycle is simply feedback from internal conditions.

The specific impact of nourishment on egg production has been reviewed previously (57, 97, 183). Mating often stimulates egg production (70), sometimes through male nutritional contributions (17). Nonreproductive processes can inhibit oogenesis through competition for energy or raw materials. For example, large-scale movements, such as migratory flight, can deplete materials that would otherwise be used for egg production. Various aspects of the interactions between reproduction and migration have been reviewed (50, 152, 158).

The social insects, through their caste systems, exhibit extreme reproductive specializations seen nowhere else in the insects. Female fertility ranges from complete sterility to hyperfecundity. The developmental processes determining caste and the physiological processes determining reproductive performance are dominated by regulation of nutrition (209). Pheromones can serve as behavioral and physiological signals that govern patterns of food distribution in colonies (63).

In this review, I adopt the perspective that nutrition is the primary requirement for egg development and do not cover the effects of season (18) or of resorption after oogenesis (15). I also consider primarily the effects of nourishment in the adult stage, even though larval nutrition can certainly have an impact on adult fecundity (209).

NOURISHMENT

Eggs of oviparous organisms generally contain all the nutritional resources required to support embryonic development. Therefore, acquiring nourishment has a central role in the reproductive biology of females in most insect species, and lack of nourishment inhibits egg development. Nourishment can be gathered during the larval stage, during the adult stage, or both. In holometabolous insects, the dissimilarity between the adult and larval environments can set up a great disparity in the nutritional resources available to the two life stages. In many hemimetabolous insects, however, larvae and adults generally share the same habitat and therefore similar availability of nourishment in both stages. Starvation in adult females tends to result in the immediate suppression of egg production. Suppression is mediated through modulation of juvenile hormone (JH) titers; JH induces yolk synthesis by fat body and uptake by oocytes (57).

Orthoptera and Blattaria

The regulation of oogenesis, including the effects of nourishment, has been studied extensively in the hemimetabolous orders Orthoptera and Blattaria. Control of JH biosynthesis by the brain is a major control point for JH-mediated processes in adult insects (103). Many studies have focused on the effect of nutrition on migratory locusts because of the importance of nutrition in phase development and migratory behavior. In the desert locust *Schistocerca gregaria*, starvation causes an accumulation of stainable neurosecretory material in the corpora cardiaca (CC) that is associated with a low and constant volume in the corpora allata (CA). The accumulated neurosecretory material is rapidly depleted when feeding resumes, and the volume of the CA increases about threefold (83). Additional evidence that the effect of starvation is modulated through the CA is provided by JH biosynthesis data. After 1–3 days of starvation, JH synthesis by CA in vitro declines in both *S. gregaria* and in the migratory locust, *Locusta migratoria* (42, 197).

Starvation also suppresses oocyte maturation through reduced JH biosynthesis in the cockroaches *Periplaneta americana* (205, 207), *Leucophaea maderae* (1), *Blattella germanica* (176), and *Diploptera punctata* (218). CA inhibition appears to result from both neural and humoral factors. Nerve transection partly alleviates starvation-induced inhibition of JH synthesis in *P. americana* (205) and *D. punctata* (218). However, in *B. germanica*, starvation-induced suppression of the CA is reversed by nerve transection less fully in protein-deprived females than in well-fed females, indicating that humoral factors correlated with nutritional state may also affect JH biosynthesis (176).

Lepidoptera

In contrast to members of the Orthoptera and Blattaria, holometabolous larvae and adults frequently experience major differences in the type and amount of nourishment available. In general, Lepidoptera use reserves accumulated during the larval stage to supply egg production. In ecological terms, this pattern can be seen as greater extraction of nitrogen and other requisite nutrients from the larval, rather than the adult, environment. Conversion of larval reserves to egg yolk can occur before, during, or after adult emergence, depending on the taxonomic group.

Lepidopteran females fall into three categories with respect to egg development at emergence. In the first, reserves are depleted and all eggs are mature before adult eclosion. The very short adult life of such insects consists of mating and ovipositing to depletion. In some groups using this strategy, the adults have only rudimentary mouthparts (Saturninae, Bombycinae). In the second category, egg development is under way at the time of adult eclosion but not yet complete. Many moths fall in this category. And in the third category, which includes most butterflies, females emerge with undeveloped ovaries and a stock of stored reserves (54).

Most Lepidopteran adults take some nourishment, especially nectar, during adult life (16). Nectar is regarded primarily as a flight fuel, but it does contain a small percentage of amino acids (9) that could contribute directly to yolk production. Adults additionally feed on such materials as decaying organic matter, blood, eye exudates, plant saps, honeydew, and bird droppings (5, 69 and references therein). In males, much of this activity appears to satisfy a need for sodium (7). Some of the resources accumulated by males may be passed to the female during mating (19) (see section on mating).

Even though substantial protein is not acquired after adult eclosion (but see discussion of Heliconiids and mating, below), adult carbohydrate ingestion can profoundly affect egg development. For example, in the speckled wood butterfly, *Pararge aegeria* (Satyridae), egg number and weight are approximately four times higher after sugar feeding than after water feeding (106). In the checkerspot butterfly, *Euphydryas editha*, sugar feeding almost doubles egg production relative to that with no feeding or water feeding. In this species, the addition of amino acids to sugar does not improve lifetime egg production but does prevent the normal decline in egg size seen in later broods (139). For the female Mormon fritillary, *Speyeria mormonia* (Nymphalidae), egg production was halved by the intake of a 25% honey water solution to half the maximal intake, and egg production was reduced to 27% of normal when the intake was cut to 1/3 the ad lib level (20). In general, whether the carbohydrate is used as a source of energy to drive the use of stored reserves or is transferred to eggs as part of the yolk is not clear.

Heliconiid butterflies represent a major exception to the rule that protein is not acquired during the adult stage. Adults of both sexes feed on pollen, from which they extract amino acids. Studies have shown that labeled free amino acids imbibed by females are incorporated into eggs. This novel feeding habit is correlated with longer-lived adults (68). The ability to gather resources as adults must liberate the females from some of the constraints of managing a limited supply of larval reserves.

In the Lepidoptera, the degree to which ovaries have developed at adult emergence reflects differences in the hormonal regulation of oogenesis. Species in which oogenesis is part of metamorphosis and is therefore well under way at emergence should have regulation systems different from those of species in which development is initiated only after feeding. Much of what we know about regulation of oogenesis in Lepidoptera has been learned from species whose ovaries are already developed at emergence, such as the tobacco hornworm, *Manduca sexta*, and the silkworms, *Hyalophora cecropia* and *Bombyx mori*. We know little, however, about the influence of adult nutrition on oogenesis in species whose ovaries do not begin to develop until the adult stage. Some important pest species fall into this category. Noctuids, for example, often migrate as young adults in their prereproductive period (215). Oogenesis is initiated following migration, but the relationship between this initiation and adult feeding is poorly understood.

In Lepidoptera that do not feed as adults, neither the CA nor JH appears to have an effect on egg development [e.g. in *H. cecropia* (146) and *B. mori* (66)] (see 57, p. 150). Hormonal control of vitellogenin synthesis has been studied in the gypsy moth, *Lymantria dispar*. As JH titers decline during the last larval stage, vitellogenin synthesis begins and titers reach peak levels. Patency and uptake, however, do not begin until the pupal stage (120). Supplementary JH analogue suppresses vitellogenin mRNA levels and delays accumulation of the yolk protein (44, 60, 85).

In *M. sexta*, eggs are already in the process of taking up vitellogenin at the time of adult eclosion. JH is not required for synthesis or uptake of this protein but is necessary for the final steps of egg maturation. The CA are activated at eclosion. During the first 24 h after emergence, the oldest follicles complete maturation, including chorion formation. If females are neither mated nor fed, release of JH continues for approximately 2 days and then ceases (141). Both feeding and mating increase CA activity (173).

In contrast, JH affects vitellogenin synthesis and/or uptake in Lepidoptera species whose egg development takes place primarily after emergence. These species include the cabbage butterfly, *Pieris brassicae* (104, 105); the mourning cloak butterfly, *Nymphalis antiopa* (80); the angle-wing, *Polygonia c-aureum* (56); and the monarch butterfly, *Danaus plexippus* (12, 147). Furthermore, in *D. plexippus*, JH treatment does not reverse inhibition of vitellogenesis induced by neck ligation at emergence, which suggests vitellogenesis requires an additional head factor unrelated to JH synthesis and release (12). In a similar experiment on *N. antiopa*, however, JH did reverse the effects of neck ligature on ovarian development (80).

Regrettably, species for which the effects of nectar feeding on fecundity are known (see above) are not the species for which the endocrine control of egg development is best understood. A subject for future study would be to determine whether the brains of species that require nectar feeding for egg development have inhibitory control of the CA or other targets.

Diptera

In contrast to Lepidoptera, egg production in most Diptera completely depends on protein feeding by adult females. The interplay of JH and ecdysteroids as mediators of feeding and nutrition are best understood in mosquitoes (77, 78). Both hormones appear to be involved in these processes throughout the Diptera.

Females of most mosquito species must take a blood meal for egg development to take place (anautogeny). However, several populations, strains, and species have evolved the ability to produce eggs from larval reserves, either exclusively (obligate autogeny) or prior to a blood meal (facultative autogeny) (41, 124, 143, 187). The considerable information available on mosquito oogenesis and ecology provides an unusually complete picture of the ecological endocrinology of egg development.

In mosquitoes, nutrition modulates oogenesis at two points. First, in anautogenous mosquitoes such as *Aedes aegypti,* inadequate larval nutrition apparently prevents an early peak of JH that, in well-fed females, prepares the ovary and fat body for later stimulation by 20-hydroxyecdysone (HE) (55, 74, 78). In females that emerge undernourished, at least one meal, which can be sugar or protein, is necessary to bring fat body and ovaries to competency (58).

The second point of modulation involves another assessment of nutritional state; if internal resource levels are above threshold, the neurosecretory factor that stimulates ecdysteroid synthesis by the ovary is released from the brain. This mechanism appears to initiate vitellogenic egg development in both anautogenous and autogenous mosquitoes (74, 77, 78, 110, 111, 123). The difference is that the threshold requirement of nutrition is met without a blood meal in autogenous mosquitoes. The degree of autogeny is probably a combination of differences in the nutritional state at emergence owing to larval history and differences in the absolute level of resources required at the threshold to release neurosecretory factors. Evidence that the availability of raw materials, especially amino acids, can be monitored comes from studies of the mosquito *Culex pipiens pallens.* Infusing the hemocoel with amino acids can cause this anautogenous species to become artificially autogenous (199, 200). These results recall Engelmann's hypothesis that the "internal nutritional milieu" could affect the CA directly, bypassing the central nervous system (57).

Higher Diptera generally require a protein meal to develop eggs [e.g. *Phormia regina* (145), *Lucilia* spp. (126), *Calliphora erythrocephala* (190), *Calliphora vomitoria* (51)]. However, some laboratory strains (e.g. of the house fly, *Musca domestica*) do exhibit autogeny (6), and the trait may actually be moderately common (47). Autogeny in higher flies has not been well characterized endocrinologically.

As in mosquitoes, in higher Diptera, JH and ecdysteroids play roles in regulating egg production. However, the interplay between neuroendocrines, JH, and ecdysteroids is not as well understood. We do know that the link between protein feeding and oogenesis is mediated by the effects of neurosecretory factors on JH and ecdysteroid levels. In the stable fly, *Stomoxys calcitrans,* JH may be important in mediating the response to protein. A protein meal is necessary to induce vitellogenic follicles; ovaries of sugar-fed flies contain no more than a trace of yolk. Allatectomy prevents vitellogenic growth; allatectomized flies receiving a protein meal can be rescued by implanted CA (134). Ecdysteroids are also important in oogenesis (36), but how the action of these hormones is integrated with the response described for JH is not known.

In house flies, lack of protein prevents egg development past the previtellogenic stages (2), as in anautogenous mosquitoes. A protein meal given 4 h after eclosion stimulates a rise in ecdysteroid and vitellogenin levels. This response is prevented by removal of the CA-CC complex (3, 4).

In *P. regina,* the protein meal induces secretion of a head factor and then a rise in ecdysteroid titer (219–221). Neck ligation shortly after the meal results in low to undetectable titers of ecdysteroid; this effect is not reversed by treatment with the JH mimic methoprene (219). JH plays a role in vitellogenin uptake into oocytes but not in stimulating vitellogenin synthesis (189).

In *Drosophila melanogaster,* starvation of adult females prevents normal growth of vitellogenic-stage ovaries. This response has been characterized at the molecular level to a greater degree in *D. melanogaster* than in other insects. Females starved from eclosion have a low, basal rate of transcription of yolk-peptide genes. Their transcription of yolk peptides is selectively reduced, compared with that of flies fed normally. After feeding, the rate of transcription increases quickly. Suppression of transcription caused by starvation can be partially reversed by an exogenous JH analogue or ecdysteroid (30). Methoprene application induces a transient ecdysteroid pulse that is followed by near normal growth of vitellogenic oocytes (179). In addition to the suppression of yolk-peptide synthesis, starvation appears to directly affect uptake by the ovary (30). This effect occurs first and is followed by a decrease in the levels of transcription from yolk-protein (YP) genes (29). Interestingly, the flanking region of the *YP-1* gene that responds to changes in diet does not interact directly with either JH or ecdysteroids (27, 30).

Yolkless Eggs

In some insect species, females do not provide sufficient nourishment within the egg to support embryonic development. For example, some parasitic insects produce microtype, hydropic eggs that are small (20–200 μm) and relatively free of yolk. As a result, embryos must develop within hosts to acquire nourishment for development (61, 62). The reduced number of chorionic layers in hydropic eggs is believed to facilitate nutrient absorption from the host (113). Eggs of the braconid wasp, *Microplitis croceipes,* lack vitellogenin (196), and protein synthesis during embryogenesis is supported by free amino acids in host hemolymph (59).

Oogenesis of microtype eggs requires little in terms of raw materials from the mother, so the cost of each egg to her is relatively small. Some parasites have numerous ovarioles; hyperparasites in the trigonalid genus *Poecilogonalus* have 450 ovarioles per ovary (94) and can produce more than 10,000 eggs in 14 days (40). Similarly, tachinid flies with microtype eggs can have over 400 ovarioles per ovary and lay thousands of eggs on the food of the host, but closely related species that lay yolky eggs directly on the host produce only hundreds of eggs (84). The hormonal coordination of host-related stimuli and oogenesis by species with microtype eggs requires exploration, as does the role JH may play in species that neither synthesize nor endocytose vitellogenin.

The production of yolkless eggs has also been reported in some viviparous insects, but nutrients for embryonic development are obtained from the mother. In Strepsiptera and some Cecidomyidae, chorion-free eggs without yolk are dispersed in the hemocoel, where they absorb nutrients directly from the blood. In other viriparous insects, developing embryos in the genital tract absorb nutrients from pseudoplacental maternal tissue (57, 76, 161).

MATING

As a rule, repeated mating is necessary for full fecundity and fertility in polyandrous female insects (162). For example, in *D. plexippus* (142) and the swallowtail, *Papilio xuthus* (204), multiple matings increase the number of eggs laid. In a bruchid beetle, *Callosobruchus maculatus,* multiple mating increases female survivorship, egg production, and to a lesser extent egg size and larval survivorship (64). Regardless of whether insects mate once or several times, mating is a complex event in which males provide neural input and various substances with physiological effects, including hormone-like compounds and nutrients, as well as sperm (138, 181). Any or all of these can stimulate egg production. Gillott & Friedel (70) reviewed "fecundity enhancing substances" in male insects that, by various means, stimulate oviposition. However, the cited studies are usually not detailed enough to determine whether mating stimulates deposition of previously made eggs or egg production itself. The best-understood case of an oviposition-stimulating substance is the sex peptide identified in several species of *Drosophila.* This peptide, produced by the male accessory glands and transferred during copulation, stimulates ovulation, specifically, the movement of eggs from the ovary into the uterus (37, 119, 177).

Physiological data on male stimulation of and contributions to egg development and oviposition are consistent with much of the literature on the evolution of mating systems in general and on paternal investments in particular (222). In stimulation of egg-production without a nutritional contribution, the male benefits because the number of eggs that his sperm will fertilize is increased. This effect is common, and males usually do not make a costly investment in nutrients unless additional benefits will accrue, such as increased acceptance by females or better survival of his progeny (195, 222).

Mating can stimulate egg development in both aedine and anopheline mosquitoes. In unfed females of autogenous strains of the saltmarsh mosquito, *Aedes taeniorhynchus,* mating increases the proportion of females maturing eggs (144). Injection of male accessory-gland fluid into females stimulates vitellogenin synthesis. This response requires the presence of the head (24). In *A. aegypti,* the percentage of mated females maturing eggs after only a small blood meal is higher than that in unmated groups. The number of eggs matured per microliter of blood did not differ between the two groups, however, which suggests the same amount of internal resources is used to produce the clutch. Injection of male accessory-gland homogenates produced a similar effect (115, 116). In *Aedes albopictus,* the amount of protein reserves is the same in mated and unmated groups of females (114, 117). As the amount of larval reserves and number of eggs produced are similar regardless of mating, Klowden (114) hypothesized that the male contributes a substance that stimulates the initiation of oogenesis, perhaps by mobilizing existing reserves for egg development.

In some anopheline species, mating appears to be required for egg development. Males of 10 species of *Anopheles* (*Nyssorhynchus*) show a 10-fold range in accessory gland size that is negatively correlated with egg development in virgin females. Thus, females of a given species are less likely to mature eggs before mating if males of that species have large accessory glands. The effect of the glands may be purely stimulatory, as in the *Aedes* spp. described above, but the possibility exists that males of species with larger accessory glands contribute more nutrients (125).

In most species of cockroaches, mating stimulates oocyte development. In general, virginity, as well as food, appears to inhibit the CA (176). Egg development is suppressed unless mating occurs in *D. punctata* (188), *Nauphoeta cinerea* (122), *P. americana* (154, 155, 206), *Supella longipalpa* (38, 185), and *Leucophaea maderae* (1). Only in *B. germanica* does mating, as distinct from social contact, not have a positive effect on oocyte development (67).

The endocrine basis for the mating stimulus has been explored in several species of cockroaches and appears to be based on modulation of the CA. Mating increases JH synthesis in *L. maderae* (1). Among virgin *P. americana,* JH synthesis rates drop if the females do not mate during the initial period of receptivity (206). Transection of the nervus corporis cardiaci 1 disinhibits oocyte growth in such virgin females (155), which demonstrates that the effect of mating acts through these nerves. The spermatophore is important in transmitting the mating stimulus; if it is removed immediately after mating, the proportion of ovipositing individuals decreases until it equals that of unmated females (153).

In some insect groups, including Orthoptera, Blattaria, Mecoptera, Diptera, and Lepidoptera, males transfer substantial amounts of nutrients to females, in which they are absorbed or ingested and become available for maintenance or reproduction (222). However, even in these groups, male donations do not always influence female fecundity [e.g. in *D. melanogaster* (34) and tettigoniid flies (208)], and incorporation of such material into the female and her eggs does not necessarily increase fecundity [e.g. in *Plodia interpunctella* (71)].

Numerous studies have documented that substances accumulated in the male reproductive system and transferred to the female during mating often are assimilated into eggs or the female's body. Proteins in the reproductive tracts of males fed labeled amino acids have been traced to female hemolymph in Acrididae (32, 65) and Coleoptera (87); to the ovaries or eggs of Lepidoptera (18, 21, 71), bushcrickets (26, 75), and Diptera (127, 211); and to other parts of the female's body in Lepidoptera (21) and Diptera (127, 156). In cockroaches, labeled urates are incorporated into oothecae and used during embryogenesis (136, 175). In a skipper, *Thymelicus lineola,* approximately 1/3 of the male's abdominal sodium is transferred to the female. Sodium affects male mating frequency, fertility of the eggs laid, and drought resistance of the eggs (157).

Male nutritional contributions often have an impact on female fecundity under nutrient-stress conditions (28, 32, 75, 128). Female *B. germanica* on low-protein diets incorporate a greater proportion of male-derived nutrients in their bodies and their oothecae than do well-fed females (136, 137). Schal & Bell (175) have related the importance of urates in accessory glands in cockroaches to the ecology of nitrogen-poor environments. Tallamy (191) presents a general hypothesis that paternal investment, including male nutritional assistance to the female, is favored by sexual selection when females' nutritional resources are limited. Male-donated nutrients are a paternal investment because a male increases the number of his progeny by increasing female fecundity (148). Boggs has modeled and reviewed the impact of male-donated nutrients on females (17).

FLIGHT AND MIGRATION

Migration and dispersal allow insects to escape adverse conditions and to colonize new areas, but both egg development and flight require substantial amounts of energy as well as an investment of protein, specifically yolk proteins for eggs and muscle proteins for flight muscles. Whether the processes are antagonistic depends on the size of carbohydrate, lipid, and amino acid pools and the demands placed on them. If the sum of the demand for any of the major nutrient categories is greater than its supply, there will be a trade-off. By this argument, the antagonism between migration and reproduction can be ameliorated by increasing the size of the pools and by reducing or shifting demands. A comparison of flight-fecundity trade-offs in the crickets *Gryllus rubens* and *Gryllus firmus* provides physiological evidence for the importance of differential resource allocation and consumption in determining the extent of the antagonism (130, 131). Insects that show no trade-off between flight activity and female reproduction are not unusual (50, 158, 159, 174, 182), but their performance should not be used as evidence against a fundamental antagonism between the two resource-intensive processes.

In some noctuid moths, trade-offs between oogenesis and flight seem to be based on lipid reserves. In the African armyworm moth (*Spodoptera exempta*), strains selected for prolonged flight contain larger reserves of lipid and allocate less of it to reproduction than do unselected strains (72). Moths that fly deplete their lipid reserves and have reduced fecundity, which can be completely reversed if the moths consume 10% sucrose after flight (73). In the soybean looper, *Pseudoplusia includens,* carbohydrate deprivation, coupled with 2-h flights, also reduces egg production (131). And in the tobacco bud worm, *Heliothis virescens,* supplementary feeding with sucrose reduces the impact of flight on the reproductive capacity of females (212).

Butterflies generally accumulate all of their protein reserves during the larval period, but the adult stages frequently require carbohydrates to develop large numbers of eggs. Therefore, the use of carbohydrate as a flight fuel could affect the amount of energy available for egg production (see previous section).

A reduction in oogenesis resulting from the use of carbohydrates for flight competition should be relatively easy to reverse if energy (e.g. nectar) is available in the environment. Amino acids are not obtained so easily and may represent a more serious constraint. Species with both short-winged (flightless) and long-winged forms and species that break down flight muscles after migration or dispersal offer an opportunity to study the competition for amino acids between flight and oogenesis. Wing dimorphism may affect fertility or the schedule of egg production (167, 186). For example, the flight apparatus of long-winged crickets becomes fully functional after emergence, but short-winged individuals do not develop flight muscles. During the first few weeks of adult life, short-winged forms begin laying eggs earlier and lay more eggs (166, 194, 224).

Proteins can be a shared resource for flight muscles and for reproduction. Generally, protein is first allocated to the flight muscles and then to reproduction. Also, in general, the muscles are subsequently histolyzed and return their components to the nutrient pool. Natural dealation and flight-muscle breakdown has been observed in several species of crickets and may be common in this group (203). Artificial removal of wings in three cricket species prevents the development of flight muscles after emergence and induces earlier oocyte development (192, 194). The number of both mature oocytes and eggs oviposited is higher than in intact long-winged females (193). In the house cricket, *Acheta domestica,* cessation of flight activity is followed quickly by wing-muscle degeneration and yolk uptake (39).

Ants are especially interesting. The queens fly only during the mating flight and, immediately after mating, begin to break down flight muscle. In many species, the nutritional constraints on queens starting new colonies are severe, in that they must lay eggs and rear their first set of workers entirely on body reserves (86, 98).

Reversible flight-muscle breakdown occurs in some Coleoptera. For example, the Colorado potato beetle, *Leptinotarsa decemlineata,* histolyzes its flight muscles before diapause and rebuilds them afterward (46, 49). Scolytid bark beetles also have reversible flight-muscle degeneration. Adults in galleries containing brood have greatly reduced flight muscles and, in females, large ovaries extending into the thoracic cavity. The reduction appears to be temporary, because reserves and flight muscle are later built up again (33).

Little direct evidence, however, supports the hypothesis that histolyzed flight-muscle components are used in oogenesis. Certainly flight muscles generate breakdown products that could be used for oogenesis. Eggs of the cotton stainer, *Dysdercus cingulatus,* contain four proteins antigenically identical to ones found in flight-muscle extracts, which indicates that total breakdown to individual amino acids is not always required (140).

The building and maintenance of the flight apparatus reduce the availability of not only amino acids to the reproductive system, but also energy, thereby reducing fecundity. In *G. rubens,* the long-winged form converts food into biomass less efficiently than the short-winged form; this difference appears to result from the energetic costs of maintaining the flight apparatus (132). The long-winged form of *G. firmus* also suffers such costs but compensates for them by eating and assimilating more nutrients (133).

The nervous and endocrine systems integrate nutritional state, flight activity, and reproduction, but specific configurations of regulation are shaped by the ecological demands on each species. Insects in which flight results in diminished fecundity might be expected to coordinate inhibition of one process with stimulation of the other. And in species where flight increases egg production, the same signals might stimulate both processes (152, 158).

JH levels and CA function are correlated with lipid storage in many insects; the absence of the CA may result in excessive accumulation of lipid in fat body (reviewed in 14). Another hormone that may be important in lipid allocation is adipokinetic hormone (AKH), which mobilizes lipids during flight. The action of AKH is best known in the migratory locust, although flight does not appear to reduce fecundity in these insects (82). Even so, the inhibitory effect of AKH on protein synthesis suggests a mechanism by which lipid mobilization and protein synthesis could be regulated antagonistically. In this insect, AKH-I increases at the end of the first ovarian cycle in vivo and almost completely inhibits protein synthesis by fat body in vitro. The dose of exogenous hormone that inhibits protein synthesis is about ten times less than that which stimulates lipid mobilization, which suggests that AKH-mediated lipid mobilization for flight could shut down the synthesis of vitellogenin, as well as of other proteins (135).

JH can be involved in coordination of oogenesis with flight activity through the maintenance or histolysis of flight muscle (152, 158). Typically, both oogenesis and flight-muscle breakdown are stimulated by JH. For example, flight-muscle histolysis and yolk uptake in *Dysdercus* spp. are induced by implantation of retrocerebral complexes or by treatment with synthetic JH (43, 53). Similarly, implanting CA-CC complexes or CA alone induces flight-muscle breakdown and egg development in crickets (39, 192). In the fire ant, *Solenopsis invicta*, flight-muscle histolysis is blocked by allatectomy and can be induced by exogenous treatment with JH (11). Furthermore, JH can stimulate egg production in this species (10). In bark beetles, topical application of JH analogue reduces flight-muscle volume (8, 22, 160) and increases acid phosphatase activity, an indicator of histolysis (172). The case of the Colorado potato beetle is unusual and interesting because breakdown of the flight muscles is associated with the absence of JH under prediapause conditions, whereas regeneration and ovarian activation are associated with increased JH levels (46).

GROUPS AND SOCIAL INSECTS

A new set of factors influences oogenesis when insects live in conspecific groups, and the increased number of interrelated stimuli affecting access to resources makes the regulatory web complex. In general, gregarious conspecifics can influence each other through repeated tactile stimulation or through pheromonal signals. Depending on the insect, group living can have positive, negative, or no effects on egg development. For example, oocyte growth is inhibited in both isolated *S. gregaria* and crowded *L. migratoria* (82). And in *B. germanica,* grouping appears to disinhibit the brain, allowing the stimulation of JH secretion and subsequent egg development (67). In *S. longipalpa,* however, grouping does not exert an effect distinct from mating (38).

The most highly evolved interactions among individuals in groups are found in the social insects (214), which include termites, ants, and the social wasps and bees. The defining feature of social insects is reproductive division of labor, with some females taking a greater role in egg production and others providing nourishment and defense for all members of the colony.

Oogenesis in a social context can be influenced by all the factors typically important for its regulation in solitary insects: nutrition, activity, mating, season, and conspecifics. It is the profound influence of conspecifics on reproduction that distinguishes the social insects. The effects of conspecifics can be mediated through regulatory factors common in nonsocial settings, such as nutrition and activity levels. The effects of conspecifics can be tied to features characteristic of social systems such as dominance interactions and pheromones (63, 109).

Considering the general importance of nourishment in regulating oogenesis in insects, we should not be surprised that social systems built on reproductive asymmetries can to a large extent be regulated through the flow and allocation of nourishment. Hunt & Nalepa (90) suggest that insect sociality cannot evolve without unequal partitioning of nourishment. Hunt (88, 89) argues that nourishment, especially the conflict between brood and adult, was critical in the evolution of sociality in wasps. Regardless of its role early on, the allocation of nutrition among colony members of higher eusocial groups plays a crucial role in generating different castes and in supporting the hyperfecundity of queens.

For insects reproducing and developing within groups, minor differences in food allocation as larvae can determine physiological and behavioral differences as adults. Moreover, behavioral differences, especially differences in activity levels, can reinforce the physiological differences. Adults that are more active and/or receive less nourishment will have less resources for reproduction, and the impact of this conflict will be greatest when food is limited. The phenomenon is similar to the oogenesis vs activity syndrome, described above, except now these trade-offs occur in a social setting, and colony members can influence each other's nutritional state.

The importance of nutrition in regulating worker reproduction is illustrated by the ant *Camponotus festinatus.* Vitellogenin titer rises sharply after emergence if workers are in queenless groups but is suppressed in workers in colonies with queens and in workers in queenless groups with larvae. Therefore, if workers have no larvae or queen to feed, they can make egg protein. The inhibitory effect is based on nutritional and, possibly, pheromonal cues (129, 130).

Hormonal regulation of oogenesis is not well understood for any social insect. Until we have evidence to the contrary, we must assume that regulation is similar to that found in solitary ancestors, and that as in most insects, JH plays a role in coordinating the influx of nourishment with oogenesis.

In the social insects, conspecifics can inhibit egg development. Some evidence suggests that such inhibition works through interference with CA function. In the wasp *Polistes dominulus,* small groups of similar females overwinter together and establish a new nest in the spring. These founders emerge from hibernation and, through behavioral interactions, establish a dominance hierarchy. Dominance interactions seem to affect oogenesis through CA function, and pheromones are not known to play a role in dominance interactions in this social wasp. Foundress females emerge from hibernation with similar CA sizes, but 24 h later, individuals that become dominant have larger CA and/or ovaries. Hormone injections can change the outcome of these dominance interactions (169).

Some ponerine ants have lost the ability to produce true queens, and mated workers are solely responsible for reproduction. In contrast to workers from other ant taxa, ponerine workers typically have spermathecae and can therefore mate and produce fertilized eggs (149). In various species lacking true queens, oogenesis can be suppressed through dominance interactions, as in *P. dominulus.* These interactions affect ovarian development in *Diacamma* sp. (151) and *Pachycondyla sublaevis* (93, 150).

In ants with true queens, the average fecundity of grouped queens may be less than that of single queens (107, 201). Mutual inhibition among queens is not always observed, however (25, 163). Apparent inhibition may result from direct pheromonal inhibition of oogenesis (201) or from lower levels of attention paid to each queen. Moreover, each queen in a multiply queened colony would presumably receive less food (107). In fire ants, virgin queens do not develop eggs in the presence of colony queens. Oogenesis can be stimulated by removing the virgins from the colony or by treating them with a JH analogue (201).

In the bumble bee *Bombus terrestris,* workers do not mate, and their ovaries regress in the presence of the mated egg-layer (the α-female). This inhibition is mediated through the effect of a pheromone from the α-female's mandibular gland on the CA of workers. Workers emerging in the presence of the mated egg-layer have JH levels much lower than do those emerging in her absence. Low JH levels are correlated with lack of egg development (168, 170).

The fact that ovary development in honey bee workers is suppressed by the queen and brood has been known for many years (45, 81, 95, 96, 118), but which pheromonal factors actually produce ovarian inhibition remain unknown. Consumption of extracts from whole queens suppresses ovary development in workers (202), and treatment with the queen-pheromone component 9-keto-(E)-2-decenoic acid (9-ODA) can have a similar effect (31). However, the complete mix of queen-pheromone components from the mandibular gland, which are highly important in emergency queen rearing and reproductive swarming (216, 217), does not inhibit ovary development in workers at any dose (213).

9-ODA does inhibit synthesis of JH by the CA in young queenless workers (102). JH, especially in low doses, stimulates vitellogenin synthesis in workers (92, 171). JH levels rise during the life of adult workers, and changes in titers correlate with changes in foraging behavior and ovarian physiology; higher titers accompany foraging and ovarian regression (reviewed in 165). Starvation in both nursing and foraging bees stimulates a rapid increase in JH synthesis (101). In conclusion, worker oogenesis in young queenless workers is regulated to some degree by JH synthesis, which in turn can be responsive to 9-ODA. However, the gaps in our understanding are highlighted by the facts that high JH titers are correlated with ovarian regression and that mandibular queen pheromones do not inhibit ovary development in workers.

Queens in the higher social insects, which are anatomically specialized as egg-producers, have a greater number of ovarioles than workers. The rate of egg production by α-females and queens in social insects spans a wide range. Reproductive workers in ponerines may lay only a single egg every 2–3 days (149), whereas queens of some species of army ants and termites can produce tens of thousands of eggs per day. Management of nutritional resources is one of the major foci of regulation of oogenesis, and nowhere is the throughput, the rate of conversion from raw materials to eggs, higher than in social insects. The extraordinarily fecund queens of social insects are long-lived [routinely living 5–15 years (86)] and generally rely on workers to channel resources to them. As a result, both short-term and life-time outputs can be extremely high.

Termite queens of the fungus-growing species *Macrotermes michaelseni* lay up to 40,000 eggs per day ($\approx$4 g) (121). Similarly high rates may exist in the most fecund army ants, *Dorylus wilverthi,* which reportedly lay 1 million eggs per batch over approximately 10 days (178). In such prolific species, and perhaps to some extent in all social insects, the oviposition rate must be limited by the rate at which nourishment can be passed to the queen and turned over into eggs.

The role of JH in regulating oogenesis in hyperfecund queens is not clear. In honey bee queens, JH does not appear to have a major role. JH synthesis by CA is no higher than that of nonforaging workers (165), and JH does not affect in vitro synthesis of vitellogenin by the queen's fat body (99, 100). Similarly, in *M. michaelseni,* JH titers and CA size in physogastric queens vary widely and do not seem to be correlated with in vitro synthesis, titer, or fresh weight (121).

Evidence that egg-production rate is limited by nutrition is found in several higher social insects. In honey bees, the amount of pollen feeding by a colony is directly related to the queen's egg-laying rate (35). In several ant species, food flow to the queen has been implicated in her fecundity. For example, queens of the Pharaoh ant, *Monomorium pharaonis,* obtain their food directly from larvae. If queens are isolated from larvae, egg laying stops within 24 h (23). Larvae are also an important factor in the fecundity of queens of *S. invicta.* The fecundity-enhancing substance in this case appears to be associated with the meconium, which is presumably passed to the queen by workers. The queen's fecundity is tied to the number and presence of large, last-instar larvae. Newly captured queens from the field can lay more than 2 eggs/min. But when isolated from larvae, egg production drops to about 1/3 of the initial rate after 24 h and to almost none after 48 h (198). In *Iridomyrmex humilis,* workers produce trophic eggs that are given to queens and larvae. The queen's oviposition rate is directly correlated with her trophic-egg consumption (13).

In *M. michaelseni,* the queen begins to become physogastric approximately 7–8 months after colony foundation, when the worker force has reached about 100 and a fungus comb has become well established. At this time, a suite of physiological and anatomical changes associated with physogastry take place (79). As the worker force continues to increase, so does queen weight and egg production (180).

Vitellogenesis in queens is generally enhanced by behavioral and anatomical specializations that direct massive amounts of nourishment toward her and provide her hypertrophied reproductive system with raw materials for conversion into eggs. This type of open-throttle oogenesis may be facilitated by the virtual absence of hormonal control. In contrast, egg production in specialized workers is suppressed by behavioral and anatomical specializations that restrict the flow of nourishment toward them, provide them with less reproductive equipment, and are greatly influenced by inhibitory cues, including social ones.

Upper limits to the rates of oogenesis are set by a combination of nourishment and the size of the reproductive system. Part of the nutritional limitation may be environmental; in central-place foragers, which includes most social insects, the amount of nourishment that can be collected in the foraging area may represent a constraint. Ants and termites with phenomenal rates of oogenesis dodge this constraint by using fungus gardens to digest plant material or by giving up a central nest and adopting a nomadic, marauding existence.

Literature Cited

1. Aclé D, Brookes VJ, Pratt GE, Feyereisen R. 1990. Activity of the corpora allata of adult female *Leucophaea maderae:* effects of mating and feeding. *Arch. Insect Biochem. Physiol.* 14:121–29

2. Adams TS, Gerst JW. 1991. The effect of pulse-feeding a protein diet on ovar-

ian maturation, vitellogenin levels, and ecdysteroid titre in houseflies, *Musca domestica*, maintained on sucrose. *Int. J. Invertebr. Reprod. Dev.* 20:49–57

3. Adams TS, Gerst JW. 1992. Interaction between diet and hormones on vitellogenin levels in the housefly, *Musca domestica*. *Int. J. Invertebr. Reprod. Dev.* 21:91–98

4. Adams TS, Gerst JW. 1993. Effect of diet on vitellogenin, vitellin and ecdysteroid levels during the second cycle of oogenesis in the housefly, *Musca domestica*. *J. Insect Physiol.* 39:835–43

5. Adler PH. 1982. Soil- and puddle-visiting habits of moths. *J. Lepid. Soc.* 36:161–73

6. Agui N, Takahashi M, Wada Y, Izumi S, Tomino S. 1985. The relationship between nutrition, vitellogenin, vitellin and ovarian development in the housefly, *Musca domestica*. *J. Insect Physiol.* 31:715–22

7. Arms K, Feeny P, Lederhouse R. 1974. Sodium: stimulus for puddling behavior by tiger swallowtail butterflies, *Papilio glaucus*. *Science* 185:372–74

8. Atkins MD, Farris SH. 1962. A contribution to the knowledge of flight muscle changes in the Scolytidae. *Can. Entomol.* 94:25–32

9. Baker HG, Baker I. 1973. Amino acids in nectar and their evolutionary significance. *Nature* 241:543–45

10. Barker JF. 1978. Neuroendocrine regulation of oocyte maturation in the imported fire ant *Solenopsis invicta*. *Gen. Comp. Endocrinol.* 35:234–37

11. Barker JF. 1979. Endocrine basis of wing casting and flight muscle histolysis in the fire ant *Solenopsis invicta*. *Experientia* 35:552–54

12. Barker JF, Herman WS. 1973. On the neuroendocrine control of ovarian development in the Monarch butterfly. *J. Exp. Zool.* 183:1–10

13. Bartels PJ. 1988. Reproductive caste inhibition by argentine ant queens: new mechanisms of queen control. *Insectes Soc.* 35:70–81

14. Beenakkers A, Van der Horst DJ, Van Marrewijk WJA. 1985. Biochemical processes directed to flight muscle metabolism. See Ref. 112, 10:451–86

15. Bell WJ, Bohm MK. 1975. Oosorption in insects. *Biol. Rev.* 50:373–96

16. Boggs CL. 1987. Ecology of nectar and pollen feeding in Lepidoptera. See Ref. 184, pp. 369–91

17. Boggs CL. 1990. A general model of the role of male-donated nutrients in female insects' reproduction. *Am. Nat.* 136:598–617

18. Boggs CL, Gilbert LE. 1979. Male contribution of egg production in butterflies: evidence for transfer of nutrients at mating. *Science* 206:83–84

19. Boggs CL, Jackson LA. 1991. Mud puddling is not a simple matter. *Ecol. Entomol.* 16:123–27

20. Boggs CL, Ross CL. 1993. The effect of adult food limitation on life history traits in *Speyeria mormonia* (Nymphalidae). *Ecology* 74:433–41

21. Boggs CL, Watt WB. 1981. Population structure of pierid butterflies. IV. Genetic and physiological investment in offspring by male *Colias*. *Oecologia* 50:320–24

22. Borden JH, Slater CE. 1968. Induction of flight muscle degeneration by synthetic juvenile hormone in *Ips confusus*. *Z. Vgl. Physiol.* 61:366–68

23. Borgesen LW. 1989. A new aspect of the role of larvae in the Pharaoh's ant society (*Monomorium pharaonis*): producer of fecundity-increasing substances to the queen. *Insectes Soc.* 36:313–27

24. Borovsky D. 1985. The role of the male accessory gland fluid in stimulating vitellogenesis in *Aedes taeniorhynchus*. *Archs. Insect Biochem. Physiol.* 2:405–13

25. Bourke AFG. 1993. Lack of experimental evidence for pheromonal inhibition of reproduction among queens in the ant *Leptothorax acervorum*. *Anim. Behav.* 45:501–9

26. Bowen BJ, Codd CG, Gwynne DT. 1984. The katydid spermatophore: male nutrient investment and its fate in the mated female. *Aust. J. Zool.* 32:23–31

27. Bownes M, Blair M. 1986. The effects of a sugar diet and hormones on the expression of the *Drosophila* yolk-protein genes. *J. Insect Physiol.* 32:493–501

28. Bownes M, Partridge L. 1987. Transfer of molecules from ejaculate to females in *Drosophila melanogaster* and *Drosophila pseudoobscura*. *J. Insect Physiol.* 33:941–47

29. Bownes M, Reid G. 1990. The role of the ovary and nutritional signals in the regulation of fat body yolk protein gene expression in *Drosophila melanogaster*. *J. Insect Physiol.* 36:471–79

30. Bownes M, Scott A, Shirras A. 1988. Dietary components modulate yolk protein gene transcription in *Drosophila melanogaster*. *Development* 103:119–28

31. Butler CG, Fairey EM. 1963. The role of the queen in preventing oogenesis in worker honeybees. *J. Apic. Res.* 2:14–18

32. Butlin RK, Woodhatch CW, Hewitt

GM. 1987. Male spermatophore investment increases female fecundity in a grasshopper. *Evolution* 41:221–25

33. Chapman JA. 1956. Flight-muscle changes during adult life in a scolytid beetle. *Nature* 177:1183

34. Chapman T, Trevitt S, Partridge L. 1994. Remating and male-derived nutrients in *Drosophila melanogaster*. *J. Evol. Biol.* 7:51–69

35. Chauvin R. 1956. Les facteurs qui gouverment la ponte chez la reine des abeilles. *Insectes Soc.* 3:499–504

36. Chen AC, Kelly TJ. 1993. Correlation of ecdysteroids with ovarian development and yolk protein–synthesis in the adult stable fly, *Stomoxys calcitrans*. *Comp. Biochem. Physiol. A* 104:485–90

37. Chen PS, Stumm-Zollinger E, Aigaki T, Balmer J, Bienz M, Bohlen P. 1988. A male accessory gland peptide that regulates reproductive behavior of female *D. melanogaster*. *Cell* 54:291–98

38. Chon T-S, Liang D, Schal C. 1990. Effects of mating and grouping in oocyte development and pheromone release activities in *Supella longipalpa*. *Environ. Entomol.* 19:1716–21

39. Chudakova IV, Bocharova-Messner OM. 1968. Endocrine regulation of the condition of the wing musculature in the imago of the house cricket (*Acheta domestica*). *Dokl. Akad. Nauk SSSR* 179: 157–59

40. Clausen CP. 1940. *Entomophagous Insects*. London: McGraw-Hill

41. Clements AN. 1992. *The Biology of Mosquitoes*, Vol. 1. London: Chapman & Hall

42. Couillard F, Girardie A. 1985. Control of C-16 juvenile hormone biosynthesis in active corpora allata of female African locusts. *Int. J. Invertebr. Reprod. Dev.* 8:303–15

43. Davis NE. 1975. Hormonal control of flight muscle histolysis in *Dysdercus fulvoniger*. *Ann. Entomol. Soc. Am.* 68: 710–14

44. Davis RE, Kelly TJ, Masler EP, Fescemyer HW, Thyagaraja BS, Borkovec AB. 1990. Hormonal control of vitellogenesis in the gypsy moth, *Lymantria dispar:* suppression of haemolymph vitellogenin by the juvenile hormone analogue, methoprene. *J. Insect Physiol.* 36:231–38

45. de Groot AP, Voogd S. 1954. On the ovary development in queenless worker bees *(Apis mellifica)*. *Experientia* 10: 384–85

46. de Kort CAD. 1969. Hormones and the structural and biochemical properties of the flight muscles in the Colorado potato beetle. *Meded. Landbouwhogsch. Wageningen* 69(2):1–69

47. Denlinger DL. 1971. Autogeny in the flesh fly *Sarcophaga argyrostoma*. *Ann. Entomol. Soc. Am.* 64:961–62

48. Deleted in proof

49. DeWilde J, DeBoer JA. 1969. Humoral and nervous pathways in photoperiodic induction of diapause in *Leptinotarsa decemlineata*. *J. Insect Physiol.* 15:661–75

50. Dingle H. 1985. Migration. See Ref. 112, 9:375–415

51. Duve H, Thorpe A, Yabi KJ, Yu CG, Tobe SS. 1992. Factors affecting the biosynthesis and release of juvenile hormone bisepoxide in the adult blow fly *Calliphora vomitoria*. *J. Insect Physiol.* 38:575–85

52. Eder J, Rembold H, eds. 1987. *Chemistry and Biology of Social Insects*. Munich: Peperny

53. Edwards FJ. 1970. Endocrine control of flight muscle histolysis in *Dysdercus intermedius*. *J. Insect Physiol.* 16:2027–31

54. Eidmann H. 1931. Morphologische und physiologische Untersuchungen am weiblichen Genitalapparat der Lepidopteren. II. Physiologischer Teil. *Z. Angew. Entomol.* 18:57–112

55. El-Akad AS, Humphreys JG. 1990. Effects of larval and adult diet plus mating/insemination upon ovarian development of laboratory-reared *Anopheles pharoensis* in Egypt. *J. Am. Mosq. Control Assoc.* 6:96–98

56. Endo K. 1970. Relation between ovarian maturation and activity of the corpora allata in seasonal forms of the butterfly *Polygonia c-aureum*. *Dev. Growth Diff.* 11:297–304

57. Engelmann F. 1970. *The Physiology of Insect Reproduction*. Oxford: Pergamon

58. Feinsod FM, Spielman A. 1980. Nutrient-mediated juvenile hormone secretion in mosquitoes. *J. Insect Physiol.* 26:113–17

59. Ferkovich SM, Dillard CR. 1986. A study of uptake of radiolabeled host proteins and protein synthesis during development of eggs of the endoparasitoid, *Microplitis croceipes*. *Insect Biochem.* 16:337–45

60. Fescemeyer HW, Masler EP, Davis RE, Kelly TJ. 1992. Vitellogenin synthesis in female larvae of the gypsy moth, *Lymantria dispar*—suppression by juvenile hormone. *Comp. Biochem. Physiol. B* 103:533–42

61. Fisher RC. 1971. Aspects of the physiology of endoparasitic Hymenoptera. *Biol. Rev.* 46:243–78

62. Flanders SE. 1942. Oosorption and ovulation in relation to oviposition in the parasitic Hymenoptera. *Ann. Entomol. Soc. Am.* 35:251–66

63. Fletcher DJC, Ross KG. 1985. Regulation of reproduction in eusocial Hymenoptera. *Ann. Rev. Entomol.* 30:319–43

64. Fox CW. 1993. The influence of maternal age and mating frequency on egg size and offspring performance in *Callosobruchus maculatus* (Coleoptera: Bruchidae). *Oecologia* 96:146

65. Friedel T, Gillott C. 1977. Contribution of male-produced proteins to vitellogenesis in *Melanoplus sanguinipes*. *J. Insect Physiol.* 23:145–51

66. Fukuda S. 1944. The hormonal mechanism of larval molting and metamorphosis in the silkworm. *J. Fac. Sci. Imp. Univ. Tokyo Sect. 4* 6:477–532

67. Gadot M, Burns E, Schal C. 1989. Juvenile hormone biosynthesis and oocyte development in adult female *Blattella germanica:* effects of grouping and mating. *Arch. Insect Biochem. Physiol.* 11:189–200

68. Gilbert LE. 1972. Pollen feeding and reproductive biology of *Heliconius* butterflies. *Proc. Natl. Acad. Sci. USA* 69:1403–7

69. Gilbert LE, Singer MC. 1975. Butterfly ecology. *Annu. Rev. Ecol. Syst.* 6:365–97

70. Gillott C, Friedel T. 1977. Fecundity-enhancing and receptivity-inhibiting substances produced by male insects: a review. *Adv. Invertebr. Reprod.* 1:199–218

71. Greenfield MD. 1982. The question of paternal investment in Lepidoptera: male-contributed proteins in *Plodia interpunctella*. *Int. J. Invertebr. Reprod. Dev.* 5:323–30

72. Gunn A, Gatehouse AG. 1993. The migration syndrome in the African armyworm moth, *Spodoptera exempta:* allocation of resources to flight and reproduction. *Physiol. Entomol.* 18:149–59

73. Gunn A, Gatehouse AG, Woodrow KP. 1988. Trade-off between flight and reproduction in the African armyworm moth, *Spodoptera exempta*. *Physiol. Entomol.* 14:419–27

74. Gwadz RL, Spielman A. 1973. Corpus allatum control of ovarian development in *Aedes aegypti*. *J. Insect Physiol.* 19:1441–48

75. Gwynne DT. 1984. Courtship feeding increases female reproductive success in bush crickets. *Nature* 307:361–63

76. Hagan HR. 1951. *Embryology of the Viviparous Insects*. New York: Ronald

77. Hagedorn HH. 1983. The role of ecdysteroids in the adult insect. In *Invertebrate Endocrinology*, ed. RGH Downer, H Laufer, 1:271–304. New York: Liss

78. Hagedorn HH. 1994. The endocrinology of the adult female mosquito. In *Advances in Disease Vector Research*, ed. KF Harris, 10:109–48. New York: Springer-Verlag

79. Han SH, Bordereau C. 1992. From colony foundation to dispersal flight in a higher fungus-growing termite, *Macrotermes subhyalinus*. *Sociobiology* 20:219–31

80. Herman WS, Bennett DC. 1975. Regulation of oogenesis, female-specific protein production, and male and female reproductive gland development by juvenile hormone in the butterfly, *Nymphalis antiopa*. *J. Comp. Physiol.* 99:331–38

81. Hess G. 1942. Über den Einfluss der Weisellosigkeit und des Fruchtbarkeitsvitamins E auf die Ovarien der Bienenarbeiterin. *Schweiz. Bienen-Ztg.* 1:33–110

82. Highnam KC, Haskell PT. 1964. The endocrine systems of isolated and crowded *Locusta* and *Schistocerca* in relation to oocyte growth, and the effects of flying upon maturation. *J. Insect Physiol.* 10:849–64

83. Highnam KC, Hill L, Mordue W. 1966. The endocrine system and oocyte growth in *Schistocerca* in relation to starvation and frontal ganglionectomy. *J. Insect Physiol.* 12:977–94

84. Hinton HE. 1981. *The Biology of Insect Eggs*. Oxford: Pergamon

85. Hiremath S, Jones D. 1992. Juvenile hormone regulation of vitellogenin in the gypsy moth, *Lymantria dispar*—suppression of vitellogenin messenger RNA in the fat body. *J. Insect Physiol.* 38:461–74

86. Hölldobler B, Wilson EO. 1990. *The Ants*. Cambridge, MA: Harvard Univ. Press

87. Huignard J. 1983. Transfer and fate of male secretion deposited in the spermatophore of females of *Acanthoscelides obtectus* (Coleoptera: Bruchidae). *J. Insect Physiol.* 29:55–63

88. Hunt JH. 1991. Nourishment and the evolution of social Vespidae. In *The Social Biology of Wasps*, ed. KG Ross, RW Matthews, pp. 426–50. Ithaca, NY: Cornell Univ. Press

89. Hunt JH. 1994. Nourishment and evolution in wasps *sensu lato*. See Ref. 91, pp. 211–44

90. Hunt JH, Nalepa CA. 1994. Nourish-

ment, evolution and insect sociality. See
Ref. 91, pp. 1–19

91. Hunt JH, Nalepa CA, eds. 1994. *Nourishment and Evolution in Insect Societies.* Boulder: Westview

92. Imboden L, Wille H, Gerig L, Lüscher M. 1976. Die Vitellogeninsynthese bei der Bienen-Arbeiterin *(Apis mellifera)* und ihre Abhängigkeit von Juvenilhormon. *Rev. Suisse Zool.* 83:928–33

93. Ito F, Higashi S. 1991. A linear dominance hierarchy regulating reproduction and polyethism of the queenless ant *Pachycondyla sublaevis. Naturwissenschaften* 78:80–82

94. Iwata K. 1955. The comparative anatomy of the ovary in Hymenoptera. Part I. Aculeata. *Mushi* 29:17–33

95. Jay SC. 1970. The effect of various combinations of immature queen and worker bees on the ovary development of worker honey bees in colonies with and without queens. *Can. J. Zool.* 48:163–73

96. Jay SC. 1972. Ovary development of worker honeybees when separated from worker brood by various methods. *Can. J. Zool.* 50:661–64

97. Johansson AS. 1964. Feeding and nutrition in reproductive processes in insects. *R. Entomol. Soc. London Symp.* 2:43–55

98. Jones RG. 1979. *The structure, development and generation of the flight muscles in the imported fire ant* Solenopsis invicta: *an ultrastructural investigation.* PhD thesis. Texas A&M Univ. 297 pp.

99. Kaatz H-H. 1987. Changes in ecdysteroid and juvenile hormone hemolymph titers of bee queens during pupal and adult development. See Ref. 52, p. 312 (Abstr.)

100. Kaatz H-H. 1987. Regulation of vitellogenin synthesis in honey bee queens. See Ref. 52, p. 317 (Abstr.)

101. Kaatz H-H, Eichmüller S, Kreissl S. 1994. Stimulatory effect of octopamine on juvenile hormone biosynthesis in honey bees *(Apis mellifera)*: physiological and immunocytological evidence. *J. Insect Physiol.* 40:865–72

102. Kaatz H-H, Hildebrandt H, Engels W. 1992. Primer effect of queen pheromone on juvenile-hormone biosynthesis in adult worker honey bees. *J. Comp. Physiol. B* 162:588–92

103. Kahn MA. 1988. Brain-controlled synthesis of juvenile hormone in adult insects. *Entomol. Exp. Appl.* 46:3–17

104. Karlinsky A. 1963. Effets de l'ablation des corpora allata imaginaux sur le développement ovarien de *Pieris brassicae. C. R. Acad. Sci. Paris* 256:4101–3

105. Karlinsky A. 1967. Reprise de la vitellogénèse après implantation de corpora allata chez *Pieris brassicae. C. R. Acad. Sci. Paris* 264:1735–38

106. Karlsson B, Wickman PO. 1990. Increase in reproductive effort as explained by body size and resource allocation in the speckled wood butterfly, *Pararge aegeria. Funct. Ecol.* 4:6009–17

107. Keller L. 1988. Evolutionary implications of polygyny in the Argentine ant, *Iridomyrmex humilis:* an experimental study. *Anim. Behav.* 36:159–65

108. Keller L, ed. 1993. *Queen Number and Sociality in Insects.* Oxford: Oxford Univ. Press. 439 pp.

109. Deleted in proof

110. Kelly TJ, Fuchs MS. 1980. *In vivo* induction of ovarian development in decapitated *Aedes atropalpus* by physiological levels of 20-hydroxyecdysone. *J. Exp. Zool.* 213:25–32

111. Kelly TJ, Fuchs MS, Kang S-H. 1981. Induction of ovarian development in autogenous *Aedes atropalpus* by juvenile hormone and 20-hydroxyecdysone. *Int. J. Invertebr. Reprod. Dev.* 3:101–12

112. Kerkut GA, Gilbert LI. eds. 1985. *Comprehensive Insect Physiology, Biochemistry and Pharamacology.* Oxford: Pergamon

113. King PE, Ratcliffe NA, Copland MJW. 1969. The structure of egg membranes in *Apanteles glomeratus* (Hymenoptera: Braconidae). *Proc. R. Entomol. Soc. Ser. A* 44:137–42

114. Klowden MJ. 1993. Mating and nutritional state affect the reproduction of *Aedes albopictus* mosquitoes. *J. Am. Mosq. Control Assoc.* 9:169–73

115. Klowden MJ, Chambers GM. 1989. Ovarian development and adult mortality in *Aedes aegypti* treated with sucrose, juvenile hormone and methoprene. *J. Insect Physiol.* 35:513–17

116. Klowden MJ, Chambers GM. 1991. Male accessory gland substances activate egg development in nutritionally stressed *Aedes aegypti* mosquitoes. *J. Insect Physiol.* 37:721–26

117. Klowden MJ, Chambers GM. 1992. Reproductive and metabolic differences between *Aedes aegypti* and *Ae. albopictus. J. Med. Entomol.* 29:467–71

118. Kropacova S, Haslbachova H. 1971. The influence of queenlessness and of unsealed brood on the development of ovaries in worker honeybees. *J. Apic. Res.* 10:57–61

119. Kubli E. 1992. The sex-peptide. *Bioessays* 14:779–84

120. Lamison CD, Ballarino J, Ma M. 1991.

Temporal events of gypsy moth vitellogenesis and ovarian development. *Physiol. Ecol.* 16:201–9

121. Lanzrein B, Gentinetta V, Fehr R. 1983. Titres of juvenile hormone and ecdysteroids in reproductives and eggs of *Macrotermes michaelseni:* relation to caste determination? In *Caste Differentiation in Social Insects,* ed. JAL Watson, BM Okot-Kotber, CH Noirot, pp. 307–27. Oxford: Pergamon

122. Lanzrein B, Wilhelm R, Buschor J. 1981. On the regulation of the corpora allata activity in adult females of the ovoviviparous cockroach *Nauphoeta cinerea.* In *Juvenile Hormone Biochemistry,* ed. GE Pratt, GT Brooks, pp. 147–60. Amsterdam: Elsevier

123. Lea AO. 1970. Endocrinology of egg maturation in autogenous and anautogenous *Aedes taeniorhynchus. J. Insect Physiol.* 16:1689–96

124. Lehane MJ. 1991. *Biology of Blood-Sucking Insects.* London: Harper-Collins Academic. 288 pp.

125. Lounibos LP. 1994. Variable egg development among *Anopheles (Nyssorhynchus):* control by mating? *Physiol. Entomol.* 19:51–57

126. Mackerras MJ. 1933. Observations on the life-histories, nutritional requirements, and fecundity of blowflies. *Bull. Entomol. Res.* 24:353–65

127. Markow TA, Ankney PF. 1988. Insemination reaction: found in species whose males contribute material to oocytes before fertilization. *Evolution* 42:1097–101

128. Markow TA, Gallagher PD, Krebs RA. 1990. Ejaculate-derived nutritional contribution and female reproductive success in *Drosophila mojavensis. Funct. Ecol.* 4:67–73

129. Martinez T, Wheeler DE. 1991. Effect of the queen, brood and worker caste on haemolymph vitellogenin titres in *Camponotus festinatus* workers. *J. Insect Physiol.* 37:347–52

130. Martinez T, Wheeler DE. 1994. Storage proteins in adult ants: roles in colony founding by queens and in larval rearing by workers. *J. Insect Physiol.* 40:723–29

131. Mason LJ, Johnson SJ, Woodring JP. 1989. Influence of carbohydrate deprivation and tethered flight on stored lipid, fecundity, and survivorship of the soybean looper (Lepidoptera: Noctuidae). *Environ. Entomol.* 18:1090–94

132. Mole S, Zera AJ. 1993. Differential allocation of resources underlies the dispersal-reproductive trade-off in the wing-dimorphic cricket, *Gryllus rubens. Oecologia* 93:121–27

133. Mole S, Zera AJ. 1994. Differential resource consumption obviates a potential flight-fecundity trade-off in the sane cricket (*Gryllus firmus*). *Funct. Ecol.* 8:573–50

134. Moobola S, Cupp EW. 1978. Ovarian development in the stable fly, *Stomoxys calcitrans,* in relation to diet and juvenile hormone control. *Physiol. Entomol.* 3:317–21

135. Moshitzky P, Applebaum SW. 1990. The role of adipokinetic hormone in the control of vitellogenesis in locusts. *Insect Biochem.* 20:319–23

136. Mullins D, Kiel C. 1980. Paternal investment of urates in cockroaches. *Nature* 283:567–69

137. Mullins D, Keil C. 1992. Maternal and paternal nitrogen investment in *Blatella germanica. J. Exp. Biol.* 162:55–72

138. Mundall E, Engelmann F. 1977. Endocrine control of vitellogenin synthesis and vitellogenesis in *Triatoma protracta. J. Insect Physiol.* 23:825–36

139. Murphy DD, Launer AE, Ehrlich PR. 1983. The role of adult feeding in egg production and population dynamics of the checkerspot butterfly *Euphydryas editha. Oecologia* 56:257–63

140. Nair CRM, Prabhu VKK. 1985. Entry of proteins from degenerating flight muscle into oocytes in *Dysdercus cingulatus. J. Insect Physiol.* 31:383–88

141. Nijhout MM, Riddiford LM. 1974. The control of egg maturation by juvenile hormone in the tobacco hornworm moth, *Manduca sexta. Biol. Bull.* 146:377–92

142. Oberhauser KS. 1989. Effects of spermatophores on male and female monarch butterfly reproductive success. *Behav. Ecol. Sociobiol.* 25:237–46

143. O'Meara GF. 1987. Nutritional ecology of blood-feeding Diptera. See Ref. 184, pp. 741–64

144. O'Meara GF, Evans DG. 1977. Autogeny in saltmarsh mosquitoes induced by a substance from the male accessory gland. *Nature* 167:342–44

145. Orr CWM. 1964. The influence of nutritional and hormonal factors on the chemistry of the fat body, blood, and ovaries of the blowfly *Phormia regina. J. Insect Physiol.* 10:103–219

146. Pan ML. 1977. Juvenile hormone and vitellogenin synthesis in the cecropia silkworm. *Biol. Bull.* 153:336–45

147. Pan ML, Wyatt GR. 1976. Control of vitellogenin synthesis in the monarch butterfly by juvenile hormone. *Dev. Biol.* 54:127–34

148. Parker GA, Simmons LW. 1989. Nuptial feeding in insects: theoretical mod-

els of male and female interests. *Ethology* 82:3–26

149. Peeters C. 1993. Monogyny and polygyny in ponerine ants with or without queens. See Ref. 108, pp. 234–61

150. Peeters C, Higashi S, Ito G. 1991. Reproduction in ponerine ants without queens: monogyny and exceptionally small colonies in the Australian *Pachycondyla sublaevis*. *Ethol. Ecol. Evol.* 3:145–52

151. Peeters C, Tsuji K. 1993. Reproductive conflict among ant workers in *Diacamma* sp. from Japan: dominance and oviposition in the absence of the gamergate. *Insectes Soc.* 40:119–36

152. Pener MP. 1985. Hormonal effects on flight and migration. See Ref. 112, 8: 491–550

153. Pipa RL. 1982. Neural influence on corpus allatum activity and egg maturation in starved virgin *Periplaneta americana*. *Physiol. Entomol.* 7:449–55

154. Pipa RL. 1985. Effects of starvation, copulation, and insemination on oocyte growth and oviposition by *Periplaneta americana*. *Ann. Entomol. Soc. Am.* 78: 284–90

155. Pipa RL. 1986. Disinhibition of oocyte growth in adult, virgin *Periplaneta americana* by corpus allatum denervation: age dependency and relatedness to mating. *Arch. Insect Biochem. Physiol.* 3:471–83

156. Pitnick S, Markow TA, Riedy MF. 1991. Transfer of ejaculate and incorporation of male-derived substances by females in the Nannoptera species group (Drosophilidae). *Evolution* 45:774–80

157. Pivnick KA, McNeil JN. 1987. Puddling in butterflies: sodium affects reproductive success in *Thymelicus lineola*. *Physiol. Entomol.* 12:461–72

158. Rankin MA, Burchsted JCA. 1992. The cost of migration in insects. *Annu. Rev. Entomol.* 37:533–59

159. Rankin MA, McAnelly ML, Bodenhamer JE. 1986. The oogenesis-flight syndrome revisited. In *Insect Flight: Disperal and Migration*, ed. W Danthanarayana, pp. 27–48. Berlin/Heidelberg: Spring-Verlag

160. Reid RW. 1962. Biology of the mountain pine beetle, *Dendroctonus monticolae* in the East Hootenay region of British Columbia. II. Behavior in the host, fecundity, and internal changes in the female. *Can. Entomol.* 94:605–13

161. Retnakaran A, Percy J. 1985. Fertilization and special modes of reproduction. See Ref. 112, 1:281–85

162. Ridley M. 1988. Mating frequency and fecundity in insects. *Biol. Rev.* 63:509–49

163. Rissing SW, Pollock GB, Higgins MR, Hagen RH, Smith DR. 1989. Foraging specialization without relatedness or dominance among co-founding ant queens. *Nature* 338:420–22

164. Deleted in proof

165. Robinson GE, Strambi C, Strambi A, Feldlaufer MF. 1991. Comparison of juvenile hormone and ecdysteroid haemolymph titres in adult worker and queen honey bees *(Apis mellifera)*. *J. Insect Physiol.* 37:929–35

166. Roff DA. 1984. The cost of being able to fly: a study of wing polymorphism in two species of crickets. *Oecologia* 63:30–37

167. Roff DA. 1986. The evolution of wing dimorphism in insects. *Evolution* 40: 1009–20

168. Röseler PF. 1977. Juvenile hormone control of oogenesis in bumblebee workers, *Bombus terrestris*. *Insect Physiol.* 23:985–92

169. Röseler PF, Röseler I, Strambi A, Augier R. 1984. Influence of insect hormones on the establishment of dominance hierarchies among foundresses of the paper wasp, *Polistes gallicus*. *Behav. Ecol. Sociobiol.* 15:133–42

170. Röseler PF, Röseler I, Van Honk CGJ. 1981. Evidence for inhibition of corpora allata activity in workers of *Bombus terrestris* by a pheromone from the queens' mandibular glands. *Experientia* 37:348–51

171. Rütz WL, Gerig L, Wille H, Lüscher M. 1976. The function of juvenile hormone in adult worker honey bees. *J. Insect Physiol.* 20:897–909

172. Sahota TS. 1975. Effect of juvenile hormone on acid phosphatases in the degenerating flight muscles of the Douglas-Fir beetle, *Dendroctonus pseudotsugae*. *J. Insect Physiol.* 21:471–78

173. Sasaki M, Riddiford LM. 1984. Regulation of reproductive behavior and egg maturation in the tobacco hawk moth, *Manduca sexta*. *Physiol. Entomol.* 9: 315–27

174. Sappington TW, Showers WB. 1992. Reproductive maturity, mating status, and long-duration flight behavior of *Agrotis ipsilon* (Lepidoptera: Noctuidae) and the conceptual misuse of the oogenesis flight syndrome by entomologists. *Environ. Entomol.* 21:677–88

175. Schal C, Bell W. 1982. Ecological correlates of paternal investment of urates in a tropical roach. *Science* 281:171–73

176. Schal C, Chiang A-S, Burns EL, Gadot M, Cooper RA. 1993. Role of the brain

in juvenile hormone synthesis and oocyte development: effects of dietary protein in the cockroach *Blattella germanica. J. Insect Physiol.* 39:303–13

177. Schmidt T, Choffat Y, Schneider M, Hunziker P, Fuyama Y, Kubli E. 1993. *Drosophila suzukii* contains a peptide homologous to the *Drosophila melanogaster* sex-peptide and functional in both species. *Insect Biochem. Mol. Biol.* 23:571–79

178. Schneirla TC. 1971. *Army Ants.* San Francisco: Freeman. 349 pp.

179. Schwartz MB, Kelly TJ, Imberski RB, Rubenstein EC. 1985. The effects of nutrition and methoprene treatment on ovarian ecdysteroid synthesis in *Drosophila melanogaster. J. Insect Physiol.* 31:947–57

180. Sieber R, Leuthold RH. 1982. Development of physogastry in the queen of the fungus-growing termite *Macrotermes michaelseni. J. Insect Physiol.* 28:979–85

181. Simmons LW. 1988. The contribution of multiple mating and spermatophore consumption to the lifetime reproductive success of female field crickets (*Gryllus bimaculatus*). *Ecol. Entomol.* 13:57–69

182. Slansky F Jr. 1980. Food consumption and reproduction as affected by tethered flight in female milkweed bugs (*Oncopeltus fasciatus*). *Entomol. Exp. Appl.* 28:277–86

183. Slanksy F Jr. 1981. Insect nutrition: an adaptationist's perspective. *Fla. Entomol.* 65:45–71

184. Slansky F Jr, Rodriguez JG, eds. 1987. *Nutritional Ecology of Insects, Mites, Spiders, and Related Invertebrates.* New York: Wiley

185. Smith AF, Yagi K, Tobe SS, Schal C. 1989. *In vitro* juvenile hormone biosynthesis in adult virgin and mated female brown-banded cockroaches, *Supella longipalpa. J. Insect Physiol.* 35:781–85

186. Solbreck C. 1986. Wing and flight muscle polymorphism in a lygaeid bug, *Horvathiolus gibbicollis:* determinants and life history consequences. *Ecol. Entomol.* 11:435–44

187. Spielman A. 1971. Bionomics of autogenous mosquitoes. *Annu. Rev. Entomol.* 16:231–48

188. Stay B, Tobe SS. 1977. Control of juvenile hormone biosynthesis during the reproductive cycle of a viviparous cockroach. I. Activation and inhibition of corpora allata. *Gen. Comp. Endocrinol.* 33:531–40

189. Stoffolano JG, Zou BX, Yin CM. 1992. Vitellogenin uptake, not synthesis, is dependent on juvenile hormone in adults of *Phormia regina. J. Insect Physiol.* 38:839–45

190. Strangways-Dixon J. 1961. The relationship between nutrition, hormones and reproduction in the blowfly *Calliphora erythrocephala.* I. Selective feeding in relation to the reproductive cycle, the corpus allatum volume and fertilization. *J. Exp. Biol.* 38:225–35

191. Tallamy D. 1994. Nourishment and the evolution of paternal investment in subsocial arthropods. See Ref. 91, pp. 21–56

192. Tanaka S. 1986. De-alation, flight muscle histolysis, and oocyte development in the striped ground cricket, *Allonemobius fasciatus. Physiol. Entomol.* 11:453–58

193. Tanaka S. 1991. De-alation and its influences on egg production and flight muscle histolysis in a cricket (*Velarifictorus parvus*) that undergoes inter-reproductive migration. *J. Insect Physiol.* 37:517–23

194. Tanaka S. 1993. Allocation of resources to egg production and flight muscle development in a wing dimorphic cricket, *Modicogryllus confirmatus. J. Insect Physiol.* 39:493–98

195. Thornhill R, Alcock J. 1983. *The Evolution of Insect Mating Systems.* Cambridge, MA: Harvard Univ. Press

196. Tilden RL, Ferkovich SM. 1987. Regulation of protein synthesis during egg development of the parasitic wasp, *Microplitis croceipes. Insect Biochem.* 17:783–92

197. Tobe SS, Chapman CS. 1979. The effects of starvation and subsequent feeding on juvenile hormone synthesis and oocyte growth in *Schistocerca americana gregaria. J. Insect Physiol.* 25:701–8

198. Tschinkel WR. 1988. Social control of egg-laying rate in queens of the fire ant, *Solenopsis invicta. Physiol. Entomol.* 13:327–50

199. Uchida K. 1993. Balanced amino acid composition essential for infusion-induced egg development in the mosquito (*Culex pipiens pallens*). *J. Insect Physiol.* 39:615–21

200. Uchida K, Ohmori D, Yamakura G, Suzuki K. 1992. Mosquito (*Culex pipiens pallens*) egg development induced by infusion of amino acids into the haemocoel. *J. Insect Physiol.* 38:953–59

201. Vargo EL. 1992. Mutual pheromonal inhibition among queens in polygyne colonies of the fire ant *Solenopsis invicta. Behav. Ecol. Sociobiol.* 31:205–10

202. Voogd S. 1956. The influence of the queen on the ovary development in worker bees. *Experientia* 12:199–201

203. Walker TJ. 1972. Deciduous wings in crickets: a new basis for wing dimorphism. *Psyche* 79:311–13

204. Watanabe M. 1988. Multiple matings increase the fecundity of the yellow swallowtail butterfly, *Papilio xuthus* in summer generations. *J. Insect Behav.* 1:17–29

205. Weaver RJ. 1984. Effects of food and water availability and of NCA-1 section upon juvenile hormone biosynthesis and oocyte development in *Periplaneta americana. J. Insect Physiol.* 30:831–38

206. Weaver RJ, Pratt GE. 1977. The effect of enforced virginity and subsequent mating on the activity of the corpus allatum of *Periplaneta americana* measured in vitro, as related to changes in the rate of ovarian maturation. *Physiol. Entomol.* 2:59–76

207. Weaver RJ, Pratt GE. 1981. Effects of starvation and feeding upon corpus allatum activity and oocyte growth in adult female *Periplaneta americana. J. Insect Physiol.* 27:75–83

208. Wedell N. 1993. Mating effort or paternal investment? Incorporation rate and cost of male donations in the wartbiter. *Behav. Ecol. Sociobiol.* 32:239–46

209. Wheeler DE. 1994. Nourishment in ants: patterns in individuals and societies. See Ref. 91, pp. 245–78

210. Wigglesworth VB. 1970. *Insect Hormones.* San Francisco: Freeman

211. Wiklund C, Kaitala A, Virpi L, Abenius J. 1993. Polyandry and its effect on female reproduction in the green-veined white butterfly *(Pieris napi). Behav. Ecol. Sociobiol.* 33:25–33

212. Willers JL, Schneider JC, Ramaswamy SB. 1987. Fecundity, longevity and caloric patterns in female *Heliothis virescens:* changes with age due to flight and supplemental carbohydrate. *J. Insect Physiol.* 33:803–8

213. Willis LG, Winston ML, Slessor KN. 1990. Queen honey bee mandibular pheromone does not affect worker ovary development. *Can. Entomol.* 122:1093–99

214. Wilson EO. 1971. *The Insect Societies.* Cambridge, MA: Harvard Univ. Press

215. Wilson K, Gatehouse AG. 1992. Migration and genetics of prereproductive period in the moth, *Spodoptera exempta* (African armyworm). *Heredity* 69:255–62

216. Winston ML, Higo HA, Colley SJ, Pankiw T, Slessor KN. 1991. The role of the queen mandibular pheromone and colony congestion in honey bee reproductive swarming. *J. Insect Behavior* 4:649–59

217. Winston ML, Higo HA, Slessor KN. 1990. Effect of various dosages of queen mandibular gland pheromone on the inhibition of queen rearing in the honey bee. *Ann. Entomol. Soc. Am.* 83:234–38

218. Woodhead AP, Stay B. 1989. Neural inhibition of corpora allata in protein-deprived *Diploptera punctata. J. Insect Physiol.* 35:415–21

219. Yin C-M, Duan H, Stoffolano JG Jr. 1993. Hormonal stimulation of the brain for its control of oogenesis in *Phormia regina. J. Insect Physiol.* 39:165–71

220. Yin C-M, Zou BX, Stoffolano JG Jr. 1989. Dietary induced hormonal control of terminal oocyte development in *Phormia regina. Proc. Symp. Host Regulated Developmental Mechanisms in Vector Arthropods, 2nd,* ed. D Borovsky, A Spielman, pp. 81–88. Vero Beach, FL: Univ. of Florida

221. Yin C-M, Zou BX, Yi S-X, Stoffolano JG Jr. 1990. Ecdysteroid activity during oogenesis in the black blowfly, *Phormia regina. J. Insect Physiol.* 36:375–82

222. Zeh DW, Smith RL. 1985. Paternal investment by terrestrial arthropods. *Am. Zool.* 25:785–805

223. Deleted in proof

224. Zera AJ, Rankin MA. 1989. Wing dimorphism in *Gryllus rubens:* genetic basis of morph determination and fertility differences between morphs. *Oecologia* 80:249–55

225. Deleted in proof

226. Zera AJ, Tiebel KC. 1988. Brachypterizing effect of group rearing, juvenile hormone III and methoprene on winglength development in the wing-dimorphic cricket, *Gryllus rubens. J. Insect Physiol.* 34:489–98

Annu. Rev. Entomol. 1996. 41:433–50

TYMPANAL HEARING IN INSECTS

R. R. Hoy and D. Robert

Section of Neurobiology & Behavior, Seeley Mudd Hall, Cornell University, Ithaca, New York 14853-2702

KEY WORDS: tympanum, acoustic insect, auditory evolution, acoustic parasitism, chordotonal organ

ABSTRACT

Specialized hearing organs, known as tympanal organs, have evolved in at least seven different orders of insects. Tympanal organs are usually defined by the presence of a tympanal membrane (or eardrum). They are backed by an air-filled space or cavity and are innervated by a chordotonal sensory organ. In some insects, however, a recognizable tympanal membrane may not be easily identified by visual inspection, yet may possess tympanal hearing organs. In insects that possess them, tympanal hearing organs may mediate the detection of predators, prey, and potential mates and rivals. Unlike the ears of vertebrates, which are localized to cranial segments, the ears of insects may be found in a bewildering variety of locations on their bodies, depending on the species. The embryological and evolutionary origins of tympanal organs are related to ancestral states as proprioceptive chordotonal organs.

INTRODUCTION

An insect, like any animal, depends on specialized organs for remote sensing of other animals: potential predators, prey, and mates or rivals. To see, hear, smell, or feel the presence of others, insects use eyes, auditory organs, olfactory organs, and internal proprioceptors and cuticular hairs sometimes borne on specialized appendages such as cerci or antennae. Compared with vision and olfaction, the ability of insects to hear may at first seem to be relatively uncommon. No one doubts that cicadas, crickets, and katydids have good hearing, because of their conspicuously loud calls, which mediate their conspecific reproductive and territorial interactions (5, 37, 51), but among the most speciose of insect taxa, such as the Coleoptera, Hymenoptera, and Diptera, reports of conspicuous sound emissions are rare. Nevertheless, even these groups appear to emit acoustic signals in various behavioral contexts (12, 50,

433

0066-4170/96/0101-0433$08.00

52). However, whether they have the means to detect such signals is not always clear.

An acoustic signal is generated by the vibrations of a sound-producing structure, and its reception requires the detection of those vibrations by appropriate mechanoreceptive organs. When an insect produces sound, part of the vibrational energy may be propagated through the substrate upon which the sender is standing and part through the air. Substrate-borne signals are generally omitted from the domain of acoustic signals, and for the sake of brevity, we do not discuss them either, although they are deserving of a thorough review (19, 44, 45). However, substrate signals can be considered part of a general class of signals that includes acoustic (airborne) signals, and when considering the structure, function, and evolution of the organs for signal production and reception, we ought to view them together. Certainly, if the discussion includes those insects that produce substrate signals and are sensitive to them, we are no longer talking about an apparent small minority of insects (46).

Even if we restrict ourselves to the domain of hearing organs that are sensitive to the airborne component of a vibrational sound source, we face the problem of defining an insect "ear" (85). An airborne sound signal has two components that affect its reach or effective communication range. The effective range, in turn, depends on several factors, including distance from the sound source, such as the size of the sound radiator relative to the wavelength of the sound (10, 11); the physical features of the sound itself, such as frequency and initial intensity of the signal; environmental factors; and the sensitivity of the receiver (49, 70). Very near the sound source, within a radius measuring one wavelength of the emitted signal, the adjacent air molecules experience relatively large translational movements that represent a significant proportion of the total intensity in the sound signal (this proportion decreases sharply as the distance to the sound source increases and rapidly becomes negligible from the sensory point of view). Within this nearby space of signal propagation, a detector is said to be in the near-field of the sound source (49).

Examples of near-field detectors include the cercal organs of cockroaches, the Johnston's organs of mosquitoes, and the aristae of drosophilid flies (46). Although these are indeed examples of auditory organs, they lack eardrums and are not tympanal organs. Depending on species and behavioral context, these organs are generally restricted in their range of communication to short distances—a few body lengths in *Drosophila melanogaster* (12) to one meter for male mosquitoes (44). In general, the range of frequency sensitivity of near-field auditory receivers is also restricted to sounds containing relatively low frequencies—a few hundred (75–500) Hertz. For example, *D. melanogaster* (12) and mosquitoes (44, 45) respond to sounds corresponding to the wingbeat frequencies of their mates, and caterpillars respond to the wingbeat frequencies of predatory wasps (78, 79). Sensory hairs borne on the insect's body or on specialized appendages, such as the cerci of orthoperans or the antennae of mosquitoes, also serve as receptors for sound vibrations (46).

Although near-field sound receptors are certainly interesting, they are not discussed further in this review because they are not tympanal ears and are discussed in several excellent reviews (23, 46, 51, 70).

Farther from the sound source, the air molecules experience the propagation of the sound as a pressure wave. The term ears, whether applied to insects, frogs, elephants, or humans, usually refers to auditory detectors sensitive to the pressure wave of the sound field. In insects, such auditory detectors are sensitive to a wide range of high frequencies (2 to more than 100 kHz) and that, depending on species, permit the detection of sounds over relatively long distances, up to tens of meters. In this article, we restrict ourselves to these far-field acoustic detectors, which are commonly designated tympanal hearing organs.

This review covers (*a*) the definition and taxonomic distribution of tympanal hearing organs (i.e. ears); (*b*) the morphology and function of tympanal ears in selected insects, including sexually dimorphic ears; (*c*) the developmental origins of some tympanal organs; and (*d*) recent studies on possible evolutionary origins of tympanal organs.

TYMPANAL HEARING ORGANS IN INSECTS

Tympanal ears are morphologically characterized by three features: a tympanal membrane; an air-filled sac or tracheal expansion, upon which the tympanum is appressed; and an associated chordotonal sensory organ. The tympanum (or eardrum) is a specially differentiated region of cuticle that is thinner than the surrounding cuticle and often appears silvery because of the way it reflects light. The chordotonal organ is a complex cellular unit consisting of a bipolar sensory neuron, the dendrite of which inserts distally into a scolopale cell. The dendrite and the scolopale cell are connected to several glial and support cells and together form the scolopophorous or scolopidial organ (46, 54). Airborne sound waves from the emitting source impinge upon the tympanum and make it vibrate. These vibrations are then sensed by the chordotonal sensory organ. In many insects, such as crickets, katydids, cicadas, moths, and locusts, the chordotonal organ may be excited not only by sound impinging directly on the external surface of the tympanum, but also by sound waves impinging on the tympanum's internal surface. For instance, in crickets, an airsac associated with the tympanum is part of the spiracular tracheal system and provides a pathway for sound to reach the internal side of the tympanum. The ability to sense the difference in sound pressure on both sides of the tympanum is the key to the directional properties of some insect auditory systems, which are known as pressure-difference receivers (51). The complex interaction between the acoustic excitation of the tympana and the anatomy of the internal tracheal pathways in pressure-difference auditory receivers exemplifies the diversity of (fundamentally different) hearing mechanisms found in insects (51).

Tympanal hearing has been well documented in seven orders of insects (Table 1). Other workers have compiled similar lists (e.g. 85). The number of orders has grown in the past few years, and not only are other orders likely to be added to that list in the near future, but more examples within each order

Table 1   The occurence of tympanal hearing organs in seven orders of insects

Order		Superfamily/ family	Species	Ear location	Reference
Neuroptera	1	Chrysopidae	*Chrysopa carnea*	Wing base	53
Lepidoptera[a]	2	Geometroidea/ Geometridae	*Larentia tristata*	Abdomen	18
	3	Noctuoidea/ Noctuidae		Metathorax	20a
		Notodontidae	*Phalera bucephala*	Metathorax	62b
		Arctiidae	*Cycnia tenera*	Metathorax	73
	4	Papilionoidea/ Nymphalidae	*Heliconius erato*	Base of fore- or hind-wing	77
	5	Hedyloidea/ Hedylidae	*Macrosoma hyacinthina*	Fore-wing base	72a
Coleoptera	6	Cicindelidae	*Cicindela marutha*	Abdomen	76
	7	Scarabaeidae	*Euetheola humilis*	Cervical membranes	26
Dictyoptera	8	Mantodea	*Mantis religiosa*	Ventral metathorax	88
		Blattoidea[b]	*Periplaneta americana*	Metathoracic leg	75
Orthoptera	9	Acrididae	*Locusta migratoria*	First abdominal segment	30a
	10	Gryllidae	*Gryllus bimaculatus*	Prothoracic leg	72
Hemiptera	11	Cicadidae	*Cystosoma saundersii*	Abdomen	91
	12	Corixidae	*Corixa punctata*	Mesothorax	62a
Diptera	13	Tachinidae	*Ormia ochracea*	Ventral prosternum	66
	14	Sarcophagidae	*Colcondamyia auditrix*	Ventral prosternum	75a[c]

[a] These four superfamilies are considered to have independently evolved tympanal organs.
[b] Absence of well-differentiated tympanal membrane.
[c] Evidence of phonotactic behavior.

will certainly be discovered. In recent years, tympanal hearing has been demonstrated in the speciose orders Coleoptera (76, 89) and Diptera (39, 65).

Does the lack of a recognizable tympanal membrane preclude audition in the far field? In a few examples of audition reported from Dictyoptera and Gryllidae, no well-defined tympanal membrane was associated with the chordotonal sensory organ. Even if the putative tympanal membrane does not consist of a transparent or translucent delimited area of cuticle, it may still be thinner than the adjacent cuticle, as in praying mantises (87, 88). Other insects also lack obvious tympana. Some cockroaches, notably the Madagascar hissing cockroach, *Gromphadorhina portentosa,* use acoustic signals for courtship and territorial displays (55, 57). The auditory organs are tibial subgenual organs (28, 56), but no obvious tympanal region or thin cuticle was found. Recently, Shaw (75) reported that the cockroach *Periplaneta americana* can detect airborne sounds. The auditory tuning curve displays a best-frequency of about 1.8 kHz, with thresholds in the range of 55–60 dB sound pressure level , which is comparable to the threshold range of other tympanate insects (51). The putative auditory sensory organ of *P. americana* is the metathoracic tibial subgenual organ (75).

Shaw used isolated cockroach legs. Such dissection quite possibly alters the acoustic characteristics of the leg tracheal system. Intact preparations are preferable, as are tuning curves from single unit recordings. However, Ritzmann et al (63) had previously reported evidence of hearing (also in *P. americana*), based on the acoustic responsiveness of identified interneurons with a striking resemblance to auditory interneurons from other insects, as discussed below.

Thus, in two distantly related genera of cockroaches, the tibial subgenual organ, which is usually thought of as a detector of substrate vibrations, has auditory sensitivity and consequently serves a dual function. Yet both species lack any obvious overlying region of thin cuticle corresponding to a tympanum. Nonetheless, two of the three structures that typically compose a tympanal organ—a scolopidial organ (the subgenual organ) and closely apposed tracheae—are present. Auditory function has also been reported for an atympanate tettigoniid, *Phasmodes ranatriformes* (38). This insect lacks any recognizable tympanal membranes and tracheal expansions in the tibiae of its forelegs. Nonetheless, neural recordings showed that a putative homologue of the auditory organ (the crista acustica) was excited by sound, although it exhibited considerably less sensitivity than that associated with a tympanum. *P. ranatriformes* looks more like a stick insect than a bushcricket or katydid, and Lakes & Schikorsky (38) posit that it may be a primitive tettigoniid.

What are the implications of these studies? For one, auditory function may be even more widespread in insects than previously reported, especially if functional hearing can occur without a clearly differentiated external tympanal membrane.

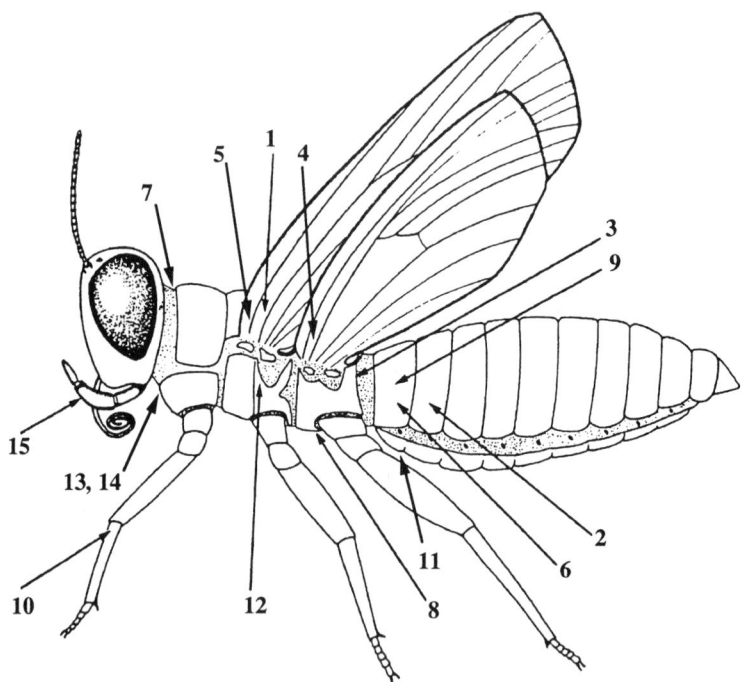

Figure 1 Diagrammatic view of a "generalized" insect, which shows the locations on the body where tympanal hearing organs have been described in various species of insects. The numbers refer to the second column of Table 1, except for number 15, which refers to the pilifer organ of hawkmoths (Lepidoptera: Sphingidae). (Modified from Reference 85.)

Tympana, whether cryptic or conspicuous, may be found at various body locations, from the head to the abdomen and on appendages, including mouthparts (68), wings (53), and legs (72), depending on species. Among insects within an order but from different families, tympanal ears may have different locations (Figure 1) (18, 30a, 73, 84). In many insects, such as in the Orthoptera and at least ten families of Lepidoptera, the tympana are anatomically conspicuous (73), a silvery eardrum can be identified on the tibial segment (crickets, bushcrickets, and katydids) or on the body (moths and grasshoppers). In other groups, e.g. the praying mantis, the tympanum is inconspicuous or hidden from view (87). It may also be cryptic, as in some heliconid butterflies (77) and the green lacewing (53), in which the effective tympanum is associated with the wing veins.

Thus, the identification of tympanate insects on the basis of morphology is not straightforward. Strictly speaking, should only those insects that possess a frankly differentiated tympanal membrane be included? Must the tympanal membrane be obvious to (human) visual inspection? These questions arise because some insects that lack a visually obvious tympanum perform similar acoustic behaviors as insects with tympana—namely, detect high-frequency sounds at long distance in the acoustic far-field—and also possess scolopophorous sensory organs associated with acoustic tracheae. Such insects, e.g. the praying mantis, possess a functionally equivalent tympanum, i.e. a thin (1–10 μm) cuticle that overlies the acoustic tracheae and the scolopophorous organ.

Table 1 and Figure 1 indicate the impressive taxonomic and locational diversity of tympanal organs among insects. Indeed, no other group of animals exhibits such a bewildering diversity of ear placement. One can find tympanal ears in almost any segment of the insect body!

THE BEHAVIORAL FUNCTION OF TYMPANAL HEARING

In the conspicuously tympanate insects, such as cicadas (91), crickets (4, 26, 37), and katydids (31, 69, 80), conspecific communication related to reproduction and territoriality is clearly mediated by loud songs. However, in insects, hearing involves not only mates and rivals, but also predators and prey. In particular, a surprising number of night-flying insects possess tympanal ears that are tuned to ultrasound (36). This is presumably an evolutionary response to predation by insectivorous bats, which detect their prey by means of echolocation using ultrasonic biosonar (24, 67). A brief summary of acoustic behavior follows in which we present a few examples out of the many hundreds in the literature.

Calling, Courtship, and Territorial Signals

Long-distance mate calling is characteristic of insects in several insectan orders. In the most common pattern, the male produces the calling song and the female usually responds by either flying or walking to his location (23, 37). The Orthoptera (crickets, katydids, grasshoppers) and Hemiptera (cicadas) include the most conspicuous of the acoustically signaling insects. In some grasshopper signaling systems, the female produces a stridulation in response to the male's call, resulting in a duet (43, 80a, 80b). Cicadas are among the loudest of daytime singing insects, and their hearing is mediated by a tympanal hearing organ (91). Among the Orthoptera, the songs of mole crickets (Gryllotalpidae) are especially loud because males sing from subterranean burrows dug in the shape of an exponential horn; the males actively adjust the tuning of this burrow (10, 80). The male's call is thus amplified and broadcast high into the air, where it attracts females on the wing (10, 27).

In field crickets, once the male has attracted a female onto his territory with his calling song, his courtship includes a stridulatory act called the courtship song. This song, which is distinct from the calling song, is initiated by mutual antennation (1, 80b) and is strictly a short-range signal. In field crickets, courtship song influences mating success under laboratory conditions (15, 42). The rivalry song that accompanies male-male aggression is also emitted at short range (2, 4, 80b).

Predator Detection

In many insects, particularly the nocturnally active, flying insects, ultrasonic frequencies elicit an acoustic startle or escape response. In general, a series of suprathreshold tone pulses with frequencies ranging from 20 to 50 kHz, although the upper frequency limit may exceed 100 kHz (24), is sufficient to elicit the startle responses (25, 36, 67). These reactions are presumably to insectivorous bats, which hunt insects on the wing by detecting them with the aide of ultrasonic biosonar (32). The best-known example is bat predation upon nocturnally flying moths (67). But since the publication of Roeder's pioneering work (67), many examples have been documented (36), and ultrasound-mediated startle responses have been demonstrated in laboratory experiments on five orders of insects (Lepidoptera, Dictyoptera, Neuroptera, Orthoptera, and Coleoptera). Whether hunting bats actually elicit such startle responses in all cases has not been rigorously demonstrated by field observations; however, predatory acts are rarely observed in nature, and most bat-insect interactions are especially difficult to observe because they occur in darkness, high above the ground.

In insects such as moths preyed upon by bats, ultrasound-sensitive ears appear to have evolved directly as a result of predation (67). In other insects, such as crickets and short- and long-horned grasshoppers, ultrasound sensitivity may have been added onto preexisting auditory capabilities that had mediated social interactions (36, 41, 64).

Acoustic Parasitism

Acoustic parasitism was first demonstrated by Cade (16), who reported that the field cricket *Gryllus integer* is parasitized by the tachinid fly *Euphasiopteryx ochracea*. The larviparous females of this fly, now reclassified as *Ormia ochracea* (83), localize their hosts, male crickets, by hearing and homing in on the mating calls of these crickets (16). Live larvae are then deposited upon or near the singing male cricket, which becomes infested (3). The auditory organ of *O. ochracea* females was described recently (65), and a similar organ was reported in females of another tachinid parasitoid, *Therobia leonidei* (39). The hearing organs of *O. ochracea* and *T. leonidei* were shown to be a typical tympanal ear; it has a well-defined membrane, backed by an air-filled acoustic chamber, and a scolopophorous sensory organ is attached directly to the membrane (65, 66). The ears of female *O. ochracea* appear to be most sensitive to sound in the range 4–6 kHz, which encompasses the dominant frequency of its host's song (4.8 kHz). However, the fly also exhibits sensitivity in the ultrasonic region, 20 to at least 60 kHz. Males of *O. ochracea* also have tympanal hearing organs, but these ears are quite different from those of the females, in both form and function (65, 66). They are relatively insensitive to the 5-kHz frequencies that characterize cricket calls, but they are just as sensitive to ultrasounds as the female fly's ears. Because these parasitoids have adopted the activity cycle of a nocturnally active cricket, and like crickets, they fly at night, they presumably must also be vulnerable to the cricket's predators, such as insectivorous bats that hunt on the wing by echolocation. Therefore, the ultrasound sensitivity in the ears of male and female *O. ochracea* may have arisen from the need to detect and avoid predation by bats.

Sexual Dimorphism in Tympanal Organs

Reports of sexual dimorphism in the auditory systems of insects have been relatively rare, although dimorphism was recently observed in species from four different orders of tympanate insects: a tettigonid (6), the gypsy moth (17), several praying mantis (86, 88), several species of cicadas (89a), and two tachinid parasitoid flies (39, 66). Interestingly, the dimorphism often takes the form of differences in auditory sensitivity, instead of in the gross morphology of the hearing organ itself (17, 88). In at least some of these examples, e.g. mantises and tachinid flies, differences may be more closely related to predator or prey detection than to intraspecific communication. Nonetheless, sometimes following a lead from functional investigations, researchers have established a firm morphological basis for the dimorphism in tachinid flies (66), tettigonids (6), uraniid moths (18), and praying mantises (86).

DEVELOPMENTAL ORIGINS OF TYMPANAL ORGANS

Introduction

The tympanal organs most thoroughly studied from a developmental viewpoint are those of the acoustically active Orthoptera, in particular those of the Caelifera (grasshoppers and locusts) and the Ensifera (katydids and crickets). The Orthoptera undergo hemimetabolous development, and so the ontogenesis of the auditory organ takes place both embryonically, before the first-instar larva hatches from the egg, and postembryonically, as development proceeds through a variable number of larval instars (up to 11 in some field crickets) (13, 90). Because fully differentiated tympanal organs are not usually expressed until the imago stage, postembryonic development must be considered an important part of their ontogenesis. Detailed studies of the auditory organs during both the embryonic and postembryonic stages have been made mainly on acridids and katydids (48, 71) and crickets (7). The following account therefore focuses primarily on the development of the various components of ensiferan organs.

Development of the Tympanal Membrane

In crickets, definitive eardrums do not appear on the surface of the tibia of the foreleg until the adult stage (90). In the katydid *Ephippiger ephippiger,* the foreleg tympana are recessed within tympanal cavities and not visible on the surface, but as in crickets, the tympana are not fully morphologically differentiated until the adult molt (71). In the Australian field cricket, *Teleogryllus commodus,* the first sign of the tympanum is a thinning of hair sensilla on the region of the cuticle that will become the eardrum; this thinning does not occur until the eighth instar (three molts from the adult). By the tenth instar (one molt to adulthood), an area devoid of hair sensilla clearly has taken on the elliptical shape of the posterior tympanal membrane (90). At adulthood, the thin, translucent tympanal membrane takes on a silvery hue because it overlies the air-filled trachea within the leg. This is the membrane upon which sound impinges, and it serves as part of a complex sound-receiving system.

Development of Leg Tracheae

Sound is conducted through leg tracheae in crickets and katydids. These air-filled tubes or chambers are crucial elements of insect audition and are associated with both eardrums and the sensory scolopidial organs (40). In *T. commodus,* the tracheae of the forelegs develop over the course of the 11 instars. As development proceeds postembryonically, the tracheae branch and, during the last few instars, increase considerably in diameter. By the adult stage, these prothoracic acoustic tracheae are proportionally larger than their serial homologues in the meso- and metathoracic tibiae (90). The acoustic tracheae in the forelegs of *E. ephippiger* similarly take on their adult form over the course of six larval instars, becoming adult-like in the fourth instar (71).

Development of the Scolopidial Sensory Organ

The scolopidial sensory organ undergoes an embryonic and a postembryonic stage of development. Meier & Reichert (48) conducted the most definitive study of embryonic development in the Ensifera using two tettigoniids, *Metaplastes ornatus* and *Poecilimon affinis.* In the prothoracic limb buds of embryos 40% into their development, a cluster of ectodermal cells invaginates the lumen of the tibia. It then separates from the epithelium of the leg, migrates, and eventually differentiates into the subgenual organ and the auditory sensory organ, i.e. the crista acustica. The characteristic linear array of scolopidia, which in the adult is correlated with tonotopic representation of frequency (61, 92), is apparently already established in embryos 50% into their development. Moreover, the use of molecular markers for sensory cell identity has shown that, in katydids, the full complement of scolopidia is generated during embryonic development (48), a finding confirmed by postembryonic studies (71). However, whether the embryonic development of these organs occurs in crickets is questionable. In *T. commodus* (7), light microscopic examination of paraffin-embedded and stained legs taken from different larval instars, showed that although the full adult complement of scolopidia in the subgenual organs had appeared by the first instar, no scolopidia in the crista acustica were present until the third instar. During postembryonic development, scolopales are systematically added in a proximal-to-distal order, and the shape of the scolopidia and associated support cells changes as the instars progress toward the adult stage.

Functionality of the Developing Tympanal Organ

Evidence indicates that the immature ear in both crickets and katydids, although lacking frank tympanal membranes, is already sensitive to sound. Hill & Boyan (34) demonstrated that auditory interneurons in the last instar of *T. commodus* had a similar tuning curve to those in adults. By recording compound action potentials from larval tympanal nerves, Rössler (71) demonstrated that the auditory spectral sensitivity is already adult-like by the fourth instar. The intensity thresholds of the earlier-instar organs tended to be considerably higher than those of the adult by at least 20 dB.

In the adult katydid, the proximal-to-distal linear array of scolopidia has functional significance. The frequency sensitivity of the tympanal organ ranges from low in the proximal regions of the crista acustica to high in the distal regions. This tonotopic organization of scolopidia within the katydid crista acustica was inferred from extracellular recordings (92) and shown directly from intracellular recordings from single cells in the scolopidial complex (60). A very similar tonotopic organization occurs in gryllids and was also confirmed at the receptor cell level (61, 93).

EVOLUTIONARY ORIGINS OF TYMPANAL ORGANS

The early histological studies by Friedrich of the anatomy of insect tympanal organs suggested that they arose from proprioceptive chordotonal organs (29, 30). Based on his studies on postembryonic instars and adults, he proposed that the crista acusticae in crickets and katydids shared a common origin with the subgenual organ. The argument that the auditory organ and the subgenual organ are in fact homologous is now confirmed by unequivocal evidence for the common embryonic origin of both subgenual and auditory organs in katydids (48). This embryological study also established that the subgenual organs in meso- and metathoracic legs develop in the same way as the specialized auditory organs in the prothoracic legs, which demonstrates that all of these organs share segmental homology. In moths, Yack & Fullard (84) present evidence that the metathoracic tympanal organ of noctuid moths may have evolved from a chordotonal stretch receptor that serves other functions in atympanate moths. For example, in the tobacco hornworm, *Manduca sexta,* it measures movements of the wings during flapping flight. Meier & Reichert (48) argue that the pair of tympanal organs of locusts, which are located in the first abdominal segment, also evolved from a chordotonal stretch receptor. These authors noted that the other abdominal segments contain pleural chordotonal organs that are serially homologous to the specialized pleural tympanal organ. Their argument is compelling, as it is based on developmental evidence. In tachinid flies, the auditory chordotonal organ is apparently also derived from a chordotonal proprioceptive organ, which has an unknown function in various outgroups of closely and distantly related higher flies (21).

In the past decade, findings on the developing nervous system of the locust, combined with anatomical data on the auditory systems of several insect species, have led to interesting general hypotheses about the functional and evolutionary relationships of tympanal hearing in insects. In locusts, the lineage of a given neuron in the adult can be traced back to its origins in the neuroblasts of the early embryo. This work was begun by the pioneering American entomologist WM Wheeler (81), and its modern-day revival has reinvigorated the studies of homology and evolutionary biology in insect neural systems (9, 62), in particular the developmental and evolutionary origins of tympanal organs in insects.

New ways of thinking about the auditory system in insects have come from the developmental work of Bastiani et al (8), Doe & Goodman (20), and Meier & Reichert (48), combined with a large body of literature on the structure and function of several species' tympanal organs and their neural projections in the central nervous system. In particular, a recent review by Boyan (14) provides an important synthesis of this information from disparate sources. These authors argue that all tympanal organs arise developmentally from a common set of segmentally homologous precursors that have mechanoreception, such as stretch sensitivity, as a primitive function. Moreover, the sensory afferent nerves of these chordotonal organs, whether auditory or not, project into specific areas of ganglionic neuropiles, where they possibly form synapses with at least some common second-order interneurons that in turn project ascending or descending axons in the central nervous system. This arrangement

implies that at least some chordotonal organs in insects have a modular structure, if not multiple functions.

The generality of this hypothesis rests on lineage studies done on the central
and peripheral sensory systems in two species, the grasshopper, *Schistocerca
gregaria* (e.g. 8, 48) and the fruit fly, *D. melanogaster* (47). The notion that
the central networks subserving auditory or chordotonal function contain homologous interneurons related by lineage certainly needs more investigation,
but tantalizing data in support of it have been reported from cockroaches (63),
praying mantises (88), and locusts (8, 14). Certainly, the data supporting a
modularity hypothesis based on serial homologies of receptor and ganglionic
interneuronal systems appear to be well founded. The assertion of phylogenetic
homology of elements between species is always problematic and remains so
also for chordotonal sensory and modular interneuronal systems. However,
species-by-species investigations into the embryological origins of identified
sensory and central neurons would certainly provide a developmental basis on
which to establish homologies and would enhance the analysis of the homology
question.

CONCLUDING REMARKS

Among terrestrial animals, only vertebrates and insects are widely endowed
with a sense of hearing. In vertebrates, hearing is associated with the possession
of an inner ear embedded in the skull. It consists of fluid-filled chambers
containing sensory hair cells and various thin membranes that overlie or are
associated with the hair cells. This organ system appears to have evolved only
once in vertebrate evolution (58), the organs being located in the head and
nowhere else. However, in insects, the possession of hearing organs is not
related to their position. Insect ears have evolved independently from taxon to
taxon and even severally within a taxon (e.g. the Lepidoptera). The reason(s)
for this diversity of ears may be related to the roles hearing plays in the life
of the insect. Thus, field crickets, bushcrickets, and katydids use hearing not
only to receive conspecific communication signals, but also to detect the
ultrasound echolocation signals of predaceous bats.

In these orthopteroids, the same hearing organ serves both these functions,
and we speculate that ultrasound sensitivity was added on to preexisting
sensitivity for lower-frequency conspecific signals. Given that the Orthoptera
can be traced back nearly 200 million years (74), whereas the fossil record
for bats can be traced back only 65 million years (59), this is not an
unreasonable proposal. However, the order Lepidoptera is relatively younger
[about 100 million years (22)], and given that many of the acoustically
sensitive lepidopteran species have ears that are sensitive exclusively to
ultrasound, the predatory pressures from insectivorous bats might have

comprised the primary force in auditory evolution. As pointed out above, the auditory receptive apparatus of many insects appears to be derived from a ubiquitous and anatomically widely distributed sets of paired pleural chordotonal receptors. In acoustically sensitive moths, this origin would account for the diversity of ear locations with (a pair of) tympana on the thorax or on one of several abdominal segments.

Tympanal hearing is at one end of a continuum in the insectan mechanoreceptive-based communication system for detecting remote signal sources. This continuum also includes near-field reception, whereas substrate-borne reception is at the other end of the continuum. We can speculate that the most primitive form of mechanoreceptive communication was substrate signaling, given the ubiquitous occurrence of vibration-sensitive chordotonal, campaniform, and sensory hair receptors in insects.

Finally, we believe that the examples of tympanal hearing we have presented above are just the tip of the iceberg with respect to the actual occurrences of tympanal hearing organs in the Insecta. The recent discoveries of tympanal hearing organs in the Coleoptera and the Diptera point to future discoveries in these speciose taxa. We should not be at all surprised by reports of tympanal hearing in the Hymenoptera. An insect's ability to hear may not be necessarily surmised from an anatomical examination of its surface anatomy. More important are the functional attributes; however, these can be difficult to demonstrate, given that many insects use signals with frequencies and amplitudes beyond (both above and below) those detectable by humans and currently available electronic instrumentation, which is designed around human hearing. The fact that some insect auditory behavior may be performed in the dark of night and high in the air provides an additional challenge for the study of hearing in insects.

Literature Cited

1. Adamo SA, Hoy RR. 1994. Courtship behavior of the field cricket *Gryllus bimaculatus* and its dependence on social and environmental cues. *Anim. Behav.* 47:857–68
2. Adamo SA, Hoy RR. 1995. Agonistic behaviour in male and female crickets (*Gryllus bimaculatus*) and how behavioral context influences its expression. *Anim. Behav.* 49:1491–1501
3. Adamo SA, Robert D, Hoy RR. 1995. The effect of the tachinid parasitoid, *Ormia ochracea*, on the behavior and

reproduction of its host, the field cricket. *J. Insect Physiol.* 41:269–77
4. Alexander RD. 1961. Aggressiveness, territoriality, and sexual behavior in field crickets (Orthoptera: Gryllidae). *Behaviour* 17:130–223
5. Bailey WJ. 1990. The ear of the bushcricket. See Ref. 5a, pp. 217–47
5a. Bailey WJ, Rentz DCF, eds. 1990. *The Tettigoniidae.* New York: Springer-Verlag
6. Bailey WJ, Römer H. 1991. Sexual differences in auditory sensitivity: mis-

match of hearing threshold and call frequency in a tettigoniid (Orthoptera, Tettigoniidae: Zaprochilinae). *J. Comp. Physiol.* 169:349–53

7. Ball EE, Young D. 1974. Structure and development of the auditory system in the prothoracic leg of the cricket *Teleogryllus commodus* (Walker). II. Postembryonic development. *Z. Zellforsch.* 147:313–24

8. Bastiani MJ, Raper JA, Goodman CS. 1984. Pathfinding by neuronal growth cones in grasshopper embryos. III. Selective affinity of the G growth cone for the P cells within the A/P fascicle. *J. Neurosci.* 4:2311–28

9. Bate CM. 1976. Embryogenesis of an insect nervous system. I. A map of the thoracic and abdominal neuroblasts in *Locusta migratoria. J. Embryol. Exp. Morphol.* 35:107–23

10. Bennet-Clark HC. 1970. The mechanism and efficiency of sound production in mole crickets. *J. Exp. Biol.* 52:619–52

11. Bennet-Clark HC. 1989. Songs and the physics of sound production. See Ref. 37, pp. 227–61

12. Bennet-Clark HC, Ewing AW. 1970. The love song of the fruit fly. *Sci. Am.* 223:84–92

13. Bentley DR, Hoy RR. 1970. Postembryonic development of adult motor patterns in crickets: a neural analysis. *Science* 170:1409–11

14. Boyan G. 1993. Another look at insect audition: tympanic receptors as an evolutionary specialization of the chordotonal system. *J. Insect Physiol.* 39: 187–200

15. Burk T. 1983. Male aggression and female choice in a field cricket *(Teleogryllus oceanicus)*: the importance of courtship song. In *Orthopteran Mating Systems*, ed. DT Gwynne, GU Morris, pp. 97–119. Boulder, CO: Westview

16. Cade WH. 1975. Acoustically orienting parasitoids: fly phonotaxis to cricket song. *Science* 190:1312–13

17. Cardone B, Fullard JH. 1988. Auditory characteristics and sexual dimorphism in the gypsy moth. *Physiol. Entomol.* 13:9–14

18. Cook MA, Scoble MJ. 1992. Tympanal organs of geometrid moths: a review of their morphology, function, and systematic importance. *Syst. Entomol.* 17:219–32

19. Dambach M. 1989. Vibrational responses. See Ref. 37, pp. 178–97

20. Doe CQ, Goodman CS. 1985. Early events in insect neurogenesis: development and segmental differences in the pattern of neuronal precursor cells. *Devl. Biol.* 111:193–205

21. Edgecomb RS, Robert D, Read MP, Hoy RR. 1994. Evolution of a dipteran tympanal hearing organ. *Soc. Neurosci. Abstr.* 20:777

22. Evans HE. 1984. *Insect Biology.* Reading, MA: Addison-Wesley

23. Ewing AW. 1989. *Arthropod Bioacoustics.* Ithaca, NY: Comstock Cornell Univ. Press

24. Fenton MB, Fullard JH. 1979. The influence of moth hearing on bat echolocation strategies. *J. Comp. Physiol.* 132: 77–86

25. Fenton MB, Fullard JH. 1981. Moth hearing and the feeding strategies of bats. *Am. Sci.* 69:266–75

26. Forrest TG. 1994. From sender to receiver: propagation and environmental effects on acoustic signals. *Am. Zool.* 34:644–54

27. Forrest TG, Green DM. 1991. Sexual selection and female choice in mole crickets (*Scapteriscus:* Gryllotalpidae): modelling the effects of intensity and male spacing. *Bioacoustics* 5:93–109

28. Fraser J, Nelson MC. 1984. Communication in the courtship of a Madagascan hissing cockroach. II. Effects of deantennation. *Anim. Behav.* 32:203–9

29. Friedrich H. 1927. Untersuchungen über die tibialen Sinnesapparate in den mitteleren und hinteren Extremitäten von Locustiden. II. *Zool. Anz.* 73:42–48

30. Friedrich H. 1928. Untersuchungen über die tibialen Sinnesapparate in den mittleren und hinteren Extremitäten von Locustiden. *Zool. Anz.* 75:86–94

30a. Fullard JH, Yack JE. 1993. *Trends Ecol. Evol.* 8:248–52

30b. Gray EG. 1960. The fine structure of the insect ear. *Phil. Trans. R. Soc. London B* 243:75–94

31. Greenfield MD. 1994. Synchronous and alternating choruses in insects and anurans: common mechanisms and diverse functions. *Am. Zool.* 34:605–15

32. Griffin DR. 1958. *Listening in the Dark.* New Haven: Yale Univ. Press

33. Deleted in proof

34. Hill KG, Boyan GS. 1977. Sensitivity to frequency and direction of sound in the auditory system of crickets (Gryllidae). *J. Comp. Physiol.* 121:79–97

35. Deleted in proof

36. Hoy RR. 1992. The evolution of hearing in insects as an adaptation to predation from bats. See Ref. 80c, pp. 115–29

37. Huber F, Moore TE, Loher W. 1989. *Cricket Behavior and Neurobiology.* Ithaca: Cornell Univ. Press

37a. Kerkut GA, Gilbert LI, eds. 1985. *Comprehensive Insect Physiology, Biochem-*

istry and Pharmacology, Vols. 1–13. New York: Pergamon

38. Lakes R, Schikorsky T. 1990. Neuroanatomy of tettigoniids. See Ref. 5a, pp. 166–90
39. Lakes-Harlan R, Heller K-G. 1992. Ultrasound-sensitive ears in a parasitoid fly. *Naturwissenschaften* 79:224–26
40. Larsen ON, Kleindienst H-U, Michelsen A. 1989. Biophysical aspects of sound reception. See Ref. 37, pp. 364–90
41. Libersat F, Hoy RR. 1991. Ultrasound startle behaviour in bushcrickets (Orthoptera: Tettigoniidae). *J. Comp. Physiol.* 169:507–14
42. Libersat F, Murray JA, Hoy RR. 1994. Frequency as a releaser in the courtship song of two crickets: *Gryllus bimaculatus* (de Geer) and *Teleogryllus oceanicus*: a neuroethological analysis. *J. Comp. Physiol. A* 174:485–94
43. Loher W, Huber F. 1966. Nervous and endocrine control of sexual behavior in a grasshopper (*Gomphocercus rufus* L., Acrididae). *Symp. Soc. Exp. Biol.* 20: 381–400
44. Markl H. 1973. Leistungen des Vibrationssinnes bei Wirbellosen Tieren. *Fortschr. Zool.* 21:100–20
45. Markl H. 1983. Vibrational communication. In *Neuroethology and Behavioural Physiology*, ed. F Huber, H Markl, pp. 332–53. Berlin: Springer-Verlag
46. McIver SB. 1985. Mechanoreception. See Ref. 37a, 6:71–132
47. Meier T, Chabaut F, Reichert H. 1991. Homologous patterns in the embryonic development of the peripheral nervous system in the grasshopper *Schistocerca gregaria* and the fly *Drosophila melanogaster*. *Development* 112:241–53
48. Meier T, Reichert H. 1990. Embryonic development and evolutionary origin of the orthopteran auditory organs. *J. Neurobiol.* 21:592–610
49. Michelsen A. 1992. Hearing and sound communication in small animals: evolutionary adaptations to the laws of physics. See Ref. 80c, pp. 61–78
50. Michelsen A, Andersen BB, Kirchner WH, Lindauer M. 1989. Honeybees can be recruited by a mechanical model of a dancing bee. *Naturwissenschaften* 76: 277–80
51. Michelsen A, Larsen ON. 1985. Hearing and sound. See Ref. 37a, 6:494–556
52. Michelsen A, Towne WF, Kirchner WH, Kryger P. 1987. The acoustic near field of the dancing honeybee. *J. Comp. Physiol.* 161:633–43
53. Miller LA. 1970. Structure of the green lacewing tympanal organ (*Chrysopa carnea*). *J. Morphol.* 131:359–82

54. Moulins M. 1976. Ultrastructure of chordotonal organs. In *Structure and Function of Proprioceptors in Invertebrates*, ed. PJ Mill, pp. 387–426. London: Chapman & Hall
55. Nelson MC. 1979. Behavioral and physiological aspects of sound reception in the cockroach, *Gromphadorhina portentosa*. *Soc. Neurosci. Abstr.* 5:1596
56. Nelson MC. 1980. Are subgenual organs "ears" for hissing cockroaches? *Soc. Neurosci. Abstr.* 6:198.5
57. Nelson MC, Fraser J. 1980. Sound production in the cockroach, *Gromphadorhina portentosa:* evidence for communication by hissing. *Behav. Ecol. Sociobiol.* 6:305–14
58. Northcutt RG. 1985. The brain and sense organs of the earliest vertebrates: reconstruction of a morphotype. In *Evolutionary Biology of Primitive Fishes*, ed. RE Foreman, A Gorbman, JM Dodd, R Olssen. New York: Plenum
59. Novacek MJ. 1985. Evidence for echolocation in the oldest known bats. *Nature* 315:140–41
60. Oldfield BP. 1982. Tonotopic organisation of auditory receptors in Tettigoniidae (Orthoptera, Ensifera). *J. Comp. Physiol.* 147:461–69
61. Oldfield BP. 1985. The tuning of auditory receptors in bushcrickets. *Hearing Res.* 17:27–35
62. Pearson KG, Boyan GS, Bastiani M, Goodman CS. 1985. Heterogeneous properties of segmentally homologous interneurons in the ventral nerve cord of locusts. *J. Comp. Neurol.* 233:133–45
62a. Prager 1976. Das mesothorakale Tympanalorgan von *Corixa punctata* III. (Heteroptera, Corixidae). *J. Comp. Physiol.* 110:33–50
62b. Richards 1933. Comparative skeletal morphology of the noctuid tympanum. *Entomol. Am.* 13:1–43
63. Ritzmann R, Pollack AJ, Hudson SE, Hyvonen A. 1991. Convergence of multi-modal sensory signals at thoracic interneurons of the escape system of the cockroach *Periplaneta americana*. *Brain Res.* 563:175–83
64. Robert D. 1989. The auditory behaviour of flying locusts. *J. Exp. Biol.* 147:279–301
65. Robert D, Amoroso J, Hoy RR. 1992. The evolutionary convergence of hearing in a parasitoid fly and its cricket host. *Science* 258:1135–37
66. Robert D, Read MP, Hoy RR. 1994. The tympanal hearing organ of the parasitoid fly *Ormia ochracea* (Diptera, Tachinidae, Ormiini). *Cell Tiss. Res.* 275:63–78

67. Roeder KD. 1967. *Nerve Cells and Insect Behavior.* Cambridge, MA: Harvard Univ. Press
68. Roeder KD. 1972. Acoustic and mechanical sensitivity of the distal lobe of the pilifer in choerocampine hawkmoths. *J. Insect Physiol.* 18:1249–64
69. Römer H, Bailey WJ. 1986. Insect hearing in the field. II. Male spacing behavior and correlated acoustic cues in the bushcricket *Mygalopsis marki. J. Comp. Physiol.* 159:627–38
70. Römer H, Tautz J. 1992. Invertebrate auditory receptors. *Adv. Comp. Environ. Physiol.* 10:185–212
71. Rössler W. 1992. Postembryonic development of the complex tibial organ in the foreleg of the bushcricket *Ephippiger ephippiger* (Orthoptera, Tettigoniidae). *Cell Tissue Res.* 269:181–88
72. Schwabe J. 1906. Beiträge zur Morphologie und Histologie der tympanalen Sinnesapparate der Orthopteren. *Zoologica* 20:1–154
72a. Scoble MJ. 1986. The structure and affinities of the Hedyloidea: a new concept of the butterflies. *Bull. Br. Museum (Nat. Hist.) (Entomol.)* 53:251–86
73. Scoble MJ. 1992. *The Lepidoptera: Form, Function, and Diversity.* New York: Oxford Univ. Press
74. Sharov A. 1971. *Phylogeny of the Orthopteroidea.* Jerusalem: Israel Program Sci. Publ.
75. Shaw SR. 1994. Detection of airborne sound by a cockroach "vibration detector": a possible missing link in insect auditory evolution. *J. Exp. Biol.* 193:13–47
75a. Soper RS, Shewell GE, Tyrrell D. 1976. *Colcondamyia auditrix* nov. sp. (Diptera: Sarcophagidae), a parasite which is attracted by the mating song of its host, *Okanagana rinosa* (Homoptera: Cicadidae). *Can. Entomol.* 108:61–68
76. Spangler HG. 1988. Hearing in tiger beetles (Cicindelidae). *Physiol. Entomol.* 13:447–52
77. Swihart SL. 1967. Hearing in butterflies (Nymphalidae: *Heliconius, Ageronia*). *J. Insect Physiol.* 13:469–76
78. Tautz J. 1977. Reception of medium vibration by thoracal hairs of caterpillars of *Barathra brassicae* L. (Lepidoptera, Noctuidae). I. Mechanical properties of the receptor hairs. *J. Comp. Physiol.* 118:13–31
79. Tautz J. 1979. Reception of particle oscillation in a medium—an unorthodox sensory capacity. *Naturwissenschaften* 66:452–61
80. Ulagaraj SM, Walker TJ. 1973. Phonotaxis of crickets in flight: attraction of male and female crickets to male calling songs. *Science* 182:1278–79
80a. von Helversen D. 1972. Gesang des Männchens und Lautschema des Weibchens bei der Feldheuschrecke *Chorthippus biguttulus* (Orthoptera, Acrididae). *J. Comp. Physiol.* 81:382–422
80b. von Hörmann-Heck S. 1957. Untersuchungen über den Erbgang einiger Verhaltensweisen bei Grillenbastarden (*Gryllus campestris* L. vs. *Gryllus bimaculatus* DeGeer). *Z. Tierpsychol.* 14:137–83
80c. Webster DB, Fay RR, Popper AN, eds. 1992. *The Evolutionary Biology of Hearing.* New York: Springer-Verlag
81. Wheeler WM. 1893. A contribution to insect embryology. *J. Morphol.* 8:1–160
82. Deleted in proof
83. Wood DM. 1987. *Tachinidae 10.* Manual of Nearctic Diptera, Vol. 2. Research Branch Agriculture Canada Monograph 28, eds. JF McAlpine, BV Peterson,GE Shewell, HJ Teskey, VR Vockeroth, DM Wood
84. Yack JE, Fullard JH. 1990. The mechanoreceptive origin of insect tympanal organs: a comparative study of similar nerves in tympanate and atympanate moths. *J. Comp. Neurol.* 300: 523–34
85. Yack JE, Fullard JH. 1993. What is an insect ear? *Anal. Entomol. Soc. Am.* 86:677–82
86. Yager DD. 1990. Sexual dimorphism of auditory function in praying mantises. *J. Morphol.* 221:517–37
87. Yager DD, Hoy RR. 1986. The cyclopean ear: a new sense for the praying mantis. *Science* 231:727–29
88. Yager DD, Hoy RR. 1989. Audition in the praying mantis, *Mantis religiosa* L.: identification of an interneuron mediating ultrasonic hearing. *J. Comp. Physiol. A* 165:471–93
89. Yager DD, Spangler HG. 1995. Characterization of auditory afferents in the tiger beetle, *Cicindela marutha* Dow. *J. Comp. Physiol. A.* 176:587–99
89a. Young D. 1990. Do cicadas radiate sound through their ear-drums? *J. Exp. Biol.* 151:41–56
90. Young D, Ball E. 1974. Structure and development of the auditory system in the prothoracic leg of the cricket *Teleogryllus commodus* (Walker). *Z. Zellforsch. Mikrosk. Anat.* 147:293–312
91. Young D, Hill KG. 1977. Structure and function of the auditory system of the cicada, *Cystosoma saundersii. J. Comp. Physiol.* 117:23–45
92. Zhantiev RD, Korsunovskaya OS. 1978. Morphological organization of

tympanal organs in *Tettigonia cantans* (Orthoptera, Tettigoniidae). *Zool. J.* 57: 1012–16

93. Zhantiev RD, Tshukanov VS. 1972. Frequency characteristics of tympanal organs of the cricket *Gryllus bimaculatus* Deg. (Orthoptera, Gryllidae). *Vestn. Moscow Univ.* 2:3–8 (In Russian)

Annu. Rev. Entomol. 1996. 41:451–72

BACILLUS SPHAERICUS TOXINS: Molecular Biology and Mode of Action

J.-F. Charles, C. Nielsen-LeRoux, and A. Delécluse

Bactéries Entomopathogènes, Institut Pasteur, 25 rue du Dr. Roux, 75724 Paris Cedex 15, France

KEY WORDS: crystal toxin, Mtx toxins, toxin genes, receptor, resistance

---

ABSTRACT

Bacillus sphaericus is a spore-forming aerobic bacterium, several strains of which are pathogenic for mosquito larvae. During sporulation, the most active strains produce a crystal toxin with a high degree of larvicidal activity. The toxin is composed of two proteins of 51.4 and 41.9 kDa, which are encoded by highly conserved chromosomal genes. After *B. sphaericus* is ingested, these proteins are released in the larva's midgut, and, in susceptible mosquito species, bind to a specific receptor present on midgut brush-border membranes. The resulting damages to the midgut cells leads to the mosquitoes' death. During vegetative growth, some *B. sphaericus* strains also synthesize mosquito larvicidal proteins of 100 and 30.8 kDa (Mtx toxins), the mode of action of which is still unknown. The mechanism of acquisition of the recessive mosquito resistance to the crystal toxin varies with selection conditions.

PERSPECTIVES AND OVERVIEW

The use of microorganisms as a source of biological compounds for insect pest control started after the discovery of the highly insecticidal bacteria *Bacillus thuringiensis*. The discovery of the strain *B. thuringiensis* serovar *israelensis* (35, 45) made possible efficient microbiological control of Diptera Nematocera vectors of diseases, such as mosquitoes (Culicidae) and black flies (Simuliidae).

The first reported *Bacillus sphaericus* strain active against mosquito larvae was isolated from moribund mosquito larvae (51). The larvicidal activity of this isolate was so low that their use in mosquito control would never have been considered (Table 1). The identification of the strain SSII-1 in India (73)

451

renewed the search for more active strains. But only after the isolation (in Indonesia from dead mosquito larvae) of strain 1593—which exhibited a much higher level of mosquitocidal activity (74) was the potential of *B. sphaericus* as a biological-control agent for mosquitoes taken seriously (75).

B. sphaericus (57) is an aerobic bacterium that produces terminal spherical spores. It lacks several biochemical pathways and thus cannot use sugars as metabolites. The findings of various genetic and biochemical studies indicate that this species is heterogenous. The increasing number of isolated strains has made differentiating between toxic and atoxic strains, and between the toxic strains themselves, difficult.

Numerous methods have been used to classify this very heterogenic species. For example, relationships between strains of *B. sphaericus* have been examined in terms of DNA homology (53). Based on the percentage of homology, five groups were identified, and group II was further subdivided into IIA (related by over 79% homology and containing all toxic strains) and IIB. Classification via two other systems, bacteriophage typing (85, 87) and sero-typing using flagellar antigens (36, 38), yielded a similar grouping for the group IIA mosquito pathogenic strains. Currently, nine serotypes (listed in Table 1) are known to contain active strains. Recently developed techniques, such as numerical classification based on taxonomy of phenotypic features (2, 46), cellular fatty acid composition analysis (42), ribotyping (3), and random amplified polymorphic DNA analysis (84), indicate that most of the pathogenic strains are recovered in a few groups.

None of these methods allows us to predict the level of toxicity of a given strain. The three mosquitocidal activity levels do not correspond to the groups defined by any other method of classification (Table 1). The genes encoding various toxins have been cloned (see below), and they might be used in hybridization experiments to predict activity (4). However, some strains react positively with toxin genes but display only weak larvicidal activity. Therefore, to identify potentially valuable strains and to further elucidate the nature of the mosquito-larvicidal activity, we must continue to determine toxicity levels by using mosquito larvae.

B. sphaericus strains are generally highly active against larvae of *Anopheles* and *Culex* species and are poorly or not toxic to larvae of *Aedes* species. However, susceptibility appears to depend on the species of mosquito and can thus vary within a genus (14). Toxicity levels vary among the larvicidal serotypes and even within the same serotype (82). Therefore, the relative potency of each strain is currently evaluated and described in terms of specific activity titers on the different mosquito species and in terms of activity ratios derived from such titers (82).

Like *B. thuringiensis* against Lepidoptera, Singer (73) first suggested that *B. sphaericus* acts by toxemia rather than septicemia. A few years later,

Table 1 Comparative properties of some mosquitocidal strains of *B. sphaericus*[a]

Strain	Origin	Source	Flagellar serotype	Phage type	DNA group	RNA (R) group	Fatty acid group	Mosquito-cidal activity[b]	Crystal genes	Mtx genes mtx	Mtx genes mtx2
Kellen K	USA	*Culiseta incidens*	1a	1	IIA	NA	AIII	L	–	+	NA
Kellen Q	USA	*Culiseta incidens*	1a	1	IIA	NA	AI	L	–	+	+
BS-197	India	NA	1a	NA	NA	NA	NA	H	+	NA	NA
SSII-1	India	*Culex fatigans*	2a2b	2	IIA	RIIA	AII	M	–	+	+
IAB 881	Ghana	*Culex* breeding site	3	–[c]	NA	NA	AV	H	+	–	NA
LP1-G	Singapore	Breeding site	3	8	IIA	NA	NA	M	+	–	NA
1593M	Indonesia	*Culex fatigans*	5a5b	3	IIA	RIIA	AIII	H	+	+	+
2362	Nigeria	*Simulium damnosum*	5a5b	3	IIA	RIIA	AIII	H	+	+	+
2317-3	Thailand	NA	5a5b	–	NA	NA	NA	H	+	+	+
1691	El Salvador	*Anopheles albimanus*	5a5b	3	IIA	NA	NA	H	+	+	+
IAB 59	Ghana	*Anopheles gambiae*	6	3	NA	NA	NA	H	+	+	+
31	Turkey	Breeding site	9a9c	8	NA	NA	NA	L	–	NA	NA
2314-2	Thailand	NA	9a9b	–	IIA	RIIA	AV	L	–	NA	NA
2297 (MR4)	Sri Lanka	*Culex pipiens*	25	4	IIA	NA	AIII	H	+	+	+
2173 (ISPC5)	India	*Culex fatigans*	26a26b	3	IIA	NA	NA	M	–	–	–
IAB 872	Ghana	*Culex* breeding site	48	3	NA	NA	NA	H	+	NA	NA
ATCC 14577[d]	USA	NA	4	–	I	RI	AIV	–	–	NA	NA

[a] Data compiled from References 2, 3, 38, 42, 46, 53, 80, 87; I Thiéry, AA Yousten, A Porter, and C Berry (personal communications); and from the IEBC (Institut Pasteur). +, present; –, absent; NA, no information available

[b] Based on the lethal concentration 50% after 48 h on fourth-instar larvae of *Culex pipiens* (LC$_{50}$): L, $\geq 1 \times 10^{-3}$ sporulated culture dilution; M, LC$_{50} \approx 1 \times 10^{-4}$ dilution; H, LC$_{50} \leq 1 \times 10^{-6}$ dilution.

[c] Strains not responding to any of the bacteriophages tested.

[d] *B. sphaericus* type strain.

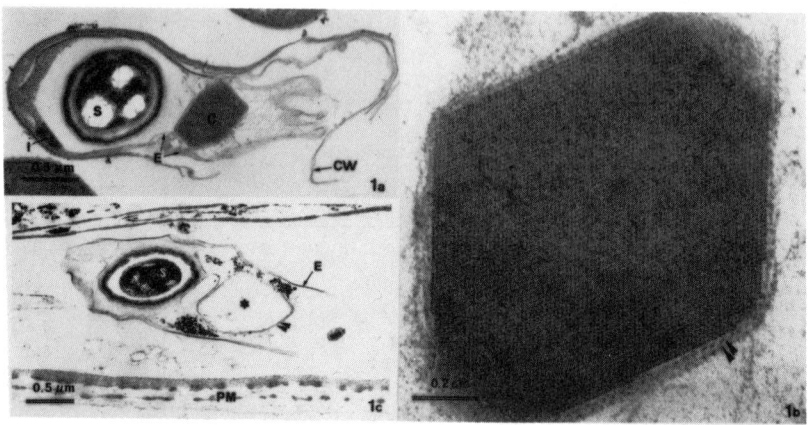

Figure 1 (*a*) Liberation of the *B. sphaericus* 2297 spore-crystal complex during sporangium lysis. C, crystal, CW, cell wall, E, exosporium, I, noncrystalline inclusion, S, mature spore. (*b*) Detail of crystal lattice. Arrowheads indicate the double envelope surrounding the protein crystal matrix. (*c*) Dissolution of *B. sphaericus* crystal matrix (*star*) in the anterior stomach of *Aedes aegyptyi* larvae. E, exosporium, PM, peritrophic membrane, S, spore core. Arrowheads indicate the empty crystal envelope.

Davidson & Myers (28) reported the presence of parasporal inclusions, which they suspected participate in the toxic action of *B. sphaericus*. Indeed, all of the most toxic strains produce parasporal inclusions during sporulation. These inclusions have a crystalline ultrastructure and are released into the medium along with the spore after the completion of sporulation (Figure 1). The relationship between sporulation, crystal formation, and mosquitocidal activity was first clearly established for strains belonging to serotypes H5a5b (18, 56), H25 (37, 49, 86), and H6 (39). Later, the partial purification of crystals toxic to mosquito larvae (64), and the study of mutants blocked at early stages of sporulation that fail to make crystals and are not mosquitocidal (24), confirmed the toxic nature and protein composition of the crystals. More recently, the identification of a group of toxins different from the crystal toxins, the Mtx toxins, renewed interest in *B. sphaericus* as a mosquitocidal agent (80, 81).

Two other recent reviews discuss the genetics of *B. sphaericus* toxins and their mode of action (11, 66). The increasing application of *B. sphaericus* in the field has recently led to cases of resistance. To restrict the development of resistance, we must understand the nature and mode of action of *B. sphaericus* toxins, as well as the mechanism of resistance. These issues have been extensively investigated, and an overview of each is given below.

BIOCHEMISTRY AND GENETICS OF *B. SPHAERICUS* TOXINS

Two kinds of toxins (crystal and Mtx toxins) seem to account for the larvicidal activity of *B. sphaericus*. They differ both in their composition and time of synthesis. The crystal toxins are present in all highly active strains and are produced during sporulation. The Mtx toxins are responsible for the activity of most of the weakly active strains. Mtx proteins seem to be synthesized only during the vegetative phase.

Crystal Toxins

The crystal toxin is composed of two proteins that are synthesized in equimolar amounts and assembled in crystal structures visible at about stage III of sporulation (12, 49, 86). The proteins are designated as P51 and P42 on the basis of their predicted molecular masses of 51.4 and 41.9 kDa (5, 9, 10, 13, 15, 47, 48). The genes encoding both proteins have been cloned from several high-toxicity strains. They appear to be organized in an operon with a 174– to 176–base pair (bp) intergenic region. A stem-loop structure (characteristic of transcription terminators) lies downstream from the P42 gene (Figure 2a) (9, 47). No sequence similar to *Bacillus subtilis* sporulation promoters has been found upstream from the gene encoding P51. However, both the P42 and P51

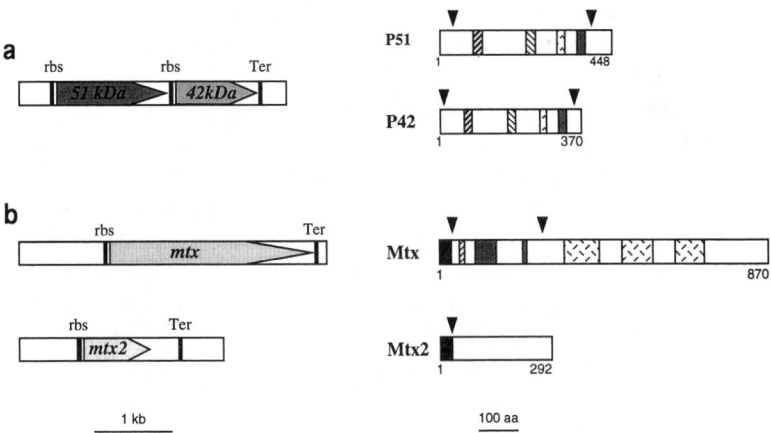

Figure 2 Genes encoding the *B. sphaericus* toxins (*left*) and schematic representation of corresponding polypeptides (*right*). (*a*) Crystal toxin. The homologous regions shared by P51 and P42 are represented by identical patterns. (*b*) Mtx toxins. The black boxes indicate the potential signal sequences; the diagonally striped box denotes the potential transmembrane sequence; the two shaded regions are homologous with ADP-ribosyltransferase toxins; and the crosshatched box signifies internal repeated sequences. Arrows point to the tryptic cleavage sites determined for each toxin. Potential ribosome binding sites and terminator sequences are also indicated for each toxin.

genes are expressed only during sporulation in *B. subtilis* (7). Moreover, *lacZ* fusions to the promoter of the crystal protein genes indicate that, in *B. sphaericus*, transcription begins immediately before the end of exponential growth and continues into stationary phase (1). Thus, enough protein for crystal formation accumulates before stage III.

The genes for P51 and P42 are chromosomal, at least in strains 1593, 2362, and 2297, although these strains contain plasmids (3). Both genes have been reported in diverse strains, and all highly toxic strains tested contain similar sequences as assessed using DNA hybridization (4). The reverse is not true: Moderately toxic strains such as LP1-G also contain these genes (Table 1). The toxin-coding and flanking regions of strains 1593, 2362, and 2317-3 are identical over a span of 3479 nucleotides, whereas the sequences in strains IAB59 and 2297 differ by 7 and 25 nucleotides, respectively (15). The resulting differences in the amino acid sequences of P51 are 5 and 3 amino acids, respectively and those for P42 are 1 and 5. These variations appear to be responsible for the differences in specificity of these strains (see below). Interestingly, LP1-G is only weakly active, although it produces the crystal toxin (Table 1). The sequence of crystal-toxin genes from this strain may indicate which regions of the protein are involved in the loss of toxicity.

The amino acid sequences of P51 and P42 are not similar to those of any other bacterial toxins, including those produced by *B. thuringiensis* serovar *israelensis* (Bti). However, P51 and P42 share four segments of sequence similarity (9) (Figure 2), the significance of which remains unclear. Therefore, P51 and P42 of *B. sphaericus* constitute a separate family of insecticidal toxins (9).

The aggregation of both P51 and P42 has been analyzed using crystal-toxin components expressed separately or together in homologous or heterologous *Bacillus* hosts. When expressed independently in *B. subtilis* and *B. sphaericus* or in *B. thuringiensis* crystal-negative hosts, the proteins form amorphous inclusions (23, 59); in contrast, crystals similar to those produced by naturally occurring, highly toxic *B. sphaericus* strains are produced when the two genes are simultaneously expressed in either *B. sphaericus* or *B. thuringiensis* (23, 59). No crystal can be detected in *B. subtilis* even when both genes are present, unless the genes are fused, eliminating the intergenic region (23). These results suggest that *B. sphaericus* and *B. thuringiensis* contain factors that help stabilize the proteins and subsequent crystallization of the toxin.

In vivo, P42 is slowly converted into a stable protein of ~39 kDa, whereas P51 is rapidly converted into a stable fragment of ~43 kDa (12, 19). In vitro deletion analysis led to the delineation of the minimal active fragments of both proteins, which indeed correspond to the activated fragments (20, 26, 62, 70). At the N and C terminus of P51, 32 and 53 amino acids, respectively, can be eliminated without loss of toxicity (26); deletions of 10 and 17 amino acids at the N and C terminus of P42, respectively, result in a protein similar to the 39-kDa activated fragment (20) (Figure 2a).

Subcloning experiments have shown that P42 alone is toxic for mosquito larvae (*Culex pipiens pipiens*), although the activity is weaker than that of crystals containing both proteins (59). In contrast, P51 alone is not toxic, but its presence enhances the larvicidal activity of P42, suggesting synergy between the polypeptides (17, 40, 59, 62). In vitro assays on mosquito cell lines also revealed that the activated form of P42 alone is toxic to *Culex pipiens quinquefasciatus* cells, whereas P51 appears to be inactive; however, no synergy between the components was observed in vitro (8). Although simultaneous presence of both proteins appears necessary for full toxicity, the activity toward various mosquito species displayed by the different *B. sphaericus* strains depends on the origin of P42, as shown by analysis of toxins mutated in vitro (14). For example, when amino acids were substituted in a region centered around position 100 of P42, the IAB59 and 2297 P42 proteins became similar to the protein from strain 2362: The activity and specificity of the mutant toxins toward *Culex* and *Aedes* larvae were comparable, unlike those of the wild-type. This finding indicates that this region is involved in specificity. Together, these observations suggest that P42 is the most important determinant of specificity and activity.

Mtx Toxins

Two types of Mtx toxins have been described to date: Mtx and Mtx2, with molecular masses of 100 and 30.8 kDa, respectively; genes for both these toxins have been cloned from the medium-toxicity strain SSII-1 (80, 81). In contrast to the crystal toxin genes, *mtx* and *mtx2* genes are expressed during the vegetative growth phase, and sequences resembling the vegetative promoters of *B. subtilis* have been found upstream from each gene (80, 81). Fusions of the *lacZ* gene to the *mtx* promoter also confirmed vegetative expression, because β-galactosidase activity was restricted to the early exponential phase, at least in *B. sphaericus* (1). Both proteins possess short N-terminal leader sequences, which is characteristic of gram-positive bacterial signal peptides (80, 81). However, both toxins have been found associated with the cell membrane of *B. sphaericus*, which indicates little or no cleavage of the signal sequence. The Mtx toxin can be further processed by gut proteases into two fragments of 27 and 70 kDa that correspond to the N- and C-terminal regions, respectively (79). The 70-kDa fragment from Mtx possesses three repeated regions of ~90 amino acids each (Figure 2b), the function of which is unknown; the 27-kDa fragment contains a short region corresponding to a transmembrane domain. The Mtx and Mtx2 toxins do not display any similarity to each other or to the crystal proteins or any other insecticidal proteins. However, the

27-kDa part of Mtx shares weak similarity with the catalytic domain of several ADP-ribosylating toxins (Figure 2b) (79, 80). Mtx2 has 34 and 29% similarity to two toxins active against mammalian cells, namely the ε-toxin from *Clostridium perfringens* and the cytotoxin from *Pseudomonas aeruginosa*, respectively (81).

The genes encoding Mtx and Mtx2 are widespread among various *B. sphaericus* strains (55, 81), including those with low, medium, and high toxicity (Table 1). The toxicity of Mtx has been measured using fusion proteins synthesized in *Escherichia coli:* It is highly active against mosquito larvae (79), with an LC_{50} value comparable to that of the two-component crystal proteins (15 ng protein/ml). Deletion analysis suggests that the 27-kDa fragment from Mtx can ADP-ribosylate itself, whereas the 70-kDa fragment is responsible for toxicity to cultured *C. pipiens quinquefasciatus* cells; however, both regions are necessary for toxicity to mosquito larvae (78).

MODE OF ACTION OF THE CRYSTAL TOXIN

The mode of action of *B. sphaericus* crystal toxin has only been studied in mosquito larvae. However, in one report, the toxin was active in adults of *C. pipiens quinquefasciatus,* but not in *A. aegypti* adults, after introduction by enema into the midgut (77).

After ingestion of the spore-crystal complex by mosquito larvae, the protein-crystal matrix quickly dissolves in the lumen of the anterior stomach (21, 25, 86) through the combined action of midgut proteinases and the high pH (22, 27). *B. sphaericus* crystals release the toxin in all species, even in nonsusceptible species such as *A. aegypti* (Figure 1c). Indeed, some studies report that the differences in susceptibility to *B. sphaericus* between mosquito species do not result from differences in solubilization and/or activation of the crystal toxin (12, 31, 58).

Cytopathology and Physiological Effects

Bti toxins completely break down the larval midgut epithelium, whereas *B. sphaericus* toxins do not. Nevertheless, midgut alterations start as soon as 15 min after ingestion of the *B. sphaericus* spore-crystal complex (21, 28, 50, 76). Midgut damages in *Culex pipiens* are the same after ingestion of spore crystals of either strain 1593 or strain 2297 (21, 28, 76). In contrast, the symptoms of intoxication produced by these two strains differ in other mosquito species. Large vacuoles (and/or cytolysosomes) appear in *C. pipiens* midgut cells, whereas large areas of low electron density appear in *Anopheles stephensi* midgut cells (compare Figures 3b and 3c with the control, Figure 3a). A generally occurring symptom is mitochondrial swelling, described for *C. pipi-*

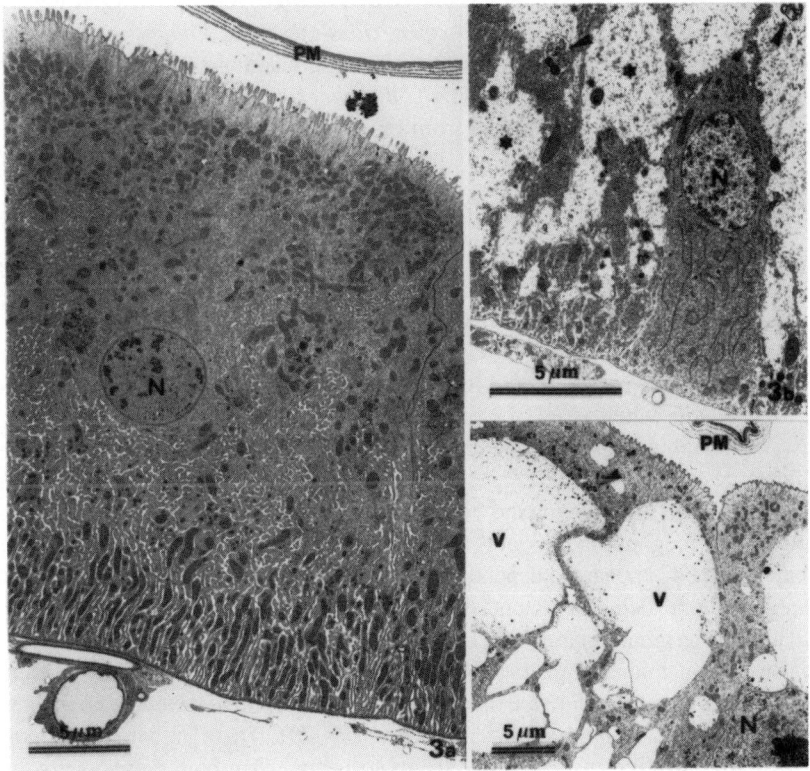

Figure 3 (*a*) Midgut cell of untreated *A. stephensi* larva. (*b*) Midgut cell of *A. stephensi* after feeding on *B. sphaericus* 2297. (*c*) Midgut cell of *C. pipiens pipiens* after feeding on *B. sphaericus* 2297. N, nucleus, PM, peritrophic membrane, V, vacuoles. Arrows indicate cytolysosomes; stars indicate areas of low electron density.

ens pipiens and *A. stephensi*, as well as for *A. aegypti* when intoxicated with a very high dose of spore crystals (21). The midgut cells, especially those of the posterior stomach and the gastric caecae, are the cells most severely damaged by the toxin, and Singh & Gill (76) also report late damage in neural tissue and in skeletal muscles.

Ultrastructural effects have been reported in cultured cells of *C. pipiens quinquefasciatus* within a few minutes of treatment with soluble and activated *B. sphaericus* toxin. These alterations consisted mainly of swelling of mitochondrial cristae and endoplasmic reticula, followed by enlargement of vacuoles and condensation of the mitochondrial matrix (33).

The physiological effects of *B. sphaericus* crystal toxin have been poorly documented. Lakshmi Narasu & Gopinathan (54) report that oxygen uptake by mitochondria isolated from *B. sphaericus*–treated *C. pipiens quinquefasciatus* larvae is inhibited, as is the activity of larval choline acetyl transferase in the presence of toxin. In addition, *B. sphaericus* toxin decreases oxygen uptake by mitochondria isolated from rat liver (54).

Binding to a Specific Receptor in the Brush-Border Membrane

As the differences in susceptibility between mosquito species do not seem related to the ability to solubilize and/or activate the binary toxin, this variation presumably results from differences at the cellular level. Indeed, studies report the binding of fluorescently labeled toxin to the gastric caecae and the posterior stomach only in very susceptible *Culex* species. P51 does not bind to the gut of *A. aegypti*, whereas P42 binds weakly and nonspecifically in this species. No regional binding was observed in *Anopheles* spp. (29, 30, 34, 63). Furthermore, these studies showed that, in *C. pipiens quinquefasciatus*, only P51 binds specifically to the caecum and posterior stomach, whereas P42 binds nonspecifically throughout the midgut (Figure 4). When the proteins are together, the binding of P42 appears to depend on the binding of P51. In vivo binding studies using nontoxic deletion mutants of the crystal toxin (63) support the hypotheses that the N-terminal region of P51 is involved in regional binding of this protein in the larval midgut and that the C-terminal region of P51 and the N- and C-terminal regions of P42 are involved in the interaction of the two components that causes P42 to bind in the same regions as P51. These authors also report that internalization of toxin only seems to occur when both components are present. However, further intracellular investigations are required to elucidate whether one or both components are internalized.

The hypothesis that a specific receptor was involved in the toxin binding was confirmed by in vitro binding assays using ^{125}I-labeled activated crystal toxin and midgut brush-border membrane fractions (BBMFs) isolated from either susceptible or nonsuceptible mosquito larvae (60). Indeed, direct binding experiments with *C. pipiens pipiens* BBMFs indicated that the toxin binds to a single class of specific receptor. The toxin-receptor binding characteristics include a dissociation constant (K_d) of 20 ± 5 nM toxin and a receptor concentration (R_t) of 7 ± 4 pmol toxin/mg of BBMF protein (Figure 5a). No significant specific binding was detected with BBMFs from *A. aegypti* (Figure 5b), which was consistent with the lack of specific binding in fluorescence-labeling studies (30, 32). Both crystal toxin components were bound to the membranes of the susceptible species, but the linearity of the Scatchard representation clearly confirmed that only one of them bound to a receptor (Figure 5a, inset).

However, binding studies exposing radiolabeled P42 and P51 separately to

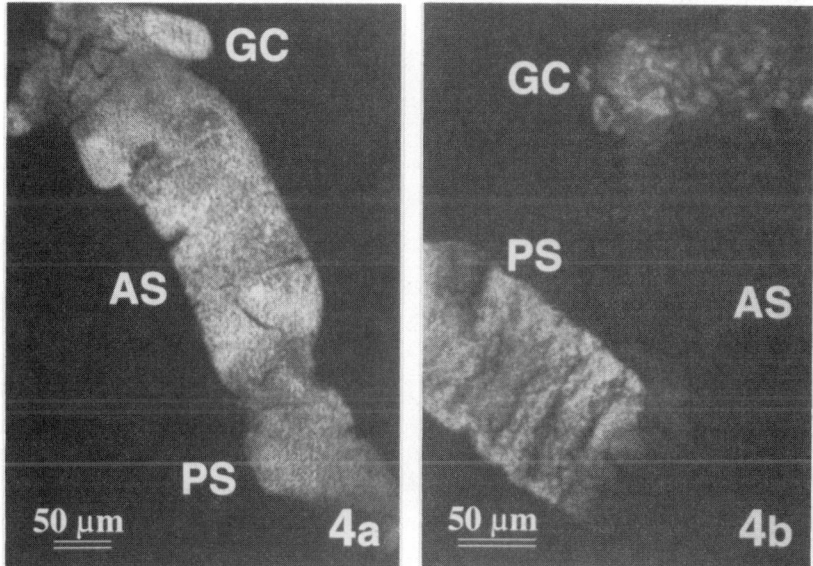

Figure 4 (a) Midgut of *C. pipiens quinquefasciatus* larva fed with fluorescently labeled P42. This protein binds over the entire midgut. (b) Larva fed with fluorescently labeled P42 and unlabeled P51; binding of P42 is regional, restricted to the gastric caecae and posterior stomach. AS, anterior stomach, GC, gastric caecae, PS, posterior stomach. Micrographs kindly provided by Dr. C Berry, Department of Biochemistry, University of Wales, College of Cardiff.

BBMF might give valuable information about toxin-receptor interactions and the kinetics for each component. If we assume that the P42 component is the toxic (active) moiety (59, 70) and P51 is the binding component, the *B. sphaericus* crystal toxin is more likely to be similar to an A/B toxin than to a binary toxin. The nature of the receptor is still unknown. Sugars including *N*-acetylamino-D-galactose or *N*-acetylamino-D-glucose do not inhibit the binding of the toxin to the receptor (60). Therefore, these sugar moieties are probably not involved in the recognition and binding of the toxin in *C. pipiens pipiens* larvae.

INSECT RESISTANCE TO *B. SPHAERICUS* CRYSTAL TOXIN

The risk of emergence of resistance should be considered when designing application strategies. Furthermore, an understanding of the mechanism of resistance might also lend insight into the mode of action of the toxin.

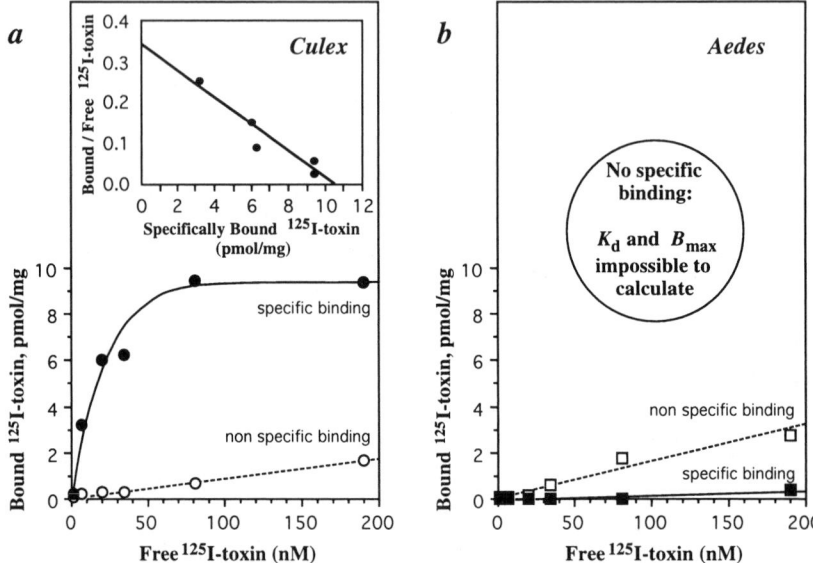

Figure 5 Binding of ^{125}I-labeled *B. sphaericus* crystal toxin to midgut brush-border membranes of (*a*) *C. pipiens pipiens* and (*b*) *A. aegypti*. Inset, specific binding in Scatchard coordinates.

Five cases of *C. pipiens* larval resistance to *B. sphaericus* crystal toxin have been reported (Table 2) in both laboratory-selected populations (43, 69) and field populations after treatment with *B. sphaericus* (67, 71, 72). The mechanism of resistance to *B. sphaericus* crystal toxin has been studied extensively in only two *C. pipiens quinquefasciatus* populations: a highly resistant laboratory-selected population (43) and a field-treated population with a low level of resistance (71).

Laboratory-Selected Resistance

A highly resistant *C. pipiens quinquefasciatus* (*Cpq*R) population was obtained under intense selection pressure by exposing a large number of early fourth-instar larvae in 12 successive generations to *B. sphaericus* toxin under a high selection pressure (43; GP Georghiou et al, unpublished data). Tests of the toxicity of *B. sphaericus* SPH88 (strain 2362) against susceptible (*Cpq*S) and resistant (*Cpq*R) larvae, their F_1 progeny (*Cpq*S × *Cpq*R), and the backcross (BC) offspring (F_1 × *Cpq*R) showed that the LC_{50} of the resistant strain was more than 100,000-fold higher than that of the susceptible strain, and that their dose-response slopes were equal. Likewise, the susceptibility of the F_1 was

Table 2 Laboratory- and field-selected *B. sphaericus* resistance in *C. pipiens quinquefasciatus* and *C. pipiens pipiens* populations[a]

C. pipiens populations	*B. sphaericus* strains or formulations used for selection	Resistance selected	Level of resistance[b]	Conditions of selection			Stability of resistance without selection pressure
				Generation	Selection pressure	Number of individuals per generation	
C. pipiens quinquefasciatus USA (California)	Strain 2362	In the laboratory (field colony)	100,000	F_{12}	LC_{94-99}	10,000–20,000	Lab: yes
C. pipiens quinquefasciatus USA (California)	Strain 2362	In the laboratory Field colony Laboratory colony	27 37	F_{80} F_{80}	LC_{70-80}	1,500–2,500	Lab: yes Lab: yes
C. pipiens quinquefasciatus India (Kochi)	Biocide-S® strain 1593M Strain 1593M	In the field In the laboratory (field colony)	150 6,200	 F_{18}	2 years 35 treatments		Field: no ND
C. pipiens quinquefasciatus Brazil (Recife)	Spherimos® strain 2362	In the field	10		2 years 37 treatments		Field: no
C. pipiens pipiens France (Port Saint-Louis)	Spherimos® strain 2362	In the field	16,000		8 years 18 treatments		Field: no Lab: yes

[a] Data compiled from References 43, 61, 67, 69, 71, 72; G Sinègre and N Pasteur (personal communications).
[b] Ratio of the LC_{50} of the resistant population to the LC_{50} of the susceptible population.

very close to that of the susceptible parent, indicating recessive resistance. The susceptibility of BC larvae showed no linear relationship to toxin concentration.

Binding experiments (61) showed that all available receptors in the parental *CpqS* population were saturable (Figure 6*a*). The corresponding Scatchard plot was linear, indicating that the radiolabeled toxin binds to a single class of specific binding sites, as previously found for the *B. sphaericus* receptor from *C. pipiens pipiens* (60). In the *CpqR* population, the toxin concentrations used did not result in specific binding to the BBMFs, whereas the nonspecific binding was equivalent to the one occurring in *CpqS* (Figure 6*b*). The BBMFs from the F_1 progeny had a single class of binding sites (Figure 6*c*) with characteristics similar to those of the susceptible strain. The binding sites of BBMFs from the BC progeny, *CpqBC*, were not saturated by toxins in the range of concentrations used, and the total amount of bound toxin was much lower than for the *CpqS* strain (Figure 6*d*). The LIGAND computer program analysis (55a) of the binding data obtained with the BC progeny suggests that the experimental data fit a two-site model better than a one-site model.

The results of binding experiments with labeled *B. sphaericus* binary toxin agreed with bioassay data. The resistance is recessive, and the two classes of binding sites found in the BC populations suggest genetic heterogeneity.

The extremely high resistance level to *B. sphaericus* that results after intense selection pressure (43) does not appear under lower selection pressure (69). This finding might indicate that the selection pressure, the number of individuals treated per generation, and the gene frequency of the resistant alleles are important elements in the potential development of high-level resistance by the mosquito populations.

Field-Selected Resistance

The efficacy of *B. sphaericus* treatment for control of the *C. pipiens quinquefasciatus,* which is the vector for Bancroftian filariasis, was tested for two years in Recife (Brazil). Good control of the vector was obtained, but the susceptibility of the field-treated population to *B. sphaericus* decreased significantly (68). Bioassays indicated that the resistance level was 6.5-fold after 33 treatments and 10.2 fold after 37 treatments (71). In vitro binding experiments using radiolabeled *B. sphaericus* toxin and BBMFs from the treated or control larval populations indicated no change in the affinity for the toxin of the midgut receptor of the treated population ($K_d \approx 11$ nM). Only one class of binding sites was found in the two populations, and only a slight decrease in receptor concentration was observed in the resistant population ($R_t \approx 7.5$ from 9.2 pmol/mg). Thus, the mechanism of resistance involved in the field-selected resistance did not appear to involve receptor functionality.

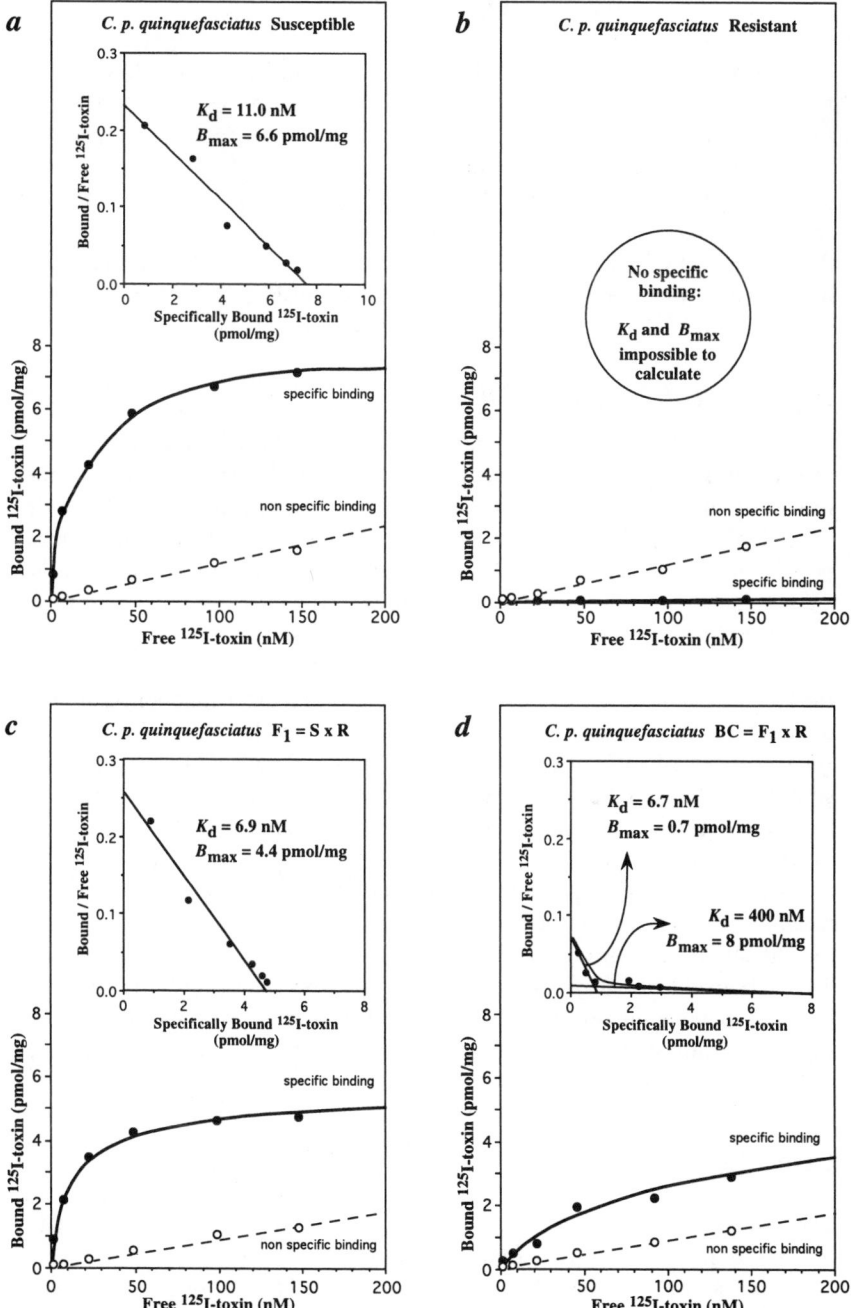

Figure 6 Binding of ^{125}I-labeled *B. sphaericus* binary toxin to *C. pipiens quinquefasciatus* midgut brush-border membranes from (*a*) susceptible (*CpqS*), (*b*) resistant (*CpqR*), (*c*) cross S × R (*CpqF1*), and (*d*) backcross F₁ × R (*CpqBC*) populations.

The resistance levels of 150-fold reported from India (67) and 16,000-fold from France (72) clearly show that high levels of resistance can emerge under field conditions. In both regions, treatments were repeated, generation turnover was high, and mixing with nonselected populations was low.

Recent results from in vitro binding studies using BBMFs from the *B. sphaericus*–resistant *C. pipiens pipiens* population from France suggest that the mechanism of resistance does not involve a receptor modification at least in its binding function (C Nielsen-LeRoux, unpublished data). Therefore, the mechanism of field resistance requires further investigation.

Physiological Function of the Receptor

Because resistance is stable in the absence of selection pressure and the resistant population appears to have good fitness in the wild, mutation of the receptor molecule probably does not suppress the expression of the receptor's normal physiological function. Alternatively, the selected individuals may possess another molecule with the same function. If the toxin receptor on the membrane has an important function in the cell, as does the aminopeptidase-N that is the receptor for *B. thuringiensis* CryIA(c) in the tobacco hornworm, *Manduca sexta* (52), the modification of one receptor could be compensated for by other molecules that have a similar physiological function. This hypothesis is supported by the observation that field populations of *B. sphaericus*–resistant *C. pipiens quinquefasciatus* in Brazil and India were more susceptible than the normal populations to Bti (two- and sevenfold, respectively) (67, 71). Experiments have shown that labeled *B. sphaericus* toxin and nonlabeled toxins from Bti crystal proteins do not compete for binding sites (C Nielsen-LeRoux, unpublished data), indicating that they have different receptors. An increase in the synthesis of other molecules, such as those able to bind Bti toxins, may compensate for the loss of *B. sphaericus* susceptibility. The high-level laboratory-selected *B. sphaericus*–resistant *C. pipiens quinquefasciatus* population was as susceptible to Bti as the unselected population; this observation suggests that laboratory- and field-selected resistance involve different mechanisms.

CONCLUSIONS

The sequences of the genes encoding P42 and P51, and of the regions important for the specificity and activity of the components, have been almost elucidated. Both P42 and P51 are needed for full activity of the *B. sphaericus* crystal toxin: P42 alone is weakly toxic, and P51 appears to be responsible for the regional binding of the toxin to a highly specific and saturable midgut membrane receptor. The nature of the receptor, the intracellular interactions, and the target molecules of the crystal toxin remain unclear, and further work is required to elucidate the mode of action. The Mtx toxins have been less extensively studied than the crystal toxins, and their modes of action probably differ.

Despite the reports of resistance, the future of *B. sphaericus* in the control of mosquito larvae is promising. Indeed, resistance in the field seems to decline very quickly when treatments are suspended (R Reuben & L Regis, personal communication), probably because of recessive gene expression. G Sinègre (personal communication) found resistance in only one breeding site in an area treated for eight years (two or three treatments a year). In addition, cross resistance between strains of *B. sphaericus* harboring different sequences has not been found to occur (43), which indicates that the different toxins may bind to different receptors. Baseline susceptibility information on target mosquito populations should be collected before treatment is started, and their susceptibility should be monitored continually to detect any significant shifts in baseline toxicity. Treatment should then be stopped before the resistance becomes a problem. Georghiou et al (44) showed that the natural variation in the susceptibility of *C. pipiens quinquefasciatus* to *B. sphaericus* was much higher than the variation in susceptibility to Bti. Thus, the probability of selecting for resistance to *B. sphaericus* is probably higher than that for Bti.

The use of *B. sphaericus* strains belonging to different serotypes and displaying no cross resistance (43) could minimize the risk of resistance. This alternative requires a search for new strains with crystal toxins whose amino acid sequences vary from that deduced for the commonly used strain, 2362. This approach seems promising, as strains such as 2297 and IAB59 produce crystal toxins different from the 2362 toxin (see section on biochemistry and genetics of the toxins). Novel crystal toxins could also be produced by construction of mutant toxins. Mutation experiments performed by Berry et al (14) suggest that this is also a feasible approach.

The best way to prevent resistance may be to produce bacterial strains that simultaneously express different toxins binding to different receptors. In addition to the crystal toxins, other toxins from *B. sphaericus* or other insecticidal microorganisms could be produced. The Mtx toxins, which do not seem to be expressed at the same time as the crystal toxins, are potential alternatives; expression of Mtx genes during sporulation of *B. sphaericus,* i.e. under the control of a sporulation promoter, could allow diversification of toxins. Similarly, Bti mosquitocidal toxin genes can be combined with *B. sphaericus* genes; such efforts include the introduction of the genes encoding CryIVB or CryIVD from Bti into toxic *B. sphaericus* strains (6, 65, 83). Conversely, *B. sphaericus* crystal toxin genes have been introduced into toxic Bti (16). Although these studies did not demonstrate any increase in toxicity against *Anopheles* and *Culex* species, such recombinants may delay insect resistance and should be studied in this regard. Several screening programs have now been established to isolate novel mosquitocidal toxins that could be combined with both Bti and *B. sphaericus* toxins. New toxins identified from one *Clostridium bifermentans* strain and several *B. thuringiensis* strains appear to be good candidates for use in combination with *B. sphaericus* toxins (41). The challenge now is to express stably and simultaneously these toxin genes in the same host and to study the development of resistance.

ACKNOWLEDGMENTS

We are very grateful to all our collegues who helped us prepare this review by providing unpublished data, micrographs, and/or helpful comments, especially Drs. C Berry, University of Wales; G Georghiou and MS Mulla, University of California, Riverside; A Porter, National University of Singapore; F Priest, Heriot-Watt University; L Regis, Fiocruz, Recife; R Reuben, CRME, Madurai; and AA Yousten, Virginia Polytechnic Institute. We are also indebted to Dr. I Thiéry and V Patricio (Institut Pasteur) for mosquito rearing, S Hamon (Institut Pasteur) for mosquito bioassays, H Ohayon and C Rollin (Institut Pasteur) for technical assistance, and A Edelman for critical reading of the manuscript. A large part of our work has been supported by the United Nations Development Program/World Bank/World Health Organization Special Program for Research and Training in Tropical Diseases, and by grants from the Commission of the European Communities SCIENCE Program, from the Danish International Development Assistance, from the University of California Mosquito Research Program, and from the CAPES and the CNPq (Brazil).

Literature Cited

1. Ahmed HK, Mitchell WJ, Priest FG. 1995. Regulation of mosquitocidal toxin synthesis in *Bacillus sphaericus. Appl. Microbiol. Biotechnol.* 43:310–14
2. Alexander B, Priest FG. 1990. Numerical classification and identification of *Bacillus sphaericus* including some strains pathogenic for mosquito larvae. *J. Gen. Microbiol.* 136:367–76
3. Aquino de Muro M, Mitchell WJ, Priest FG. 1992. Differentiation of mosquito-pathogenic strains of *Bacillus sphaericus* from non-toxic varieties by ribosomal RNA gene restriction patterns. *J. Gen. Microbiol.* 138:1159–66
4. Aquino de Muro M, Priest FG. 1994.

A colony hybridization procedure for the identification of mosquitocidal strains of *Bacillus sphaericus* on isolation plates. *J. Invertebr. Pathol.* 63:310–13
5. Arapinis C, de la Torre F, Szulmajster J. 1988. Nucleotide and deduced amino acid sequence of the *Bacillus sphaericus* 1593M gene encoding a 51.4 kD polypeptide which acts synergistically with the 42 kD protein for expression of the larvicidal toxin. *Nucleic Acids Res.* 16:7731
6. Bar E, Lieman-Hurwitz J, Rahamim E, Keynan A, Sandler N. 1991. Cloning and expression of *Bacillus thuringiensis*

israelensis δ-endotoxin DNA in *B. sphaericus. J. Invertebr. Pathol.* 57: 149–58
7. Baumann L, Baumann P. 1989. Expression in *Bacillus subtilis* of the 51- and 42- kilodalton mosquitocidal toxin genes of *Bacillus sphaericus. Appl. Environ. Microbiol.* 55:252–53
8. Baumann L, Baumann P. 1991. Effects of components of the *Bacillus sphaericus* toxin on mosquito larvae and mosquito-derived tissue culture–grown cells. *Curr. Microbiol.* 23:51–57
9. Baumann L, Broadwell AH, Baumann P. 1988. Sequence analysis of the mosquitocidal toxin genes encoding 51.4- and 41.9-kilodalton proteins from *Bacillus sphaericus* 2362 and 2297. *J. Bacteriol.* 170:2045–50
10. Baumann P, Baumann L, Bowditch RD, Broadwell AH. 1987. Cloning of the gene for the larvicidal toxin of *Bacillus sphaericus* 2362: evidence for a family of related sequences. *J. Bacteriol.* 169: 4061–67
11. Baumann P, Clark MA, Baumann L, Broadwell AH. 1991. *Bacillus sphaericus* as a mosquito pathogen: properties of the organism and its toxins. *Microbiol. Rev.* 55:425–36
12. Baumann P, Unterman BM, Baumann L, Broadwell AH, Abbene SJ, et al. 1985. Purification of the larvicidal toxin of *Bacillus sphaericus* and evidence for high-molecular-weight precursors. *J. Bacteriol.* 163:738–47
13. Berry C, Hindley J. 1987. *Bacillus sphaericus* strain 2362: identification and nucleotide sequence of the 41.9 kDa toxin gene. *Nucleic Acids Res.* 15:5591
14. Berry C, Hindley J, Ehrhardt AF, Grounds T, de Souza I, et al. 1993. Genetic determinants of host ranges of *Bacillus sphaericus* mosquito larvicidal toxins. *J. Bacteriol.* 175:510–18
15. Berry C, Jackson-Yap J, Oei C, Hindley J. 1989. Nucleotide sequence of two toxin genes from *Bacillus sphaericus* IAB59: sequence comparisons between five highly toxinogenic strains. *Nucleic Acids Res.* 17:7516
16. Bourgouin C, Delécluse A, de la Torre F, Szulmajster J. 1990. Transfer of the toxin protein genes of *Bacillus sphaericus* into *Bacillus thuringiensis* subsp. *israelensis* and their expression. *Appl. Environ. Microbiol.* 56:340–44
17. Broadwell AH, Baumann L, Baumann P. 1990. Larvicidal properties of the 42 and 51 kilodalton *Bacillus sphaericus* proteins expressed in different bacterial hosts: evidence for a binary toxin. *Curr. Microbiol.* 21:361–66
18. Broadwell AH, Baumann P. 1986. Sporulation-associated activation of *Bacillus sphaericus* larvicide. *Appl. Environ. Microbiol.* 52:758–64
19. Broadwell AH, Baumann P. 1987. Proteolysis in the gut of mosquito larvae results in further activation of the *Bacillus sphaericus* toxin. *Appl. Environ. Microbiol.* 53:1333–37
20. Broadwell AH, Clark MA, Baumann L, Baumann P. 1990. Construction by site-directed mutagenesis of a 39-kilodalton mosquitocidal protein similar to the larva-processed toxin of *Bacillus sphaericus* 2362. *J. Bacteriol.* 172: 4032–36
21. Charles J-F. 1987. Ultrastructural midgut events in Culicidae larvae fed with *Bacillus sphaericus* 2297 spore/crystal complex. *Ann. Inst. Pasteur/Microbiol.* 138:471–84
22. Charles J-F, de Barjac H. 1981. Variation du pH de l'intestin moyen d'*Aedes aegypti* en relation avec l'intoxication par les cristaux de *Bacillus thuringiensis* var. *israelensis* (sérotype H 14). *Bull. Soc. Pathol. Exot.* 74:91–95
23. Charles J-F, Hamon S, Baumann P. 1993. Inclusion bodies and crystals of *Bacillus sphaericus* mosquitocidal proteins expressed in various bacterial hosts. *Res. Microbiol.* 144:411–16
24. Charles J-F, Kalfon A, Bourgouin C, de Barjac H. 1988. *Bacillus sphaericus* asporogenous mutants: ultrastructure, mosquito larvicidal activity and protein analysis. *Ann. Inst. Pasteur/Microbiol.* 139:243–59
25. Charles J-F, Nicolas L. 1986. Recycling of *Bacillus sphaericus* 2362 in mosquito larvae: a laboratory study. *Ann. Inst. Pasteur/Microbiol.* 137B:101–11
26. Clark MA, Baumann P. 1990. Deletion analysis of the 51-kilodalton protein of the *Bacillus sphaericus* 2362 binary mosquitocidal toxin: construction of derivatives equivalent to the larva-processed toxin. *J. Bacteriol.* 172:6759–63
27. Dadd RH. 1975. Alkalinity within the midgut of mosquito larvae with alkaline-active digestive enzymes. *J. Insect Physiol.* 21:1847–53
28. Davidson EW. 1981. A review of the pathology of bacilli infecting mosquitoes, including an ultrastructural study of larvae fed *Bacillus sphaericus* 1593 spores. *Dev. Ind. Microbiol.* 22:69–81
29. Davidson EW. 1988. Binding of the *Bacillus sphaericus* (Eubacteriales: Bacillaceae) toxin to midgut cells of mosquito (Diptera: Culicidae) larvae: relationship to host range. *J. Med. Entomol.* 25:151–57

30. Davidson EW. 1989. Variation in binding of *Bacillus sphaericus* toxin and wheat germ agglutinin to larval midgut cells of six species of mosquitoes. *J. Invertebr. Pathol.* 53:251–59

31. Davidson EW, Bieger AL, Meyer M, Shellabarge RC. 1987. Enzymatic activation of the *Bacillus sphaericus* mosquito larvicidal toxin. *J. Invertebr. Pathol.* 50:40–44

32. Davidson EW, Oei C, Meyer M, Bieber AL, Hindley J, et al. 1990. Interaction of the *Bacillus sphaericus* mosquito larvicidal proteins. *Can. J. Microbiol.* 36:870–78

33. Davidson EW, Titus M. 1987. Ultrastructural effects of the *Bacillus sphaericus* mosquito larvicidal toxin on cultured mosquito cells. *J. Invertebr. Pathol.* 50:213–20

34. Davidson EW, Yousten AA. 1990. The mosquito larval toxin of *Bacillus sphaericus*. See Ref. 38a, pp. 237–55

35. de Barjac H. 1978. Une nouvelle variété de *Bacillus thuringiensis* très toxique pour les moustiques: *B. thuringiensis* var. *israelensis* sérotype H14. *C. R. Acad. Sci. Paris Sér. D* 286:797–800

36. de Barjac H. 1990. Classification of *Bacillus sphaericus* strains and comparative toxicity to mosquito larvae. See Ref. 38a, pp. 228–36

37. de Barjac H, Charles J-F. 1983. Une nouvelle toxine active sur les moustiques, présente dans des inclusions cristallines produites par *Bacillus sphaericus*. *C. R. Acad. Sci. Paris Sér. III* 296:905–10

38. de Barjac H, Larget-Thiéry I, Cosmao Dumanoir V, Ripouteau H. 1985. Serological classification of *Bacillus sphaericus* strains in relation with toxicity to mosquito larvae. *Appl. Microbiol. Biotechnol.* 21:85–90

38a. de Barjac H, Sutherland DJ, eds. 1990. *Bacterial Control of Mosquitoes and Blackflies*. New Brunswick, NJ: Rutgers Univ. Press

39. de Barjac H, Thiéry I, Cosmao Dumanoir V, Frachon E, Laurent P, et al. 1988. Another *Bacillus sphaericus* serotype harbouring strains very toxic to mosquito larvae: serotype H6. *Ann. Inst. Pasteur/Microbiol.* 139:363–77

40. de la Torre F, Bennardo T, Sebo P, Szulmajster J. 1989. On the respective roles of the two proteins encoded by the *Bacillus sphaericus* 1593M toxin genes expressed in *Escherichia coli* and *Bacillus subtilis*. *Biochem. Biophys. Res. Commun.* 164:1417–22

41. Delécluse A, Barloy F, Thiéry I. 1995. Mosquitocidal toxins from various *Bacillus thuringiensis* and *Clostridium bifermentans*. In *Bacillus thuringiensis Biotechnology and Environmental Benefits*, ed. HS Yuan, pp. 99–114. Taipei, Taiwan.

42. Frachon E, Hamon S, Nicolas L, de Barjac H. 1991. Cellular fatty acid analysis as a potential tool for predicting mosquitocidal activity of *Bacillus sphaericus* strains. *Appl. Environ. Microbiol.* 57:3394–98

43. Georghiou GP, Malik JI, Wirth M, Sainato K. 1992. Characterization of resistance of *Culex quinquefasciatus* to the insecticidal toxins of *Bacillus sphaericus* (strain 2362). In *Univ. Calif., Mosq. Control Res., Annu. Rep. 1992*, pp. 34–35. Division of Agriculture and Natural Resources

44. Georghiou GP, Wirth MC, Ferrari J, Tran H. 1991. Baseline susceptibility and analysis of variability toward biopesticides in California populations of *Culex quinquefasciatus*. In *Univ. Calif., Mosq. Control Res., Annu. Rep. 1991*, pp. 25–27. Division of Agriculture and Natural Resources

45. Goldberg LH, Margalit J. 1977. A bacterial spore demonstrating rapid larvicidal activity against *Anopheles sergentii*, *Uranotaenia unguiculata*, *Culex univitatus*, *Aedes aegypti* and *Culex pipiens*. *Mosq. News* 37:355–58

46. Guerineau M, Alexander B, Priest FG. 1991. Isolation and identification of *Bacillus sphaericus* strains pathogenic for mosquito larvae. *J. Invertebr. Pathol.* 57:325–33

47. Hindley J, Berry C. 1987. Identification, cloning and sequence analysis of the *Bacillus sphaericus* 1593 41.9 kD larvicidal toxin gene. *Mol. Microbiol.* 1:187–94

48. Hindley J, Berry C. 1988. *Bacillus sphaericus* strain 2297: nucleotidic sequence of a 41.9kDa toxin gene. *Nucleic Acids Res.* 16:4168

49. Kalfon A, Charles J-F, Bourgouin C, de Barjac H. 1984. Sporulation of *Bacillus sphaericus* 2297: an electron microscope study of crystal-like inclusion biogenesis and toxicity to mosquito larvae. *J. Gen. Microbiol.* 130:893–900

50. Karch S, Coz J. 1983. Histopathologie de *Culex pipiens* Linné (Diptera, Culicidae) soumis à l'activité larvicide de *Bacillus sphaericus* 1593-4. *Cah. ORSTOM Sér. Entomol. Méd. Parasitol.* 21:225–30

51. Kellen W, Clark T, Lindergren J, Ho B, Rogoff M, et al. 1965. *Bacillus sphaericus* Neide as a pathogen of mosquitoes. *J. Invertebr. Pathol.* 7:442–48

52. Knight PJK, Crickmore N, Ellar DJ. 1994. The receptor for *Bacillus thuringiensis* CryIA(c) delta-endotoxin in the brush border membrane of the lepidopteran *Manduca sexta* is aminopeptidase N. *Mol. Microbiol.* 11:429–36

53. Krych VK, Johnson JL, Yousten AA. 1980. Deoxyribonucleic acid homologies among strains of *Bacillus sphaericus*. *Int. J. Syst. Bacteriol.* 30: 476–82

54. Lakshmi Narasu M, Gopinathan KP. 1988. Effect of *Bacillus sphaericus* 1593 toxin on choline acetyl transferase and mitochondrial oxidative activities of the mosquito larvae. *Ind. J. Biochem. Biophys.* 25:253–56

55. Liu JW, Hindley J, Porter AG, Priest FG. 1993. New high-toxicity mosquitocidal strains of *Bacillus sphaericus* lacking a 100-kilodalton-toxin gene. *Appl. Environ. Microbiol.* 59:3470–73

55a. Munson PJ, Rodbard D. 1980. LIGAND: a versatile computerized approach for characterization of ligand-binding systems. *Anal. Biochem.* 107: 220–39

56. Myers P, Yousten AA, Davidson EW. 1979. Comparative studies of the mosquito-larval toxin of *Bacillus sphaericus* SSII-1 and 1593. *Can. J. Microbiol.* 25:1227–31

57. Neide E. 1904. Botanische Beschreibung einiger sporenbildenden Bacterien. *Zentralbl. Bakteriol. Parasitenkd. Infektionskr. Hyg. Abt.* (pt. 1) 12:1–11

58. Nicolas L, Lecroisey A, Charles J-F. 1990. Role of the gut proteinases from mosquito larvae in the mechanism of action and the specificity of the *Bacillus sphaericus* toxin. *Can. J. Microbiol.* 36: 804–7

59. Nicolas L, Nielsen-LeRoux C, Charles J-F, Delécluse A. 1993. Respective role of the 42- and 51-kDa component of the *Bacillus sphaericus* toxin overexpressed in *Bacillus thuringiensis*. *FEMS Microbiol. Lett.* 106:275–80

60. Nielsen-LeRoux C, Charles J-F. 1992. Binding of *Bacillus sphaericus* binary toxin to a specific receptor on midgut brush-border membranes from mosquito larvae. *Eur. J. Biochem.* 210:585–90

61. Nielsen-LeRoux C, Charles J-F, Thiéry I, Georghiou GP. 1995. Resistance in a laboratory population of *Culex quinquefasciatus* (Diptera: Culicidae) to *Bacillus sphaericus* binary toxin is due to a change in the receptor on midgut brush-border membranes. *Eur. J. Biochem.* 228:206–10

62. Oei C, Hindley J, Berry C. 1990. An analysis of the genes encoding the 51.4-

and 41.9-kDa toxins of *Bacillus sphaericus* 2297 by deletion mutagenesis: the construction of fusion proteins. *FEMS Microbiol. Lett.* 72:265–74

63. Oei C, Hindley J, Berry C. 1992. Binding of purified *Bacillus sphaericus* binary toxin and its deletion derivatives to *Culex quinquefasciatus* gut: elucidation of functional binding domains. *J. Gen. Microbiol.* 138:1515–26

64. Payne JM, Davidson EW. 1984. Insecticidal activity of crystalline parasporal inclusions and other components of the *Bacillus sphaericus* 1593 spore complex. *J. Invertebr. Pathol.* 43:383–88

65. Poncet S, Delécluse A, Guido A, Klier A, Rapoport G. 1994. Transfer and expression of the *cryIVB* and *cryIVD* genes of *Bacillus thuringiensis* subsp. *israelensis* in *Bacillus sphaericus* 2297. *FEMS Microbiol. Lett.* 117:91–96

66. Porter AG, Davidson EW, Liu JW. 1993. Mosquitocidal toxins of bacilli and their genetic manipulation for effective biological control of mosquitoes. *Microbiol. Rev.* 57:838–61

67. Rao DR, Mani TR, Rajendran R, Joseph ASJ, Gajanana A, et al. 1995. Development of a high level of resistance to *Bacillus sphaericus* in a field population of *Culex quinquefasciatus* from Kochi, India. *J. Am. Mosq. Contr. Assoc.* 11:1–5

68. Regis L, Silva-Filha MHNL, de Oliveiro CMF, Rios EM, da Silva SB, et al. 1995. Integrated control measures against *Culex quinquefasciatus*, the vector of filariasis in Recife. *Mem. Inst. Oswaldo Cruz* 90:115–19

69. Rodcharoen J, Mulla MS. 1994. Resistance development in *Culex quinquefasciatus* (Diptera: Culicidae) to the microbial agent *Bacillus sphaericus*. *J. Econ. Entomol.* 87:1133–40

70. Sěbo P, Bennardo T, de la Torre F, Szulmajster J. 1990. Delineation of the minimal portion of the *Bacillus sphaericus* 1593M toxin required for the expression of larvicidal activity. *Eur. J. Biochem.* 194:161–65

71. Silva-Filha M-H, Regis L, Nielsen-LeRoux C, Charles J-F. 1995. Low-level resistance to *Bacillus sphaericus* in a field-treated population of *Culex quinquefasciatus* (Diptera: Culicidae). *J. Econ. Entomol.* 88:525–30

72. Sinègre G, Babinot M, Quermel J-M, Gavon B. 1994. First field occurrence of *Culex pipiens* resistance to *Bacillus sphaericus* in southern France. *Eur. Meet. Soc. Vector Ecol., 8th, Barcelona,* p. 17 (Abstr.)

73. Singer S. 1973. Insecticidal activity of

recent bacterial isolates and their toxins against mosquito larvae. *Nature* 244: 110–11

74. Singer S. 1974. Entomogenous bacilli against mosquito larvae. *Dev. Ind. Microbiol.* 15:187–94

75. Singer S. 1977. Isolation and development of bacterial pathogens in vectors. In *Biological Regulation of Vectors, DHEW Publ. No. (NIH) 77–1180,* pp. 3–18. Bethesda, MD: NIH

76. Singh GJP, Gill SS. 1988. An electron microscope study of the toxic action of *Bacillus sphaericus* in *Culex quinquefasciatus* larvae. *J. Invertebr. Pathol.* 52:237–47

77. Stray JE, Klowden MJ, Hurlbert RE. 1988. Toxicity of *Bacillus sphaericus* crystal toxin to adult mosquitoes. *Appl. Environ. Microbiol.* 54:2320–21

78. Thanabalu T, Berry C, Hindley J. 1993. Cytotoxicity and ADP-ribosylating activity of the mosquitocidal toxin from *Bacillus sphaericus* SSII-1: possible roles of the 27-kilodalton and 70-kilodalton peptides. *J. Bacteriol.* 175:2314–20

79. Thanabalu T, Hindley J, Berry C. 1992. Proteolytic processing of the mosquitocidal toxin from *Bacillus sphaericus* SSII-1. *J. Bacteriol.* 174:5051–56

80. Thanabalu T, Hindley J, Jackson-Yap J, Berry C. 1991. Cloning, sequencing, and expression of a gene encoding a 100-kilodalton mosquitocidal toxin from *Bacillus sphaericus* SSII-1. *J. Bacteriol.* 173:2776–85

81. Thanabalu T, Porter AG. 1995. *Bacillus sphaericus* gene encoding a novel class of mosquitocidal toxin with homology to *Clostridium* and *Pseudomonas* toxins. *Gene.* In press

82. Thiéry I, de Barjac H. 1989. Selection of the most potent *Bacillus sphaericus* strains, based on activity ratios determined on three mosquito species. *Appl. Microbiol. Biotechnol.* 31:577–81

83. Trisrisook M, Pantuwatana S, Bhumiratana A, Panbangred W. 1990. Molecular cloning of the 130-kilodalton mosquitocidal δ-endotoxin gene of *Bacillus thuringiensis* subsp. *israelensis* in *Bacillus sphaericus. Appl. Environ. Microbiol.* 56:1710–16

84. Woodburn MA, Yousten AA, Hilu KH. 1995. Random amplified polymorphic DNA fingerprinting of mosquito-pathogenic and nonpathogenic strains of *Bacillus sphaericus. Int. J. Syst. Bacteriol.* 45:212–17

85. Yousten AA. 1984. Bacteriophage typing of mosquito pathogenic strains of *Bacillus sphaericus. J. Invertebr. Pathol.* 43:124–25

86. Yousten AA, Davidson EW. 1982. Ultrastructural analysis of spores and parasporal crystals formed by *Bacillus sphaericus* 2297. *Appl. Environ. Microbiol.* 44:1449–55

87. Yousten AA, de Barjac H, Hedrick J, Cosmao Dumanoir V, Myers P. 1980. Comparison between bacteriophage typing and serotyping for the differentiation of *Bacillus sphaericus* strains. *Ann. Microbiol.* 131B:297–308

Annu. Rev. Entomol. 1996. 41:473–94
Copyright © 1996 by Annual Reviews Inc. All rights reserved

SEXUAL RECEPTIVITY IN INSECTS

John Ringo

Department of Zoology, University of Maine, Orono, Maine 04469

KEY WORDS: sexual behavior, remating, corpus allatum, juvenile hormone, sex peptide

ABSTRACT

Sexual receptivity is female behavior that allows or helps a male to fertilize her eggs; through this behavior, females play an active role in reproduction. Multiple signals may be used for receptivity or unreceptivity. Insect species exhibit three ontogenetic patterns of receptivity: cyclic, in which females alternately become receptive and unreceptive; brief, in which females mate during one short developmental period; and continuous. Primary (initial) receptivity may be stimulated or inhibited by diet, ovarian development, or juvenile hormone. In species with cyclic receptivity, remating may be inhibited by copulation itself, the presence of eggs, sperm stored in spermathecae, or seminal factors—usually peptides—secreted by the male accessory glands. In many species, there is substantial genetic variation for both primary receptivity and speed of remating. Several single-gene mutations reduce female receptivity; most of these mutations also impair sensory functioning.

PERSPECTIVES AND OVERVIEW

Insects that reproduce sexually—which includes most insects (34)—use internal fertilization. Sperm are transferred from donating male to receiving female either through mating or, in the Apterygota, via an indirect process. Females are generally reluctant to mate (14, 164) and do so only after appropriate stimulation by males. Forced copulation is relatively uncommon in insects, or at least is difficult to demonstrate (161); hence, the probability of successful fertilization generally depends upon a female's receptivity, or willingness to receive sperm. Accordingly, receptivity is a major feature of the reproductive biology of most insects.

A female indicates her receptivity by allowing or helping the male to fertilize her eggs. Thus, fertilization seems at least partly controlled by female behavior.

473

For sexually mature, conspecific females and males to mate, they must exchange signals. The signals comprise either sexual advertisement (long-range signals) or courtship (short-range signals given just prior to mating). In species using sexual advertisement, either males or females broadcast chemical or auditory signals to attract potential mates from a distance. Because even sexually mature females that advertise may still be initially reluctant to mate when males approach, all insect species that mate use some type of courtship. This process can be as simple as moving to an appropriate location and displaying species-specific chemicals on the cuticle, or it can be a long, complex, dance-like ritual using signals in all the sensory modalities. Despite numerous examples of insect mating with little or no discernible courtship (42, 162, 164), prospective mates cannot locate each other without premating signals, however cryptic or short-lived they may be. Except in forced copulation, the transition between courtship and copulation requires the female's acceptance of the male. In some taxa, acceptance amounts to the absence of rejection.

The ontogeny of receptivity is generally coordinated with the temporal pattern of mating. In monocoitic species (i.e. those mating only once), females are receptive before mating but not afterwards. Females of most multicoitic species go through cycles of receptivity and unreceptivity. They usually start adult life unreceptive and later develop primary (initial) receptivity, which is lost after mating. Secondary (postmating) receptivity then develops, after which mating receptivity again declines sharply. The physiological mechanisms underlying these behavioral changes are far from being understood, but several factors influencing receptivity have been identified, including control by portions of the central nervous system (CNS), diet, hormones, and chemical messages donated along with sperm. A few specific genes have been identified that influence receptivity.

BEHAVIOR ASSOCIATED WITH RECEPTIVITY

The human observer can assess the receptivity of a female insect only by analyzing her behavior before and during insemination. This is an easy matter in species whose females exert firm control over insemination but otherwise may prove difficult. In the Apterygota, the male deposits spermatophores on the substrate, and a female acquires sperm by placing the vulva over one (for review, see 44); clearly, a female engaging in such behavior is receptive. Even in Apterygota species whose males court, coaxing or pushing the female toward their spermatophores, the females are clearly in control (44). At the other extreme, one cannot generally observe mating, let alone female responses to males, within a mating swarm (150, 162); perhaps, however, it is only the receptive females that enter the swarm. The mosquito *Aedes aegypti* provides an example of cryptic receptivity. The *A. aegypti* female exerts strong post-mating control; she may copulate repeatedly, perhaps being forced by the male, but she appears to expel sperm until she is ready for insemination (87, 179).

Male Behavior That Increases Receptivity

During courtship, male insects typically use several different signals, often in more than one sensory modality, and repeat or continuously emit the signals (for reviews, see 14, 164). The female presumably integrates male courtship stimuli in the CNS (e.g. 99, 103). The probability of female acceptance is influenced by the amount or intensity of male courtship. For example, female houseflies (*Musca domestica*) mated at a higher rate if they were exposed to male sex pheromone just before mating tests (140). Female scorpionflies (*Hylobittacus apicalis*) prefer males offering nuptial gifts (captured insects) larger than 16 mm^3 (163). To manipulate the intensity of the auditory component of courtship of *Drosophila melanogaster*, Ewing (46) removed portions of the male's wings; courtship success was linearly related to the remaining wing area. In this species and its sibling *D. simulans*, stimulating females with artificial or recorded components of male courtship songs induced females to mate faster (86, 172). *D. melanogaster* females homozygous for sensory mutations mated more slowly than did normal females (105).

Signaling Rejection

In many species, females give ritualized, stereotyped, rejection signals; males respond to such signals. These signals may involve the overall posture or movements of the wings, legs, and ovipositor (6, 27, 31, 71, 156). Kicking by females is a favorite means of rejection, as is simply running or flying away (27, 37, 73, 93, 97, 153). In those few species for which forced copulation (i.e. rape) is a common occurrence (161), a female may struggle actively in attempting to escape the male. The best-documented example of forced copulation is described in riveting detail by Thornhill (161), who observed apparent forced copulation in 7 of 18 species of *Panorpa* scorpionflies in the field, and in most of these 18 species in the laboratory. Usually, the female is appeased during mating by a nuptial gift presented by the male. A male without a nuptial gift may lunge at a female, grasping her with genital claspers and the notal organ (161). The female struggles violently, successfully freeing herself in ~80% of the forcing attempts (161). When the notal organs were covered, forced copulation was attempted at the same rate, but was never successful (161). Females of other species also struggle to escape a male that is copulating or attempting copulation without having received the appropriate acceptance signal (37, 38, 138).

Signaling Acceptance

Receptivity is obvious and unambiguous in only a few species, such as the katydid *Requena verticalis,* whose females fight for access to males (98). The more control females exert in the mating process, the more conspicuous their acceptance signals. In some of the cockroaches, a female need only fail to give a critical signal to interrupt courtship (58). In contrast, some of the Apoidea and Acalyptratae have no ritualized courtship in either sex, and acceptance signals are not apparent to a human observer (42, 153). Females respond to the males' signals with various countersignals: exposing the vagina (168) or opening the vaginal plates (36, 58, 151), stridulation (71, 80), vibrating or fanning the wings (119, 181), and laying eggs (95). In some species, the female may mount the male or climb on his back (41, 69, 145, 171a). A remarkable example occurs in the giant water bug *Abedus herberti.* The female climbs on the male's back just before copulation (145) and afterward lays one or two eggs on the male's back. Then, courtship begins again. The pair have multiple copula, mating an average of 1.4 times per egg (145).

ONTOGENY

Receptivity develops in one of three patterns in the adult female (164): (*a*) She is receptive during a brief period; (*b*) her receptivity is cyclic; or (*c*) she is continuously receptive. Regardless of the pattern of ontogeny, a phenomenon that deserves further research is whether receptivity is switched on or off in a discontinuous fashion or varies continuously in response to developmental stage, external stimuli, and physiology. In *D. melanogaster,* primary receptivity increases rapidly during the first 48 h of adult life. At first it appears to be off, but then it constantly changes (28). Receptivity usually declines in aged virgin insects (56, 82, 103) and may decrease as a function of the number of matings (118, 128).

Brief Receptivity

Many orders are receptive only for a short time; this pattern has been most frequently reported in the Hymenoptera and Diptera (for reviews, see 40, 42, 127, 164), as well as the Ephemeroptera (150). A good example is the solitary bee *Centris pallida.* Pupae of this species live underground; adult males, which emerge before females, dig down to meet emerging bees (4). If the emerging bee is a female, the male initiates courtship on the ground, and the pair usually fly to a nearby bush where courtship continues. The courting male stimulates the female with complex tactile and possibly auditory behavior, and the female may respond with a sound. Mating takes place a few minutes later, but only when the female is receptive (her acceptance signal is unknown). Females mate only once; indeed, subsequent male-female encounters never proceed to courtship, presumably because females signal unreceptivity (4). Mating is restricted to a particular setting associated with a set of stimuli not encountered later in life. This restriction also applies to insects that mate in swarms (40, 76, 150).

Some species, such as *C. pallida,* that have a brief period of receptivity are monocoitic whereas others are multicoitic; here, these terms refer only to females. Cole (25) and Ridley (127) reviewed the mating behavior of nearly 235 species of insects, of which these authors claimed ~25% are monocoitic. Unfortunately, the definition of monocoitic remains controversial. Taken literally, the term would mean that all females of a monocoitic species would mate at most once in nature, but many investigators call a species monocoitic if most of the females studied were observed to mate only once (127, 164). For example, *M. domestica* is widely cited as a monocoitic species (127, 128, 164), yet in a large and careful study (128), 3–52% of mature females remated, depending on experimental conditions, such as (and especially) the age of the male and whether the female had already oviposited extensively.

Cyclic Receptivity

This pattern is widespread in the Odonata, Mecoptera, Orthoptera, Heteroptera, Diptera, Hymenoptera, and Coleoptera (for reviews, see 7, 44, 164, 168). Well-documented examples of cyclic receptivity and of strong female control of mating come from studies of ovoviviparous cockroaches (6, 58, 135, 136). In *Blabera craniifer,* a male is attracted to a female emitting a sex pheromone. After antennal fencing between the female and male, the male raises his wings and wingsheaths and the female moves toward his abdominal tergites, from which the male secretes a liquid. The female advances, straddling the male abdomen, and licks the male secretions. The male slides underneath the female and hooks the edge of the female subgenital plate with his phallomere. She then turns 180°, opens her vaginal plates, and allows the male to insert the phallus. The female may withdraw to stop the courtship at several points: at initial contact, before straddling the male, before licking the male, and before turning. A male does not continue courtship if the female fails to give the next positive signal (58). In quick succession after emergence, the female becomes receptive for a brief period, mates, and oviposits; she then carries the eggs in an ootheca or brood sac for many days. After parturition, she enters another cycle of receptivity, mating, oviposition, and egg-carrying. In contrast, many other cyclic species exhibit temporal patterns of oviposition that are not clearly related to the temporal pattern of receptivity.

Continuous Receptivity

This pattern is less common than the other two but has been observed in the Orthoptera, Heteroptera, and Hymenoptera (for review, see 164). Female green stink bugs (*Nezara viridula*) are multicoitic and, after becoming sexually mature, spend a relatively high proportion of their time in copula (66). During courtship, the male repeatedly and rapidly butts his head against the female's abdomen, then spins around to face away from her and attempts copulation. The receptive female stands with abdomen raised, awaiting intromission (66). Sexually mature females spend 10–70% of their time in copula, yet the duration of copulation and latency do not correlate with the likelihood of the next mating (66). The adaptive reasons for continuous receptivity of this bug are unknown, but the high ratio of hours spent in copula to number of eggs laid is striking.

PHYSIOLOGICAL FACTORS CONTROLLING PRIMARY RECEPTIVITY

Diet and Egg Production

The activity of the corpora allata and the growth of ovaries are often related to diet (8, 21, 44, 45, 51, 135), and not surprisingly, receptivity can also be connected to diet. For example, starved females of several species of blaberid cockroach (135, 137), of the bugs *Pyrrhocoris apterus* (182) and *Dysdercus sidae* (44), and of the rabbit flea *Spilopsyllus cuniculi* (107) are less receptive. Futhermore, starvation partially inhibits receptivity in *Haematosiphon inodorus* (88). In contrast, starvation has no effect upon receptivity in the cockroach *Nauphoeta cinerea* (135) and the Colorado potato beetle, *Leptinotarsa decemlineata* (160). Dietary protein stimulates the development of female receptivity the dipterans *D. melanogaster* (20), *Musca autumnalis* (21), *Lucilia cuprina* (8), and *A. aegypti* (147). Receptivity in the blowfly (*L. cuprina*) was also strongly stimulated by the insect growth regulator Altosid® (8). Protein-fed females of *D. melanogaster* were somewhat more receptive than sugar-fed females in one study (32) but not in another (103). Female *D. melanogaster* given access to food remated at a higher rate than females denied food (67). In *R. verticalis,* virgin females fed a high-protein diet rejected males more often than females fed a low-protein diet; the opposite was true of once-mated females (139).

Oviposition also affects receptivity. In the beetle *Callosobruchus maculatus* (48), the flies *M. domestica* (128) and *D. melanogaster* (167), and the mosquito *Culex tarsalis* (180), females that are allowed to lay eggs after mating regain receptivity faster than females not so allowed.

Endocrine Control

The corpora allata produce juvenile hormone (JH), which stimulates oocyte growth, vitellogenesis, and absorption of yolk proteins by the oocyte (125). Because the development of sexual receptivity, ovarian maturation, and corpus allatum activity are temporally related in many species (44, 81, 177), JH seems likely to play a role in the development of receptivity. This role varies from species to species. Adultoids and supernumerary larvae of *P. apterus* induced by JH analogues exhibit normal sexual behavior, including female receptivity. JH or at least the corpus allatum can promote the development of primary receptivity or may have no measurable effect (see also earlier reviews, 7, 44, 168). The influence of the endocrine hormones on secondary receptivity in species whose primary receptivity is stimulated by JH is largely unknown.

One must be cautious in claiming a requirement for JH in the development of receptivity. First, such a claim is based upon negative evidence in a limited set of observations. For example, Engelmann (43) reported that only 30% of females of allatectomized cockroaches (*Leucophaea maderae*) were receptive, whereas Roth & Barth (136) performed the same experiment and observed normal receptivity. When Engelmann & Barth (45) later replicated the experiment, allatectomized females had reduced receptivity. The discrepancy may be accounted for by the use of different strains in the two experiments and by higher densities used in the mating tests of Roth & Barth (136) than in those of Engelmann (43). Second, the observation that allatectomy reduces but does not abolish primary receptivity in a species does not imply that the corpus allatum is inessential in that species because some JH may have been produced before the allatectomy. In *M. domestica,* 11% of females allatectomized 2.5 h after eclosion were receptive, compared with 86% of the operated but not allectomized controls (1). Perhaps allatectomy of *M. domestica* at an earlier age would block the onset of receptivity in all females.

The corpora allata appear to control the onset of primary receptivity in a few species (87, 94, 157, 166). For example, allatectomized females of the cricket *Gomphocerus rufus* did not attain primary receptivity and actively rejected male courtship; reimplanting the corpora allata rendered these females receptive (94). Allatectomy also prevented the development of receptivity in the grasshopper *Schistocerca gregaria* (157). *A. aegypti* females are multicoitic soon after emergence, but no sperm can be transferred until they are at least 24–60 h old (64, 87). Females that were allatectomized at 1 h after emergence mated repeatedly 72 h later but were not inseminated (87). Implanting corpora allata from donors restored the ability to be inseminated in 94% of the allatectomized females, and severing the ventral nerve cord did not inhibit receptivity (87). The fly *Calliphora vomitoria* gradually develops primary receptivity from 3 to 6 days posteclosion; nearly all females allatectomized at 3 h posteclosion were unreceptive on day 5 (166). Topical application of methoprene accelerated the onset of primary receptivity in intact females (166).

In some species, the corpora allata or JH stimulates the development of primary receptivity but may not be essential in that process (1, 12, 43, 63, 84, 102, 112, 132, 173). In allatectomized *M. domestica,* which have low primary receptivity, treatment with a JH analogue partially restored receptivity (1). Implanting active corpora allata–corpora cardiaca complexes in pupae accelerated the onset of primary receptivity in *D. melanogaster* (102), as did topical application of JH or methoprene (12). Methoprene also markedly accelerated the development of receptivity in *Drosophila grimshawi* (132). Receptivity of female milkweed bugs (*Oncopeltus fasciatus*) increased somewhat when they were fed juvenoids but remained unaffected when they were fed precocene II, an antijuvenoid (172).

In several species, receptivity develops independently of the corpora allata (5, 35, 47, 96, 117, 126, 136, 155, 183). For example, receptivity of *L. decemlineata* is neither reduced by allatectomy nor enhanced by application of JH III or a JH analogue (160). Allatectomized female crickets of *Teleogryllus commodus* became receptive and produced eggs regardless of whether the allatectomy was performed at the beginning, middle, or end of the penultimate larval stadium (96). Receptivity is also normal in allatectomized members of another cricket species, *Gryllus bimaculatus* (117). These experiments are compelling because mating in all three species, *L. decemlineata, T. commodus,* and *G. bimaculatus,* depends upon active responses of females.

The corpora allata can also inhibit receptivity. Allatectomized grasshoppers of *Chorthippus curtipennis* became receptive for life, although the effect seemed to be mediated by the ovaries (70). JH injections inhibited receptivity in the cockroach *Blattella germanica;* however, JH-treated females remained attractive to males (124). Rasmaswamy & Gupta (124) suggested that the JH treatment may have induced a sensory deficit in these females.

The production and dispersion of sex pheromones by female insects is often strongly correlated with primary receptivity and, like receptivity, may be influenced by the corpora allata (5). Allatectomized females of the black cutworm, *Agrotis ipsilon,* did not undergo ovarian maturation, call, or mate; treating intact females with fenoxycarb, a JH analogue, accelerated the development of both calling and first mating (55). Allatectomy abolished calling in *B. germanica,* but treatment of such females with fenoxycarb restored calling (92). JH inhibited calling in two moths, *Plodia interpunctella* (115) and *Platynota stultana* (174). In the moth *Antheria pernyi* allatectomized females attracted and mated with males, whereas females were rendered unattractive and did not mate after removal of the corpora cardiaca (126); receptivity per se was probably not involved. However, in the tobacco hornworm, *Manduca sexta,* removal of the corpora allata and corpora cardiaca had no effect on calling behavior (78).

Roth & Barth (136) suggested that receptivity in cockroaches may be under neurosecretory control. This attractive hypothesis may be generally true for the class, but it has received little attention from experimentalists. Females of *L. maderae* become permanently unreceptive after electrocoagulation of their pars intercerebralis (45), a cluster of giant neurosecretory cells in the brain essential to ovarian development in many species (125). In *D. melanogaster,* destruction of the pars intercerebralis inhibited both oviposition and primary receptivity (13). Ablating the medial neurosecretory cells of *A. aegypti* had no effect upon primary receptivity (87).

The ovaries, like the corpus allatum, may stimulate, inhibit, or have no effect upon primary receptivity. Ovariectomy decreased primary receptivity in *C. vomitoria* (166) and in the grasshoppers *Omocestus viridulus, Chorthippus brunneus,* and *Chorthippus parallelus* (72). In contrast, *Chorthippus antipennis* become permanently receptive after ovariectomy (70). Ovariectomy had no effect upon receptivity in *L. maderae* (43), *G. rufus* (94), and *M. autumnalis* (21). A mutant strain of *D. melanogaster* that lacked ovaries, *fs(2)B,* had more or less normal receptivity when reared on a normal diet but exhibited low receptivity when reared on sugar water (32).

Neural Control

The role of the brain can be assessed in a crude way by observing the receptivity, or lack thereof, in decapitated females. Most females of *M. domestica* in nature are monocoitic; decapitation resulted in repeated matings (90). To copulate, the female inserts her ovipositor into the genital opening of the male, and decapitated females managed this with no apparent difficulty. Decapitating the female after a first or even a second mating still led to numerous additional matings (90). In evaluations of virgin females of ten *Drosophila* species (62, 152), which are cyclic maters, those of several species responded in the opposite way to decapitation, becoming much less receptive (152). The highest rate of copulation among decapitated *Drosophila* was seen in *D. auraria* (69%), while headless *D. pseudoobscura* and *D. hydei* did not mate (152). Females of all ten species gave strong rejection signals to males (62, 152), suggesting that these rejection signals can operate reflexively.

Although a head is not a necessary prerequisite for the control of mating behavior, the brain does play an important role in the intact insect, as we have seen already. The behavior of gynandromorphs, which are part female, part male, offers clues to which portion of the insect controls the decision to mate or not. The female portion of a gynandromorph exhibits female characteristics, including receptivity, while the male portion exhibits none (65). In nature, gynandromorphs are rare, but they can be produced in *D. melanogaster* through genetic manipulation (65, 77, 114, 165). A portion of the dorsal brain must be female (have two X chromosomes) for the fly to be receptive (165), which suggests that a "receptivity center" is located in the brain of *D. melanogaster* Gynandromorphs that were male on either the left or right side of the critical group of cells, and female on the other, were unreceptive, i.e. primary receptivity requires female-specific genes functioning on both sides of this structure (165). Stimulation of abdominal neural structures influences receptivity in some insects (58, 134) but not others (64). In *A. aegypti*, the terminal ganglion maintains postmating unreceptivity (64).

FACTORS THAT INHIBIT REMATING

After mating, primary receptivity declines sharply in most insect species. The search for mechanisms has been conducted mostly in the Diptera and the Orthoptera, and little is known about the causes of the decline in other orders.

The presence of the ootheca in the brood sac of *N. cinerea* (134, 135) and other cockroaches (58) inhibits mating, and its removal leads to the return of receptivity in this insect (135). Whether the loss of receptivity is the direct result of sensory stimulation, or whether neuroendocrine factors intervene, is not known (44, 58), although transection of the ventral nerve cord partially restored receptivity in *B. craniifer* (58). In *C. curtipennis*, mating is not followed by loss of receptivity in ovariectomized females. Implanting glass "eggs" in the oviduct of this species reduced receptivity, whereas implanting a spermatophore had no effect (68). These experiments suggest that mechanical stimulation of the oviduct by eggs contributes directly to the refusal by *C. curtipennis* to remate. Both ovariectomy and allatectomy of *M. domestica* increase secondary receptivity, even though both treatments decrease primary receptivity (1). Implanting glass beads in the uteri of the tsetse, *Glossina morsitans*, similarly reduces receptivity (57). In *A. aegypti*, the terminal ganglion, not the brain or subesophageal ganglion, processes the sensory stimuli leading to loss of receptivity (64). In mated female *A. aegypti*, removing the terminal ganglion increases the rate of remating, while its removal in virgin females lowers receptivity (64). Decapitation of mated *M. domestica* females restored receptivity; 72% of decapitated females remated three times, vs 0% in intact controls (90). The presence of sperm in the spermathecae reduces the receptivity of *D. melanogaster* (103, 142). Both sperm and prolonged copulation inhibit remating in the melon fly, *Bactrocera* (= *Dacus*) *cucurbitae* (85).

Male accessory glands produce many seminal factors that are donated to females during copulation (89), some of which inhibit remating (22). Injecting male accessory-gland material or extracts into females reduced the receptivity of *A. aegypti* (33); the onion fly, *Delia antiqua* (146); *M. domestica* (2, 128, 129); *D. melanogaster* (109); *G. morsitans* (57); and the stable fly, *Stomoxys calcitrans* (111). Peptides and small proteins produced in the accessory glands enter the hemolymph from the vagina after mating (22, 57, 90, 110, 159). The sites of receptors for antiaphrodisiac peptides are unknown, but the brain seems a likely target (22, 90, 141).

In *M. domestica,* the antiaphrodisiac factor is a sex peptide that consists of very few amino acids (113, 159). The antiaphrodisiac factor of the *D. melanogaster* male accessory-gland secretion is a 36–amino acid sex peptide, and in *Drosophila funebris,* it is a 27–amino acid peptide (9, 10, 22–24). The sex peptide of each species both inhibits remating and stimulates oviposition (10, 22, 23). In *A. aegypti,* the inhibitory factor is matrone, which consists of a pair of proteins, α- and β-matrone (49, 50). Females injected with both α- and β-matrone became unreceptive (i.e. insemination is prevented), but females injected with either α- or β-matrone alone remained receptive (75). Coinjection of β-matrone and the sex peptide of *D. melanogaster* also rendered females unreceptive, which suggests that the *Drosophila* sex peptide resembles α-matrone. Oviposition was stimulated by injection of α-matrone alone, α+β-matrone, or *Drosophila* sex peptide plus β-matrone, but not by β-matrone alone.

GENETIC FACTORS

Questions regarding the evolution and the proximal causes of female receptivity can best be answered by studying the genetics of the insect. For example, evolutionary studies have shown the potential of populations to undergo evolutionary change in primary receptivity and in speed of remating. Investigators have also sought to discover genetically correlated fitness traits by detecting and analyzing naturally occurring, usually polygenic, variation. The genetic study of neural, hormonal, and behavioral control mechanisms has been done via a thoroughly different approach, that of analyzing the role of single genes on the one hand and of dissecting physiological processes by manipulating the genotype on the other. Most of the genetic studies of receptivity have used *Drosophila,* especially *D. melanogaster.*

Natural Genetic Variation in Primary Receptivity

Despite the many powerful methods available for detecting genetic variation, most geneticists studying primary receptivity have restricted themselves to two of the less effective approaches: artificial selection, and comparing and crossing mildly inbred strains (but see 170, 171). A response to selection implies that the selected trait is heritable in the population from which the selection line was established, i.e. that this population contains additive genetic variation for the trait and therefore can evolve under natural selection. A key question is, what trait was selected in an experiment? So far, no experiments have selected for female receptivity per se.

Selection for fast and slow maters has been documented in *D. melanogaster* (100), *D. simulans* (104), *D. pseudoobscura* (83, 143), *D. persimilis* (149), and *D. ananassae* (144). Analysis of each line after selection suggests that some alteration of female receptivity occurred in all of these experiments. The mating speed of couples was selected for in the above studies; in five other studies (29, 77, 101, 121, 154), selection was for mating speed or for the onset of primary receptivity in females only. Each of the latter studies demonstrated altered receptivity in the females.

In Cook's (29) experiment, he selected for an increased rate of mating between females and wingless males of *D. melanogaster*. After selection, females of the selected line were less active than control females in the presence of wingless males (low activity normally precedes mating), and after antennectomy, the selected females were less active in the presence of winged males (30). This finding suggests that the courtship song performed by winged males decreased receptivity of the selected females.

In some instances, the onset of receptivity differed in the selected lines from that of controls (77, 121, 143). Using *D. melanogaster*, Hudak & Gromko (77) selected for mating that occurred either earlier or later in development instead of mating speed. The responses to this selection were rapid and large, which indicated a heritability of mating speed. After selection, in two early-maturing lines and one late-maturing line, female mating speed (tested with control males) differed significantly from the mating speed of control females (77). This result suggests that female receptivity in *D. melanogaster* may be genetically correlated with the rate of sexual maturation.

Comparisons of several strains and lines has revealed genetic variation for receptivity within species (26, 148, 175, 178). Isofemale lines (each established from a single, wild-caught, inseminated female) of *D. melanogaster* and *D. simulans* varied significantly in receptivity (15, 18); receptivity of *D. melanogaster* was positively correlated with speed of sexual maturation (16) and with willingness to hybridize with *D. simulans* males (18). Directional dominance for the speed of female onset of receptivity was found in crosses between isofemale lines of *D. melanogaster*. In another study, two strains of *A. aegypti* differed in the age at which females became receptive; applying a JH analogue to the slow-mating strain accelerated its development of receptivity (63). In a novel approach, Veuille & Mazeau (170, 171) sampled chromosomes from a wild population and constructed heterozygotes from a random set of these genes, so that the resulting individuals were highly heterozygous, yet genetically identical with respect to any given chromosome being manipulated. There was genetic variation for female mating frequency (apparently primary receptivity), but female and male "eagerness to mate" were not correlated. This suggests a lack of genetic correlation between female receptivity and male mating propensity, in contrast to some other studies (e.g. 59).

When a strain is deliberately inbred, or when it passes through bottlenecks in population size, its genetic variation is reduced. In some cases, primary receptivity declines as well (108, 130, 131). Receptivity was reduced in all-female parthenogenetic strains of *Drosophila mercatorum,* compared with that of females in sexual strains (19, 158). However, in some species even extreme inbreeding may have no effect on receptivity (120).

Natural Genetic Variation in Remating Frequency

Many populations of *Drosophila* show substantial additive genetic variation in the speed of remating, as determined from selection experiments (59, 60, 123). In four lines selected for remating speed and four others selected for primary receptivity, higher receptivity was associated with faster remating, based on line means (61, 77). Subsequent analysis proved that the variation in remating frequency was polygenic; in one population, all three major chromosomes contributed significantly, while in another, chromosome 2 had the largest effect (52, 53). In this case, a major effect was mapped to the right arm of chromosome 2 (54).

Mutations Affecting Receptivity

Two kinds of genes affect sexual receptivity, those essential to the development or functioning of sensory systems and those acting on some other physiological substratum of receptivity.

A female's willingness to mate depends upon her response to male signals, and sensory mutants often mate more slowly than normal females, despite being courted equally by normal, wild-type males. In genetic experiments, mutations can have pleiotropic effects; i.e. a mutant may have both a sensory defect and low receptivity without the latter's being caused by the former. Such effects make interpreting the results difficult. All mutants discussed in this section are *D. melanogaster* unless stated otherwise. Primary receptivity was sharply reduced in congenitally deaf *aristaeless; thread* females (105), and was also low in females rendered partially anosmic by the *smellblind* mutation (105). Receptivity depends less upon vision than upon chemoreception or audition in *D. melanogaster;* females blind because of the *norp*^{A} mutation mated at 60% of the rate of sighted flies in a stringent test (105). However, Willmund & Ewing (176) found, using *sevenless* and *dipp6*, respectively, that both the high-acuity visual system (retinal cells R7–8) and the high-sensitivity system (retinal cells R1–6) affect female receptivity. The light-independent (*lin*) strain of the normally light-dependent species *Drosophila subobscura* does not require light to mate (122). Females of this strain are more receptive than wild-type females and also mate with *D. pseudoobscura* males at a higher frequency.

Mutations known to lower or increase receptivity, and that apparently do not act via the sensory system, comprise a short list: *yellow, ebony, hypoactiveC, inactive, apterous, fs(2)B*, and *fs(1)M72*. Females carrying the *yellow* mutation are less receptive than normal females (39, 106). Antennectomized *yellow* females have even lower receptivity, which suggests that receptivity is not reduced by stimuli received via the antennae (106). Female-sterile mutations generally have no effect on receptivity, except for *fs(2)B* (32) and some alleles of *apterous* (133). Some mutant alleles of *apterous* lower receptivity and reduce JH production; receptivity is partially restored in *apterous*[56f] by the topical application of methoprene (133). However, *apterous* mutations cause sterility through an independent mechanism unrelated to female receptivity (133). A mutation that strongly reduces receptivity but has no other known effects is *fs(1)M72*, which is incorrectly designated as a female-sterile mutation (82). This mutation does not induce any known sensory or motor defects, nor does it measurably affect ovarian development or postmating fertility (82). The X-linked mutations *hypoactiveC* and *inactive* (two alleles thereof) cause flies to be less active than normal and to develop primary receptivity slowly (116); the causes of reduced receptivity are not known. One mutation, *ebony*, reportedly speeds up the development of primary receptivity; the mechanism may be related to β-alanine metabolism (79).

Molecular Genetics

The molecular analysis of genes and their products, now used to study the actions of sex peptide (SP) in *D. melanogaster*, holds promise for the study of other species for which classical genetics has not yielded much information. The gene encoding SP cannot be approached with ease genetically because attempts to find variant alleles (e.g. induced mutations) have not succeeded (24). Rather, its chromosomal location and its structure have been discovered via reverse genetics, in which a cDNA copy of the *SP* gene was isolated using a synthetic DNA probe whose sequence was deduced from the *SP* primary structure (24). The *SP* gene contains a tiny intron, and the transcript is translated into a 55–amino acid precursor with a 19–amino acid signal peptide (141). Transformants were made using a hybrid gene consisting of *SP* cDNA fused either to a heat-shock protein (*hsp-70*) promoter or to the enhancer of a yolk protein (*yp1*) gene (3). In *hsSP* transformants, SP synthesis can be induced by heat shock, whereas in transformants carrying the *yp1* enhancer and the genomic *SP* sequence, synthesis is constitutive in the adult fat body. Virgin *hsSP* transformants become unreceptive within an hour after heat shock; receptivity returns to normal after about 24 h (3). Transformants carrying the *yp1* enhancer are constitutively unreceptive. SP-induced unreceptivity is characterized by the same rejection behavior of normally mated females, namely high rates of ovipositor extrusion (3). The active portion of SP includes all but the seven N-terminal amino acids and must contain a disulfide bridge. However, the post-translational modifications of amino acids found in natural SP is inessential for function (141). These determinations were made by injecting synthetic SP into virgin females (141).

IMPLICATIONS FOR APPLIED ENTOMOLOGY

The sexual receptivity of insects is of more than academic interest. For example, the release of sterile males to control the rate of reproduction of pest species has been successful against several pests, but it works only if (*a*) females of the target population are receptive to the sterile males and (*b*) mating with sterile males inhibits subsequent mating. As we have seen, however, genetic experiments have repeatedly documented substantial amounts of natural genetic variation for both primary receptivity and the control of remating. Therefore, either of these two characteristics of receptivity can evolve under natural selection. An intriguing property of the sex peptides of male accessory glands is the ability of some to reduce receptivity in females of distantly related taxa (22, 75). If the functional portions of these molecules have been relatively conserved in evolution, molecular-genetic methods could be used to introduce the genes into wild populations of pest insects, rendering females conditionally unreceptive. In other words, the switch gene introduced into a population could be conditional upon a signal supplied exogenously. However, a fundamental problem with any scheme to introduce antifitness genes into a population is that they will be immediately and strongly selected against.

Literature Cited

1. Adams TS, Hintz AM. 1969. Relationship of age, ovarian development, and the corpus allatum to mating in the house-fly, *Musca domestica*. *J. Insect Physiol*. 15:201–15
2. Adams TS, Nelson DR. 1968. Bioassay of crude extracts for the factor that prevents second matings in female *Musca domestica*. *Ann. Entomol. Soc. Am*. 61:112–16
3. Aigaki T, Fleischmann I, Chen PS, Kubli E. 1991. Ectopic expression of sex peptide alters reproductive behavior of female *D. melanogaster*. *Neuron* 7:557–63

4. Alcock J, Jones CE, Buchmann SL. 1976. Location before emergence of the female bee, *Centris pallida*, by its male (Hymenoptera: Anthophoridae). *J. Zool. London* 179:189–99
5. Barth RH. 1962. The endocrine control of mating behavior in the cockroach *Byrsotria fumigata* (Guerin). *Gen. Comp. Endocrinol*. 2:53–69
6. Barth RH. 1964. The mating behaviour of *Byrsotria fumigata* (Guerin) (Blattidae, Blaleninae). *Behaviour* 23:1–30
7. Barth RH, Lester LJ. 1973. Neuro-hormonal control of sexual behavior in insects. *Annu. Rev. Entomol*. 18:445–72

8. Barton Browne L, Bartell RJ, van Gerwen ACM, Lawrence LA. 1976. Relationship between protein ingestion and sexual receptivity in females of the Australian sheep blowfly *Lucilia cuprina*. *Physiol. Entomol.* 1:235–40

9. Baumann H. 1974. The isolation, partial characterization, and biosynthesis of the paragonial substances, PS-1 and PS-2, of *Drosophila funebris*. *J. Insect Physiol.* 20:2181–94

10. Baumann H. 1974. Biological effects of paragonial substances PS-1 and PS-2, in females of *Drosophila funebris*. *J. Insect Physiol.* 20:2347–62

11. Deleted in proof

12. Bouletreau-Merle J. 1973. Receptivite sexuelle et vitellogenese chez les femelles de *Drosophila melanogaster:* effets d'une application d'hormone juvenile et de deux analogues hormonaux. *C. R. Acad. Sci. Paris* 277: 2045–48

13. Bouletreau-Merle J. 1976. Destruction de la pars intercerebralis chez *Drosophila melanogaster:* effet sur la fecondite et sur sa stimulation par accouplement. *J. Insect Physiol.* 22:933–40

14. Cade WH. 1985. Insect mating and courtship behaviour. In *Comprehensive Insect Physiology, Biochemistry, and Pharmacology*, ed. GA Kerkut, LI Gilbert, 9:591–620. Oxford: Pergamon

15. Carracedo MC, Casares P. 1987. Sexual isolation between *Drosophila melanogaster* females and *Drosophila simulans* males. I. Relation between homospecific and heterospecific mating sucess. *Genet. Sel. Evol.* 19:21–36

16. Carracedo MC, Casares P, Izquierdo JI, Pineiro R. 1991. Receptivity and sexual maturation of *Drosophila melanogaster* females in relation to hybridization with *Drosophila simulans* males: a populational analysis. *Anim. Behav.* 42:201–8

17. Deleted in proof

18. Carracedo MC, Garcia-Florez L, San Miguel E. 1989. Sexual maturation in *Drosophila melanogaster* females and hybridization with *Drosophila simulans* males: a study of inheritance modes. *J. Hered.* 80:57–58

19. Carson HL, Teramoto LT, Templeton AR. 1977. Behavioral differences among isogenic strains of *Drosophila mercatorum*. *Behav. Genet.* 7:189–97

20. Chapman T, Trevitt S, Partridge L. 1994. Remating and male-derived nutrients in *Drosophila melanogaster*. *J. Evol. Biol.* 7:1–69

21. Chaudhury MFB, Ball HJ. 1973. The effect of age, nutritional factors, and gonadal development on the mating behaviour of the face fly, *Musca autumnalis*. *J. Insect Physiol.* 19:57–64

22. Chen PS. 1984. The functional morphology and bichemistry of insect male accessory glands and their secretions. *Annu. Rev. Entomol.* 29:233–55

23. Chen PS, Buhler R. 1970. Paragonial substance (sex peptide) and other free ninhydrin-positive components in male and female adults of *Drosophila melanogaster*. *J. Insect Physiol.* 16:615–27

24. Chen PS, Strumm-Zollinger E, Aigaki T, Balmer J, Bienz M, Bohlen P. 1988. A male accessory gland peptide that regulates reproductive behavior of female *D. melanogaster*. *Cell* 54:291–98

25. Cole BJ. 1983. Multiple mating and the evolution of social behavior in the Hymenoptera. *Behav. Ecol. Sociobiol.* 12:191–201

26. Connolly K, Burnet B, Kearney M, Eastwood L. 1974. Mating speed and courtship behaviour of inbred strains of *Drosophila melanogaster*. *Behaviour* 48:61–74

27. Connolly K, Cook R. 1973. Rejection responses by female *Drosophila melanogaster:* their ontogeny, causality, and effects upon the behaviour of the courting male. *Behaviour* 44:142–66

28. Cook RM. 1973. Physiological factors in the courtship processing of *Drosophila melanogaster*. *J. Insect Physiol.* 19:397–406

29. Cook RM. 1973. Courtship processing in *Drosophila melanogaster*. I. Selection for receptivity to wingless males. *Anim. Behav.* 21:338–48

30. Cook RM. 1973. Courtship processing in *Drosophila melanogaster*. II. An adaptation to selection for receptivity to wingless males. *Anim. Behav.* 21:349–58

31. Cook RM. 1975. Courtship of *Drosophila melanogaster:* rejection without extrusion. *Behaviour* 52:155–71

32. Cook RM, Connolly K. 1976. Sexual behaviour of a female-sterile mutant of *Drosophila melanogaster*. *J. Insect Physiol.* 22:1727–35

33. Craig GB Jr. 1967. Mosquitoes: female monogamy induced by male accessory gland substance. *Science* 156:1499–501

34. Davey KG. 1965. *Reproduction in the Insects*. Edinburgh/London: Oliver & Boyd. 96 pp.

35. Davis NT. 1965. Studies of the reproductive physiology of Cimicidae (Hemiptera). III. The seminal stimulus. *J. Insect Physiol.* 11:1199–211

36. De Villier PS, Hanrahan SA. 1991.

Sperm competition in the Namib desert beetle, *Onymacris unguicularis*. *J. Insect Physiol.* 37:1–8

37. Dickinson JL. 1986. Prolonged mating in the milkweed leaf beetle *Labidomera clivicollis* (Coleoptera: Chrysomelidae): a test of the "sperm-loading" hypothesis. *Behav. Ecol. Sociobiol.* 18: 331–38

38. Dodson G. 1987. Biological observations on *Aciurina trixa* and *Valentibulla dodsoni* (Diptera: Tephritidae) in New Mexico. *Ann. Entomol. Soc. Am.* 80: 494–500

39. Dow M. 1976. The genetic basis of receptivity of *yellow* mutant *Drosophila* females. *Behav. Genet.* 6:141–43

40. Downes JA. 1969. The swarming and mating flight of Diptera. *Annu. Rev. Entomol.* 14:271–98

41. Eggert AK, Sakaluk SK. 1994. Sexual cannibalism and its relation to male mating success in sagebrush crickets, *Cyphoderris strepitans* (Haglidae: Orthoptera). *Anim. Behav.* 47:1171–77

42. Eickwort GC, Ginsberg HS. 1980. Foraging and mating behavior in Apoidea. *Annu. Rev. Entomol.* 25:421–46

43. Engelmann F. 1960. Hormonal control of mating in an insect. *Experientia* 16: 69–70

44. Engelmann F. 1970. *The Physiology of Insect Reproduction.* New York: Pergammon. 307 pp.

45. Engelmann F, Barth RH. 1968. Endocrine control of female receptivity in *Leucophaea maderae* (Blattaria) *Ann. Entomol. Soc. Am.* 61:503–5

46. Ewing AW. 1964. The influence of wing area on the courtship behaviour of *Drosophila melanogaster. Anim. Behav.* 12:316–20

47. Fockler CE, Borden JH. 1973. Mating activity and ovariole development of *Trypodendron lineatum:* effect of juvenile hormone analogs. *Ann. Entomol. Soc. Am.* 66:509–12

48. Fox CW, Hickman DL. 1994. Influence of oviposition substrate on female receptivity to multiple mating in *Callosobruchus maculatus* (Coleoptera: Bruchidae). *Ann. Entomol. Soc. Am.* 87: 395–98

49. Fuchs MS, Craig GB Jr, Despommier DD. 1969. The protein nature of the substance inducing female monogamy in *Aedes aegypti. J. Insect Physiol.* 15: 701–9

50. Fuchs MS, Hiss EA. 1970. The partial purification and separation of the protein components of matron from *Aedes aegypti. J. Insect Physiol.* 16:931–39

51. Grillou H. 1973. A study of sexual receptivity in *Blaber craniifer* Burm. (Blattaria). *J. Insect Physiol.* 19:173–93

52. Fukui HH, Gromko MH. 1991. Genetic basis for remating in *Drosophila melanogaster*. IV. A chromosome substitution analysis. *Behav. Genet.* 21: 169–82

53. Fukui HH, Gromko MH. 1991. Genetic basis for remating in *Drosophila melanogaster*. V. Biometrical and planned comparisons analysis. *Behav. Genet.* 21:183–97

54. Fukui HH, Gromko MH. 1991. Genetic basis for remating in *Drosophila melanogaster*. VI. Recombination analysis. *Behav. Genet.* 21:199–209

55. Gadenne C. 1993. Effects of fenoxycarb, juvenile hormone mimetic, on female sexual behaviour of the black cutworm, *Agrotis ipsilon* (Lepidoptera: Noctuidae). *J. Insect Physiol.* 39:25–29

56. Giebultowicz JM, Raina AK, Uebel EC. 1990. Mated-like behaviour in senescent virgin females of gypsy moth *Lymantria dispar. J. Insect Physiol.* 36:495–98

57. Gillott C, Langley PA. 1981. The control of receptivity and ovulation in the tsetse fly, *Glossina morsitans. Physiol. Entomol.* 6:269–81

58. Grillou H. 1973. A study of sexual receptivity in *Blabera craniifer* Burm. (Blattaria). *J. Insect Physiol.* 19:173–93

59. Gromko MH. 1992. Genetic correlation of male and female mating frequency: evidence from *Drosophila melanogaster. Anim. Behav.* 43:176–77

60. Gromko MH, Newport MEA. 1988. Genetic basis for remating in *Drosophila melanogaster*. II. Response to selection based on the behavior of one sex. *Behav. Genet.* 18:621–32

61. Gromko MH, Newport MEA. 1988. Genetic basis for remating in *Drosophila melanogaster*. III. Correlated responses to selection for female remating speed. *Behav. Genet.* 18:633–43

62. Grossfield J. 1972. Decapitated females as a tool in the analysis of *Drosophila* behaviour. *Anim. Behav.* 20:243–51

63. Gwadz RW. 1970. Monofactorial inheritance of early sexual receptivity in the mosquito, *Aedes atropalpus. Anim. Behav.* 18:358–61

64. Gwadz RW. 1972. Neurohormonal regulation of sexual receptivity in female *Aedes aegypti. J. Insect Physiol.* 18: 259–66

65. Hall JC. 1977. Portions of the central nervous system controlling reproductive behavior in *Drosophila melanogaster. Behav. Genet.* 7:291–312

66. Harris VE, Todd JW. 1980. Temporal and numerical patterns of reproductive

behavior in the Southern green stink bug, *Nezara viridula* (Hemiptera: Pentatomidae). *Entomol. Exp. Appl.* 27: 105–16

67. Harshman LG, Hoffmann AA, Prout T. 1988. Environmental effects on remating in *Drosophila melanogaster*. *Evolution* 42:312–21

68. Hartmann R, Loher W. 1974. Control of sexual behavior pattern 'secondary defence' in the female grasshopper *Chorthippus curtipennis*. *J. Insect Physiol.* 20:1713–28

69. Hartman HB, Roth LM. 1967. Stridulation by the cockroach *Nauphoeta cinerea* during courtship behaviour. *J. Insect Physiol.* 13:579–86

70. Hartmann R, Wolf W, Loher W. 1972. The influence of the endocrine system on reproductive behavior and development in grasshoppers. *Gen. Comp. Endocrinol. Suppl.* 3:518–28

71. Haskell PT. 1958. Stridulation and associated behaviour in certain Orthoptera. 2. Stridulation of females and their behaviour with males. *Anim. Behav.* 6: 27–42

72. Haskell PT. 1960. Stridulation and associated behaviour in certain Orthoptera. 3. The influence of the gonads. *Anim. Behav.* 8:76–81

73. Heady SE. 1993. Factors affecting female sexual receptivity in the planthopper, *Prokelisia dolus*. *Physiol. Entomol.* 18:263–70

74. Deleted in proof

75. Hiss EA, Fuchs MS. 1972. The effect of matrone on oviposition in the mosquito, *Aedes aegypti*. *J. Insect Physiol.* 18:2217–27

76. Hölldobler B. 1976. The behavioral ecology of mating in harvester ants (Hymenoptera: Formicidae: Pogomyrmex). *Behav. Ecol. Sociobiol.* 1:405–23

77. Hudak MJ, Gromko MH. 1989. Response to selection for early and late development of sexual maturity in *Drosophila melanogaster*. *Anim. Behav.* 38:344–51

78. Itagaki H, Conner WE. 1986. Physiological control of pheromone release behaviour in *Manduca sexta* L. *J. Insect Physiol.* 32:657–64

79. Jacobs ME. 1978. Influence of beta-alanine on mating and territorialism in *Drosophila melanogaster*. *Behav. Genet.* 8:487–502

80. Jansson A. 1973. Stridulation and its significance in the genus *Cenocorixa* (Hemiptera, Corixidae). *Behaviour* 46: 1–36

81. Kambysellis MP, Craddock EM. 1991. Insemination patterns in Hawaiian *Drosophila* species (Diptera: Drosophilidae) correlate with ovarian development. *J. Insect Behav.* 4:83–100

82. Kerr CL. 1994. *Characterization of a sex-linked female mutant of* Drosophila melanogaster *discovered with low receptivity*. MS thesis. Univ. Maine, Orono. 60 pp.

83. Kessler S. 1968. The genetics of *Drosophila* mating behaviour. I. Organization of mating speed in *Drosophila pseudoobscura*. *Anim. Behav.* 16:485–91

84. Koudele K, Stout JF, Reichert D. 1987. Factors which influence female crickets' (*Achaeta domesticus*) phonotactic and sexual responsiveness to males. *Physiol. Entomol.* 12:67–80

85. Kuba H, Ito Y. 1993. Remating inhibition in the melon fly, *Bactrocera* (=*Dacus*) *cururbitae* (Diptera: Tephritidae): copulation with spermless males inhibits female remating. *J. Ethol.* 11: 23–28

86. Kyriacou CP, Hall JC. 1982. The function of courtship song rhythms in *Drosophila*. *Anim. Behav.* 30:794–801

87. Lea AO. 1968. Mating without insemination in virgin *Aedes aegypti*. *J. Insect Physiol.* 14:305–8

88. Lee RD. 1955. The biology of the Mexican chicken bug *Haematosiphon inodorus* (Duges) (Hemiptera: Cimicidae). *Pan-Pac. Entomol.* 31:47–61

89. Leopold RA. 1976. The role of male accessory glands in insect reproduction. *Annu. Rev. Entomol.* 21:199–221

90. Leopold RA, Terranova AC, Swilley EM. 1971. Mating refusal in *Musca domestica*: effects of repeated mating and decerebration upon frequency and duration of mating. *J. Exp. Zool.* 176: 353–60

91. Deleted in proof

92. Liang D, Schal C. 1994. Neural and hormonal regulation of calling behavior in *Blattella germanica* females. *J. Insect Physiol.* 40:251–58

93. Linley JR, Mook MS. 1975. Behavioural interaction between sexually experienced *Culicoides melleus* (Coquillett) (Diptera: Ceratopogonidae). *Behaviour* 54:97–110

94. Loher W. 1962. Die Kontrolle des Weibchengesanges von *Gomphocerus rufus* L. (Acridinae) durch die Corpora Allata. *Naturwissenschaften* 49:406

95. Loher W. 1981. The effect of mating on female sexual behavior of *Teleogryllus commodus* Walker. *Behav. Ecol. Sociobiol.* 9:219–25

96. Loher W, Giannakakis A. 1990. Allatectomy in the penultimate larval instar of *Teleogryllus commodus*: morphological,

physiological, and behavioral consequences. *Zool. Jahrb. Abt. Physiol.* 94: 167–79

97. Loher W, Huber F. 1964. Experimentelle Untersuchungen am Sexualverhalten des Weibchens der Heuschrecke *Gomphocerus rufus* L. (Acridinae). *J. Insect Physiol.* 10:13–36

98. Lynam AJ, Morris S, Gwyne DT. 1992. Differential mating success of virgin female katydids *Requena verticalis* (Orthopetra: Tettigomidae). *J. Insect Behav.* 5:51–59

99. Manning A. 1960. The sexual behaviour of two sibling *Drosophila* species. *Behaviour* 15:123–45

100. Manning A. 1961. The effects of artificial selection for mating speed in *Drosophila melanogaster. Anim. Behav.* 9: 82–92

101. Manning A. 1963. Selection for mating speed in *Drosophila melanogaster* based on the behaviour of one sex. *Anim. Behav.* 11:116–20

102. Manning A. 1966. Corpus allatum and sexual receptivity in female *Drosophila melanogaster. Nature* 211:1321–22

103. Manning A. 1967. The control of sexual receptivity in female *Drosophila. Anim. Behav.* 15:239–50

104. Manning A. 1968. The effects of artificial selection for slow mating in *Drosophila simulans.* 1. The behavioural changes. *Anim. Behav.* 16:108–13

105. Markow TA. 1987. Behavioral and sensory basis of courtship success in *Drosophila melanogaster. Proc. Natl. Acad. Sci. USA* 84:6200–4

106. Mayr E. 1950. The role of the antennae in the mating behavior of female *Drosophila. Evolution* 4:149–54

107. Mead-Briggs AR. 1964. The reproductive biology of the rabbit flea *Spilopsyllus cuniculi* (Dale) and the dependence of this species upon the breeding of its host. *J. Exp. Biol.* 41: 371–402

108. Meffert LM, Bryant EH. 1991. Mating propensity and courtship behavior in serially bottlenecked lines of the housefly. *Evolution* 45:293–306

109. Merle J. 1968. Fonctionnement ovarien et receptivite sexuelle de *Drosophila melanogaster* apres implantation de fragments de l'appareil genital male. *J. Insect Physiol.* 14:1159–68

110. Monsma SA, Harada HA, Wolfner MF. 1990. Synthesis of two *Drosophila* male accessory gland proteins and their fate after transfer into the female during mating. *Dev. Biol.* 142:465–75

111. Morrison PE, Venkatesh K, Thompson B. 1982. The role of male accessory gland substance on female reproduction with some observations of spermatogenesis in the stable fly. *J. Insect Physiol.* 28:607–14

112. Muller HP. 1965. Zur Frage der Steuerung des Paarungs Verhaltens und der Eireifung bei der Feldheuschrecke *Enthystira brachyptera* Oisk. unter besonderer Beruchsichtigung der Rolle der Corpora Allata. *Z. Vgl. Physiol.* 50:447–97

113. Nelson DR, Adams TS, Pomonis JG. 1969. Initial studies on the extraction of the active substance inducing monocoitic behavior in house flies, black blow flies, and screwworm flies. *J. Econ. Entomol.* 62:634–39

114. Nissani M. 1977. Gynandromorph analysis of some aspects of sexual behaviour in *Drosophila melanogaster. Anim. Behav.* 25:555–66

115. Oberlander H, Sower L, Silhacek D. 1975. Mating behavior of *Plodia interpunctella* reared on juvenile-hormone treated diet. *J. Insect Physiol.* 21:681–85

116. O'Dell K, Burnet B, Jallon J-M. 1969. Effects of the *hypoactive* and *inactive* mutations on mating success in *Drosophila melanogaster. Heredity* 62:373–81

117. Orshan L, Pener MP. 1991. Effects of the corpora allata on sexual receptivity and completion of oocyte development in females of the cricket, *Gryllus bimaculatus. Physiol. Entomol.* 16:231–42

118. Otronen M. 1989. Female mating behaviour and multiple matings in the fly, *Dryomyza anilis. Behaviour* 111:77–97

119. Parker GA. 1968. The sexual behaviour of the blowfly, *Protophormia terrae-novae* R.-D. *Behaviour* 32:291–308

120. Parsons PA. 1965. The determinants of mating speeds in *Drosophila melanogaster* for various combinations of inbred lines. *Experientia* 21:478

121. Pineiro R, Carracedo MC, Izquinierdo JI, Casares P. 1993. Bidirectional selection for female receptivity in *Drosophila melanogaster. Behav. Genet.* 23:77–83

122. Pinsker W, Doschek E. 1980. Courtship and rape: the mating behavior of *Drosophila subobscura* in light and darkness. *Z. Tierpsychol.* 54:57–70

123. Pyle DW, Gromko MH. 1981. Genetic basis for repeated mating in *Drosophila melanogaster. Am. Nat.* 117:133–46

124. Ramaswamy SB, Gupta AP. 1981. Effects of juvenile hormone on sense organs involved in mating behaviour of *Blattella germanica* L. (Dictyoptera: Blattellidae). *J. Insect Physiol.* 27:601–8

125. Raabe M. 1989. *Recent Developments*

in *Insect Neurohormones.* New York: Plenum. 503 pp.

126. Riddiford LM, Williams CM. 1971. Role of the corpora cardiaca in the behavior of saturniid moths. I. Release of sex pheromone. *Biol. Bull.* 140:1–7

127. Ridley M. 1989. The timing and frequency of mating in insects. *Anim. Behav.* 37:535–45

128. Riemann JG, Moen DJ, Thorson BJ. 1967. Female monogamy and its control in houseflies. *J. Insect Physiol.* 13:407–8

129. Riemann JG, Thorson B. 1969. Effect of male accessory gland material on oviposition and mating by female house flies. *Ann. Entomol. Soc. Am.* 62:828–34

130. Ringo J. 1986. The effect of successive founder events on mating propensity of *Drosophila.* In *Evolutionary Genetics of Invertebrate Behavior: Progress and Prospects,* ed. MD Huettel, 8:79–88. New York: Plenum. 335 pp.

131. Ringo J, Barton K, Dowse H. 1986. The effect of genetic drift on mating propensity, courtship behaviour, and postmating fitness in *Drosophila simulans. Behaviour* 97:226–33

132. Ringo JM, Pratt NR. 1978. A juvenile hormone analog induces precocious sexual behavior in *Drosophila grimshawi* (Diptera: Drosophilidae). *Ann. Entomol. Soc. Am.* 71:264–66

133. Ringo J, Werczberger R, Altaratz M, Segal D. 1991. Female sexual receptivity is defective in juvenile-hormone deficient mutants of the *apterous* gene of *Drosophila melanogaster. Behav. Genet.* 21:453–69

134. Roth LM. 1962. Hypersexual activity induced in females of the cockroach *Nauphoeta cinerea. Science* 138:1267–69

135. Roth LM. 1964. Control of reproduction in female cockroaches with special reference to *Nauphoeta cinerea.* I. First pre-oviposition period. *J. Insect Physiol.* 10:915–45

136. Roth LM, Barth RH. 1964. The control of sexual receptivity in female cockroaches. *J. Insect Physiol.* 10:965–75

137. Roth LM, Stay B. 1962. Oocyte development in *Blattella germanica* and *Blattella vaga* (Blattaria). *Ann. Entomol. Soc. Am.* 55:633–42

138. Rowe L. 1992. Convenience polyandry in a water strider: foraging conflicts and female control of copulation frequency and guarding duration. *Anim. Behav.* 44:189–202

139. Schatral A. 1993. Diet influences male-female interactions in the bush cricket *Requena verticalis* (Orthoptera: Tettigoniidae). *J. Insect Behav.* 6:379–88

140. Schein Y, Galun R. 1984. Male housefly (*Musca domestica* L.) genital system as a source of mating pheromone. *J. Insect Physiol.* 30:175–77

141. Schmidt T, Choffat Y, Klauser S, Kubli E. 1993. The *Drosophila melanogaster* sex-peptide: a molecular analysis of structure-function relationships. *J. Insect Physiol.* 39:361–68

142. Scott D. 1987. The timing of the sperm effect on female *Drosophila melanogaster* receptivity. *Anim. Behav.* 35:142–49

143. Sherwin RN. 1975. Selection for mating activity in two chromosomal arrangements of *Drosophila pseudoobscura. Evolution* 29:519–30

144. Singh BN, Chatterjee S. 1988. Selection for high and low mating propensity in *Drosophila ananassae. Behav. Genet.* 18:357–69

145. Smith RL. 1979. Paternity assurance and altered roles in the mating behaviour of a giant water bug, *Abedus herberti* (Heteroptera: Belostomatidae). *Anim. Behav.* 27:716–25

146. Spencer JL, Bush GL Jr, Keller JE, Miller JR. 1992. Modification of female onion fly, *Delia antiqua* (Meigen), reproductive behavior by male paragonial extracts (Diptera: Anthomyiidae). *J. Insect Behav.* 5:689–97

147. Spielman A, Leahy MG, Skaff V. 1969. Failure of effective insemination of young female *Aedes aegypti* mosquitoes. *J. Insect Physiol.* 15:1471–79

148. Spiess EB. 1968. Courtship and mating time in *Drosophila pseudoobscura. Anim. Behav.* 16:470–79

149. Spiess EB, Stankewych AJ. 1973. Mating speed selection and egg chamber correlation in *Drosophila persimilis. Egypt. J. Genet. Cytol.* 2:177–94

150. Spieth HT. 1940. Studies on the biology of the Ephemeroptera. II. The nuptial flight. *J. NY Entomol. Soc.* 48:379–90

151. Spieth HT. 1952. Mating behavior within the genus *Drosophila. Bull. Am. Mus. Nat. Hist.* 99:395–474

152. Spieth HT. 1966. Drosophilid mating behaviour: the behaviour of decapitated females. *Anim. Behav.* 14:226–35

153. Spieth HT. 1974. Courtship behavior in *Drosophila. Annu. Rev. Entomol.* 19:385–405

154. Stamenkovic-Radak M, Partridge L, Andjelkovic M. 1992. A genetic correlation between the sexes for mating speed in *Drosophila melanogaster. Anim. Behav.* 43:389–96

155. Stout JF, Gerard G, Hasso S. 1976. Sexual responsiveness mediated by the

corpora allata and its relationship to phonotaxis in the female cricket, *Achaeta domesticus* L. *J. Comp. Physiol. A* 108:1–9

156. Stride GO. 1958. Further studies on the courtship behaviour of African mimetic butterflies. *Anim. Behav.* 6:224–30

157. Strong L, Amerasinghe FP. 1977. Allatectomy and sexual receptivity in females of *Schistocerca gregaria*. *J. Insect Physiol.* 23:131–35

158. Takenaka-Dacanay JH, Carson HL. 1991. Sexual behavior in laboratory strains of *Drosophila mercatorum* that have spontaneously adopted parthenogenesis. *Behav. Genet.* 21:305–16

159. Terranova AC, Leopold RA, Degrugliellier ME, Johnson Jr. 1972. Electrophoresis of the male accessory gland secretion and its fate in the mated female. *J. Insect Physiol.* 18:1573–91

160. Thibout E. 1982. La comportement sexuel du doryphore, *Leptinotarsa decemlineata* Say et son possible controle par l'hormone juvenile et les corps allates. *Behaviour* 80:199–217

161. Thornhill R. 1980. Rape in *Panorpa* scorpionflies and a general rape hypothesis. *Anim. Behav.* 28:52–59

162. Thornhill R. 1980. Sexual selection within mating swarms of the lovebug, *Plecia nearctica* (Diptera: Bibionidae). *Anim. Behav.* 28:405–12

163. Thornhill R. 1980. Mate choice in *Hylobittacus apicalis* (Insecta: Mecoptera) and its relation to some models of female choice. *Evolution* 34:519–38

164. Thornhill R, Alcock J. 1983. *The Evolution of Insect Mating Systems*. Cambridge, MA: Harvard Univ. Press. 547 pp.

165. Tompkins L, Hall J. 1983. Identification of brain sites controlling female receptivity in mosaics of *Drosophila melanogaster*. *Genetics* 103:179–95

166. Trabalon M, Campan M. 1984. Etude de la receptivite sexuelle de la femelle de *Calliphora vomitoria* (Dipters, Calliphoridae) an cours du premier cycle gonadotrope. I. Approches comportmentale et physiologuque. *Behaviour* 90:241–58

167. Trevitt S, Fowler K, Partridge L. 1988. An effect of egg-deposition on the subsequent fertility and remating frequency of female *Drosophila melanogaster*. *J. Insect Physiol.* 34:821–28

168. Truman JW, Riddiford LM. 1974. Hormonal mechanisms underlying insect behaviour. *Adv. Insect Physiol.* 10:297–352

169. Deleted in proof

170. Veuille M, Mazeau S. 1986. Variation in sexual behavior and negative assortative mating in *Drosophila melanogaster*. *Behav. Genet.* 16:307–17

171. Veuille M, Mazeau S. 1988. Genetic variability of sexual behavior in a natural population of *Drosophila melanogaster*. *Behav. Genet.* 18:389–403

171a. von Helversen D, von Helversen O. 1991. Premating sperm removal in the bushcricket *Metaplastes ornatus* Ramme 1931 (Orthoptera, Tettigonoidea, Phaneropteridae). *Behav. Ecol. Sociobiol.* 28:391–96

172. von Schilcher F. 1976. The function of pulse song and sine song in the courtship of *Drosophila melanogaster*. *Anim. Behav.* 24:622–25

173. Walker WF. 1978. Mating behaviour in *Oncopeltus fasciatus* (Dallas): effects of diet, photoperiod, juvenoids, and precocene II. *Physiol. Entomol.* 3:147–55

174. Webster RP, Cardé RT. 1984. The effects of mating, exogenous juvenile hormone, and a juvenile hormone analogue on pheromone titre, calling, and oviposition in the omnivorous leafroller moth (*Platynota stultana*). *J. Insect Physiol.* 30:113–18

175. Wellbergen P, Spruijt BM, van Dijken FR. 1992. Mating speed and the interplay between female and male courtship responses in *Drosophila melanogaster* (Diptera: Drosophilidae). *J. Insect Behav.* 5:229–44

176. Willmund R, Ewing A. 1982. Visual signals in the courtship of *Drosophila melanogaster*. *Anim. Behav.* 30:209–15

177. Wong TTY, McInnis DO, Nishimoto JI. 1986. Melon fly (Diptera: Tephritidae): sexual maturation rates and mating responses of laboratory-reared and wild flies. *Ann. Entomol. Soc. Am.* 79:605–9

178. Yacher TH, Spiess EB. 1973. The development of mating propensity in two karyotypes of *Drosophila persimilis*. *Anim. Behav.* 21:359–70

179. Young ADM, Downe AER. 1982. Renewal of sexual receptivity in mated female mosquitoes, *Aedes aegypti*. *Physiol. Entomol.* 7:467–71

180. Young ADM, Downe AER. 1983. Influence of mating on sexual receptivity and oviposition in the mosquito, *Culex tarsalis*. *Physiol. Entomol.* 8:213–17

181. Zagatti P. 1981. Comportement sexuel de la pyrale de la canne a sucre *Eldana saccharina* (Wilk.) lie a deux phero-

mones emise par la male. *Behaviour* 78:81–98

182. Zdarek J. 1970. Mating behaviour in the bug *Pyrrhocoris apterus* L. (Heteroptera): ontogeny and its environmental control. *Behaviour* 37:253–68

183. Zdarek J, Slama K. 1968. Mating activity in adultoids or supernumerary larvae induced by agents with high juvenile hormone activity. *J. Insect Physiol.* 14: 563–67

SUBJECT INDEX

CUMULATIVE INDEXES

CONTRIBUTING AUTHORS, VOLUMES 32–41

CHAPTER TITLES, VOLUMES 32–41